w/CD Rom

ST. OLAF COLLEGE
SCIENCE LIBRARY

OCT 03 2002

The Finite Element Method Using MATLAB
Second Edition

CRC MECHANICAL ENGINEERING SERIES

Series Editor Frank Kreith

Published

Entropy Generation Minimization
  Adrian Bejan
Finite Element Method Using MATLAB
  Young W. Kwon and Hyochoong Bang
Fundamentals of Environmental Discharge Modeling
  Lorin R. Davis
Intelligent Transportation Systems: New Principles and Architectures
  Sumit Ghosh and Tony Lee
Mathematical and Practical Modeling of Materials Processing Operations
  Olusegun Johnson Ileghus, Manabu Iguchi, and Walter E. Wahnsiedler
Mechanics of Composite Materials
  Autar K. Kaw
Mechanics of Fatigue
  Vladimir V. Bolotin
Nonlinear Analysis of Structures
  M. Sathyamoorthy
Practical Inverse Analysis in Engineering
  David M. Trujillo and Henry R. Busby
Viscoelastic Solids
  Roderic S. Lakes

To be Published

Distributed Generation: The Power Paradigm for the New Millenium
  Marie Borbely and Jan F. Kreider
Engineering Experimentation
  Euan Somerscales
Energy Audit of Building Systems: An Engineering Approach
  Moncef Krarti
Fluid Power Circuits and Controls: Fundamentals and Applications
  John S. Cundiff
Introduction Finite Element Method
  Chandrakant S. Desai and Tribikram Kundu
Mechanics of Solids and Shells
  Gerald Wempner and Demosthenes Talaslidis
Mechanism Design: Enumeration of Kinematic Structures According to Function
  Lung-Wen Tsai
Principles of Solid Mechanics
  Rowland Richards, Jr.
Thermodynamics for Engineers
  Kau-Fui Wong

CHAPTER 4. DIRECT APPROACH WITH SPRING SYSTEM   71

    4.0   Chapter Overview   71
    4.1   Linear Spring   71
    4.2   Axial Member   74
    4.3   Torsional Member   76
    4.4   Other Systems   78
           Problems   81

CHAPTER 5. LAPLACE'S AND POISSON'S EQUATIONS   85

    5.0   Chapter Overview   85
    5.1   Governing Equation   85
    5.2   Linear Triangular Element   88
    5.3   Bilinear Rectangular Element   92
    5.4   Boundary Integral   94
    5.5   Transient Analysis   98
    5.6   Time Integration Technique   100
    5.7   Axisymmetric Analysis   103
    5.8   Three-Dimensional Analysis   105
    5.9   MATLAB Application to 2-D Steady State Analysis   109
    5.10 MATLAB Application to Axisymmetric Analysis   121
    5.11 MATLAB Application to Transient Analysis   126
    5.12 MATLAB Application to 3-D Steady State Analysis   150
           Problems   155

CHAPTER 6. ISOPARAMETRIC ELEMENTS   159

    6.0   Chapter Overview   159
    6.1   One-Dimensional Elements   159
    6.2   Quadrilateral Elements   163
    6.3   Triangular Elements   171
    6.4   Gauss Quadrature   172
    6.5   MATLAB Application to Gauss Quadrature   179
    6.6   MATLAB Application to Laplace's Equation   186
           Problems   195

CHAPTER 7. TRUSS STRUCTURES   199

    7.0   Chapter Overview   199
    7.1   One-Dimensional Truss   199
    7.2   Plane Truss   201
    7.3   Space Truss   205
    7.4   MATLAB Application to Static Analysis   206
    7.5   MATLAB Application to Eigenvalue Analysis   218
    7.6   MATLAB Application to Transient Analysis   226
           Problems   234

TABLE OF CONTENTS

CHAPTER 1. INTRODUCTION TO MATLAB — 1

1.0	Chapter Overview	1
1.1	Finite Element Method	1
1.2	Overview of the Book	2
1.3	About MATLAB	2
1.4	Vector and Matrix Manipulations	3
1.5	Matrix Functions	4
1.6	Data Analysis Functions	10
1.7	Tools for Polynomials	12
1.8	Making Complex Numbers	15
1.9	Nonlinear Algebraic Equations	15
1.10	Solving Differential Equations	17
1.11	Loop and Logical Statements	18
1.12	Writing Function Subroutines	20
1.13	File Manipulation	22
1.14	Basic Input-Output Functions	23
1.15	Plotting Tools	24

CHAPTER 2. APPROXIMATION TECHNIQUES — 31

2.0	Chapter Overview	31
2.1	Methods of Weighted Residual	31
2.2	Weak Formulation	34
2.3	Piecewise Continuous Trial Function	35
2.4	Galerkin's Finite Element Formulation	37
2.5	Variational Method	42
2.6	Rayleigh-Ritz Method	42
2.7	Rayleigh-Ritz Finite Element Method	43
	Problems	46

CHAPTER 3. FINITE ELEMENT PROGRAMMING — 51

3.0	Chapter Overview	51
3.1	Overall Program Structure	51
3.2	Input Data	52
3.3	Assembly of Element Matrices and Vectors	54
3.4	Application of Constraints	54
3.5	Example Programs	56
	Problems	70

Library of Congress Cataloging-in-Publication Data

Catalog record is available from the Library of Congress.

This book contains information obtained from authentic and highly regarded sources. Reprinted material is quoted with permission, and sources are indicated. A wide variety of references are listed. Reasonable efforts have been made to publish reliable data and information, but the author and the publisher cannot assume responsibility for the validity of all materials or for the consequences of their use.

Neither this book nor any part may be reproduced or transmitted in any form or by any means, electronic or mechanical, including photocopying, microfilming, and recording, or by any information storage or retrieval system, without prior permission in writing from the publisher.

The consent of CRC Press LLC does not extend to copying for general distribution, for promotion, for creating new works, or for resale. Specific permission must be obtained in writing from CRC Press LLC for such copying.

Direct all inquiries to CRC Press LLC, 2000 N.W. Corporate Blvd., Boca Raton, Florida 33431.

Trademark Notice: Product or corporate names may be trademarks or registered trademarks, and are used only for identification and explanation, without intent to infringe.

© 2000 by CRC Press LLC

No claim to original U.S. Government works
International Standard Book Number 0-8493-0096-7
Printed in the United States of America 1 2 3 4 5 6 7 8 9 0
Printed on acid-free paper

The Finite Element Method Using MATLAB

Second Edition

Young W. Kwon
Hyochoong Bang

CRC Press
Boca Raton   London   New York   Washington, D.C.

CHAPTER 8. BEAM AND FRAME STRUCTURES — 237

8.0	Chapter Overview	237
8.1	Euler-Bernoulli Beam	237
8.2	Timoshenko Beam	244
8.3	Beam Elements With Only Displacement Degrees of Freedom	248
8.4	Mixed Beam Element	252
8.5	Hybrid Beam Element	255
8.6	Composite Beams	259
8.7	Two-Dimensional Frame Element	262
8.8	Three-Dimensional Frame Element	265
8.9	MATLAB Application to Static Analysis	268
8.10	MATLAB Application to Eigenvalue Analysis	283
8.11	MATLAB Application to Transient Analysis	287
8.12	MATLAB Application to Modal Analysis of Undamped System	292
8.13	MATLAB Application to Modal Analysis of Damped System	299
8.14	MATLAB Application to Frequency Response Analysis	302
	Problems	307

CHAPTER 9. ELASTICITY PROBLEM — 311

9.0	Chapter Overview	311
9.1	Plane Stress and Plane Strain	311
9.2	Force Vector	315
9.3	Energy Method	317
9.4	Three-Dimensional Solid	319
9.5	Axisymmetric Solid	322
9.6	Dynamic Analysis	324
9.7	Thermal Stress	325
9.8	MATLAB Application to 2-D Stress Analysis	327
9.9	MATLAB Application to Axisymmetric Analysis	339
9.10	MATLAB Application to 3-D Stress Analysis	351
	Problems	361

CHAPTER 10. PLATE AND SHELL STRUCTURES — 365

10.0	Chapter Overview	365
10.1	Classical Plate Theory	365
10.2	Classical Plate Bending Element	368
10.3	Shear Deformable Plate Element	370
10.4	Plate Element With Displacement Degrees of Freedom	373
10.5	Mixed Plate Element	377
10.6	Hybrid Plate Element	383
10.7	Shell Made of Inplane and Bending Elements	386
10.8	Shell Degenerated from 3-D Solid	389

10.9	MATLAB Application to Plates	394
10.10	MATLAB Application to Shells	401
	Problems	424

CHAPTER 11. CONTROL OF FLEXIBLE STRUCTURES 427

11.0	Chapter Overview	427
11.1	Introduction	427
11.2	Stability Theory	429
11.3	Stability of Multiple Degrees of Freedom Systems	432
11.4	Analysis of a Second Order System	436
11.5	State Space Form Description	441
11.6	Transfer Function Analysis	451
11.7	Control Law Design for State Space Systems	461
11.8	Linear Quadratic Regulator	472
11.9	Modal Control for Second Order Systems	482
11.10	Dynamic Observer	485
11.11	Compensator Design	490
11.12	Output Feedback Design by Using Collocated Sensor/Actuator	494
	Problems	509

CHAPTER 12. SPECIAL TOPICS 513

12.0	Chapter Overview	513
12.1	Stationary Singular Elements	513
12.2	Quarter-Point Singular Elements	519
12.3	Moving Singular Elements	522
12.4	Semi-Infinite Element	528
12.5	Thermal Stress in Layered Beams	531
12.6	Buckling Analysis	534
12.7	Nonlinear Analysis	537
12.8	MATLAB Application to Buckling Problem	539
12.9	MATLAB Application to Nonlinear Problem	541

REFERENCES 557

APPENDIX A: FEA MATLAB FUNCTION FILES 563

APPENDIX B: EXAMPLES OF PRE- AND POST-PROCESSOR 591

INDEX 599

PREFACE TO THE SECOND EDITION

This second edition has the same objectives as the first edition to serve the same purpose. This edition expanded the previous one to include more diverse problems in the application of the finite element method along with some organizational change. With the expansion, this book may be used for a two-semester course as a textbook or used for a more extensive reference for practicing/research engineers/scientists.

One of the major topics included is analysis of shell structures because it is one of the most important structural applications. Both formulations and MATLAB example programs are included. Two different formulations are discussed along with their programs. One is the formulation based on the combination of inplane elements and plate bending elements. The other formulation is based on degeneration from 3-D solids.

A new chapter (Chapter 12) is added for special topics for finite element applications. It includes analysis of cracks using elements with singularity, analysis of semi-infinite or infinite domains, analysis of buckling, thermal stress analysis, and analysis of nonlinear differential equations. Various crack tip elements are presented for stationary and moving cracks. Thermal analysis is presented for multi-layered structures. Buckling analysis includes both static and dynamic bucking. Three different linearization techniques are discussed to solve nonlinear differential equations. Some MATLAB example programs are also provided at the end of the chapter. Finally, some MATLAB programs are presented for the pre- and post-processor in the appendix as illustrative examples for a simple shape of domain.

The organizational change is inclusion of the chapter overview in the beginning of each chapter. Each chapter overview will provide readers with the subject matter to be discussed in each chapter and some logic of why materials are presented in the given sequence within the chapter. This will be useful to give a big picture of each chapter.

The authors gratefully acknowledge comments and encouragements from many colleagues after publishing the first edition even if we do not list their names here. The authors also acknowledge help by the staff in CRC in preparing the manuscript. We also acknowledge preparing the index by Elliot and Soonja.

April, 2000
Y. W. Kwon
H. C. Bang

PREFACE TO THE FIRST EDITION

The finite element method has become one of the most important and useful engineering tools for engineers and scientists. This book presents introductory and some advanced topics of the Finite Element Method (FEM). Finite element theories, formulations, and various example programs written in MATLAB[1] are presented. The book is written as a textbook for upper level undergraduate and lower level graduate courses, as well as a reference book for engineers and scientists who want to write quick finite element analysis programs.

Understanding basic program structures of the Finite Element Analysis (FEA) is an important part for better comprehension of the finite element method. MATLAB is especially convenient to write and understand finite element analysis programs because a MATLAB program manipulates matrices and vectors with ease. These algebraic operations constitute major parts of the FEA program. In addition, MATLAB has built-in graphics features to help readers visualize the numerical results in two- and/or three-dimensional plots. Graphical presentation of numerical data is important to interpret the finite element results. Because of these benefits, many examples of finite element analysis programs are provided in MATLAB.

The book contains extensive illustrative examples of finite element analyses using MATLAB program for most problems discussed in the book. Subroutines (MATLAB functions) are provided in the appendix and a computer diskette which contains all the subroutines and example problems is also provided.

Chapter 1 has a brief summary of useful MATLAB commands which can be used in programming FEA. Readers may refer to MATLAB manuals for additional information. However, this chapter may be a good start for readers who have no experience with MATLAB.

Subsequent chapters are presented in a logical order. Chapter 2 discusses the weighted residual method which is used for the formulation of FEA in the remaining chapters. Initially, continuous trial functions are used to obtain approximate solutions using the weighted residual method. Next, piecewise continuous functions are selected to achieve approximate solutions. Then, FEM is introduced from the concept of piecewise continuous functions. Finally, classical variational formulations are compared with the weighted residual formulations.

Chapter 3 shows the basic program structure of FEA using ordinary differential equations for a one-dimensional system. MATLAB programs are provided to explain the programming. Both program input and output as well as internal program structure are fully discussed. A direct FEM approach using simple mechanics is presented in Chapter 4. This chapter presents the basic concept of FEM using an intuitive and physical approach.

---

[1]MATLAB is a registered trademark of The MathWorks, Inc. For additional information contact:
The MathWorks, Inc.
3 Apple Hill Drive
Natick, MA 01760
phone: (508) 647-7000, fax: (508) 647-7001

Finite element formulations for partial differential equations are presented in Chapter 5. This chapter explains not only domain integration for computation of the finite element matrices but also boundary integration to compute column vectors. Applications of Laplace's equation to two- and three-dimensional domains as well as an axially symmetric domain are presented for both steady-state and transient problems.

Chapter 6 shows concepts and programming of isoparametric finite elements. Because a complex shape of domain with curved boundary can be easily handled using isoparametric finite elements, these elements are very useful and common in FEA. Both one-dimensional and two-dimensional isoparametric elements are presented. A numerical technique and its programming concept are also discussed. As a program example, Laplace's equation is solved using isoparametric elements.

Chapters 7 and 8 discuss truss and frame structures. Static, dynamic, and eigenvalue problems are solved. In addition, one-, two- and three-dimensional structures are considered. As a result, coordinate transformation from local to global axes is explained. In particular, various formulations for the beam structure are compared; the relative advantages and disadvantages of each are cited. Modeling of laminated beams with embedded cracks is also discussed. Further, Chapter 8 presents the modal analysis and Fast Fourier Transform.

Elasticity is studied in Chapter 9. Plane stress/strain, axisymmetric and three dimensional problems are included. Both static and dynamic analyses are presented. The finite element formulations are presented in terms of the weighted residual method. However, an energy method is also discussed for comparison. Plate bending is given in Chapter 10. Similar to beam formulations, different plate bending formulations are presented for comparison.

Finally, structural control using FEM is presented in Chapter 11. This chapter is intended to provide a broad understanding of the basic concepts of control law in conjunction with FEM. Due to limited space, only a few major control theories are presented. It is assumed that readers are already familiar with fundamentals of linear dynamic systems analysis.

This book contains more material than can be covered in one semester. Thus, materials may be selected depending on course objectives. For an introductory FEM course, Chapters 2 through 9 are recommended. Depending on the desired course contents, some sections may be deleted.

We would like to thank individuals who have contributed to this book. The authors would like to express our appreciation to Professor Aleksandra Vinogradov for reviewing the manuscript and providing us with many useful suggestions. We are also indebted to the staff of CRC Press for their professional guidance in the production of this book. Finally but not lastly the authors sincerely appreciate the lifelong support and encouragement by their parents.

March, 1996
Y. W. Kwon
H. C. Bang

To Our Dedicated Parents!

CHAPTER ONE

INTRODUCTION TO MATLAB

1.0 Chapter Overview

This chapter describes what the finite element method is and how this book is organized. Because MATLAB is used as the programming tool, an introduction to MATLAB and its commands useful for the finite element method are provided in this chapter.

1.1 Finite Element Method

In order to analyze an engineering system, a mathematical model is developed to describe the system. While developing the mathematical model, some assumptions are made for simplification. Finally, the governing mathematical expression is developed to describe the behavior of the system. The mathematical expression usually consists of differential equations and given conditions.

These differential equations are usually very difficult to obtain solutions which explain the behavior of the given engineering system. With the advent of high performance computers, it has become possible to solve such differential equations. Various numerical solution techniques have been developed and applied to solve numerous engineering problems in order to find their approximate solutions. Especially, the finite element method has been one of the major numerical solution techniques. One of the major advantages of the finite element method is that a general purpose computer program can be developed easily to analyze various kinds of problems. In particular, any complex shape of problem domain with prescribed conditions can be handled with ease using the finite element method.

The finite element method requires division of the problem domain into many subdomains and each subdomain is called a *finite element*. Therefore, the problem domain consists of many finite element patches.

1.2 Overview of the Book

This book is written as a textbook for engineering students as well as a reference book for practicing engineers and researchers. The book consists of two parts: theory and program. Therefore, each chapter has initial sections explaining fundamental theories and formulations of the finite element method, and subsequent sections showing examples of finite element programs written in the MATLAB program. The collection of MATLAB function files (i.e., *m-files*) used in the example programs is summarized in Appendix A and provided in a separate computer disc.

A brief summary of some MATLAB commands is provided in the following sections for readers who are not familiar with them. Those are the commands which may be used in finite element programs. Especially, the MATLAB commands for matrix operation and solution are most frequently used in the programs. For visualization of the finite element solution, some plotting commands are also explained.

1.3 About MATLAB

MATLAB is an interactive software which has been used recently in various areas of engineering and scientific applications. It is not a computer language in the normal sense but it does most of the work of a computer language. Writing a computer code is not a straightforward job, typically boring and time consuming for beginners. One attractive aspect of MATLAB is that it is relatively easy to learn. It is written on an intuitive basis and it does not require in-depth knowledge of operational principles of computer programming like compiling and linking in most other programming languages. This could be regarded as a disadvantage since it prevents users from understanding the basic principles in computer programming. The interactive mode of MATLAB may reduce computational speed in some applications.

The power of MATLAB is represented by the length and simplicity of the code. For example, one page of MATLAB code may be equivalent to many pages of other computer language source codes. Numerical calculation in MATLAB uses collections of well-written scientific/mathematical subroutines such as LINPACK and EISPACK. MATLAB provides Graphical User Interface (GUI) as well as three-dimensional graphical animation.

In general, MATLAB is a useful tool for vector and matrix manipulations. Since the majority of the engineering systems are represented by matrix and vector equations, we can relieve our workload to a significant extent by using MATLAB. The finite element method is a well-defined candidate for which MATLAB can be very useful as a solution tool. Matrix and vector manipulations are essential parts in the method. MATLAB provides a *help* menu so that we can type the *help* command when we need help to figure out a command. The *help* utility is quite convenient for both beginners and experts.

1.4 Vector and Matrix Manipulations

Once we get into MATLAB, we meet a prompt >> called the MATLAB prompt. This prompt receives a user command and processes it providing the output on the next line. Let us try the following command to define a matrix.

>> $A = [1, 3, 6; 2, 7, 8; 0, 3, 9]$

Then the output appears in the next line as shown below.

$$A = \begin{array}{ccc} 1 & 3 & 6 \\ 2 & 7 & 8 \\ 0 & 3 & 9 \end{array}$$

Thus, a matrix is entered row by row, and each row is separated by the semicolon(;). Within each row, elements are separated by a space or a comma(,). Commands and variables used in MATLAB are case-sensitive. That is, lower case letters are distinguished from upper case letters. The size of the matrix is checked with

>> **size**(A)

$ans = 3 \quad 3$

Transpose of a matrix In order to find the transpose of matrix A, we type

>> A'

The result is

$$ans = \begin{array}{ccc} 1 & 2 & 0 \\ 3 & 7 & 3 \\ 6 & 8 & 9 \end{array}$$

Column or row components MATLAB provides columnwise or rowwise operation of a matrix. The following expression

>> $A(:, 3)$

yields

$$ans = \begin{array}{c} 6 \\ 8 \\ 9 \end{array}$$

which is the third column of matrix A. In addition,

>> $A(1, :)$

represents the first row of A as

$ans = 1 \quad 3 \quad 6$

We can also try

>> $A(1, :) + A(3, :)$

as addition of the first and third rows of A with the result

$ans = 1\ 6\ 15$

Now let us introduce another matrix B as

>> $B = [3, 4, 5; 6, 7, 2; 8, 1, 0];$

Then there seems to be no output on the screen. MATLAB does not prompt output on the screen when an operation ends with the semicolon(;).

If we want to check the B matrix again, we simply type

>> B

The screen output will be

$$B = \begin{matrix} 3 & 4 & 5 \\ 6 & 7 & 2 \\ 8 & 1 & 0 \end{matrix}$$

Matrix addition Adding two matrices is straightforward like

>> $C = A + B$

$$C = \begin{matrix} 4 & 7 & 11 \\ 8 & 14 & 10 \\ 8 & 4 & 9 \end{matrix}$$

Thus we defined a new matrix C as the sum of the previous two matrices.

Matrix subtraction In order to subtract matrix B from matrix A, we type

>> $C = A - B$

$$C = \begin{matrix} -2 & -1 & 1 \\ -4 & 0 & 6 \\ -8 & 2 & 9 \end{matrix}$$

Note that C is now a new matrix, not the summation of A and B anymore.

Matrix multiplication Similarly, matrix multiplication can be done as

>> $C = A * B$

$$C = \begin{matrix} 69 & 31 & 11 \\ 112 & 65 & 24 \\ 90 & 30 & 6 \end{matrix}$$

1.5 Matrix Functions

Manipulation of matrices is a key feature of the MATLAB functions. MATLAB is a useful tool for matrix and vector manipulations. Collections of representative MATLAB matrix functions are listed in Table 1.5.1. Examples and detailed explanations are provided for each function below.

Section 1.5 Matrix Functions

Table 1.5.1 Basic Matrix Functions

Symbol	Explanations
inv	inverse of a matrix
det	determinant of a matrix
rank	rank of a matrix
cond	condition number of a matrix
eye(n)	the n by n identity matrix
trace	summation of diagonal elements of a matrix
zeros(n,m)	the n by m matrix consisting of all zeros

Matrix inverse The inverse of a matrix is as simple as

>> **inv**(A)

ans

$$\begin{array}{ccc} 1.8571 & -0.4286 & -0.8571 \\ -0.8571 & 0.4286 & 0.1905 \\ 0.2857 & -0.1429 & 0.0476 \end{array}$$

In order to verify the answer, we can try

>> $A*$**inv**(A);

which should be a 3 by 3 identity matrix.

Determinant of a matrix

>> $d =$**det**(A)

produces the determinant of the matrix A. That is,

$d = 21$

Rank of a matrix The rank of a matrix A, which is the number of independent rows or columns, is obtained from

>>**rank**(A);

Identity matrix

>> **eye**(3)

yields

$$ans = \begin{array}{ccc} 1 & 0 & 0 \\ 0 & 1 & 0 \\ 0 & 0 & 1 \end{array}$$

eye(n) produces an identity matrix of size n by n. This command is useful when we initialize a matrix.

Matrix of random numbers A matrix consisting of random numbers can be generated using the following MATLAB function.

>>**rand**(3, 3)

$$ans = \begin{matrix} 0.2190 & 0.6793 & 0.5194 \\ 0.0470 & 0.9347 & 0.8310 \\ 0.6789 & 0.3835 & 0.0346 \end{matrix}$$

That is, **rand**(3, 3) produces a 3 by 3 matrix whose elements consist of random numbers. The general usage is **rand**(n, m).

trace Summation of diagonal elements of a matrix can be obtained using the **trace** operator.

For example,

>> $C = [1 \ 3 \ 9; \ 6 \ 7 \ 2; \ 8 \ -1 \ -2];$

Then, **trace**(C) produces 6, which is the sum of diagonal elements of C.

zero matrix

>> **zeros**(5, 4)

produces a 5 by 4 matrix consisting of all zero elements. In general, **zeros**(n, m) is used for an n by m zero matrix.

condition number The command **cond**(A) is used to calculate the condition number of a matrix A. The condition number represents the degree of singularity of a matrix. An identity matrix has a condition number of unity, and the condition number of a singular matrix is infinity.

>>**cond**(eye(6))

$ans =$
  1

An example matrix which is near singular is

$$A = \begin{bmatrix} 1 & 1 \\ 1 & 1 + 10^{-6} \end{bmatrix}$$

The condition number is

>>**cond**(A)

$ans =$
  4.0000e+006

Further matrix functions are presented in Table 1.5.2. They do not include all matrix functions of the MATLAB, but represent only a part of the whole MATLAB

Table 1.5.2 Basic Matrix Functions (Continued)

Symbol	Explanations
expm	exponential of a matrix
eig	eigenvalues/eigenvectors of a matrix
lu	LU decomposition of a matrix
svd	singular value decomposition of a matrix
qr	QR decomposition of a matrix
\	used to solve a set of linear algebraic equations

functions. Readers can use the MATLAB Reference Guide or *help* command to check when they need more MATLAB functions.

Matrix exponential The **expm(A)** produces the exponential of a matrix A. In other words,

```
>> A =rand(3,3)
```

$$A = \begin{matrix} 0.2190 & 0.6793 & 0.5194 \\ 0.0470 & 0.9347 & 0.8310 \\ 0.6789 & 0.3835 & 0.0346 \end{matrix}$$

```
>>expm(A)
```

$$ans = \begin{matrix} 1.2448 & 0.0305 & 0.6196 \\ 1.0376 & 1.5116 & 1.3389 \\ 1.0157 & 0.1184 & 2.0652 \end{matrix}$$

Eigenvalues The eigenvalue problem of a matrix is defined as

$$A\phi = \lambda\phi$$

where λ is the eigenvalue of matrix A, and ϕ is the associated eigenvector.

```
>> e =eig(A)
```

gives the eigenvalues of A, and

```
>> [V, D] =eig(A)
```

produces the V matrix, whose columns are eigenvectors, and the diagonal matrix D whose values are eigenvalues of the matrix A.
For example,

```
>> A = [5 3 2; 1 4 6; 9 7 2];
>> [V, D] =eig(A)
```

$$V = \begin{matrix} 0.4127 & 0.5992 & 0.0459 \\ 0.5557 & -0.7773 & -0.6388 \\ 0.7217 & 0.1918 & 0.7680 \end{matrix}$$

$$D = \begin{matrix} 12.5361 & 0 & 0 \\ 0 & 1.7486 & 0 \\ 0 & 0 & -3.2847 \end{matrix}$$

LU Decomposition The LU decomposition command is used to decompose a matrix into a combination of upper and lower triangular matrices.

```
>> A = [1 3 5; 2 4 8; 4 7 3];
>> [L,U] =lu(A)
```

$$L = \begin{matrix} 0.2500 & 1.0000 & 0 \\ 0.5000 & 0.4000 & 1.0000 \\ 1.0000 & 0 & 0 \end{matrix}$$

$$U = \begin{matrix} 4.0000 & 7.0000 & 3.0000 \\ 0 & 1.2500 & 4.2500 \\ 0 & 0 & 4.8000 \end{matrix}$$

In order to check the result, we try

```
>> L * U
```

$$ans = \begin{bmatrix} 1 & 3 & 5 \\ 2 & 4 & 8 \\ 4 & 7 & 3 \end{bmatrix}$$

The lower triangular matrix L is not perfectly triangular. There is another command available

```
>> [L,U,P] =lu(A)
```

$$L = \begin{matrix} 1.0000 & 0 & 0 \\ 0.2500 & 1.0000 & 0 \\ 0.5000 & 0.4000 & 1.0000 \end{matrix}$$

$$U = \begin{matrix} 4.0000 & 7.0000 & 3.0000 \\ 0 & 1.2500 & 4.2500 \\ 0 & 0 & 4.8000 \end{matrix}$$

$$P = \begin{matrix} 0 & 0 & 1 \\ 1 & 0 & 0 \\ 0 & 1 & 0 \end{matrix}$$

Here, the matrix P is the permutation matrix such that $P * A = L * U$.

Singular value decomposition The **svd** command is used for singular value decomposition of a matrix. For a given matrix,

$$A = U\Sigma V'$$

where Σ is a diagonal matrix consisting of non-negative values. For example, we define a matrix D like

>> $D = [1\ 3\ 7;\ 2\ 9\ 5;\ 2\ 8\ 5]$

The singular value decomposition of the matrix is

>> $[U, Sigma, V] = \mathbf{svd}(D)$

which results in

$$U = \begin{bmatrix} 0.4295 & 0.8998 & -0.0775 \\ 0.6629 & -0.3723 & -0.6495 \\ 0.6133 & -0.2276 & 0.7564 \end{bmatrix}$$

$$Sigma = \begin{bmatrix} 15.6492 & 0 & 0 \\ 0 & 4.1333 & 0 \\ 0 & 0 & 0.1391 \end{bmatrix}$$

$$V = \begin{bmatrix} 0.1905 & -0.0726 & 0.9790 \\ 0.7771 & -0.5982 & -0.1956 \\ 0.5999 & 0.7980 & -0.0576 \end{bmatrix}$$

QR decomposition A matrix can be also decomposed into a combination of an orthonormal matrix and an upper triangular matrix. In other words,

$$A = QR$$

where Q is the matrix with orthonormal columns, and R is the upper triangular matrix. The QR algorithm has wide applications in the analysis of matrices and associated linear systems. For example,

$$A = \begin{bmatrix} 0.2190 & 0.6793 & 0.5194 \\ 0.0470 & 0.9347 & 0.8310 \\ 0.6789 & 0.3835 & 0.0346 \end{bmatrix}$$

Application of the **qr** operator follows as

>> $[Q, R] = \mathbf{qr}(A)$

and yields

$$Q = \begin{bmatrix} -0.3063 & -0.4667 & -0.8297 \\ -0.0658 & -0.8591 & 0.5076 \\ -0.9497 & 0.2101 & 0.2324 \end{bmatrix}$$

$$R = \begin{bmatrix} -0.7149 & -0.6338 & -0.2466 \\ 0 & -1.0395 & -0.9490 \\ 0 & 0 & -0.0011 \end{bmatrix}$$

Solution of linear equations The solution of a linear system of equations is frequently needed in the finite element method. The typical form of a linear system of algebraic equations is written as

Table 1.6.1 Data Analysis Functions

Symbol	Explanations
min(max)	minimum(maximum) of a vector
sum	sum of elements of a vector
std	standard deviation of a data collection
sort	sort the elements of a vector
mean	mean value of a vector
	used for componentwise operation of a vector

$$Ax = y$$

and the solution is obtained by

>> $x =$ **inv**$(A) * y$

or we can use \ sign as

>> $x = A \backslash y$

For example

>> $A = [1 \ 3 \ 4; \ 5 \ 7 \ 8; \ 2 \ 3 \ 5];$

and

>> $y = [10; \ 9; \ 8];$

Let us compare two different approaches.

>> $[\textbf{inv}(A) * y \quad A \backslash y]$

$$ans = \begin{array}{cc} -4.2500 & -4.2500 \\ 1.7500 & 1.7500 \\ 2.2500 & 2.2500 \end{array}$$

1.6 Data Analysis Functions

Up to now, we discussed matrix related functions and operators of MATLAB. MATLAB also has data analysis functions for a vector or a column of a matrix. In Table 1.6.1, some operators for data manipulation are listed.

Minimum (maximum) The **min (max)** finds a minimum (maximum) value of a given vector. For example,

>> $v = [11 \ 23 \ 73 \ 25 \ 49 \ 92 \ 28 \ 23];$

Section 1.6   Data Analysis Functions   11

>>min(v)

yields

    $ans =$
        11

>>max(v)
$ans =$
    92

sum The **sum** command produces the summation of elements of a vector. For example,

>> sum(v)

yields

    $ans =$
        324

Standard deviation The **std** command calculates the standard deviation of a vector. For example,

>> std([1 4 10 −5 6 9 −20])

$ans =$
    10.4517

Sort a vector The **sort** command is used to sort a vector in the ascending order.

>> sort([1 4 10 −5 6 9 −20])

$ans =$
    -20 -5 1 4 6 9 10

Mean value of a vector The **mean** calculates the mean value of a vector.

>> mean([1 4 10 −5 6 9 −20])

$ans =$
    0.7143

Vector componentwise operation Let us define two vectors

>> $v_1 = [1, 5, 6, 7]; v_2 = [0, 2, 3, 5];$

Sometimes we want to multiply components of v_1 with the corresponding components of v_2. The operation is

>> $v_3 = v_1 .* v_2$

$ans =$
    0 10 18 35

Table 1.7.1 Polynomial Functions

Symbol	Explanations
poly	converts collection of roots into a polynomial equation
roots	finds the roots of a polynomial equation
polyval	evaluates a polynomial for a given value
conv	multiply two polynomials
deconv	decompose a polynomial into a dividend and a residual
polyfit	curve fitting of a given polynomial

In other words, ( ).*( ) represents the componentwise multiplication of two vectors. Another useful operator is

>> $v_4 = v_2./v_1$

with

$v_4 = 0\ \ 0.4\ \ 0.5\ \ 0.7143$

Note that the data analysis tools explained in the above are applicable to matrices, too. Each matrix column is regarded as a vector for data analysis.

1.7 Tools for Polynomials

Polynomials are frequently used in the analysis of linear systems. MATLAB provides some tools for handling polynomials. The summary of polynomial functions is provided in Table 1.7.1.

Roots of a polynomial equation   A polynomial equation is given by

$$a_1 x^n + a_2 x^{n-1} + \cdots + a_n x + a_{n+1} = 0$$

The roots of the polynomial equation are solved using **roots** command

$$\mathbf{roots}([a_1\ \ a_2\ \ \cdots\ \ a_n\ \ a_{n+1}])$$

For example,

$$x^4 + 4x^3 - 5x^2 + 6x - 9 = 0$$

>>**roots**([1 4 −5 6 −9])

yields

ans =
 -5.2364
 1.2008
 0.0178 + 1.1963i
 0.0178 - 1.1963i

Generation of a polynomial equation using roots The **poly** command takes the roots, and converts them into a polynomial equation. For instance, if we know $[r_1, r_2, \cdots, r_n]$ in

$$(x - r_1)(x - r_2) \cdots (x - r_n) = x^n + a_1 x^{n-1} + a_2 x^{n-2} + \cdots + a_{n-1} x + a_n$$

then

>>**poly**($[r_1, r_2, \cdots, r_n]$)

provides us the coefficients($[a_1, a_2, \cdots, a_n]$) of the polynomial equation. For example,

>>**poly**($[-1 \quad -2+2*i \quad -2-2*i \quad -5+7*i \quad -5-7*i]$)

produces

ans =
 1 15 136 498 968 592

In order to check the result, we use the **roots** command again.

>>**roots**([1 15 136 498 968 592])

The result should be $[-1 \quad -2+2*i \quad -2-2*i \quad -5+7*i \quad -5-7*i]$.

Polynomial value When we want to calculate the value of a polynomial at a certain point, we can use **polyval**.

>> y =**polyval**([1 3 4 -5], 2)

ans =
 23

which evaluates the polynomial $s^4 + 3s^3 + 4s - 5$ at $s = 2$.

Multiplication of two polynomials The **conv** command is used to multiply two polynomials. For example,

$$a(s) = s^2 + 3s - 1, \qquad b(s) = s^3 - 2s^2 + 6s - 7$$

The multiplication of $a(s)$ and $b(s)$ follows as

>> c =**conv**([1 3 -1], [1 -2 6 -7])

c =
 1 1 -1 13 -27 7

In other words, we obtain the coefficient vector c of the product of $a(s)$ and $b(s)$.

Decomposition of a polynomial The **deconv** is used to decompose a polynomial as a multiplicand and a residue. Let

$$a(s) = b(s)m(s) + r(s)$$

That is, the polynomial *a(s)* is represented in terms of a multiplicand *m(s)* and a residue *r(s)* via *b(s)*. The MATLAB command is

>> $[m, r]$ =**deconv**(a, b)

where the parameters are coefficient vectors for given polynomials. An example is given by

>> $[m, r]$ =**deconv**$([1 \ -2 \ 6 \ -7], [1 \ 3 \ -1])$

m =
    1 -5

r =
    0   0   22 -12

If we change the order of polynomials,

>> $[m, r]$ =**deconv**$([1 \ 3 \ -1], [1 \ -2 \ 6 \ -7])$

m =
    0

r =
    1   3 -1

Polynomial fit The **polyfit** command is to generate a polynomial curve which fits a given set of data. The polynomial is obtained by minimizing the error between the polynomial and the given data set. The synopsis is

$$p = \mathbf{polyfit}(x, y, n)$$

where x and y are vectors of the given data set in (x, y) form, and n is the order of the desired polynomial to fit the data set. The output result is p, the coefficient vector of the fitting polynomial. An example is provided below:

>> $x = [1 \ 2 \ 3 \ 4 \ 5 \ 6]$;

>> $y = [-1 \ 3 \ 5 \ 2 \ -3 \ 1]$;

>> p =**polyfit**$(x, y, 1)$

p =
    -0.3143   2.2667

A linear curve fitting is performed for data set (x, y).

1.8 Making Complex Numbers

In order to make a complex number $2 + 3*i$, we use

>> 2 + 3*i

or

>> 2 + 3*j

MATLAB takes i and j as a pure complex number. In case i or j is defined already, we can use $\sqrt{-1}$ as

>> i = **sqrt**(−1)

$i = 0 + 1.0000i$

abs, angle For a given complex number, we use **abs** and **angle** commands to find out the magnitude (**abs**) and phase angle (**angle**) of the given complex number. For example, if

>> $c = -1 + i$;

then

>> **abs**(c)

$ans = 1.4142$

>> **angle**(c)

$ans = 2.3562$

Real, imaginary parts of a complex number The **real** and **imag** are used to take the real and imaginary parts of a complex number. For example,

>> $c = -10 + 9*i$

>> [**real**(c),**imag**(c)]

$ans = -10 \quad 9$

Conjugate The **conj** command is used to generate a complex conjugate number. For example

>> **conj**($-1 + 5*i$)

$ans = -1 - 5*i$

1.9 Nonlinear Algebraic Equations

Nonlinear algebraic equations are frequently adopted in many different areas. The nonlinear equations are different from linear equations, and there is no unique

Table 1.9.1 Functions for Nonlinear Algebraic Equations

Symbol	Explanations
fmin	finds minimum of a function of one variable
fzero	solves a nonlinear algebraic equation of one variable

analysis tool to the nonlinear equations. MATLAB is equipped with some functions which can handle nonlinear equations. The list is presented in Table 1.9.1.

Minimum of a function The MATLAB command **fmin** minimizes a function by finding out a value which minimizes the given function. The synopsis is

$$\textbf{fmin}('func', x_1, x_2)$$

where $'func'$ is the name of a function to be minimized and $x_1(x_2)$ represents a lower(upper) limit of the interval of the function argument. For example,

>>**fmin**$('x*cos(x)', -2, 2)$

produces

$ans = -0.8603$

Solution of a nonlinear algebraic equation When a nonlinear algebraic equation is written as

$$x^3 + x*cos(x) - 4x = 0$$

the MATLAB function **fzero** can be used to find a solution of the nonlinear algebraic equation. The synopsis is

>> $sol =$**fzero**$('function', x0)$

where $'function'$ is a MATLAB function subroutine and $x0$ is an initial condition vector of the variables. For the given example, we write a function subroutine $fctn.m$ as

function $[f] =$**fctn**$(x);$
$f = x\hat{\ }3 + x*cos(x) - 4x;$

Then, we use the **fzero** command as

>> $sol =$**fzero**$('fctn', -5)$

$sol = \textit{-2.1281}$

In order to check the solution

>>**fctn**(-2.1281)

$ans =$
$\textit{1.9192e-004}$

Table 1.10.1 Numerical Techniques for Differential Equations

Symbol	Explanations
ode23	solution using the 2nd/3rd order Runge-Kutta algorithm
ode45	solution using the 4th/5th order Runge-Kutta algorithm

The error is due to the numerical format error. The number is truncated for screen display purpose, even if it is calculated using double precision format inside MATLAB.

1.10 Solving Differential Equations

Linear and nonlinear differential equations can be also solved using MATLAB. A list of numerical techniques solving differential equations is in Table 1.10.1.

Runge-Kutta second and third order algorithm MATLAB uses the Runge-Kutta algorithm to solve a differential equation or a set of differential equations. The general synopsis is

$$[t, x] = \mathbf{ode23}('func', t0, tf, x0);$$

where $'func'$ is a function containing the derivative information, $t0$ (tf) is the initial (final) time, and $x0$ is an initial condition vector. The outputs are t, which contains the returned time points, and x which is the integrated output.

For example, we want to solve

$$\ddot{x} + sin(x) = 0$$

which can be rewritten as

$$\dot{x}_1 = x_2$$
$$\dot{x}_2 = -sin(x_1)$$

where $x_1 = x$ and $x_2 = \dot{x}$. The $'func'$ function should be provided as an independent function subroutine as $func.m$ in a directory, which MATLAB can locate. Now we execute the **ode23** command

$$>> [t, x] = \mathbf{ode23}('func', 0, 10, x0);$$

where

$$x0 = \begin{Bmatrix} 0 \\ 0 \end{Bmatrix}$$

is an initial condition and $func.m$ is provided as

function $[f]$ =**func**(t,x);
$[f] = zeros(2, 1);$
$f(1) = x(2);$
$f(2) = -sin(x(1));$

Table 1.11.1 Loop and Logical Statements

Symbol	Explanations
for	loop command similar to other languages
while	used for a loop combined with conditional statement
if	produces a conditional statement
elseif, else	used in conjunction with if command
break	breaks a loop when a condition is satisfied

Runge-Kutta fourth and fifth order algorithm There is another Runge-Kutta algorithm **ode45** which is more accurate than **ode23**.

$$[t,x] = \mathbf{ode45}('func', t0, tf, x0)$$

The same calling synopsis as **ode23** can be applied to make use of the **ode45** function.

1.11 Loop and Logical Statements

There are some logical statements available in MATLAB which help us in writing combinations of MATLAB commands. Furthermore, loop commands can be used as in other programming languages. In fact, we can duplicate the majority of existing programs using MATLAB commands, which significantly reduces the size of the source codes. A collection of loop and logical statements in MATLAB is presented in Table 1.11.1.

for loop The **for** is a loop command which ends with **end** command.

```
>> for i = 1 : 100
      a(i,i) = 2 * i;
   end
```

In the above example, i is a loop index which starts from 1 and ends at 100. There may be also multiple loops.

```
>> for i = 1 : 100
      for j = 1 : 50
         for k = 1 : 50
            a(i,j) = b(i,k) * c(k,j) + a(i,j);
         end
      end
   end
```

while The **while** command is useful for an infinite loop in conjunction with a conditional statement. The general synopsis for the **while** command is as follows:

>**while** *condition*
>*statements*
>**end**

For example,

>$i = 1$
>**while** $(i < 100)$
> $i = i + 1;$
>**end**

Another example of the **while** command is

>$n = 1000;$
>$var = [\ ];$
>**while** $(n > 0)$
> $n = n/2 - 1;$
> $var = [var, n];$
>**end**

The result is

>var =
>Columns 1 through 6
>
>4.9900e+002 2.4850e+002 1.2325e+002 6.0625e+001 2.9313e+001 1.3656e+001
>
>Columns 7 through 9
>5.8281e+000 1.9141e+000 -4.2969e-002

where we used [ ] in order to declare an empty matrix.

if, elseif, else The **if, elseif,** and **else** commands are conditional statements which are used in combination.

>**if** *condition #1*
>*statement #1*
>**elseif** *condition #2*
>*statement #2*
>**else**
>*statement #3*
>**end**

For example,

>$n = 100;$
>**if** $(\mathbf{rem}(n, 3) == 0)$
> $x = 0;$
>**elseif** $(\mathbf{rem}(n, 3) == 1)$
> $x = 1;$
>**else**

Table 1.11.2 Loop and Logical Statements

Symbol	Explanations
==	two conditions are equal
~=	two conditions are not equal
<= (>=)	one is less (greater) than or equal to the other
< (>)	one is less (greater) than the other
&	*and* operator - two conditions are met
~	*not* operator
|	*or* operator - either condition is met

    $x = 2$;
end

where **rem(x,y)** is used to calculate the remainder of x divided by y.

break   The **break** command is used to exit from a loop such as **if** and **while**. For example,

    for $i = 1 : 100$
      $i = i + 1$;
        $if(i == 10)$ **break**;
      end
    end

Logical and relational operators   The logical and relational operators of MATLAB are as listed in Table 1.11.2.

The above command sets are used in combination.

1.12 Writing Function Subroutines

MATLAB provides a convenient tool, by which we can write a program using collections of MATLAB commands. This approach is similar to other common programming languages. It is quite useful especially when we write a series of MATLAB commands in a text file. This text file is edited and saved for later use.

The text file should have *filename.m* format normally called *m-file*. That is, all MATLAB subroutines should end with *.m* extension, so that MATLAB recognizes them as MATLAB compatible files. The general procedure is to make a text file using any text editor. If we generate a file called *func1.m*, then the file *func1.m* should start with the following file header

$$\mathbf{function}[ov_1, ov_2, \ldots] = \mathbf{func1}(iv_1, iv_2, \ldots)$$

where $iv_1, iv_2, \ldots$ are input variables while $ov_1, ov_2, \ldots$ are output variables. The input variables are specific variables and the output variables are dummy variables, for which we can use any variables.

For example, let us solve a second order algebraic equation.

$$ax^2 + bx + c = 0$$

The solution is given in analytical form as

$$x = \frac{-b \pm \sqrt{b^2 - 4*a*c}}{2*a}$$

We want to write an *m-file* with the name *secroot.m*, which produces the analytical solution.

```
function [r1,r2]=secroot(a,b,c);
%
% Find Determinant ——— Any command in MATLAB which starts with
% % sign is a comment statement
Det = b^2 - 4*a*c;
if (Det < 0),
r1 = (-b + j*sqrt(-Det))/2/a;
r2 = (-b - j*sqrt(-Det))/2/a;
disp('The two roots are complex conjugates');
elseif(Det == 0),
r1 = -b/2/a;
r2 = -b/2/a;
disp('There are two repeated roots');
else(Det > 0)
r1 = (-b + sqrt(Det))/2/a;
r2 = (-b - sqrt(Det))/2/a;
disp('The two roots are real');
end
```

Some commands appearing in the above example will be discussed later. Once the *secroot.m* is created, we call that function as

>> [r1,r2]=secroot(3,4,5)

or

>> [p1,p2] =secroot(3,4,5)

One thing important about the function command is to set up the *m-file* pathname. The *m-file* should be in the directory which is set up by the MATLAB configuration set up stage. In the recent version of MATLAB, the set up procedure is relatively easy by simply adding a directory which we want to access in a MATLAB configuration file.

Table 1.13.1 File Manipulation Commands

Symbol	Explanations
save	save current variables in a file
load	load a saved file into MATLAB environment
diary	save screen display output in text format

Another function subroutine *fct.m* is provided below.

function $[f]$ =**fct**(x)
$f = (1 - x)\hat{\,}2;$

The above function represents $f(x) = (1-x)^2$. In the MATLAB command prompt, we call the function as

$>> y =$**fct**$(9);$

The function subroutine utility of MATLAB allows users to write their own subroutines. It provides flexibility of developing programs using MATLAB.

1.13 File Manipulation

Manipulating files is another attractive feature of MATLAB. We can save MATLAB workspace, that is, all variables used, in a binary file format and/or a text file format. The saved file can also be reloaded in case we need it later on. The list of file manipulation commands is presented in Table 1.13.1.

save The **save** command is used to save variables when we are working in MATLAB. The synopsis is as follows

$$\textbf{save} \quad filename \quad var_1 \quad var_2 \quad ...$$

where *filename* is the filename and we want to save the variables, $var_1, var_2, ...$ The filename generated by **save** command has extension of *.mat*, called a *mat-file*. If we do not include the variables name, then all current variables are saved automatically. In case we want to save the variables in a standard text format, we use

$$\textbf{save} \quad filename \quad var_1 \quad var_2 \quad .../ascii/double$$

load The **load** command is the counterpart of **save**. In other words, it reloads the variables in the file which was generated by **save** command. The synopsis is as follows

$$\textbf{load} \quad filename \quad var_1 \quad var_2 \quad ...$$

where *filename* is a *mat-file* saved by **save** command. Without the variables name specified, all variables are loaded. For example,

Table 1.14.1 Input-Output Functions

Symbol	Explanations
input	save current variables in a file
disp	load a saved file into MATLAB environment
format	check the file status in the directory

```
>> a = [1 3 4];
>> b = 3;
>>save   test
>>clear  all        % clear all variables
>>who               % display current variables being used
>>load   test
>>who
```

diary Using **diary** command, we can capture all MATLAB texts including command and answer lines which are displayed on the screen. The texts will be saved in a file, so that we can edit the file later. For example,

```
>>diary   on
>> a = 1; b = 4; c = 5;
>> [a b c ]
>> d = a * b
>> e = g * h
>>diary   off
```

Now we can use any text editor to modify the *diary* file. The **diary** command is useful displaying the past work procedures. Also, it can be used to save data in a text format.

1.14 Basic Input-Output Functions

Input/output functions in MATLAB provide users with a friendly programming environment. Some input/output functions are listed in Table 1.14.1.

input The **input** command is used to receive a user input from the keyboard. Both numerical and string inputs are available. For example,

$$>> age= \mathbf{input}('How\ old\ are\ you?')$$

$$>> name = \mathbf{input}('What\ is\ your\ name','s')$$

The $'s'$ sign denotes the input type is string.

disp The **disp** command displays a string of text or numerical values on the screen. It is useful when we write a function subroutine in a user-friendly manner. For example,

>> disp('*This is a MATLAB tutorial!*')

>> c=3*4;

>> disp('*The computed value of c turns out to be*')

>> c

format The **format** command is used to display numbers in different formats. MATLAB calculates floating numbers in the double precision mode. We do not want to, in some situations, display the numbers in the double precision format on the screen. For a display purpose, MATLAB provides the following different formats

>> $x = 1/9$

x = 0.1111

>>**format short e**

$x = 1.1111e - 001$

>>**format long**

$x = 0.111111111111111$

>>**format long e**

$x = 1.111111111111111e - 001$

>>**format hex**

$x = 3fbc71c71c71c71c$

1.15 Plotting Tools

MATLAB supports some plotting tools, by which we can display the data in a desired format. The plotting in MATLAB is relatively easy with various options available. The collection of plotting commands is listed in Table 1.15.1.

A sample plotting command is shown below.

>> $t = 0 : 0.1 : 10$;
>> $y = sin(t)$;
>> **plot**(y)
>> title('plot(y)')

The resultant plot is presented at the top of Fig. 1.15.1.

>> $t = 0 : 0.1 : 10$;
>> $y = sin(t)$;
>> **plot**(t, y)
>> title('plot(t,y)')

Table 1.15.1 Plotting Commands

Symbol	Explanations
plot	basic plot command
xlabel(ylabel)	attach label to x(y) axis
axis	manually scale x and y axes
text	place a text on the specific position of graphic screen
title	place a graphic title on top of the graphic
ginput	produce a coordinate of a point on the graphic screen
gtext	receives a text from mouse input
grid	add a grid mark to the graphic window
pause	hold graphic screen until keyboard is hit
subplot	breaks a graphic window into multiple windows

The resultant plot is presented at the bottom of Fig. 1.15.1. In the above example, $t = 0 : 0.1 : 10$ represents a vector t which starts from 0 and ends at 10 with an interval of 0.1. We can use just y or both y and t together. In the first case, the horizontal axis represents number of data, from 0 to 101. In the second case, the horizontal axis is the actual time scale t in the **plot(t,y)** command.

Plotting multiple data We plot multiple data sets as shown below.

>> $t = 0 : 1 : 100$;
>> $y1 = sin(t).*t$;
>> $y2 = cos(t).*t$;
>> **plot**$(t,y1,'-',t,y2,'-')$

where $'-'$ and $'-'$ represent line styles. The line styles, line marks, and colors are listed in Table 1.15.2.

For example, if we want to plot data in a *dashed blue* line, the command becomes

>> **plot**$(y,'-b')$;

xlabel, ylabel The **xlabel**$('text')$ and **ylabel**$('text')$ are used to label the x and y axes.

axis The **axis** command sets up the limits of axes. The synopsis is

$$\text{axis}[x_{min}, x_{max}, y_{min}, y_{max}]$$

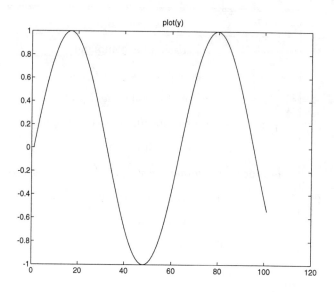

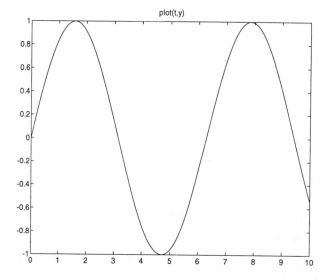

Figure 1.15.1 A Sample Plot

text The **text** command is used to write a text on the graphic window at a designated point. The synopsis is

$$\text{text}(x, y, 'text\ contents')$$

where x and y locate the (x,y) position of the $'text\ contents'$. A text can be also added in 3-D coordinates as shown below:

$$\text{text}(x, y, z, 'text\ contents')$$

Section 1.15   Plotting Tools

Table 1.15.2 Line, Mark, and Color Styles

Style		Line marks		Color	
solid	'-'	point	.	red	r
dashed	'--'	star	*	green	g
dotted	':'	circle	o	blue	b
dashdot	'-.'	plus	+	white	w
		x-mark	x	invisible	i

ginput   This command allows us to pick up any point on a graphic window. The synopsis is
$$[x, y] = \mathbf{ginput}$$
We can pick as many points as we want on the graphic screen. The vector $[x, y]$ then contains all the points.

gtext   The **gtext** command is used to place text on the graphic window using the mouse input. The synopsis is
$$\mathbf{gtext}('text')$$
Once the above command is entered or read in a function subroutine, the cursor on the graphic window is activated waiting for the mouse input, so that the $'text'$ is located at the point selected by the mouse.

grid   The **grid** command adds grids to the graphic window. It is useful when we want to clarify axis scales.

An example plot constructed using some of the commands described above is presented in Fig. 1.15.2. The following commands are used for the plot output.

```
>> t=0:0.1:20;
>> plot(t, sin(t))
>> xlabel('Time(sec)'))
>> ylabel('ydata')
>> title('This is a plot example')
>> grid
>> gtext('sin(t)')
>> axis([0 20 -1.5 1.5])
```

pause   This command is useful when we display multiple graphic windows sequentially. It allows us to display one at a time with the keyboard interrupt.

subplot   The **subplot** is used to put multiple plots on the same MATLAB figure window. The command is

```
>> subplot(pqr)
```

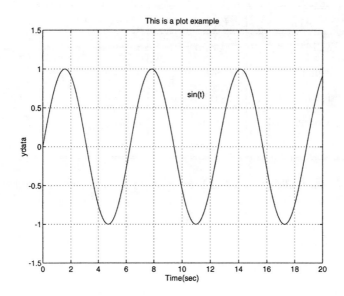

Figure 1.15.2 A Plot Example With Some Commands

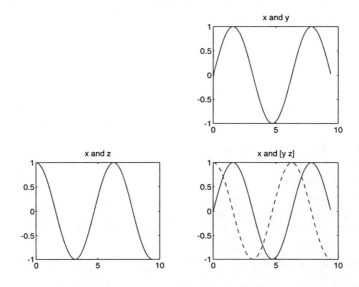

Figure 1.15.3 A Subplot Example

The plot size is adjusted by a p by q matrix on the whole size of the graphic window. Then the third index r picks one frame out of the p by q plot frames. An example subplot is presented in Fig. 1.15.3 with the following commands entered.

```
>> x = 0 : 0.1 : 3*pi; y = sin(x); z = cos(x);
>> subplot(222)
>> plot(x,y)
>> title('x and y')
>> subplot(223)
>> plot(x,z)
>> title('x and z')
>> subplot(224)
>> plot(x,y,'-',x,z,'--')
>> title('x and [y z]')
```

where *pi* is an internally defined variable equivalent to π.

CHAPTER TWO

APPROXIMATION TECHNIQUES

2.0 Chapter Overview

This chapter presents methods of weighted residual for ordinary differential equations to obtain approximate solutions. It begins with globally continuous functions as candidates for the approximate solutions. Later, piecewise continuous functions are introduced for the approximate solutions through weak formulation. Finally, the finite element approximation is discussed. Both the Galerkin finite element formulation and the Rayleigh-Ritz formulation are presented.

2.1 Methods of Weighted Residual

Methods of weighted residual are useful to obtain approximate solutions to a differential governing equation. In order to explain the methods, we consider the following sample problem:

$$\begin{cases} \frac{d^2u}{dx^2} - u = -x, & 0 < x < 1 \\ u(0) = 0, \text{ and } u(1) = 0 \end{cases} \quad (2.1.1)$$

The first step in the methods of weighted residual is to assume a trial function which contains unknown coefficients to be determined later. For example, a trial function, $\tilde{u} = ax(1-x)$, is selected as an approximate solution to Eq. (2.1.1). Here, ~ denotes an approximate solution which is usually different from the exact solution. The trial function is chosen here such that it satisfies the boundary conditions (i.e., $\tilde{u}(0) = 0$ and $\tilde{u}(1) = 0$), and it has one unknown coefficient a to be determined.

In general, accuracy of an approximated solution is dependent upon proper selection of the trial function. However, a simple form of trial function is selected for the present example to show the basic procedure of the methods of weighted residual.

Once a trial function is selected, the residual is computed by substituting the trial function into the differential equation. That is, the residual R becomes

$$R = \frac{d^2\tilde{u}}{dx^2} - \tilde{u} + x = -2a - ax(1-x) + x \tag{2.1.2}$$

Because $\tilde{u}$ is different from the exact solution, the residual does not vanish for all values of x within the domain. The next step is to determine the unknown constant a such that the chosen test function best approximates the exact solution. To this end, a test (or weighting) function w is selected and the weighted average of the residual over the problem domain is set to zero. That is,

$$\begin{aligned} I &= \int_0^1 wR\,dx = \int_0^1 w\left(\frac{d^2\tilde{u}}{dx^2} - \tilde{u} + x\right)dx \\ &= \int_0^1 w\{-2a - ax(1-x) + x\}dx = 0 \end{aligned} \tag{2.1.3}$$

The next step is to decide the test function. The resultant approximate solution differs depending on the test function. The methods of weighted residual can be classified based on how the test function is determined. Some of the methods of weighted residual are explained below. Readers may refer to Refs. [1-3] for other methods.

1. <u>Collocation Method</u>. The Dirac delta function, $\delta(x - x_i)$, is used as the test function, where the sampling point x_i must be within the domain, $0 < x_i < 1$. In other words,

$$w = \delta(x - x_i) \tag{2.1.4}$$

Let $x_i = 0.5$ and we substitute the test function into the weighted residual, Eq. (2.1.3), to find $a = 0.2222$. Then, the approximate solution becomes $\tilde{u} = 0.2222x(1-x)$.

2. <u>Least Squares Method</u>. The test function is determined from the residual such that

$$w = \frac{dR}{da} \tag{2.1.5}$$

Applying Eq.(2.1.5) to Eq. (2.1.2) yields $w = -2 - x(1-x)$. Substitution of the test function into Eq. (2.1.3) results in $a = 0.2305$. Then $\tilde{u} = 0.2305x(1-x)$.

3. <u>Galerkin's Method</u>. For Galerkin's method, the test function comes from the chosen trial function. That is,

$$w = \frac{d\tilde{u}}{da} \tag{2.1.6}$$

For the present trial function, $w = x(1-x)$. Applying this test function to Eq. (2.1.3) gives $a = 0.2272$ so that $\tilde{u} = 0.2272x(1-x)$. Comparison of these three approximate solutions to the exact solution at $x = 0.5$ is provided in Table

Table 2.1.1 Comparison of Solutions to Eq. (2.1.1) at x=0.5

Exact Solution	Collocation	Least Squares	Galerkin
0.0566	0.0556	0.0576	0.0568

Table 2.1.2 Test Functions for Methods of Weighted Residual

Method	Description
Collocation	$w_i = \delta(x - x_i), \quad i = 1, 2, ..., n$ where x_i is a point within the domain
Least Squares	$w_i = \partial R/\partial a_i, \quad i = 1, 2, ..., n,$ where R is the residual and a_i is an unknown coefficient in the trial function
Galerkin	$w_i = \partial \tilde{u}/\partial a_i, \quad i = 1, 2, ..., n$ where $\tilde{u}$ is the selected trial function

2.1.1. As seen in the comparison, all three methods result in reasonably accurate approximate solutions to Eq. (2.1.1).

In order to improve the approximate solutions, we can add more terms to the previously selected trial function. For example, another trial function is $\tilde{u} = a_1 x(1-x) + a_2 x^2(1-x)$. This trial function has two unknown constants to be determined. Computation of the residual using the present trial function yields

$$R = a_1(-2 - x + x^2) + a_2(2 - 6x - x^2 + x^3) + x \qquad (2.1.7)$$

We need the same number of test functions as that of unknown constants so that the constants can be determined properly. Table 2.1.2 summarizes how to determine test functions for a chosen trial function which has n unknown coefficients. Application of Table 2.1.2 to the present trial function results in the following test functions for each method.

$$\text{Collocation Method}: \quad w_1 = \delta(x - x_1), \quad w_2 = \delta(x - x_2) \qquad (2.1.8)$$

$$\text{Least Squares Method}: \quad w_1 = -2 - x + x^2, \quad w_2 = 2 - 6x - x^2 + x^3 \qquad (2.1.9)$$

$$\text{Galerkin's Method}: \quad w_1 = x(1-x), \quad w_2 = x^2(1-x) \qquad (2.1.10)$$

For the collocation method, x_1 and x_2 must be selected such that the resultant weighted residual, i.e. Eq. (2.1.3), can produce two independent equations to determine unknowns a_1 and a_2 uniquely. The least squares method produces a symmetric matrix regardless of a chosen trial function. Example 2.1.1 shows symmetry of the matrix resulting from the least squares method. Galerkin's method does not result in a symmetric matrix when it is applied to Eq. (2.1.1). However, Galerkin's method may produce a symmetric matrix under certain conditions as explained in the next section.

♣ **Example 2.1.1** A differential equation is written as

$$L(u) = f \tag{2.1.11}$$

where L is a linear differential operator. A trial solution is chosen such that

$$\tilde{u} = \sum_{i=1}^{n} a_i g_i \tag{2.1.12}$$

in which g_i is a known function in terms of the spatial coordinate system and it is assumed to satisfy boundary conditions. Substitution of Eq. (2.1.12) into Eq. (2.1.11) and collection of terms with the same coefficient a_i yield the residual as seen below:

$$R = \sum_{i=1}^{n} a_i h_i + p \tag{2.1.13}$$

Here, h_i and p are functions in terms of the spatial coordinate system. Test functions for the least squares method are

$$w_j = h_j, \quad j = 1, 2, \ldots, n \tag{2.1.14}$$

The weighted average of the residual over the domain yields the matrix equation

$$I = \int_\Omega w_j R \, d\Omega = \sum_{i=1}^{n} A_{ij} a_i - b_j = 0, \quad j = 1, 2, \ldots, n \tag{2.1.15}$$

where

$$A_{ij} = \int_\Omega h_i h_j \, d\Omega \tag{2.1.16}$$

Equation (2.1.16) shows that $A_{ij} = A_{ji}$ (symmetry). ‡

2.2 Weak Formulation

We consider the previous sample problem, Eq. (2.1.1), again. The formulation described in the preceding section is called the *strong formulation* of the weighted residual method. The strong formulation requires evaluation of $\int_0^1 w(\partial^2 \tilde{u}/\partial x^2)dx$, which includes the highest order of derivative term in the differential equation. The integral must have a non-zero finite value to yield a meaningful approximate solution to the differential equation. This means a trial function should be differentiable twice and its second derivative should not vanish.

So as to reduce the requirement for a trial function in terms of order of differentiability, integration by parts is applied to the strong formulation. Then Eq. (2.1.3) becomes

$$\begin{aligned} I &= \int_0^1 w\left(\frac{d^2\tilde{u}}{dx^2} - \tilde{u} + x\right)dx \\ &= \int_0^1 \left(-\frac{dw}{dx}\frac{d\tilde{u}}{dx} - w\tilde{u} + xw\right)dx + \left[w\frac{d\tilde{u}}{dx}\right]_0^1 = 0 \end{aligned} \tag{2.2.1}$$

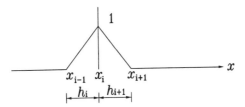

Figure 2.3.1 Piecewise Linear Functions

As seen in Eq. (2.2.1), the trial function needs the first order differentiation instead of the second order differentiation. As a result, the requirement for the trial function is reduced for Eq. (2.2.1). This formulation is called the *weak formulation*.

Weak formulation has an advantage for Galerkin's method where test functions are obtained directly from the selected trial function. If a governing differential equation is the self-adjoint operator, Galerkin's method along with the *weak formulation* results in a symmetric matrix in terms of unknown coefficients of the trial function. Using a trial function $\tilde{u} = ax(1-x)$ for the *weak formulation*, Eq. (2.2.1) results in the same solution as obtained from the *strong formulation*, as expected. However, when a piecewise function is selected as a trial function, we see the advantage of the *weak formulation* over the *strong formulation*.

2.3 Piecewise Continuous Trial Function

Regardless of the weak or strong formulation, the accuracy of an approximate solution depends greatly on the chosen trial function. However, assuming a proper trial function for the unknown exact solution is not an easy task. This is especially true when the unknown exact solution is expected to have a large variation over the problem domain, the domain has a complex shape in two-dimensional or three-dimensional problems, and/or the problem has complicated boundary conditions. In order to overcome these problems, a trial function can be described using piecewise continuous functions.

Consider piecewise linear functions in a one-dimensional domain as defined below:

$$\phi_i(x) = \begin{cases} (x - x_{i-1})/h_i & \text{for } x_{i-1} \leq x \leq x_i \\ (x_{i+1} - x)/h_{i+1} & \text{for } x_i \leq x \leq x_{i+1} \\ 0 & \text{otherwise} \end{cases} \quad (2.3.1)$$

The function defined in Eq. (2.3.1) is plotted in Fig. 2.3.1 and Example 2.3.1 illustrates the use of the function as a trial function.

♣ **Example 2.3.1** Consider the same problem as given in Eq. (2.1.1). It is rewritten here

$$\begin{cases} \frac{d^2u}{dx^2} - u = -x, & 0 < x < 1 \\ u(0) = 0, \text{ and } u(1) = 0 \end{cases} \quad (2.1.1)$$

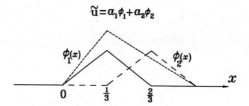

Figure 2.3.2 Piecewise Linear Trial Function

The weak formulation is also rewritten as below:

$$I = \int_0^1 w\left(\frac{d^2\tilde{u}}{dx^2} - \tilde{u} + x\right)dx$$
$$= \int_0^1 \left(-\frac{dw}{dx}\frac{d\tilde{u}}{dx} - w\tilde{u} + xw\right)dx + \left[w\frac{d\tilde{u}}{dx}\right]_0^1 = 0 \qquad (2.2.1)$$

A trial function is chosen such that $\tilde{u} = a_1\phi_1(x) + a_2\phi_2(x)$ in which a_1 and a_2 are unknown constants to be determined, and ϕ_1 and ϕ_2 are defined as below:

$$\phi_1(x) = \begin{cases} 3x, & 0 \leq x \leq \frac{1}{3} \\ 2 - 3x, & \frac{1}{3} \leq x \leq \frac{2}{3} \\ 0, & \frac{2}{3} \leq x \leq 1 \end{cases} \qquad (2.3.2)$$

$$\phi_2(x) = \begin{cases} 0, & 0 \leq x \leq \frac{1}{3} \\ 3x - 1, & \frac{1}{3} \leq x \leq \frac{2}{3} \\ 3 - 3x, & \frac{2}{3} \leq x \leq 1 \end{cases} \qquad (2.3.3)$$

$\phi_1(x)$ and $\phi_2(x)$ are plotted in Fig. 2.3.2. For the present trial function, the problem domain is divided into three subdomains and two piecewise linear functions are used. Of course, more piecewise functions can be used along with more subdomains to improve accuracy of the approximate solution. The trial function can be rewritten as

$$\tilde{u} = \begin{cases} a_1(3x), & 0 \leq x \leq \frac{1}{3} \\ a_1(2 - 3x) + a_2(3x - 1), & \frac{1}{3} \leq x \leq \frac{2}{3} \\ a_2(3 - 3x), & \frac{2}{3} \leq x \leq 1 \end{cases} \qquad (2.3.4)$$

Use of Galerkin's method yields the following test functions

$$w_1 = \begin{cases} 3x, & 0 \leq x \leq \frac{1}{3} \\ 2 - 3x, & \frac{1}{3} \leq x \leq \frac{2}{3} \\ 0, & \frac{2}{3} \leq x \leq 1 \end{cases} \qquad (2.3.5)$$

and

$$w_2 = \begin{cases} 0, & 0 \leq x \leq \frac{1}{3} \\ 3x - 1, & \frac{1}{3} \leq x \leq \frac{2}{3} \\ 3 - 3x, & \frac{2}{3} \leq x \leq 1 \end{cases} \qquad (2.3.6)$$

Averaged weighted residuals are

$$I_1 = \int_0^1 (-\frac{dw_1}{dx}\frac{d\tilde{u}}{dx} - w_1\tilde{u} + xw_1)dx = 0 \qquad (2.3.7)$$

$$I_2 = \int_0^1 (-\frac{dw_2}{dx}\frac{d\tilde{u}}{dx} - w_2\tilde{u} + xw_2)dx = 0 \qquad (2.3.8)$$

where $[w\frac{d\tilde{u}}{dx}]_0^1$ is omitted because $w_1(0) = w_1(1) = w_2(0) = w_2(1) = 0$. Substitution of both trial and test functions into Eq. (2.3.7) and Eq. (2.3.8) respectively gives

$$I_1 = \int_0^{\frac{1}{3}} [-3(3a_1) - 3x(3a_1x) + x(3x)]dx +$$

$$\int_{\frac{1}{3}}^{\frac{2}{3}} [3(-3a_1 + 3a_2) - (2 - 3x)(2a_1 - 3a_1x + 3a_2x - a_2) \qquad (2.3.9)$$

$$+ x(2 - 3x)]dx + \int_{\frac{2}{3}}^1 0 dx$$

$$= -6.222a_1 + 2.9444a_2 + 0.1111 = 0$$

$$I_2 = \int_0^{\frac{1}{3}} 0 dx + \int_{\frac{1}{3}}^{\frac{2}{3}} [-3(-3a_1 + 3a_2)$$

$$- (3x - 1)(2a_1 - 3a_1x + 3a_2x - a_2) + x(3x - 1)]dx + \qquad (2.3.10)$$

$$\int_{\frac{2}{3}}^1 [3(-3a_2) - (3 - 3x)(3a_2 - 3a_2x) + x(3 - 3x)]dx$$

$$= 2.9444a_1 - 6.2222a_2 + 0.2222 = 0$$

Solutions for a_1 and a_2 are $a_1 = 0.0448$ and $a_2 = 0.0569$ from Eq. (2.3.9) and Eq. (2.3.10). That is, the approximate solution is $\tilde{u} = 0.0448\phi_1(x) + 0.0569\phi_2(x)$. If the trial function Eq. (2.3.4) were used for the strong formulation Eq. (2.1.3), it would not give a reasonable, approximate solution because $\frac{d^2\tilde{u}}{dx^2}$ vanishes completely over the domain. ‡

2.4 Galerkin's Finite Element Formulation

As seen in the previous section, use of piecewise continuous functions for the trial function has advantages. As we increase the number of subdomains for the piecewise functions, we can represent a complex function by using the sum of simple piecewise linear functions. Later, the subdomains are called finite elements. From now on, ~ used to denote a trial function is omitted unless there is any confusion.

This section shows how to compute weighted residual in a systematic manner using finite elements and piecewise continuous functions. In the previous section, the piecewise continuous functions were defined in terms of the generalized coefficients

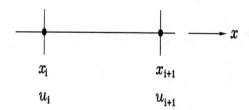

Figure 2.4.1 Two-Node Linear Element

(i.e. a_1, a_2, etc.). For a systematic formulation, the piecewise continuous functions are defined in terms of nodal variables.

Consider a subdomain or a finite element shown in Fig. 2.4.1. The element has two nodes, one at each end. At each node, the corresponding coordinate value (x_i or x_{i+1}) and the nodal variable (u_i or u_{i+1}) are assigned. Let us assume the unknown trial function to be

$$u = c_1 x + c_2 \tag{2.4.1}$$

We want to express Eq. (2.4.1) in terms of nodal variables. In other words, c_1 and c_2 need to be replaced by u_i and u_{i+1}. To this end, we evaluate u at $x = x_i$ and $x = x_{i+1}$. Then

$$u(x_i) = c_1 x_i + c_2 = u_i \tag{2.4.2}$$
$$u(x_{i+1}) = c_1 x_{i+1} + c_2 = u_{i+1} \tag{2.4.3}$$

Solving Eq. (2.4.2) and Eq. (2.4.3) simultaneously for c_1 and c_2 gives

$$c_1 = \frac{u_{i+1} - u_i}{x_{i+1} - x_i} \tag{2.4.4}$$

$$c_2 = \frac{u_i x_{i+1} - u_{i+1} x_i}{x_{i+1} - x_i} \tag{2.4.5}$$

Substitution of Eq. (2.4.4) and Eq. (2.4.5) into Eq. (2.4.1) and rearrangement of the resultant expression result in

$$u = H_1(x) u_i + H_2(x) u_{i+1} \tag{2.4.6}$$

where

$$H_1(x) = \frac{x_{i+1} - x}{h_i} \tag{2.4.7}$$

$$H_2(x) = \frac{x - x_i}{h_i} \tag{2.4.8}$$

$$h_i = x_{i+1} - x_i \tag{2.4.9}$$

Equation (2.4.6) gives an expression for the variable u in terms of nodal variables, and Eq. (2.4.7) and Eq. (2.4.8) are called linear shape functions. The shape functions are plotted in Fig. 2.4.2. These functions have the following properties:

Section 2.4 Galerkin's Finite Element Formulation

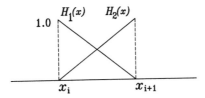

Figure 2.4.2 Linear Shape Functions

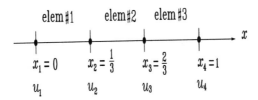

Figure 2.4.3 Finite Element Mesh With 3 Linear Elements

1. The shape function associated with node i has a unit value at node i and vanishes at other nodes. That is,

$$H_1(x_i) = 1, \quad H_1(x_{i+1}) = 0, \quad H_2(x_i) = 0, \quad H_2(x_{i+1}) = 1 \qquad (2.4.10)$$

2. The sum of all shape functions is unity.

$$\sum_{i=1}^{2} H_i(x) = 1 \qquad (2.4.11)$$

These are important properties for shape functions. The first property, Eq. (2.4.10), states that the variable u must be equal to the corresponding nodal variable at each node (i.e. $u(x_i) = u_i$ and $u(x_{i+1}) = u_{i+1}$ as enforced in Eq. (2.4.2) and Eq. (2.4.3)). The second property, Eq. (2.4.11), says that the variable u can represent a uniform solution within the element. If the solution remains constant within the element, then $u = u_i = u_{i+1}$. Substitution of this condition into Eq. (2.4.6) gives

$$u = \{H_1(x) + H_2(x)\}u_i = u_i \qquad (2.4.12)$$

Equation (2.4.12) results in the second property of shape functions, Eq. (2.4.11).

♣ **Example 2.4.1** We solve the same problem as given in Example 2.3.1 using the linear finite elements. The weighted residual can be written as

$$I = \sum_{i=1}^{n} \int_{x_i}^{x_{i+1}} \left(-\frac{dw}{dx}\frac{du}{dx} - wu + xw \right) dx + \left[u'w \right]_0^1 = 0 \qquad (2.4.13)$$

for n elements. If the problem domain is discretized into three equal-size elements, i.e. $n = 3$, Fig. 2.4.3 shows the corresponding finite element mesh. Consider the ith element (i.e. $i=1$, 2, or 3). The integral for this element is

$$\int_{x_i}^{x_{i+1}} \left(-\frac{dw}{dx}\frac{du}{dx} - wu + xw \right) dx \qquad (2.4.14)$$

The trial function u is expressed as

$$u = H_1(x)u_i + H_2(x)u_{i+1} \qquad (2.4.6)$$

and test functions for Galerkin's method are $w_1 = H_1(x)$ and $w_2 = H_2(x)$. Putting these u and w into Eq. (2.4.13) gives

$$-\int_{x_i}^{x_{i+1}} \left(\begin{Bmatrix} H_1' \\ H_2' \end{Bmatrix} [H_1' H_2'] + \begin{Bmatrix} H_1 \\ H_2 \end{Bmatrix} [H_1 H_2] \right) dx \begin{Bmatrix} u_i \\ u_{i+1} \end{Bmatrix}$$

$$+ \int_{x_i}^{x_{i+1}} x \begin{Bmatrix} H_1 \\ H_2 \end{Bmatrix} dx \qquad (2.4.15)$$

where H_i' denotes $\frac{dH_i(x)}{dx}$ and H_i is given in Eq. (2.4.7) and Eq. (2.4.8). Computation of these integrals finally yields

$$-\begin{bmatrix} \frac{1}{h_i} + \frac{h_i}{3} & -\frac{1}{h_i} + \frac{h_i}{6} \\ -\frac{1}{h_i} + \frac{h_i}{6} & \frac{1}{h_i} + \frac{h_i}{3} \end{bmatrix} \begin{Bmatrix} u_i \\ u_{i+1} \end{Bmatrix} + \begin{Bmatrix} \frac{h_i}{6}(x_{i+1} + 2x_i) \\ \frac{h_i}{6}(2x_{i+1} + x_i) \end{Bmatrix} \qquad (2.4.16)$$

For each element, Eq. (2.4.16) can be written as

Element #1

$$\begin{bmatrix} -3.111 & 2.9444 \\ 2.9444 & -3.111 \end{bmatrix} \begin{Bmatrix} u_1 \\ u_2 \end{Bmatrix} + \begin{Bmatrix} 0.0185 \\ 0.0370 \end{Bmatrix} \qquad (2.4.17)$$

Element #2

$$\begin{bmatrix} -3.111 & 2.9444 \\ 2.9444 & -3.111 \end{bmatrix} \begin{Bmatrix} u_2 \\ u_3 \end{Bmatrix} + \begin{Bmatrix} 0.0741 \\ 0.0926 \end{Bmatrix} \qquad (2.4.18)$$

Element #3

$$\begin{bmatrix} -3.111 & 2.9444 \\ 2.9444 & -3.111 \end{bmatrix} \begin{Bmatrix} u_3 \\ u_4 \end{Bmatrix} + \begin{Bmatrix} 0.1296 \\ 0.1481 \end{Bmatrix} \qquad (2.4.19)$$

As shown in Eq. (2.4.13), we need to sum Eqs. (2.4.17) through (2.14.19). Each element has different nodes associated with it. As a result, we expand each expression such that the expression has a matrix and a vector of size m which is the total number of degrees of freedom in the system. For the present problem, $m = 4$. The number of total degrees of freedom is the same as the total number of nodes because each node has one degree of freedom for the present problem. Rewriting Eq. (2.4.17) for the expanded matrix and vector gives

$$\begin{bmatrix} -3.111 & 2.9444 & 0 & 0 \\ 2.944 & -3.111 & 0 & 0 \\ 0 & 0 & 0 & 0 \\ 0 & 0 & 0 & 0 \end{bmatrix} \begin{Bmatrix} u_1 \\ u_2 \\ u_3 \\ u_4 \end{Bmatrix} + \begin{Bmatrix} 0.0185 \\ 0.0370 \\ 0 \\ 0 \end{Bmatrix} \qquad (2.4.20)$$

Section 2.4 — Galerkin's Finite Element Formulation

Similarly, Eq. (2.4.18) and Eq. (2.4.19) can be rewritten as

$$\begin{bmatrix} 0 & 0 & 0 & 0 \\ 0 & -3.111 & 2.9444 & 0 \\ 0 & 2.9444 & -3.1111 & 0 \\ 0 & 0 & 0 & 0 \end{bmatrix} \begin{Bmatrix} u_1 \\ u_2 \\ u_3 \\ u_4 \end{Bmatrix} + \begin{Bmatrix} 0 \\ 0.0741 \\ 0.0926 \\ 0 \end{Bmatrix} \quad (2.4.21)$$

$$\begin{bmatrix} 0 & 0 & 0 & 0 \\ 0 & 0 & 0 & 0 \\ 0 & 0 & -3.1111 & 2.9444 \\ 0 & 0 & 2.9444 & -3.1111 \end{bmatrix} \begin{Bmatrix} u_1 \\ u_2 \\ u_3 \\ u_4 \end{Bmatrix} + \begin{Bmatrix} 0 \\ 0 \\ 0.1296 \\ 0.1481 \end{Bmatrix} \quad (2.4.22)$$

Adding directly Eqs. (2.4.20) through (2.4.22) results in

$$\begin{bmatrix} -3.1111 & 2.9444 & 0 & 0 \\ 2.9444 & -6.2222 & 2.9444 & 0 \\ 0 & 2.9444 & -6.2222 & 2.9444 \\ 0 & 0 & 2.9444 & -3.1111 \end{bmatrix} \begin{Bmatrix} u_1 \\ u_2 \\ u_3 \\ u_4 \end{Bmatrix}$$

$$+ \begin{Bmatrix} 0.0185 - u'(0) \\ 0.1111 \\ 0.2222 \\ 0.1481 + u'(1) \end{Bmatrix} = 0 \quad (2.4.23)$$

The Neuman boundary conditions are added to the column vector from Eq. (2.4.13). For the present problem, the Dirichlet boundary conditions are provided at both ends (i.e. $u_1 = 0$ and $u_4 = 0$). Therefore, the Neumann boundary conditions (i.e. $u'(0)$ and $u'(1)$) are not provided. Equation (2.4.23) can be solved with the given boundary conditions, $u_1 = 0$ and $u_4 = 0$, to find the rest of the nodal variables and unknown Neumann boundary conditions. In actual finite element programming, Eqs. (2.4.17) through (2.4.19) are directly summed into Eq. (2.4.23) without using Eqs. (2.4.20) through (2.4.22). Equations (2.4.20) through (2.4.22) are used here only to help the conceptual understanding of the assembly process. Furthermore, in computer programming, unknown nodal values, called the primary variables, are solved first and then the unknown boundary conditions are solved later. To this end, Eq. (2.4.23) is modified with the known boundary conditions.

$$\begin{bmatrix} 1 & 0 & 0 & 0 \\ 2.9444 & -6.2222 & 2.9444 & 0 \\ 0 & 2.9444 & -6.2222 & 2.9444 \\ 0 & 0 & 0 & 1 \end{bmatrix} \begin{Bmatrix} u_1 \\ u_2 \\ u_3 \\ u_4 \end{Bmatrix} = \begin{Bmatrix} 0 \\ -0.1111 \\ -0.2222 \\ 0 \end{Bmatrix} \quad (2.4.24)$$

The first and last equations in Eq. (2.4.23) are replaced by the Dirichlet boundary conditions. From Eq. (2.4.24), the solution gives $u_1 = 0$, $u_2 = 0.0448$, $u_3 = 0.0569$, and $u_4 = 0$. These nodal solutions can be substituted into Eq. (2.4.23) to find $u'(0)$ and $u'(1)$. Once the nodal variables are determined, the solution within each element can be obtained from corresponding nodal variables and shape functions. For example, the solution within the first element $(0 \leq x \leq \frac{1}{3})$ is $u = H_1(x)u_1 + H_2(x)u_2 = 0.1344x$. ‡

2.5 Variational Method

The variational method is also commonly used to derive the finite element matrix equation. We want to derive the functional for the sample problem

$$\begin{cases} \frac{d^2u}{dx^2} - u = -x, & 0 < x < 1 \\ u(0) = 0, \text{ and } u(1) = 0 \end{cases} \tag{2.1.1}$$

The variational expression for Eq. (2.1.1) is

$$\delta J = \int_0^1 \left(-\frac{d^2u}{dx^2} + u - x\right) \delta u \, dx + \left[\frac{du}{dx} \delta u\right]_0^1 \tag{2.5.1}$$

where δ is the *variational* operator. The first term in the above equation is the differential equation and the second term is the unknown *Neumann boundary condition* (or natural boundary condition). Applying integration by parts to the first term of Eq. (2.5.1) yields

$$\delta J = \int_0^1 \left(\frac{du}{dx} \frac{d(\delta u)}{dx} + u\delta u - x\delta u\right) dx \tag{2.5.2}$$

Since the *variational* operator is commutative with both differential and integral operators (i.e. $\frac{d(\delta u)}{dx} = \delta(\frac{du}{dx})$ and $\int \delta u \, dx = \delta \int u \, dx$), Eq. (2.5.2) can be written as

$$\delta J = \delta \int_0^1 \left\{\frac{1}{2}\left(\frac{du}{dx}\right)^2 + \frac{1}{2}u^2 - xu\right\} dx \tag{2.5.3}$$

The functional is obtained from Eq. (2.5.3) as

$$J = \int_0^1 \left\{\frac{1}{2}\left(\frac{du}{dx}\right)^2 + \frac{1}{2}u^2 - xu\right\} dx \tag{2.5.4}$$

Conversely, taking the variation of Eq. (2.5.4) will result in the differential equation as given in Eq. (2.1.1). A functional represents energy in many engineering applications. For example, the total potential energy in solid mechanics is a functional. The solution to the governing equation is obtained by minimizing the functional. The *principle of minimum total potential energy* in solid mechanics is one example to determine the stable equilibrium solution [4,5]. Energy principles are discussed in later chapters. For more detailed information for *variational method*, readers may refer to Refs. [6-8].

2.6 Rayleigh-Ritz Method

The *Rayleigh-Ritz* method obtains an approximate solution to a differential equation with given boundary conditions using the functional of the equation. The procedure of this technique can be summarized in two steps as given below:

1. Assume an admissible solution which satisfies the *Dirichlet* boundary condition (or essential boundary condition) and contains unknown coefficients.
2. Substitute the assumed solution into the functional and find the unknown coefficients to minimize the functional.

♣ **Example 2.6.1** In order to solve Eq. (2.1.1) using the *Rayleigh-Ritz* method, we assume the following function as an approximate solution:

$$u = ax(1-x) \qquad (2.6.1)$$

where a is an unknown coefficient. This function satisfies the essential boundary conditions. Substituting Eq. (2.6.1) into the functional, Eq. (2.5.4), yields

$$J = \frac{1}{2}a^2 \int_0^1 [(1-2x)^2 + x^2(1-x)^2]dx - a\int_0^1 x^2(1-x)dx \qquad (2.6.2)$$

Minimizing the functional with respect to the unknown coefficient a, i.e. $\frac{dJ}{da}=0$, yields a=0.2272. Therefore, the approximate solution is $u = 0.2272x(1-x)$ which is the same as that obtained in Sec. 2.1 using Galerkin's method. In order to improve the approximate solution, we need to add more terms. For example, we may assume

$$u = a_1 x(1-x) + a_2 x^2(1-x) \qquad (2.6.3)$$

where a_1 and a_2 are two unknown coefficients. We substitute the expression into the functional and take derivatives with respect to a_1 and a_2 in order to minimize the functional.

$$\frac{\partial J}{\partial a_1} = 0 \quad \text{and} \quad \frac{\partial J}{\partial a_2} = 0 \qquad (2.6.4)$$

This operation will give solutions for unknown coefficients a_1 and a_2. ‡

2.7 Rayleigh-Ritz Finite Element Method

The *Rayleigh-Ritz* method can be applied to a problem domain using continuous piecewise functions. As a result, the problem domain is divided into subdomains of finite elements. For elements with two nodes apiece, the linear shape functions as in Eqs. (2.4.7) and (2.4.8) can be used for the *Rayleigh-Ritz* method. The following example explains the finite element procedure using the *Rayleigh-Ritz* method.

♣ **Example 2.7.1** We will solve Example 2.4.1 again using the Rayleigh-Ritz method. The problem domain and its discretization are shown in Fig. 2.4.3. The functional can be expressed for the discretized domain as

$$J = \sum_{i=1}^{n} \int_{x_i}^{x_{i+1}} \left\{ \frac{1}{2}\left(\frac{du}{dx}\right)^2 + \frac{1}{2}u^2 - xu \right\} dx \qquad (2.7.1)$$

where $n = 3$, $x_1 = 0$, $x_2 = 1/3$, $x_3 = 2/3$ and $x_4 = 1$ as shown in Fig. 2.4.3. Using the linear shape functions, the solution u for the ith element is expressed as

$$u = H_1(x)\, u_i + H_2(x)\, u_{i+1} = [H]\{u^i\} \tag{2.7.2}$$

where

$$[H] = [H_1 \quad H_2] \tag{2.7.3}$$

$$\{u^i\} = \{u_i \quad u_{i+1}\}^T \tag{2.7.4}$$

and H_1 and H_2 are given in Eqs. (2.4.7) and (2.4.8). Substituting Eq. (2.7.2) into the functional yields

$$\int_{x_i}^{x_{i+1}} \left\{ \frac{1}{2}\left(\frac{du}{dx}\right)^2 + \frac{1}{2}u^2 - xu \right\} dx = \int_{x_i}^{x_{i+1}} \left\{ \frac{1}{2}\{u^i\}^T \left[\frac{dH}{dx}\right]^T \left[\frac{dH}{dx}\right]\{u^i\} + \frac{1}{2}\{u^i\}^T [H]^T [H]\{u^i\} - \{u^i\}^T [H]^T x \right\} dx \tag{2.7.5}$$

in which

$$\left[\frac{dH}{dx}\right] = \left[\frac{dH_1}{dx} \quad \frac{dH_2}{dx}\right] \tag{2.7.6}$$

Evaluation of the integral in Eq. (2.7.5) gives

$$\frac{1}{2}\{u_i \quad u_{i+1}\} \begin{bmatrix} \frac{1}{h_i} + \frac{h_i}{3} & -\frac{1}{h_i} + \frac{h_i}{6} \\ -\frac{1}{h_i} + \frac{h_i}{6} & \frac{1}{h_i} + \frac{h_i}{3} \end{bmatrix} \left\{ \begin{array}{c} u_i \\ u_{i+1} \end{array} \right\}$$
$$- \{u_i \quad u_{i+1}\} \left\{ \begin{array}{c} \frac{h_i}{6}(x_{i+1} + 2x_i) \\ \frac{h_i}{6}(2x_{i+1} + x_i) \end{array} \right\} \tag{2.7.7}$$

Here, the matrix expression in Eq. (2.7.7) came from the first and second terms of the right-hand side of Eq. (2.7.5) while the vector expression came from the last term. Summing Eq. (2.7.7) over the total number of elements and substituting proper values give the functional

$$J = \frac{1}{2}\{u_1 \quad u_2 \quad u_3 \quad u_4\} \begin{bmatrix} 3.1111 & -2.9444 & 0 & 0 \\ -2.9444 & 6.2222 & -2.9444 & 0 \\ 0 & -2.9444 & 6.2222 & -2.9444 \\ 0 & 0 & -2.9444 & 3.1111 \end{bmatrix} \left\{ \begin{array}{c} u_1 \\ u_2 \\ u_3 \\ u_4 \end{array} \right\}$$
$$- \{u_1 \quad u_2 \quad u_3 \quad u_4\} \left\{ \begin{array}{c} 0.0185 \\ 0.1111 \\ 0.2222 \\ 0.1481 \end{array} \right\} \tag{2.7.8}$$

The summation process for Eq. (2.7.8) is the same as explained in Example 2.4.1. In order to find the solution, we need to minimize the functional with respect to the unknown nodal vector $\{u\} = \{u_1 \; u_2 \; u_3 \; u_4\}^T$. Invoking $\frac{dJ}{d\{u\}} = 0$ results in

$$\begin{bmatrix} 3.1111 & -2.9444 & 0 & 0 \\ -2.9444 & 6.2222 & -2.9444 & 0 \\ 0 & -2.9444 & 6.2222 & -2.9444 \\ 0 & 0 & -2.9444 & 3.1111 \end{bmatrix} \begin{Bmatrix} u_1 \\ u_2 \\ u_3 \\ u_4 \end{Bmatrix} - \begin{Bmatrix} 0.0185 \\ 0.1111 \\ 0.2222 \\ 0.1481 \end{Bmatrix} = 0 \qquad (2.7.9)$$

Applying the boundary conditions $u_1 = 0$ and $u_4 = 0$ to Eq. (2.7.9) yields Eq. (2.4.24) in Example 2.4.1. The solutions for nodal variables are $u_1 = 0$, $u_2 = 0.0448$, $u_3 = 0.0569$, and $u_4 = 0$ again as before. ‡

Problems

2.1 Find an approximate solution to the boundary value problem

$$\frac{d^2u}{dx^2} = x \quad 0 < x < 1$$

$$u(0) = 0 \text{ and } u(1) = 0$$

Use a trial function $u = ax(1-x)$ where a is a constant to be determined. Apply the collocation method with a collocation point located at $x = 0.5$.

2.2 Redo Prob. 2.1 using the least squares method.

2.3 Redo Prob. 2.1 using Galerkin's method.

2.4 Solve the following two-point boundary value problem using the collocation method.

$$x^2 \frac{d^2u}{dx^2} - 2u = 1 \quad 1 < x < 2$$

$$u(1) = 0 \text{ and } u(2) = 0$$

Use a trial function $u = a(x-1)(x-2)$ where a is a constant to be determined.

2.5 Redo Prob. 2.4 using the least squares method.

2.6 Redo Prob. 2.4 using Galerkin's method.

2.7 Determine an approximate solution to the differential equation

$$x^2 \frac{d^2u}{dx^2} - 2x \frac{du}{dx} + 2u = 0 \quad 1 < x < 4$$

$$u(1) = 0 \text{ and } u(4) = 12$$

Use a quadratic polynomial for the trial function and the collocation method.

2.8 Redo Prob. 2.7 using the least squares method.

2.9 Redo Prob. 2.7 using Galerkin's method.

2.10 Apply Galerkin's method to find an approximate solution to the following differential equation:

$$\frac{d^2u}{dx^2} + \frac{du}{dx} - 2u = 0 \quad 0 < x < 1$$

$$u(0) = 0 \text{ and } u(1) = 1$$

Assume (a) a quadratic polynomial and (b) a cubic polynomial as a trial function, respectively.

2.11 Redo Prob. 2.10 using the collocation method.

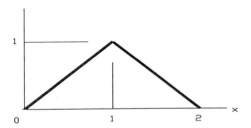

Figure P2.14 Problem 2.14

2.12 Redo Prob. 2.10 using the least squares method.

2.13 Use Galerkin's method to determine an approximate solution to

$$\frac{d^2u}{dx^2} + \frac{du}{dx} - 2u = x \quad 0 < x < 1$$

$$u(0) = 0 \quad \text{and} \quad u(1) = 1$$

Assume a quadratic polynomial as a trial function.

2.14 Solve the problem given below using Galerkin's method and piecewise linear functions. The piecewise function is shown in Fig. P.2.14.

$$\frac{d^2u}{dx^2} = 1 \quad 0 < x < 2$$

$$u(0) = 0 \quad \text{and} \quad u(2) = 0$$

2.15 Solve the boundary value problem using piecewise linear functions.

$$\frac{d^2u}{dx^2} = 1 \quad 0 < x < 3$$

$$u(0) = 0 \quad \text{and} \quad u(3) = 0$$

(a) Derive the weak formulation. (b) Develop the matrix equation using three equal sizes of subdomains. (c) What is the approximate solution at $x=1.5$?

2.16 Apply the piecewise linear functions to

$$\frac{d^2u}{dx^2} + \frac{du}{dx} - 2u = 0 \quad 0 < x < 1$$

$$u(0) = 0 \quad \text{and} \quad u(1) = 1$$

Divide the domain into three equal sizes of subdomains.

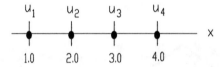

Figure P2.17 Problem 2.17

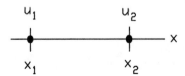

Figure P2.19 Problem 2.19

2.17 A differential equation with boundary conditions is given below:

$$x^2 \frac{d^2 u}{dx^2} + 2x \frac{du}{dx} + 2 = 0 \quad 1 < x < 4$$

$$u(1) = 1 \text{ and } u(4) = 0$$

(a) Derive the weak formulation. (b) Compute element matrices and vectors for the given mesh discretization using linear shape functions. (c) Assemble them into the global matrix and vector. (d) Apply the boundary conditions to the matrix equation. (e) Solve for the unknown nodal values.

2.18 Redo Prob. 2.16 using the linear finite elements.

2.19 (a) For a finite element with two end nodes, derive shape functions using the polynomial $u(x) = a + bx^2$. In other words, find $H_1(x)$ and $H_2(x)$ such that $u(x) = H_1(x)u_1 + H_2(x)u_2$. (b) Using the shape functions obtained from (a), compute the following integral:

$$\int_0^1 \left(\frac{du}{dx} \frac{dw}{dx} + wu \right) dx$$

where w is the test function. Use Galerkin's method.

2.20 For a two-node element, (a) develop the shape functions $H_1(x)$ and $H_2(x)$ using $u(x) = ax + bx^2$ such that $u(x) = H_1(x)u_1 + H_2(x)u_2$, and (b) compute $\int_{-1}^{1} w \frac{du}{dx} dx$ using Galerkin's method. (c) Does the element converge as the mesh is refined? Explain why.

2.21 (a) Develop the functional of the differential equation given in Prob. 2.1. (b) With the functional derived in (a), redo Prob. 2.1 using the Rayleigh-Ritz method.

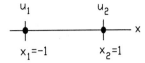

Figure P2.20 Problem 2.20

2.22 (a) Develop the functional of the differential equation given in Prob. 2.14. (b) With the functional derived in (a), redo Prob. 2.14 using the Rayleigh-Ritz method.

2.23 Redo Prob. 2.21 using the Rayleigh-Ritz finite element method along with linear elements.

2.24 Redo Prob. 2.22 using the Rayleigh-Ritz finite element method along with linear elements.

CHAPTER THREE

FINITE ELEMENT PROGRAMMING

3.0 Chapter Overview

The basic program structure of finite element analysis is presented with an example of a two-point boundary value problem. It discusses what input data are required in general, how the matrix equation is assembled, how the boundary conditions are applied to the matrix equation, and how the matrix equation is solved using examples of MATLAB programs. Complete sample programs in MATLAB are provided in the chapter.

3.1 Overall Program Structure

In order to understand fundamental concepts of the finite element method, it is very useful (or sometimes essential) to understand the skeleton of the program structure of the finite element analysis. This chapter explains the basic structure of the program. The main procedures in the finite element analysis are

1. Read input data and allocate proper array sizes.
2. Calculate element matrices and vectors for every element.
3. Assemble element matrices and vectors into the system matrix and vector.
4. Apply constraints to the system matrix and vector.
5. Solve the matrix equation for the primary nodal variables.
6. Compute secondary variables.
7. Plot and/or print desired results.

Each procedure is explained in the subsequent sections using the following second order ordinary differential equation

$$a\frac{d^2u}{dx^2} + b\frac{du}{dx} + cu = f(x), \quad 0 < x < L$$
$$u(0) = 0 \text{ and } u(L) = 0 \tag{3.1.1}$$

The weak formulation of the equation is

$$\int_0^L \left\{ -a\frac{dw}{dx}\frac{du}{dx} + bw\frac{du}{dx} + cwu \right\} dx = \int_0^L wf(x)dx - \left[aw\frac{du}{dx}\right]_0^L \qquad (3.1.2)$$

If we use the linear shape functions, Eqs. (2.4.7) and (2.4.8), the element matrix for the ith element becomes

$$[K^e] = \int_{x_i}^{x_{i+1}} \left(-a \begin{Bmatrix} H_1' \\ H_2' \end{Bmatrix} [H_1' H_2'] + b \begin{Bmatrix} H_1 \\ H_2 \end{Bmatrix} [H_1' H_2'] + c \begin{Bmatrix} H_1 \\ H_2 \end{Bmatrix} [H_1 H_2] \right) dx \qquad (3.1.3)$$

where ( )' denotes the derivative with respect to x. Evaluation of the integration gives

$$[K^e] = -\frac{a}{h_i}\begin{bmatrix} 1 & -1 \\ -1 & 1 \end{bmatrix} + \frac{b}{2}\begin{bmatrix} -1 & 1 \\ -1 & 1 \end{bmatrix} + \frac{c\,h_i}{6}\begin{bmatrix} 2 & 1 \\ 1 & 2 \end{bmatrix} \qquad (3.1.4)$$

On the other hand, the element vector is

$$F^e = \int_{x_i}^{x_{i+1}} f(x) \begin{Bmatrix} H_1 \\ H_2 \end{Bmatrix} dx \qquad (3.1.5)$$

If $f(x) = 1$, the element vector becomes

$$F^e = \frac{h_i}{2} \begin{Bmatrix} 1 \\ 1 \end{Bmatrix} \qquad (3.1.6)$$

3.2 Input Data

The major input parameters needed for the finite element analysis program for Eq. (3.1.1) are

>   the number of total nodes in the system,
>   the number of total elements in the system,
>   coordinate values of every node in terms of the global coordinate system,
>   types of every element,
>   information for boundary conditions, and
>   coefficients for Eq. (3.1.1).

Most of these input data are associated with the finite element mesh upon which a user decides. The mesh can be generated either using an automatic mesh generation program called *pre-processor* or manually. The type of element includes how many nodes per element as well as how many degrees of freedom for each node of the element. If the same type of elements is used over the whole domain, this information is needed for one element. However, if the system (or domain) has many different types of elements, this information should be supplied for all different elements. For the present problem, Eq. (3.1.1), we use the same type of finite elements for the sake of simplicity. The problem domain is discretized in Fig. 3.2.1. Here five equal size linear elements are used. Therefore, the number of total nodes in the system

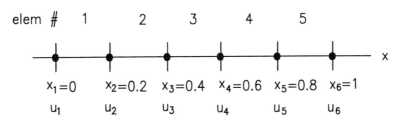

Figure 3.2.1 A Mesh With 5 Linear Elements

($nnode$) is 6 and the number of total elements in the system (nel) is 5. Since it is a one-dimensional problem, each node has only x coordinate values. If $gcoord$ denotes the array storing the coordinate values, then

$gcoord(1)=0.0$, $gcoord(2)=0.2$, $gcoord(3)=0.4$,
$gcoord(4)=0.6$, $gcoord(5)=0.8$, $gcoord(6)=1.0$

in which the index in the parentheses is the node number which varies from 1 to 6 and the size of array $gcoord$ is the same as the total node number $nnode$. The number of nodes per element ($nnel$) is 2 and the number of degrees of freedom per node ($ndof$) is 1. Then the number of degrees of freedom per system is $sdof=nnode*ndof$.

In general, information for nodal connectivity for each element is an input to the program. This is also called element topology. This information is important to evaluate element matrices and vectors as well as to assemble these matrices and vectors into the system (or global) matrix and vector. For the present one-dimensional problem using linear elements, this information can be constructed in the program in a simple way if the node numbering and element numbering are sequential from one end of the domain to the other end of the domain. It is stored in an array called *nodes*. The array is a two-dimensional array. The first index indicates the element number and the second index denotes the nodes associated with the element. For the example problem, the ith element has two nodes, i^{th} and $(i+1)^{th}$ nodes. That is,

$nodes(i,1)=i$ and $nodes(i,2)=i+1$ for $i=1,2,3,4,5$

This can be coded easily in the program and is shown in examples in a later section.

Information for boundary conditions includes the nodal degrees of freedoms where constraints and external forces (or fluxes) are applied. In order to specify the nodal degrees of freedom, we need to provide node numbers and corresponding degrees of freedom of the specified nodes. In addition, the prescribed constraint values should be provided. For the present problem, the information for constrained nodes is

$bcdof(1)=1$, $bcdof(2)=6$

where *bcdof* contains the node numbers where constraints are given. In other words, the size of array *bcdof* is 2 because there are two constrained nodes, and the first and second constrained node numbers are 1 and 6, respectively. Furthermore, the constrained values are read in *bcval* such as

$bcval(1)=0.0$ and $bcval(2)=0.0$.

Here the first value is for node 1 and the second value is for node 6.

The element matrix Eq. (3.1.4) is derived for arbitrary constants a, b, and c as in Eq. (3.1.1). As a result, the coefficients should be provided. For the present problem, let $a = 1$, $b = -1$, and $c = 2$.

3.3 Assembly of Element Matrices and Vectors

The element matrix and vector are expressed in Eqs. (3.1.4) and (3.1.6). These expressions are functions of the length of each element. As a result, the length of each element is computed from the coordinate values of the nodes associated with the element. For example, the ith element is associated with the ith and $(i+1)^{th}$ nodes. The coordinate values of the nodes are $gcoord(i)$ and $gcoord(i+1)$. As a result, the element length h_i is equal to $gcoord(i+1) - gcoord(i)$. If the element length is the same for the whole domain, the length can be provided as an input.

Once these matrices and vectors are computed, they need to be assembled into the system matrix and vector. To this end, we need the information where the element matrix and vector are to be located in the system matrix and vector. This information is obtained from the array $index$ whose size equals the number of degrees of freedom per element, i.e. 2 for the present problem. Because each node has a single degree of freedom (i.e. $nodf=1$), the size of the array $index$ is the same as that for array $nodes$.

$index(1)=i$ and $index(2)=i+1$ for the ith element

The following example shows assembly of element matrices and vectors.

♣ **Example 3.3.1**   Let k and f be the element matrix and vector for any element. In addition, kk and ff are the system matrix and vector. The array $index$ contains the degrees of freedom associated with the element. Then, k and f are stored into kk and ff in the following way. This is repeated for every element.

```
edof=nnel*ndof;      % edof=number of degrees of freedom per element
for ir = 1:edof;     % loop for element rows
    irs = index(ir);     % address for the system row
    ff(irs) = ff(irs) + f(ir);     % assembly into the system vector
    for ic = 1:edof;     % loop for element columns
        ics = index(ic);     % address for the system column
        kk(irs,ics) = kk(irs,ics) + k(ir,ic);  % assembly into system matrix
    end     % end of column loop
end     % end of row loop   ‡
```

3.4 Application of Constraints

The information for constraints or boundary conditions is provided in arrays called $bcdof$ and $bcval$ as described in the previous section. The system matrix equation is modified using this information. The size of the system matrix equation is equal to the total number of degrees of freedom in the system. Without applying the constraints to the system of equations, the matrix equation is singular so that it cannot be inverted. In context of solid/structural mechanics, this means that the matrix equation contains rigid body motions. As a result, the constraints prevent the matrix equation from being singular. If a constraint is applied to the nth degree of freedom in the matrix equation, the nth equation in the matrix is replaced by the constraint equation.

♣ **Example 3.4.1** For the present example, the system matrix equation is

$$[kk]\{u\} = \{ff\} \qquad (3.4.1)$$

The size of the matrix equation is $sdof$=6 and there are two constraints. These constraints are applied to the system matrix equation as shown below:

```
for ic = 1:2;      % loop for two constraints
    id = bcdof(ic);    % extract the degree of freedom of a constraint
    val = bcval(ic);   % extract the corresponding constrained value
        for i = 1:sdof;    % loop for number of equations in system
            kk(id,i) = 0;    % set all the id^th row to zero
        end
    kk(id,id) = 1;     % set the id^th diagonal to unity
    ff(id) = val;      % put the constrained value in the column
end  ‡
```

The algorithm shown in Example 3.4.1 destroys the symmetry if the system matrix, before applying the boundary conditions, is symmetric. If we want to maintain the symmetry even after applying the boundary conditions, the next example shows the algorithm.

♣ **Example 3.4.2** This example shows another way to apply the boundary condition without destroying the symmetry of the system matrix.

```
for ic = 1:2;      % loop for two constraints
    id = bcdof(ic);    % extract the degree of freedom for constraint
    val = bcval(ic);   % extract the corresponding constrained value
        for i = 1:sdof;    % loop for number of equations in system
            ff(i) = ff(i)-val*kk(i,id)  % modify column using constrained value
            kk(id,i) = 0;    % set all the id^th row to zero
            kk(i,id) = 0;    % set all the id^th column to zero
        end
```

```
        kk(id,id) = 1;    % set the id^th diagonal to unity
        ff(id) = val;     % put the constrained value in the column
   end  ‡
```

Once the system matrix equation is modified as shown in the above example, the modified matrix equation is solved for the primary nodal unknowns. In the MATLAB program, it can be solved as

u = kk' \ ff

where kk' denotes the modified matrix equation. Once the primary nodal variable u is determined from the matrix equation, the natural boundary conditions (i.e. the secondary variable) are found from

ff = kk*u

3.5 Example Programs

This section shows examples of finite element analysis programs. The second-order ordinary differential equations are used as governing equations.

♣ **Example 3.5.1**   We want to solve Eq. (3.1.1) using the finite element method. The coefficients in the differential equation are assumed $a = 1$, $b = -3$, and $c = 2$ while function $f(x)$ is assumed 1. The domain size is equal to 1 (i.e. $L = 1$) and five linear elements of equal size are used for the present analysis. The computer program written in MATLAB is provided below along with the results. The main program is first shown below.

```
%————————————————————————————————————
% EX3.5.1.m
% to solve the ordinary differential equation given as
% a u" + b u' + c u = 1, 0 < x < 1
% u(0) = 0 and u(1) = 0
% using 5 linear elements
%
% Variable descriptions
% k = element matrix
% f = element vector
% kk = system matrix
% ff = system vector
% index = a vector containing system dofs associated with each element
% bcdof = a vector containing dofs associated with boundary conditions
% bcval = a vector containing boundary condition values associated with
%           the dofs in bcdof
```

```
%------------------------------------------------------------
%
%------------------------------------------------
% input data for control parameters
%------------------------------------------------
nel=5;                          % number of elements
nnel=2;                         % number of nodes per element
ndof=1;                         % number of dofs per node
nnode=6;                        % total number of nodes in system
sdof=nnode*ndof;                % total system dofs
%
%------------------------------------------------
% input data for nodal coordinate values
%------------------------------------------------
gcoord(1)=0.0; gcoord(2)=0.2; gcoord(3)=0.4; gcoord(4)=0.6;
gcoord(5)=0.8; gcoord(6)=1.0;
%
%------------------------------------------------
% input data for nodal connectivity for each element
%------------------------------------------------
nodes(1,1)=1; nodes(1,2)=2; nodes(2,1)=2; nodes(2,2)=3;
nodes(3,1)=3; nodes(3,2)=4; nodes(4,1)=4; nodes(4,2)=5;
nodes(5,1)=5; nodes(5,2)=6;
%
%------------------------------------------------
% input data for coefficients of the ODE
%------------------------------------------------
acoef=1;                        % coefficient 'a' of the diff eqn
bcoef=-3;                       % coefficient 'b' of the diff eqn
ccoef=2;                        % coefficient 'c' of the diff eqn
%
%------------------------------------------------
% input data for boundary conditions
%------------------------------------------------
bcdof(1)=1;                     % first node is constrained
bcval(1)=0;                     % whose described value is 0
bcdof(2)=6;                     % 6th node is constrained
bcval(2)=0;                     % whose described value is 0
%
%------------------------------------------------
% initialization of matrices and vectors
%------------------------------------------------
ff=zeros(sdof,1);               % initialization of system force vector
kk=zeros(sdof,sdof);            % initialization of system matrix
index=zeros(nnel*ndof,1);       % initialization of index vector
%
%------------------------------------------------
% computation of element matrices and vectors and their assembly
```

```
%----------------------------------------------------------
for iel=1:nel                        % loop for the total number of elements
%
nl=nodes(iel,1); nr=nodes(iel,2);    % extract nodes for (iel)-th element
xl=gcoord(nl); xr=gcoord(nr);        % extract nodal coord values
eleng=xr-xl;                         % element length
index=feeldof1(iel,nnel,ndof);       % extract system dofs associated
%
k=feode2l(acoef,bcoef,ccoef,eleng);  % compute element matrix
f=fef1l(xl,xr);                      % compute element vector
[kk,ff]=feasmbl2(kk,ff,k,f,index);   % assemble element matrices and vectors
%
end                                  % end of loop for total elements
%
%----------------------------------
% apply boundary conditions
%----------------------------------
[kk,ff]=feaplyc2(kk,ff,bcdof,bcval);
%
%------------------------------
% solve the matrix equation
%------------------------------
fsol=kk\ff;
%
%-----------------------------
% analytical solution
%-----------------------------
c1=0.5/exp(1);
c2=-0.5*(1+1/exp(1));
for i=1:nnode
x=gcoord(i);
esol(i)=c1*exp(2*x)+c2*exp(x)+1/2;
end
%
%-------------------------------------
% print both exact and fem solutions
%-------------------------------------
num=1:1:sdof;
results=[num' fsol esol']
%----------------------------------------------------------
```

The function programs (i.e. *m-files*) used in the main program are also given below.

```
function [kk,ff]=feaplyc2(kk,ff,bcdof,bcval)
%----------------------------------------------------------
% Purpose:
% Apply constraints to matrix equation [kk]x=ff
```

```
%
% Synopsis:
% [kk,ff]=feaplybc(kk,ff,bcdof,bcval)
%
% Variable Description:
% kk - system matrix before applying constraints
% ff - system vector before applying constraints
% bcdof - a vector containing constrained dof
% bcval - a vector containing constrained value
%
% For example, there are constraints at dof=2 and 10
% and their constrained values are 0.0 and 2.5,
% respectively. Then, bcdof(1)=2 and bcdof(2)=10; and
% bcval(1)=1.0 and bcval(2)=2.5.
%----------------------------------------------------------------
%
n=length(bcdof);
sdof=size(kk);
%
for i=1:n
c=bcdof(i);
for j=1:sdof
kk(c,j)=0;
end
%
kk(c,c)=1;
ff(c)=bcval(i);
end
%----------------------------------------------------------------


function [kk,ff]=feasmbl2(kk,ff,k,f,index)
%----------------------------------------------------------------
% Purpose:
% Assembly of element matrices into the system matrix and
% Assembly of element vectors into the system vector
%
% Synopsis:
% [kk,ff]=feasmbl2(kk,ff,k,f,index)
%
% Variable Description:
% kk - system matrix
% ff - system vector
% k - element matrix
% f - element vector
% index - d.o.f. vector associated with an element
%----------------------------------------------------------------
```

```
%
edof = length(index);
for i=1:edof
ii=index(i);
ff(ii)=ff(ii)+f(i);
  for j=1:edof
  jj=index(j);
  kk(ii,jj)=kk(ii,jj)+k(i,j);
  end
end
%————————————————————————————
```


```
function [index]=feeldof1(iel,nnel,ndof)
%————————————————————————————
% Purpose:
% Compute system dofs associated with each element in one-
% dimensional problem
%
% Synopsis:
% [index]=feeldof1(iel,nnel,ndof)
%
% Variable Description:
% index - system dof vector associated with element iel
% iel - element number whose system dofs are to be determined
% nnel - number of nodes per element
% ndof - number of dofs per node
%————————————————————————————
%
edof = nnel*ndof;
start = (iel-1)*(nnel-1)*ndof;
%
for i=1:edof
index(i)=start+i;
end
%————————————————————————————
```


```
function [f]=fef1l(xl,xr)
%————————————————————————————
% Purpose:
% element vector for f(x)=1
% using linear element
%
% Synopsis:
% [f]=fef1l(xl,xr)
```

```
%
% Variable Description:
% f - element vector (size of 2x1)
% xl - coordinate value of the left node
% xr - coordinate value of the right node
%----------------------------------------------------------------
%
% element vector
%
eleng=xr-xl;                                    % element length
f=[ eleng/2; eleng/2];
%----------------------------------------------------------------
```

```
function [k]=feode2l(acoef,bcoef,ccoef,eleng)
%----------------------------------------------------------------
% Purpose:
% element matrix for (a u" + b u' + c u)
% using linear element
%
% Synopsis:
% [k]=feode2l(acoef,bcoef,ccoef,eleng)
%
% Variable Description:
% k - element matrix (size of 2x2)
% acoef - coefficient of the second order derivative term
% bcoef - coefficient of the first order derivative term
% ccoef - coefficient of the zero-th order derivative term
% eleng - element length
%----------------------------------------------------------------
%
% element matrix
%
a1=-(acoef/eleng); a2=bcoef/2; a3=ccoef*eleng/6;
k=[ a1-a2+2*a3      -a1+a2+a3;...
   -a1-a2+a3        a1+a2+2*a3];
%----------------------------------------------------------------
```

The finite element solutions are compared to the exact solutions at the nodal points.

```
results =
node #   fem sol    exact sol
1.0000   0.00000    0.00000         % solution at x=0
2.0000  -0.0621    -0.0610          % solution at x=0.2
3.0000  -0.1133    -0.1110          % solution at x=0.4
4.0000  -0.1388    -0.1355          % solution at x=0.6
5.0000  -0.1142    -0.1111          % solution at x=0.8
```

          6.0000   0.00000   0.00000   % solution at $x=1$

♣ **Example 3.5.2** The same differential equation as that in Example 3.5.1 is solved here. However, the boundary conditions are different. They are

$$u(0) = 0 \quad \text{and} \quad \frac{du(1)}{dx} = 1 \qquad (3.5.1)$$

The left end is the essential boundary condition as before while the right end is the natural boundary condition. As seen in Eq. (3.1.2), the boundary condition with a known value of $\frac{du}{dx}$ contributes to the right-hand side column vector. For example, Eq. (2.4.23) shows how the natural boundary condition is incorporated into the column vector. Because the column vector moves to the right-hand side of the matrix equation, $\frac{du(1)}{dx} = 1$ is subtracted from the right-hand side column vector. The program list is given below for completeness. Comparing the program to that given in the previous example tells the difference between the essential (i.e. $u(1) = 0$) and natural (i.e. $\frac{du(1)}{dx} = 1$) boundary conditions.

```
%————————————————————————————————
% EX3.5.2.m
% to solve the ordinary differential equation given as
% a u" + b u' + c u = 1, 0 < x < 1
% u(0) = 0 and u'(1) = 1
% using 5 or 10 linear elements
%
% Variable descriptions
% k = element matrix
% f = element vector
% kk = system matrix
% ff = system vector
% index = a vector containing system dofs associated with each element
% bcdof = a vector containing dofs associated with boundary conditions
% bcval = a vector containing boundary condition values associated with
%         the dofs in bcdof
%————————————————————————————————
%
%————————————————————————————————
% input data for control parameters
%————————————————————————————————
nel=5;               % number of elements
nnel=2;              % number of nodes per element
ndof=1;              % number of dofs per node
nnode=6;             % total number of nodes in system
sdof=nnode*ndof;     % total system dofs
```

```
%
%----------------------------------------
% input data for nodal coordinate values
%----------------------------------------
gcoord(1)=0.0; gcoord(2)=0.2; gcoord(3)=0.4; gcoord(4)=0.6;
gcoord(5)=0.8; gcoord(6)=1.0;
%
%----------------------------------------
% input data for nodal connectivity for each element
%----------------------------------------
nodes(1,1)=1; nodes(1,2)=2; nodes(2,1)=2; nodes(2,2)=3;
nodes(3,1)=3; nodes(3,2)=4; nodes(4,1)=4; nodes(4,2)=5;
nodes(5,1)=5; nodes(5,2)=6;
%
%----------------------------------------
% input data for coefficients of the ODE
%----------------------------------------
acoef=1;                              % coefficient 'a' of the diff eqn
bcoef=-3;                             % coefficient 'b' of the diff eqn
ccoef=2;                              % coefficient 'c' of the diff eqn
%
%----------------------------------------
% input data for boundary conditions
%----------------------------------------
bcdof(1)=1;                           % first node is constrained
bcval(1)=0;                           % whose described value is 0
%
%----------------------------------------
% initialization of matrices and vectors
%----------------------------------------
ff=zeros(sdof,1);                     % initialization of system force vector
kk=zeros(sdof,sdof);                  % initialization of system matrix
index=zeros(nnel*ndof,1);             % initialization of index vector
%
%----------------------------------------
% computation of element matrices and vectors and their assembly
%----------------------------------------
for iel=1:nel                         % loop for the total number of elements
%
nl=nodes(iel,1); nr=nodes(iel,2);     % extract nodes for (iel)-th element
xl=gcoord(nl); xr=gcoord(nr);         % extract nodal coord values
eleng=xr-xl;                          % element length
index=feeldof1(iel,nnel,ndof);        % extract system dofs associated
%
k=feode2l(acoef,bcoef,ccoef,eleng);   % compute element matrix
f=fefl1(xl,xr);                       % compute element vector
[kk,ff]=feasmbl2(kk,ff,k,f,index);    % assemble element matrices and vectors
%
```

```
end                                        % end of loop for total elements
%
%----------------------------------------
% apply the natural boundary condition at the last node
%----------------------------------------
ff(nnode)=ff(nnode)-1;                     % include $u'(1)=1$ in column vector
%
%----------------------------------------
% apply boundary conditions
%----------------------------------------
[kk,ff]=feaplyc2(kk,ff,bcdof,bcval);
%
%----------------------------------------
% solve the matrix equation
%----------------------------------------
fsol=kk\ff;
%
%----------------------------------------
% analytical solution
%----------------------------------------
c1=(1+0.5*exp(1))/(2*exp(2)-exp(1));
c2=-(1+exp(2))/(2*exp(2)-exp(1));
for i=1:nnode
x=gcoord(i);
esol(i)=c1*exp(2*x)+c2*exp(x)+1/2;
end
%
%----------------------------------------
% print both exact and fem solutions
%----------------------------------------
num=1:1:sdof;
results=[num' fsol esol']
%----------------------------------------
```

The solutions using 5 elements and 10 elements are shown below, respectively. Comparing the two finite element solutions to the exact solution shows the convergence of the finite element solution as the mesh is refined.

```
results for five elements=
node #    fem sol      exact sol
1.0000    0.00000      0.00000          % solution at $x=0$
2.0000   -0.0588      -0.0578           % solution at $x=0.2$
3.0000   -0.1043      -0.1024           % solution at $x=0.4$
4.0000   -0.1203      -0.1180           % solution at $x=0.6$
5.0000   -0.0802      -0.0792           % solution at $x=0.8$
6.0000    0.0586       0.0546           % solution at $x=1$

results for ten elements=
node #    fem sol      exact sol
```

1.0000	0.00000	0.00000	% solution at $x=0$
2.0000	-0.0300	-0.0298	% solution at $x=0.1$
3.0000	-0.0580	-0.0578	% solution at $x=0.2$
4.0000	-0.0829	-0.0825	% solution at $x=0.3$
5.0000	-0.1028	-0.1024	% solution at $x=0.4$
6.0000	-0.1157	-0.1151	% solution at $x=0.5$
7.0000	-0.1186	-0.1180	% solution at $x=0.6$
8.0000	-0.1080	-0.1075	% solution at $x=0.7$
9.0000	-0.0794	-0.0792	% solution at $x=0.8$
10.000	-0.0273	-0.0275	% solution at $x=0.9$
11.000	0.0556	0.0546	% solution at $x=1$

‡

♣ **Example 3.5.3** This example solves the following differential equation:

$$x^2 \frac{\partial^2 u}{\partial x^2} - 2x \frac{\partial u}{\partial x} - 4u = x^2, \quad 10 < x < 20 \qquad (3.5.2)$$

with the boundary conditions $u(10) = 0$ and $u(20) = 100$. The *weak formulation* of Eq. (3.5.2) is

$$\int_{10}^{20} \left(x^2 \frac{\partial w}{\partial x} \frac{\partial u}{\partial x} + 4xw \frac{\partial u}{\partial x} + 4wu \right) dx = -\int_{10}^{20} wx^2 dx + \left[x^2 w \frac{\partial u}{\partial x} \right]_{10}^{20} \qquad (3.5.3)$$

Discretizing the domain into a number of linear elements and evaluating the element matrix and vector using Galerkin's method yield

$$[K^e] = \frac{1}{h_e^2} \begin{bmatrix} 4x_r^2 x_l - 6x_r x_l^2 - x_l^3 + 3x_l^3 & 2x_r^2 x_l - x_r^3 - x_l^3 \\ -2x_r x_l^2 + x_r^3 + x_l^3 & 6x_r^2 x_l - 4x_r x_l^2 - 3x_r^3 + x_l^3 \end{bmatrix} \qquad (3.5.4)$$

and

$$\{F^e\} = \frac{1}{12h_e} \begin{Bmatrix} -4x_r x_l^3 + x_r^4 + 3x_l^4 \\ -4x_r^3 x_l + 3x_r^4 + x_l^4 \end{Bmatrix} \qquad (3.5.5)$$

where h_e is the length of the linear element, and x_l and x_r are nodal coordinate values of left and right nodes of the element. The MATLAB program using 10 elements is provided below along with new necessary functions.

```
%---------------------------------------------------------------
% EX3.5.3.m
% to solve the ordinary differential equation given as
% x^2 u" - 2x u' -4 u = x^2, 10 < x < 20
% u(10) = 0 and u(20) = 100
% using 10 linear elements
%
% Variable descriptions
```

```
% k = element matrix
% f = element vector
% kk = system matrix
% ff = system vector
% index = a vector containing system dofs associated with each element
% bcdof = a vector containing dofs associated with boundary conditions
% bcval = a vector containing boundary condition values associated with
%             the dofs in bcdof
%----------------------------------------------------------------
%
%----------------------------------------
% input data for control parameters
%----------------------------------------
nel=10;                             % number of elements
nnel=2;                             % number of nodes per element
ndof=1;                             % number of dofs per node
nnode=11;                           % total number of nodes in system
sdof=nnode*ndof;                    % total system dofs
%
%----------------------------------------
% input data for nodal coordinate values
%----------------------------------------
gcoord(1)=10; gcoord(2)=11; gcoord(3)=12; gcoord(4)=13;
gcoord(5)=14; gcoord(6)=15; gcoord(7)=16; gcoord(8)=17;
gcoord(9)=18; gcoord(10)=19; gcoord(11)=20;
%
%------------------------------------------------
% input data for nodal connectivity for each element
%------------------------------------------------
nodes(1,1)=1; nodes(1,2)=2; nodes(2,1)=2; nodes(2,2)=3;
nodes(3,1)=3; nodes(3,2)=4; nodes(4,1)=4; nodes(4,2)=5;
nodes(5,1)=5; nodes(5,2)=6; nodes(6,1)=6; nodes(6,2)=7;
nodes(7,1)=7; nodes(7,2)=8; nodes(8,1)=8; nodes(8,2)=9;
nodes(9,1)=9; nodes(9,2)=10; nodes(10,1)=10; nodes(10,2)=11;
%
%------------------------------------------
% input data for boundary conditions
%------------------------------------------
bcdof(1)=1;                         % first node is constrained
bcval(1)=0;                         % whose described value is 0
bcdof(2)=11;                        % 11th node is constrained
bcval(2)=100;                       % whose described value is 100
%
%------------------------------------------
% initialization of matrices and vectors
%------------------------------------------
ff=zeros(sdof,1);                   % initialization of system force vector
kk=zeros(sdof,sdof);                % initialization of system matrix
```

Section 3.5   Example Programs   67

```
index=zeros(nnel*ndof,1);            % initialization of index vector
%
%----------------------------------------------------------------
%
% computation of element matrices and vectors and their assembly
%----------------------------------------------------------------
%
for iel=1:nel                         % loop for the total number of elements
%
nl=nodes(iel,1); nr=nodes(iel,2);     % extract nodes for (iel)-th element
xl=gcoord(nl); xr=gcoord(nr);         % extract nodal coord values
eleng=xr-xl;                          % element length
index=feeldof1(iel,nnel,ndof);        % extract system dofs associated
%
k=feodex2l(xl,xr);                    % compute element matrix
f=fefx2l(xl,xr);                      % compute element vector
[kk,ff]=feasmbl2(kk,ff,k,f,index);    % assemble element matrices and vectors
%
end                                   % end of loop for total elements
%
%---------------------------------
% apply boundary conditions
%---------------------------------
[kk,ff]=feaplyc2(kk,ff,bcdof,bcval);
%
%-----------------------------
% solve the matrix equation
%-----------------------------
fsol=kk\ff;
%
%-----------------------------
% analytical solution
%-----------------------------
for i=1:nnode
x=gcoord(i);
esol(i)=0.00102*x^4-0.16667*x^2+64.5187/x;
%
%---------------------------------------
% print both exact and fem solutions
%---------------------------------------
num=1:1:sdof;
results=[num' fsol esol']
%----------------------------------------------------------------



function [k]=feodex2l(xl,xr)
%----------------------------------------------------------------
% Purpose:
% element matrix for (x^2 u" - 2x u' - 4 u)
```

```
% using linear element
%
% Synopsis:
% [k]=feodex2l(xl,xr)
%
% Variable Description:
% k - element matrix (size of 2x2)
% xl - coordinate value of the left node of the linear element
% xr - coordinate value of the right node of the linear element
%---------------------------------------------------------------
%
% element matrix
%
eleng=xr-xl;
k=(1/eleng^2)*[ (4*xr^2*xl-6*xr*xl^2-xr^3+3*xl^3) ...
(2*xr^2*xl-xr^3-xl^3);...
(-2*xr*xl^2+xr^3+xl^3) ...
(6*xr^2*xl-4*xr*xl^2-3*xr^3+xl^3)];
%---------------------------------------------------------------


function [f]=fefx2l(xl,xr)
%---------------------------------------------------------------
% Purpose:
% element vector for f(x)=$x^2$
% using linear element
%
% Synopsis:
% [f]=fefx2l(xl,xr)
%
% Variable Description:
% f - element vector (size of 2x1)
% xl - coordinate value of the left node
% xr - coordinate value of the right node
%---------------------------------------------------------------
%
% element vector
%
eleng=xr-xl;                                    % element length
f=(1/(12*eleng))*[ (-4*xr*xl^3+xr^4+3*xl^4);...
(-4*xr^3*xl+3*xr^4+xl^4)];
%---------------------------------------------------------------
```

The results are

node #	fem sol	exact sol	
1.0000	0.0000	0.0000	% solution at x=0
2.0000	0.6046	0.6321	% solution at x=0.1

3.0000	2.4650	2.5268	% solution at $x=0.2$
4.0000	5.8421	5.9280	% solution at $x=0.3$
5.0000	11.028	11.126	% solution at $x=0.4$
6.0000	18.343	18.438	% solution at $x=0.5$
7.0000	28.137	28.212	% solution at $x=0.6$
8.0000	40.785	40.819	% solution at $x=0.7$
9.0000	56.689	56.659	% solution at $x=0.8$
10.000	76.276	76.155	% solution at $x=0.9$
11.000	100.00	100.00	% solution at $x=1$

‡

Problems

3.1 Modify the MATLAB programs provided in Sec. 3.5 and solve Prob. 2.15 using the programs.

3.2 Solve Prob. 2.16 using the modified computer program.

3.3 Solve Prob. 2.17 using the modified computer program.

3.4 Redo Prob. 3.1 using twice as many elements as the number used in the problem. Compare the two finite element solutions to the exact solution.

3.5 Redo Prob. 3.2 with an increasing number of elements and observe the convergence to the exact solution.

CHAPTER FOUR

DIRECT APPROACH WITH SPRING SYSTEM

4.0 Chapter Overview

An intuitive approach for the finite element method is presented using linear springs without mathematical complication. The linear spring system can represent many practical engineering applications such as axial members, torsional members, electric circuits, heat conduction, flow along ducts, etc. The equivalency between the linear spring system and those systems are discussed so that the finite element method can be applied to those systems.

4.1 Linear Spring

Consider a linear spring as shown in Fig. 4.1.1 (a). The displacements of the two end points of the spring are u_1 and u_2 and the two points are subjected to axial forces f_1 and f_2, respectively. Both displacements and forces are assumed in the right-hand side direction which is assumed to be positive in the present finite element formulation. If the spring is in equilibrium, the sum of forces becomes zero. That is,

$$f_1 + f_2 = 0 \qquad (4.1.1)$$

As a result, $f_2 = -f_1$ and Fig. 4.1.1 (b) shows the equilibrated linear spring. The spring is compressed by these forces and the contraction of the spring is proportional to them. Using the spring constant k, the force and displacement relationship becomes

$$k(u_1 - u_2) = f_1 \qquad (4.1.2)$$

From Eqs. (4.1.1) and (4.1.2), we obtain

$$k(-u_1 + u_2) = f_2 \qquad (4.1.3)$$

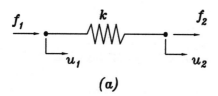

(a)

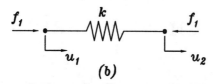

(b)

Figure 4.1.1 Linear Spring

Rewriting Eqs. (4.1.2) and (4.1.3) in matrix form yields

$$\begin{bmatrix} k & -k \\ -k & k \end{bmatrix} \begin{Bmatrix} u_1 \\ u_2 \end{Bmatrix} = \begin{Bmatrix} f_1 \\ f_2 \end{Bmatrix} \qquad (4.1.4)$$

This is the matrix equation for a linear spring. A spring is like a linear finite element. As a result, the matrix is called the *element stiffness matrix* and the right-hand side vector is called the *element force vector*. A system consisting of serial and parallel linear springs can be analyzed using the finite element analysis concept.

♣ **Example 4.1.1** Consider three springs connected in series as shown in Fig. 4.1.2. The matrix equation for each spring is similar to Eq. (4.1.4). Assembling them into the system of matrix equation yields

$$\begin{bmatrix} k_1 & -k_1 & 0 & 0 \\ -k_1 & (k_1+k_2) & -k_2 & 0 \\ 0 & -k_2 & (k_2+k_3) & -k_3 \\ 0 & 0 & -k_3 & k_3 \end{bmatrix} \begin{Bmatrix} u_1 \\ u_2 \\ u_3 \\ u_4 \end{Bmatrix} = \begin{Bmatrix} f_1 \\ f_2 \\ f_3 \\ f_4 \end{Bmatrix} \qquad (4.1.5)$$

Depending on the constraints, the system may be statically determinate or statically indeterminate. For example, if only u_1 is constrained to be zero, the system is statically determinate. On the other hand, if both u_1 and u_4 are constrained to zero, then the system becomes statically indeterminate. In terms of the finite element formulation, there is no distinction between the statically determinate and indeterminate systems because the formulation uses not only equilibrium equations but also compatibility of the displacements. As an example of a statically indeterminate system, let $k_1 = 20$ MN/m, $k_2 = 30$ MN/m and $k_3 = 10$ MN/m. In addition, an external force is applied at node 2, i.e.

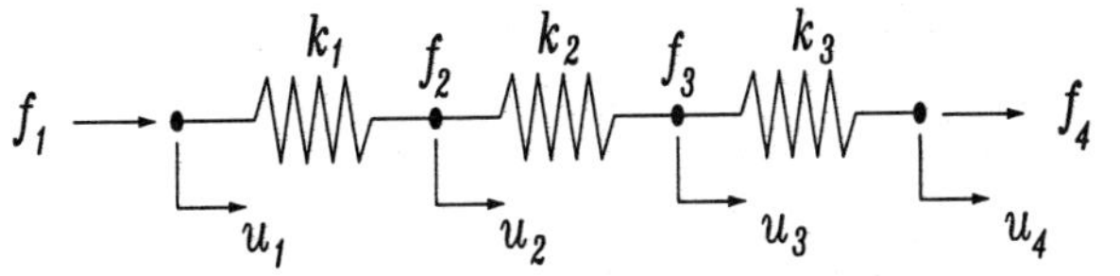

Figure 4.1.2 Linear Springs in Serial Connection

$f_2 = 1000$ N. Then, the matrix equation becomes

$$10^6 \begin{bmatrix} 20 & -20 & 0 & 0 \\ -20 & 50 & -30 & 0 \\ 0 & -30 & 40 & -10 \\ 0 & 0 & -10 & 10 \end{bmatrix} \begin{Bmatrix} u_1 \\ u_2 \\ u_3 \\ u_4 \end{Bmatrix} = \begin{Bmatrix} f_1 \\ 1000 \\ 0 \\ f_4 \end{Bmatrix} \qquad (4.1.6)$$

Note that f_3 is set to zero because there is no external force applied at the node. Applying the constraints $u_1 = 0$ and $u_4 = 0$ to the above equation results in the modified matrix equation

$$10^6 \begin{bmatrix} 1 & 0 & 0 & 0 \\ -20 & 50 & -30 & 0 \\ 0 & -30 & 40 & -10 \\ 0 & 0 & 0 & 1 \end{bmatrix} \begin{Bmatrix} u_1 \\ u_2 \\ u_3 \\ u_4 \end{Bmatrix} = \begin{Bmatrix} 0 \\ 1000 \\ 0 \\ 0 \end{Bmatrix} \qquad (4.1.7)$$

Solution of the matrix equation yields the displacements $u_2 = 36.36 \times 10^{-6}$m and $u_3 = 27.27 \times 10^{-6}$m. Substitution of these displacements along with the constrained displacements into Eq. (4.1.6) determines the reaction forces at both constrained ends, i.e. $f_1 = -727.3$N and $f_2 = -272.7$N. ‡

♣ **Example 4.1.2** Linear springs are connected as shown in Fig. 4.1.3. The rigid and massless plates are assumed to move vertically without rotation. We want to find the deflection of the system. Each spring constitutes one linear element and there are 7 elements in the system. The number of total nodes in the system is 6 because some nodes are shared by more than two elements. As a result, the number of degrees of freedom before applying the constraint is 6. Assembling these elements into the system matrix yields

$$k \begin{bmatrix} 1 & -1 & 0 & 0 & 0 & 0 \\ -1 & 7 & -2 & -1 & -3 & 0 \\ 0 & -2 & 3 & -1 & 0 & 0 \\ 0 & -1 & -1 & 4 & -2 & 0 \\ 0 & -3 & 0 & -2 & 7 & -2 \\ 0 & 0 & 0 & 0 & -2 & 2 \end{bmatrix} \begin{Bmatrix} x_1 \\ x_2 \\ x_3 \\ x_4 \\ x_5 \\ x_6 \end{Bmatrix} = \begin{Bmatrix} f_1 \\ w \\ w \\ w \\ w \\ 0 \end{Bmatrix} \qquad (4.1.8)$$

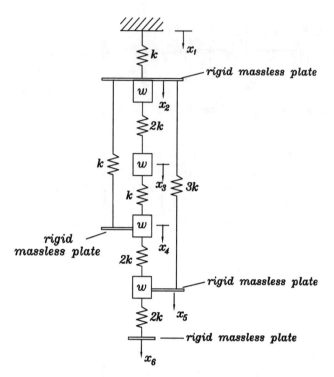

Figure 4.1.3 A Mass-Spring System

There is no weight at node 6 and the sixth component of the right-hand side column vector is zero as the result. In addition, f_1 is unknown here because $x_1 = 0$ is known. Applying the constraint to Eq. (4.1.8) gives

$$k \begin{bmatrix} 1 & 0 & 0 & 0 & 0 & 0 \\ -1 & 7 & -2 & -1 & -3 & 0 \\ 0 & -2 & 3 & -1 & 0 & 0 \\ 0 & -1 & -1 & 4 & -2 & 0 \\ 0 & -3 & 0 & -2 & 7 & -2 \\ 0 & 0 & 0 & 0 & -2 & 2 \end{bmatrix} \begin{Bmatrix} x_1 \\ x_2 \\ x_3 \\ x_4 \\ x_5 \\ x_6 \end{Bmatrix} = \begin{Bmatrix} 0 \\ w \\ w \\ w \\ w \\ 0 \end{Bmatrix} \qquad (4.1.9)$$

The matrix equation determines the displacements of the springs. ‡

4.2 Axial Member

The linear spring can represent various systems in engineering applications. One direct application is the axial member. Consider an axial member with length L, uniform cross-section A and elastic modulus E. The elongation δ of the axial member subjected to an axial force P is computed from

$$\delta = \frac{PL}{AE} \qquad (4.2.1)$$

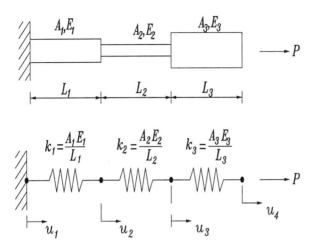

Figure 4.2.1 Axial Members Represented by Equivalent Springs

Rewriting Eq. (4.2.1) gives

$$P = \frac{AE}{L}\delta \qquad (4.2.2)$$

As a result, the equivalent spring constant for the axial bar is

$$k_{eq} = \frac{AE}{L} \qquad (4.2.3)$$

♣ **Example 4.2.1** Axial members can be represented by serial and/or parallel linear springs. For example, a bar of telescoped shape can be replaced by a series of springs as shown in Fig. 4.2.1. There are three linear spring elements and each element has the matrix expression as shown in Eq. (4.1.4) with proper spring constant k_i. Assembling them gives

$$\begin{bmatrix} k_1 & -k_1 & 0 & 0 \\ -k_1 & k_1+k_2 & -k_2 & 0 \\ 0 & -k_2 & k_2+k_3 & -k_3 \\ 0 & 0 & -k_3 & k_3 \end{bmatrix} \begin{Bmatrix} u_1 \\ u_2 \\ u_3 \\ u_4 \end{Bmatrix} = \begin{Bmatrix} f_1 \\ 0 \\ 0 \\ P \end{Bmatrix} \qquad (4.2.4)$$

where $k_i = \frac{A_i E_i}{L_i}$ and f_1 is the unknown reaction force at the left end support. Instead, $u_1 = 0$ is given as the boundary condition. Solution of Eq. (4.2.4) with this boundary condition will result in $f_1 = -P$ which can be also obtained from the static equilibrium. However, the finite element formulation already includes equations of static equilibrium so that we may not use additional equilibrium equations. These equations are redundant. ‡

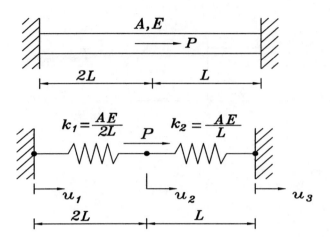

Figure 4.2.2 A Statically Indeterminate Axial Member

♣ **Example 4.2.2** Consider a statically indeterminant system as shown in Fig. 4.2.2. The axial member can be replaced by two linear springs as seen in the figure. The finite element matrix equation for this system is

$$\begin{bmatrix} 0.5k & -0.5k & 0 \\ -0.5k & 1.5k & -k \\ 0 & -k & k \end{bmatrix} \begin{Bmatrix} u_1 \\ u_2 \\ u_3 \end{Bmatrix} = \begin{Bmatrix} f_1 \\ P \\ f_3 \end{Bmatrix} \qquad (4.2.5)$$

where $k = \frac{AE}{L}$, and f_1 and f_3 are the unknown forces at the supports. Because the system is statically indeterminate, we cannot find the forces from the static equilibrium equation only. However, the finite element formulation includes not only equilibrium but also geometric compatibility (compatibility of deformation). Therefore, Eq. (4.2.5) along with boundary conditions $u_1 = u_3 = 0$ can solve the deformation as well as the reaction forces.

Another statically indeterminate system made of axial rods can be represented by a spring constant as shown in Fig. 4.2.3. ‡

4.3 Torsional Member

A circular rod subjected to a twisting moment produces an angle of twist

$$\theta = \frac{TL}{GJ} \qquad (4.3.1)$$

in which θ is the angle of twist, T is the applied torque, L is the length of the member, G is the shear modulus, and J is the polar moment of inertia of the circular cross-section. Rewriting the equation yields

$$T = \frac{GJ}{L}\theta \qquad (4.3.2)$$

Section 4.3   Torsional Member   77

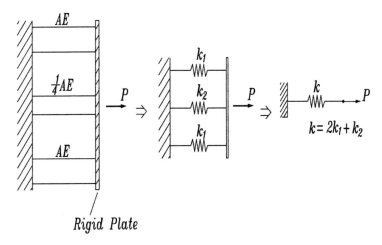

Figure 4.2.3 A Spring Representing a Statically Indeterminate System

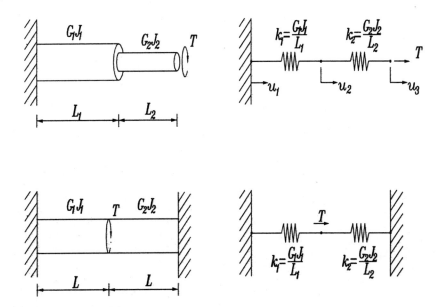

Figure 4.3.1 Torsional Members

The equivalent spring constant is $k_{eq} = \frac{GJ}{L}$. Torque T corresponds to the spring force F and angle of twist θ corresponds to the spring displacement u. As examples, both statically determinate and indeterminate torsional members are shown in Fig. 4.3.1 along with their equivalent spring systems.

4.4 Other Systems

Other frequently used engineering systems, which can be easily substituted with a spring system, are heat conduction, simple fluid flow along pipes, and electric circuits. For one-dimensional heat conduction, heat flux is proportional to temperature difference. That is, heat flux q is

$$q = -k_t \Delta T \tag{4.4.1}$$

where k_t is the heat conduction coefficient and ΔT is the temperature difference between two end points of a one-dimensional bar. The minus sign in the equation denotes that heat flux is in the opposite direction as the temperature increases along the positive axis. The equivalent spring system has the spring constant $k_{eq} = -k_t$, the spring force $F = q$, and the spring displacement $u = T$.

Fluid flow rate through a pipe of constant cross-section is proportional to the pressure difference of the two ends. That is,

$$Q = -k_p \Delta p \tag{4.4.2}$$

Here, Q is the flow rate, Δp is the pressure difference, and k_p is the proportional constant. For a laminar flow, the proportional constant can be expressed as

$$k_p = \frac{\pi d^4}{128 \mu L} \tag{4.4.3}$$

where μ is the fluid viscocity, d is the pipe diameter, and L is the pipe length. The equivalent spring system has the spring constant $k_{eq} = -k_p$, the spring force $F = Q$, and the spring displacement $u = p$.

Electric current flow i through a resistance R is

$$i = \frac{V}{R} \tag{4.4.4}$$

where V is voltage. The equivalent spring system has spring constant $k_{eq} = \frac{1}{R}$, force $F = i$, and displacement $u = V$. One may consider $V = iR$ such that $F = V$, $k_{eq} = R$ and $u = i$. Is the equivalent spring constant R or $\frac{1}{R}$? One is right and the other is wrong. One way to select the right form is to understand the nature of spring force and to find the parameter equivalent to the force. Then, the rest of the parameters can be determined accordingly.

When two springs are separated from each other, internal forces occur between the two springs. These forces are equal in magnitude and opposite in direction. When the two springs are put together, the forces cancel each other and become zero if there is no external force applied at the joint as seen in Fig. 4.4.1. This is known as *Newton's third law*. When considering electric current, its sum at the joint of resistants (i.e. the middle point in Fig. 4.4.2) is also zero like the force. However, voltage does not

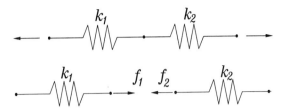

Figure 4.4.1 Mechanical Forces From Newton's Third Law

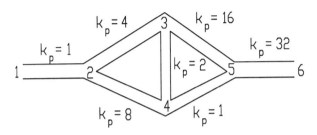

Figure 4.4.2 Flow Through a Pipe System

vanish at such a joint. As a result, electric current is similar to the mechanical force. Therefore, the equivalent spring constant for the electric circuit is $\frac{1}{R}$ but not R.

♣ **Example 4.4.1** Consider a pipe system for fluid flow as seen in Fig. 4.4.2. The pressure at the inlet is 200 while the flow rate at the outlet is 50. The proportional constants k_p in Eq. (4.4.2) are given in the figure. All units are assumed consistent. We want to determine the flow rate between node 3 and node 4.

Using the linear spring equivalency, the matrix equation becomes

$$\begin{bmatrix} -1 & 1 & 0 & 0 & 0 & 0 \\ 1 & -13 & 4 & 8 & 0 & 0 \\ 0 & 4 & -22 & 2 & 16 & 0 \\ 0 & 8 & 2 & -11 & 1 & 0 \\ 0 & 0 & 16 & 1 & -49 & 32 \\ 0 & 0 & 0 & 0 & 32 & -32 \end{bmatrix} \begin{Bmatrix} p_1 \\ p_2 \\ p_3 \\ p_4 \\ p_5 \\ p_6 \end{Bmatrix} = \begin{Bmatrix} Q_1 \\ 0 \\ 0 \\ 0 \\ 0 \\ Q_6 = 50 \end{Bmatrix} \qquad (4.4.5)$$

Applying the known pressure $p_1 = 200$ to this equation yields the following pressure at each node.

$$p_2 = 150, \quad p_3 = 142.2, \quad p_4 = 147.6, \quad p_5 = 139.6, \quad p_6 = 138$$

Therefore, the flow rate between nodes 3 and 4 is

$$Q_{3-4} = -2 \times (147.6 - 142.2) = -10.8$$

Hence, the flow is upward with a rate of 10.8. ‡

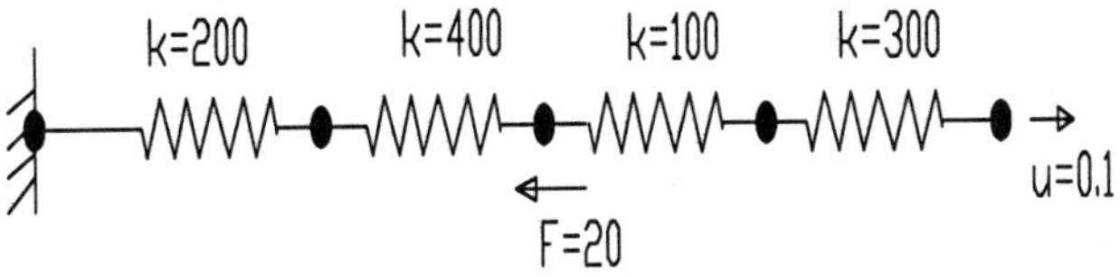

Figure P4.1 Problem 4.1

Problems

4.1 Find the system of equations for the spring system shown in Fig. P4.1. Solve the matrix equation to find the displacements of the nodal points.

4.2 A circular shaft is made of two different materials and fixed at both ends. It is subjected to torsions as shown in Fig. P4.2. The diameter of the shaft is 0.1 m. Find the angles of twist at the nodal points.

4.3 For the given mass-spring system (see Fig. P4.3), (a) develop the system mass and stiffness matrices to determine natural frequencies of the system. (b) Apply the given boundary conditions to the matrix equations.

4.4 An electric circuit is shown in Fig. P4.4. Find the current flow using the equivalent spring system.

4.5 For a laminar pipe flow, the flow rate is proportional to pressure difference. Construct the system of equations for the given pipe flow as shown in Fig. P4.5 in order to find the flow rate through each pipe.

4.6 Heat conduction through a circular pipe is expressed as

$$q = -\frac{2\pi k L}{ln(r_0/r_i)} \Delta T$$

in which k is the heat conduction coefficient, L is the length of the cylinder, and r_o and r_i are the outer and inner radii of the cylinder, respectively. Therefore, two concentric cylinders with radii $r_1 < r_2 < r_3$ can be represented by two springs in serial connection. Find the equivalent spring constants for the two linear springs in Fig. P4.6 and construct the system of equations.

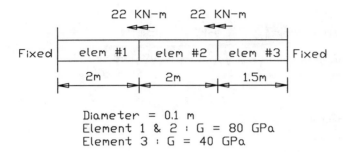

Figure P4.2 Problem 4.2

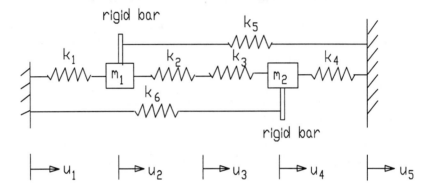

Figure P4.3 Problem 4.3

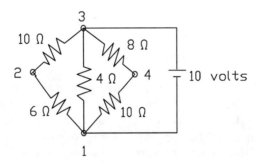

Figure P4.4 Problem 4.4

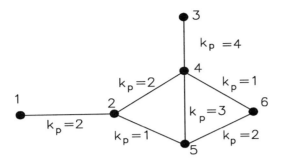

Figure P4.5 Problem 4.5

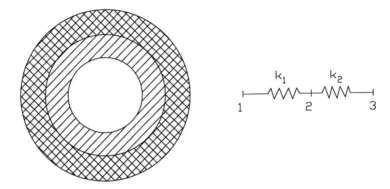

Figure P4.6 Problem 4.6

CHAPTER FIVE

LAPLACE'S AND POISSON'S EQUATIONS

5.0 Chapter Overview

This chapter presents the finite element formulations for partial differential equations. *Laplace*'s and *Poisson*'s equations are used as examples because they are common field equations in many engineering applications. Finite elements for 2-D, 3-D, and axisymmetric problems are developed. Both steady-state (boundary value) problems and transient (initial value) problems are considered. Sample MATLAB programs are provided for the formulations developed in the chapter.

5.1 Governing Equation

Laplace's and *Poisson*'s equations are common field governing equations to describe various physical natures. For example, these differential equations can represent heat conduction, potential flow, and torsion of noncircular members. Therefore, we study the finite element formulation of these equations. *Laplace*'s equation is

$$\nabla^2 u = 0 \tag{5.1.1}$$

while *Poisson*'s equation is

$$\nabla^2 u = g \tag{5.1.2}$$

Because *Poisson*'s equation is more general than *Laplace*'s equation as seen above, we consider *Poisson*'s equation in the following formulation.

Poisson's equation in terms of the Cartesian coordinate system becomes

$$\frac{\partial^2 u}{\partial x^2} + \frac{\partial^2 u}{\partial y^2} = g(x,y) \quad \text{in } \Omega \tag{5.1.3}$$

for a two-dimensional domain Ω. The boundary conditions are

$$u = \bar{u} \quad \text{on } \Gamma_e \tag{5.1.4}$$

and
$$\frac{\partial u}{\partial n} = \bar{q} \quad \text{on } \Gamma_n \qquad (5.1.5)$$

where $\bar{u}$ and $\bar{q}$ denote known variable and flux boundary conditions, and n in Eq. (5.1.5) is the outward normal unit vector at the boundary. Furthermore, Γ_e and Γ_n are boundaries for essential and natural boundary conditions, respectively. For the well-posed boundary value problem,

$$\Gamma_e \cup \Gamma_n = \Gamma \qquad (5.1.6)$$

and

$$\Gamma_e \cap \Gamma_n = \emptyset \qquad (5.1.7)$$

in which $\cup$ and $\cap$ denote sum and intersection respectively, and Γ is the total boundary of the domain Ω.

Integration of weighted residual of the differential equation and boundary condition is

$$I = \int_\Omega w \left(\frac{\partial^2 u}{\partial x^2} + \frac{\partial^2 u}{\partial y^2} - g(x,y) \right) d\Omega - \int_{\Gamma_e} w \frac{\partial u}{\partial n} d\Gamma \qquad (5.1.8)$$

In order to develop the *weak formulation* of Eq. (5.1.8), integration by parts is applied to reduce the order of differentiation within the integral. The subsequent example shows the integration by parts.

♣ **Example 5.1.1** Consider a two-dimensional domain as seen in Fig. 5.1.1. First of all, we want to evaluate the first term of Eq. (5.1.8)

$$\int_\Omega w \frac{\partial^2 u}{\partial x^2} d\Omega \qquad (5.1.9)$$

The domain integral can be expressed as

$$\int_{y_1}^{y_2} \left(\int_{x_1}^{x_2} w \frac{\partial^2 u}{\partial x^2} dx \right) dy \qquad (5.1.10)$$

where y_1 and y_2 are the minimum and maximum values of the domain in the y-axis as the strip along the x-axis moves in the y-direction as seen in Fig. 5.1.1. Integration by parts with respect to x yields

$$-\int_{y_1}^{y_2} \int_{x_1}^{x_2} \frac{\partial w}{\partial x} \frac{\partial u}{\partial x} dx dy + \int_{y_1}^{y_2} \left[w \frac{\partial u}{\partial x} \right]_{x_1}^{x_2} dy \qquad (5.1.11)$$

and rewriting the expression using the domain and boundary integrations as shown in Fig. 5.1.1 results in

$$-\int_\Omega \frac{\partial w}{\partial x} \frac{\partial u}{\partial x} d\Omega + \int_{\Gamma_2} w \frac{\partial u}{\partial x} n_x d\Gamma - \int_{\Gamma_1} w \frac{\partial u}{\partial x} n_x d\Gamma \qquad (5.1.12)$$

in which n_x is the x-component of the unit normal vector which is assumed to be positive in the outward direction as shown in Fig. 5.1.1. Finally combining the two boundary integals gives

$$-\int_\Omega \frac{\partial w}{\partial x}\frac{\partial u}{\partial x}d\Omega + \oint_\Gamma w\frac{\partial u}{\partial x}n_x d\Gamma \tag{5.1.13}$$

where the boundary integral is in the counter-clockwise direction
Similarly, the second term in Eq. (5.1.8) can be written as

$$-\int_\Omega \frac{\partial w}{\partial y}\frac{\partial u}{\partial y}d\Omega + \oint_\Gamma w\frac{\partial u}{\partial y}n_y d\Gamma \tag{5.1.14}$$

Adding Eqs. (5.1.13) and (5.1.14) produces

$$\int_\Omega w(\frac{\partial^2 u}{\partial x^2} + \frac{\partial^2 u}{\partial y^2})d\Omega =$$
$$-\int_\Omega \left(\frac{\partial w}{\partial x}\frac{\partial u}{\partial x} + \frac{\partial w}{\partial y}\frac{\partial u}{\partial y}\right)d\Omega + \oint_\Gamma w\left(\frac{\partial u}{\partial x}n_x + \frac{\partial u}{\partial y}n_y\right)d\Gamma \tag{5.1.15}$$

Since the boundary integral can be written as

$$\frac{\partial u}{\partial n} = \frac{\partial u}{\partial x}n_x + \frac{\partial u}{\partial y}n_y \tag{5.1.16}$$

we can rewrite Eq. (5.1.15) as

$$\int_\Omega w(\frac{\partial^2 u}{\partial x^2} + \frac{\partial^2 u}{\partial y^2})d\Omega =$$
$$-\int_\Omega \left(\frac{\partial w}{\partial x}\frac{\partial u}{\partial x} + \frac{\partial w}{\partial y}\frac{\partial u}{\partial y}\right)d\Omega + \oint_\Gamma w\frac{\partial u}{\partial n}d\Gamma \tag{5.1.17}$$

The symbol $\oint$ to denote the line integral around a closed boundary is replaced by $\int$ for simplicity in the following text. Equation (5.1.17) is known as *Green's theorem*. ‡

Use of Eq. (5.1.17) to Eq. (5.1.8) results in

$$I = -\int_\Omega \left(\frac{\partial w}{\partial x}\frac{\partial u}{\partial x} + \frac{\partial w}{\partial y}\frac{\partial u}{\partial y}\right)d\Omega - \int_\Omega wg(x,y)d\Omega + \int_{\Gamma_n} w\frac{\partial u}{\partial n}d\Gamma \tag{5.1.18}$$

The first volume integral becomes a matrix term while both the second volume integral and the line integral become vector terms. In the context of heat conduction, the second volume integral is related to heat source or sink within the domain, and the line integral denotes the heat flux through the natural boundary.

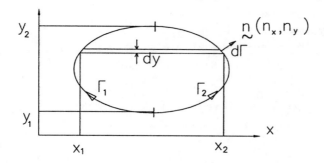

Figure 5.1.1 Two-Dimensional Domain

5.2 Linear Triangular Element

Discretization of the domain in Eq. (5.1.18) is performed using selected two-dimensional finite elements. One of the simplest two-dimensional elements is the three-noded triangular element. This is also known as the linear triangular element. The element is shown in Fig. 5.2.1. It has three nodes at the vertices of the triangle and the variable interpolation within the element is linear in x and y, as

$$u = a_1 + a_2 x + a_3 y \qquad (5.2.1)$$

or

$$u = \begin{bmatrix} 1 & x & y \end{bmatrix} \begin{Bmatrix} a_1 \\ a_2 \\ a_3 \end{Bmatrix} \qquad (5.2.2)$$

where a_i is the constant to be determined. The interpolation function, Eq. (5.2.1), should represent the nodal variables at the three nodal points. Therefore, substituting the x and y values at each nodal point gives

$$\begin{Bmatrix} u_1 \\ u_2 \\ u_3 \end{Bmatrix} = \begin{bmatrix} 1 & x_1 & y_1 \\ 1 & x_2 & y_2 \\ 1 & x_3 & y_3 \end{bmatrix} \begin{Bmatrix} a_1 \\ a_2 \\ a_3 \end{Bmatrix} \qquad (5.2.3)$$

Here, x_i and y_i are the coordinate values at the ith node and u_i is the nodal variable as seen in Fig. 5.2.1. Inverting the matrix and rewriting Eq. (5.2.3) give

$$\begin{Bmatrix} a_1 \\ a_2 \\ a_3 \end{Bmatrix} = \frac{1}{2A} \begin{bmatrix} x_2 y_3 - x_3 y_2 & x_3 y_1 - x_1 y_3 & x_1 y_2 - x_2 y_1 \\ y_2 - y_3 & y_3 - y_1 & y_1 - y_2 \\ x_3 - x_2 & x_1 - x_3 & x_2 - x_1 \end{bmatrix} \begin{Bmatrix} u_1 \\ u_2 \\ u_3 \end{Bmatrix} \qquad (5.2.4)$$

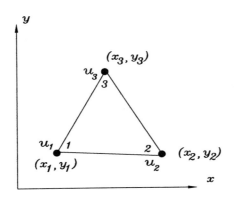

Figure 5.2.1 Linear Triangular Element

where

$$A = \frac{1}{2} \det \begin{bmatrix} 1 & x_1 & y_1 \\ 1 & x_2 & y_2 \\ 1 & x_3 & y_3 \end{bmatrix} \quad (5.2.5)$$

The magnitude of A is equal to the area of the linear triangular element. However, its value is positive if the element node numbering is in the counter-clockwise direction and negative otherwise. For the finite element computation, the element nodal sequence must be in the same direction for every element in the domain.

Substitution of Eq. (5.2.4) into Eq. (5.2.2) produces

$$u = H_1(x,y)u_1 + H_2(x,y)u_2 + H_3(x,y)u_3 \quad (5.2.6)$$

in which $H_i(x,y)$ is the shape function for the linear triangular element and it is given below:

$$H_1 = \frac{1}{2A}[(x_2 y_3 - x_3 y_2) + (y_2 - y_3)x + (x_3 - x_2)y] \quad (5.2.7)$$

$$H_2 = \frac{1}{2A}[(x_3 y_1 - x_1 y_3) + (y_3 - y_1)x + (x_1 - x_3)y] \quad (5.2.8)$$

$$H_3 = \frac{1}{2A}[(x_1 y_2 - x_2 y_1) + (y_1 - y_2)x + (x_2 - x_1)y] \quad (5.2.9)$$

These shape functions also satisfy the conditions

$$H_i(x_j, y_j) = \delta_{ij} \quad (5.2.10)$$

and

$$\sum_{i=1}^{3} H_i = 1 \quad (5.2.11)$$

Here, δ_{ij} is the *Kronecker delta*. That is,

$$\delta_{ij} = \begin{Bmatrix} 1 & \text{if } i = j \\ 0 & \text{if } i \neq j \end{Bmatrix} \quad (5.2.12)$$

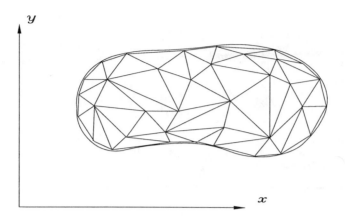

Figure 5.2.2 Finite Element Discretization

A problem domain is discretized into a number of finite elements using the linear triangular elements. An example of finite element mesh discretization is illustrated in Fig. 5.2.2. As seen in the discretization, the actual curved boundary is approximated by a piecewise linear boundary. The crude mesh in the figure may be refined for closer approximation of the actual boundary using linear triangular elements. Another alternative is to use higher order finite elements which can fit the curved boundary using a higher order of polynomial expressions.

For a linear triangular element shown in Fig. 5.2.1, the element matrix is computed as derived below.

$$[K^e]\{u^e\} = \int_{\Omega^e} \left(\frac{\partial w}{\partial x} \frac{\partial u}{\partial x} + \frac{\partial w}{\partial y} \frac{\partial u}{\partial y} \right) d\Omega = \int_{\Omega^e} \left(\begin{Bmatrix} \frac{\partial H_1}{\partial x} \\ \frac{\partial H_2}{\partial x} \\ \frac{\partial H_3}{\partial x} \end{Bmatrix} \begin{Bmatrix} \frac{\partial H_1}{\partial x} & \frac{\partial H_2}{\partial x} & \frac{\partial H_3}{\partial x} \end{Bmatrix} + \right.$$

$$\left. \begin{Bmatrix} \frac{\partial H_1}{\partial y} \\ \frac{\partial H_2}{\partial y} \\ \frac{\partial H_3}{\partial y} \end{Bmatrix} \begin{Bmatrix} \frac{\partial H_1}{\partial y} & \frac{\partial H_2}{\partial y} & \frac{\partial H_3}{\partial y} \end{Bmatrix} \right) d\Omega \{u^e\} \qquad (5.2.13)$$

where Ω^e is the element domain.

Performing integration after substituting the shape functions Eqs. (5.2.7) through (5.2.9) into Eq. (5.2.13) gives

$$[K^e] = \begin{bmatrix} k_{11} & k_{12} & k_{13} \\ k_{21} & k_{22} & k_{23} \\ k_{31} & k_{32} & k_{33} \end{bmatrix} \qquad (5.2.14)$$

in which

$$k_{11} = \frac{1}{4A}[(x_3 - x_2)^2 + (y_2 - y_3)^2] \qquad (5.2.15)$$

$$k_{12} = \frac{1}{4A}[(x_3 - x_2)(x_1 - x_3) + (y_2 - y_3)(y_3 - y_1)] \qquad (5.2.16)$$

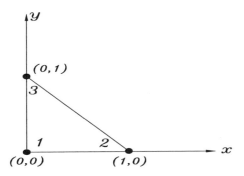

Figure 5.2.3 Triangular Domain With One Element

$$k_{13} = \frac{1}{4A}[(x_3 - x_2)(x_2 - x_1) + (y_2 - y_3)(y_1 - y_2)] \quad (5.2.17)$$

$$k_{21} = k_{12} \quad (5.2.18)$$

$$k_{22} = \frac{1}{4A}[(x_1 - x_3)^2 + (y_3 - y_1)^2)] \quad (5.2.19)$$

$$k_{23} = \frac{1}{4A}[(x_1 - x_3)(x_2 - x_1) + (y_3 - y_1)(y_1 - y_2)] \quad (5.2.20)$$

$$k_{31} = k_{13} \quad (5.2.21)$$

$$k_{32} = k_{23} \quad (5.2.22)$$

$$k_{33} = \frac{1}{4A}[(x_2 - x_1)^2 + (y_1 - y_2)^2] \quad (5.2.23)$$

Because $\frac{\partial H_i}{\partial x}$ and $\frac{\partial H_i}{\partial y}$ are constant for the linear triangular element, the integrand in Eq. (5.2.13) is constant. As a result, the integration in Eq. (5.2.13) becomes the integrand multiplied by the area of the element domain and the result is given in Eqs. (5.2.15) through (5.2.23).

♣ **Example 5.2.1**   We compute the element matrix for *Poisson's* equation for the linear triangular element shown in Fig. 5.2.3. Element node numbering is in the counter-clockwise direction. The area of the triangular element is 0.5 and the element matrix is

$$[K^e] = \begin{bmatrix} 1.0 & -0.5 & -0.5 \\ -0.5 & 0.5 & 0 \\ -0.5 & 0 & 0.5 \end{bmatrix} \quad (5.2.24)$$

‡

The other domain integral term to be evaluated in Eq. (5.1.18) is

$$\int_\Omega wg(x,y)d\Omega \tag{5.2.25}$$

This integration results in a column vector as shown below. Computation of this integral over each linear triangular element yields

$$\int_{\Omega^e} \begin{Bmatrix} H_1 \\ H_2 \\ H_3 \end{Bmatrix} g(x,y)d\Omega \tag{5.2.26}$$

Analytical integration may not be simple depending on function $g(x,y)$. Then, a numerical integration technique may be applied to compute this integral. Some numerical techniques are discussed in Chapter 6.

5.3 Bilinear Rectangular Element

The bilinear rectangular element is shown in Fig. 5.3.1. The shape functions for this element can be derived from the following interpolation function:

$$u = a_1 + a_2 x + a_3 y + a_4 xy \tag{5.3.1}$$

This function is linear in both x and y. Applying the same procedure as used in the previous section results in

$$H_1 = \frac{1}{4bc}(b-x)(c-y) \tag{5.3.2}$$

$$H_2 = \frac{1}{4bc}(b+x)(c-y) \tag{5.3.3}$$

$$H_3 = \frac{1}{4bc}(b+x)(c+y) \tag{5.3.4}$$

$$H_4 = \frac{1}{4bc}(b-x)(c+y) \tag{5.3.5}$$

where $2b$ and $2c$ are the length and height of the element, respectively.

The shape functions Eqs. (5.3.2) through (5.3.5) can be obtained by calculating the product of two sets of one-dimensional shape functions. Let the linear shape functions in the x-direction with nodes located at $x = -b$ and $x = b$ be

$$\phi_1(x) = \frac{1}{2b}(b-x) \tag{5.3.6}$$

and

$$\phi_2(x) = \frac{1}{2b}(b+x) \tag{5.3.7}$$

Section 5.3 Bilinear Rectangular Element

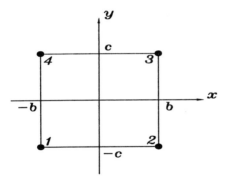

Figure 5.3.1 Bilinear Element

Similarly, the linear shape functions in the y-direction are

$$\psi_1(x) = \frac{1}{2c}(c - y) \tag{5.3.8}$$

and

$$\psi_2(x) = \frac{1}{2c}(c + y) \tag{5.3.9}$$

The product of Eqs. (5.3.6) and (5.3.7) and Eqs. (5.3.8) and (5.3.9) yields Eqs. (5.3.2) through (5.3.5). The shape functions obtained by the products as shown above are called the *Lagrange* shape functions.

♣ **Example 5.3.1** We want to compute the element matrix for *Poisson's* equation using the bilinear shape functions.

$$[K^e] = \int_{\Omega^e} \left(\begin{Bmatrix} \frac{\partial H_1}{\partial x} \\ \frac{\partial H_2}{\partial x} \\ \frac{\partial H_3}{\partial x} \\ \frac{\partial H_4}{\partial x} \end{Bmatrix} \begin{bmatrix} \frac{\partial H_1}{\partial x} & \frac{\partial H_2}{\partial x} & \frac{\partial H_3}{\partial x} & \frac{\partial H_4}{\partial x} \end{bmatrix} + \right.$$

$$\left. \begin{Bmatrix} \frac{\partial H_1}{\partial y} \\ \frac{\partial H_2}{\partial y} \\ \frac{\partial H_3}{\partial y} \\ \frac{\partial H_4}{\partial y} \end{Bmatrix} \begin{bmatrix} \frac{\partial H_1}{\partial y} & \frac{\partial H_2}{\partial y} & \frac{\partial H_3}{\partial y} & \frac{\partial H_4}{\partial y} \end{bmatrix} \right) d\Omega \tag{5.3.10}$$

where H_i is the bilinear shape function. This will be a 4×4 matrix. The first component is

$$K^e_{11} = \int_{-b}^{b} \int_{-c}^{c} \left[\frac{\partial H_1}{\partial x} \frac{\partial H_1}{\partial x} + \frac{\partial H_1}{\partial y} \frac{\partial H_1}{\partial y} \right] dy dx$$

$$= \int_{-b}^{b} \int_{-c}^{c} \frac{1}{16 b^2 c^2} \left[(y - c)^2 + (x - b)^2 \right] dy dx$$

$$= \frac{c^2 + b^2}{3bc} \tag{5.3.11}$$

Performing integrations for all terms results in the following element matrix for the bilinear rectangular element.

$$[K^e] = \begin{bmatrix} k_{11} & k_{12} & k_{13} & k_{14} \\ k_{12} & k_{22} & k_{23} & k_{24} \\ k_{13} & k_{23} & k_{33} & k_{34} \\ k_{14} & k_{24} & k_{34} & k_{44} \end{bmatrix} \quad (5.3.12)$$

in which

$$k_{11} = \frac{b^2 + c^2}{3bc} \quad (5.3.13)$$

$$k_{12} = \frac{b^2 - 2c^2}{6bc} \quad (5.3.14)$$

$$k_{13} = -\frac{b^2 + c^2}{6bc} \quad (5.3.15)$$

$$k_{14} = \frac{c^2 - 2b^2}{6bc} \quad (5.3.16)$$

$$k_{22} = k_{11} \quad (5.3.17)$$

$$k_{23} = k_{14} \quad (5.3.18)$$

$$k_{24} = k_{13} \quad (5.3.19)$$

$$k_{33} = k_{11} \quad (5.3.20)$$

$$k_{34} = k_{12} \quad (5.3.21)$$

$$k_{44} = k_{11} \quad (5.3.22)$$

‡

The other domain integral becomes

$$\int_{-b}^{b} \int_{-c}^{c} \begin{Bmatrix} H_1 \\ H_2 \\ H_3 \\ H_4 \end{Bmatrix} g(x,y) dy dx \quad (5.3.23)$$

This is similar to that in Eq. (5.2.26).

5.4 Boundary Integral

The boundary integral in Eq. (5.1.18) is

$$\int_{\Gamma_n} w \frac{\partial u}{\partial n} d\Gamma = \sum \int_{\Gamma^e} w \frac{\partial u}{\partial n} d\Gamma \quad (5.4.1)$$

Section 5.4 Boundary Integral

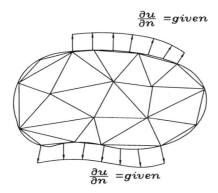

Figure 5.4.1 Elements at Boundary

where subscript n denotes the natural boundary and superscript e indicates the element boundary. Here, the summation is taken over the elements which are located at the boundary of the domain and whose element boundaries are subjected to the natural boundary condition as shown in Fig. 5.4.1.

For simplicity, we consider an element boundary which is parallel to the x-axis as seen in Fig. 5.4.2. The element boundary is subjected to a positive constant flux. That is, the flux is in the outward direction which is assumed to be positive. Since linear triangular elements are used for the domain discretization, the element boundary has two nodes as shown in Fig. 5.4.2. As a result, linear one-dimensional shape functions are used to interpolate the element boundary. The boundary integral along the element boundary becomes

$$\int_{\Gamma^e} w \frac{\partial u}{\partial n} d\Gamma =$$
$$\bar{q} \int_{x_i}^{x_j} \left\{ \begin{array}{c} \frac{x_j - x}{x_j - x_i} \\ \frac{x - x_i}{x_j - x_i} \end{array} \right\} dx = \frac{\bar{q} h_{ij}}{2} \left\{ \begin{array}{c} 1 \\ 1 \end{array} \right\} \qquad (5.4.2)$$

where
$$h_{ij} = x_j - x_i : \text{length of the element boundary} \qquad (5.4.3)$$

This column vector is added to locations associated with nodes i and j. If the element boundary is along the y-axis or is in an arbitrary orientation about the xy-axes, the same result is obtained as long as h_{ij} is the length of the element boundary.

♣ **Example 5.4.1** We consider a heat conduction problem with a triangular domain shape which is discretized into four linear triangular elements (see Fig. 5.4.3). There are six nodes in the domain. One boundary is insulated or symmetric so that there is no flux ($\frac{\partial u}{\partial n} = 0$) through the boundary. Another boundary has a constant heat flux and the third boundary has a known temperature. Find the temperature at the nodal points.

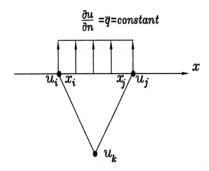

Figure 5.4.2 Triangular Element Subjected to Constant Flux

Each element matrix can be obtained from Eqs. (5.2.14) through (5.2.23). The local and global node numbering is shown for each element in Fig. 5.4.4. The global node numbering is used to identify what nodes are associated with each element while the local node numbering is related to the assignment of shape functions to the nodes. Therefore, the local node numbers are always 1, 2, and 3 for the linear triangular element. For the present elements, element matrices are the same and one of them is given below:

$$[K^e] = \begin{bmatrix} 0.5 & -0.5 & 0.0 \\ -0.5 & 1.0 & -0.5 \\ 0.0 & -0.5 & 0.5 \end{bmatrix} \tag{5.4.4}$$

If the local node numbering changes for each element, the element matrix becomes different from that in Eq. (5.4.4). Assembling the element matrices into the system matrix based on the global node numbers results in

$$[K] = \begin{bmatrix} 0.5 & -0.5 & 0.0 & 0.0 & 0.0 & 0.0 \\ -0.5 & 2.0 & -1.0 & -0.5 & 0.0 & 0.0 \\ 0.0 & -1.0 & 2.0 & 0.0 & -1.0 & 0.0 \\ 0.0 & -0.5 & 0.0 & 1.0 & -0.5 & 0.0 \\ 0.0 & 0.0 & -1.0 & -0.5 & 2.0 & -0.5 \\ 0.0 & 0.0 & 0.0 & 0.0 & -0.5 & 0.5 \end{bmatrix} \tag{5.4.5}$$

for the system nodal vector $\{\, u_1 \;\; u_2 \;\; u_3 \;\; u_4 \;\; u_5 \;\; u_6 \,\}$.

The system column vector is obtained from the boundary integration of the given flux. Element boundaries with specified nonzero flux are boundaries $4-5$ and $5-6$. Using Eq. (5.4.2), the equivalent nodal fluxes are

$$\begin{Bmatrix} F_4 \\ F_5 \end{Bmatrix} = \begin{Bmatrix} 2 \\ 2 \end{Bmatrix} \tag{5.4.6}$$

and

$$\begin{Bmatrix} F_5 \\ F_6 \end{Bmatrix} = \begin{Bmatrix} 2 \\ 2 \end{Bmatrix} \tag{5.4.7}$$

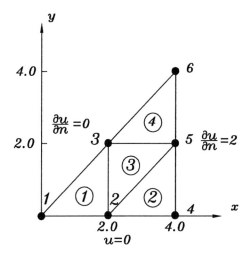

Figure 5.4.3 A Triangular Domain

On the other hand, element boundaries $1-3$ and $3-6$ have zero nodal fluxes because they are insulated. Combining all these vectors yields the system column vector

$$\{F\} = \{\,F_1 \quad F_2 \quad 0.0 \quad F_4 \quad 4.0 \quad 2.0\,\}^T \tag{5.4.8}$$

where F_1, F_2, and F_4 are unknown nodal fluxes. At node 4 the flux is known at the edge $4-5$ but it is not known at the edge $2-4$. The nodal flux at node 4 is affected by the fluxes at the both edges and one of them is not known. As a result, F_4 is unknown. The same explanation holds for F_1. Since temperature is known at nodes 1, 2 and 4, we can solve the matrix equation

$$[K]\{u\} = \{F\} \tag{5.4.9}$$

with $u_1 = 0.0$, $u_2 = 0.0$, and $u_4 = 0.0$. The solution gives $u_3 = 3.0$, $u_5 = 6.0$, and $u_6 = 10.0$. ‡

♣ **Example 5.4.2** One common boundary condition in heat transfer is heat convection at the boundary. This boundary condition is expressed as

$$\frac{\partial u}{\partial n} = -a(u - u_o) \tag{5.4.10}$$

where a is the heat convection coefficient and u_o is the ambient temperature. That is, heat flux is proportional to the temperature difference of the body surface and the environment. Rewriting this in a more general expression gives

$$\frac{\partial u}{\partial n} = -au + b \tag{5.4.11}$$

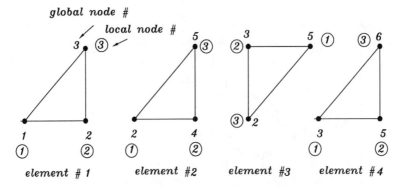

Figure 5.4.4 Elements With Local and Global Node Numbers

where a and b are known functions because u_o is a known value. Substituting Eq. (5.4.11) into the element boundary integral in Eq. (5.4.1) results in

$$\int_{\Gamma_e} w \frac{\partial u}{\partial n} d\Gamma = \int_{\Gamma_e} w\{-a(x,y)u + b(x,y)\} d\Gamma \qquad (5.4.12)$$

Whenever there is a product of test function and trial function, the term becomes a matrix while the test function only produces a vector. As a result, the first term of Eq. (5.4.12) becomes, using linear shape functions to interpolate the element boundary

$$-\int_{s_i}^{s_j} a \left\{ \begin{array}{c} \frac{s_j - s}{s_j - s_i} \\ \frac{s - s_i}{s_j - s_i} \end{array} \right\} \left\{ \begin{array}{cc} \frac{s_j - s}{s_j - s_i} & \frac{s - s_i}{s_j - s_i} \end{array} \right\} d\Gamma \left\{ \begin{array}{c} u_i \\ u_j \end{array} \right\} \qquad (5.4.13)$$

where s_i and s_j are the coordinate values of the local axis located along the element boundary as seen in Fig. 5.4.4, and u_i and u_j are the nodal variables at the element boundary. This integral results in a 2×2 matrix for an element with two nodes on the boundary. This matrix should be added to the system matrix. The remaining term in Eq. (5.4.12) can be dealt with in the same way, described in this section. ‡

5.5 Transient Analysis

The governing equation for transient heat conduction is

$$\frac{\partial u}{\partial t} = \frac{1}{a}\left(\frac{\partial^2 u}{\partial x^2} + \frac{\partial^2 u}{\partial y^2}\right) \quad \text{in } \Omega \qquad (5.5.1)$$

where t denotes time and a is a known function. Usually, for heat conduction problems with constant material properties, a is equal to $\frac{\rho c_p}{k}$ where k is the coefficient of heat conduction, ρ is density and c_p is specific heat. Here, heat generation or heat sink is neglected.

Applying the method of weighted residual to Eq. (5.5.1) in the same way as given in Section 5.1 gives

$$I = \int_\Omega w \frac{\partial u}{\partial t} d\Omega + \frac{1}{a} \int_\Omega \left(\frac{\partial w}{\partial x} \frac{\partial u}{\partial x} + \frac{\partial w}{\partial y} \frac{\partial u}{\partial y} \right) d\Omega - \frac{1}{a} \int_{\Gamma_n} w \frac{\partial u}{\partial n} d\Gamma \qquad (5.5.2)$$

As noticed here, the method of weighted residual is applied to the spatial domain but not to the temporal domain regardless of whether it is a steady state or a transient problem. As a result, the difference between the transient and steady state problems is the first term in Eq. (5.5.2). The other difference is, of course, that the variable u is a function of both space and time for the transient problem.

The variable $u = u(x, y, t)$ is interpolated within a finite element in a similar way as before using shape functions.

$$u(x, y, t) = \sum_{i=1}^{n} H_i(x, y) u_i(t) \qquad (5.5.3)$$

where $H_i(x, y)$ is the shape function and n is the number of nodes per element. One thing to be noted here is that the shape functions are used to interpolate the spatial variation within the element while the temporal variation is related with the nodal variables. Applying Eq. (5.5.3) to the first term in Eq. (5.5.2) yields

$$[M^e] = \int_{\Omega^e} \begin{Bmatrix} H_1 \\ H_2 \\ H_3 \end{Bmatrix} \{ H_1 \quad H_2 \quad H_3 \} d\Omega \begin{Bmatrix} \dot{u}_1 \\ \dot{u}_2 \\ \dot{u}_3 \end{Bmatrix} \qquad (5.5.4)$$

for a linear triangular element. On the other hand, the matrix and vector obtained from the second and third integrals of Eq. (5.5.2) are the same as those developed in the previous sections other than that a should be included in the matrix.

Computation of Eq. (5.5.4) results in

$$[M^e] = \frac{A}{12} \begin{bmatrix} 2 & 1 & 1 \\ 1 & 2 & 1 \\ 1 & 1 & 2 \end{bmatrix} \qquad (5.5.5)$$

where A is the area of the triangular element. Similarly, the bilinear rectangular element yields

$$[M^e] = \frac{A}{36} \begin{bmatrix} 4 & 2 & 1 & 2 \\ 2 & 4 & 2 & 1 \\ 1 & 2 & 4 & 2 \\ 2 & 1 & 2 & 4 \end{bmatrix} \qquad (5.5.6)$$

Therefore, the final matrix equation for Eq. (5.5.1) becomes

$$[M]\{\dot{u}\}^t + [K]\{u\}^t = \{F\}^t \qquad (5.5.7)$$

Because this equation should be true at any time, we place superscript t in Eq. (5.5.7) to denote the time when the equation is satisfied. Furthermore, matrices $[M]$ and $[K]$ are independent of time. Now, the parabolic differential equation has transformed into a set of ordinary differential equations using the finite element method. In order to solve the equations, we use the finite difference method for the time derivative. The next sections show the solution techniques.

5.6 Time Integration Technique

First of all, we explain the forward difference method for the time derivative. The forward difference is expressed as

$$\{\dot{u}\}^t = \frac{\{u\}^{t+\Delta t} - \{u\}^t}{\Delta t} \tag{5.6.1}$$

Substitution of Eq. (5.6.1) into Eq. (5.5.7) results in

$$[M]\{u\}^{t+\Delta t} = \Delta t\left(\{F\}^t - [K]\{u\}^t\right) + [M]\{u\}^t \tag{5.6.2}$$

In the above equation, all the terms defined at time t are put on the right-hand side of the equation while the term associated with time $t+\Delta t$ is at the left-hand side. Equation (5.6.2) can be solved from the given initial condition $\{u\}^0$ and known boundary conditions $\{F\}^t$, as explained below:

1. Setting $t=0$ in Eq. (5.6.2), we can find the solution for $\{u\}^{\Delta t}$ from $\{u\}^0$ and $\{F\}^0$.
2. Once $\{u\}^{\Delta t}$ is found, we can continue the previous step again by setting $t=\Delta t$ in Eq. (5.6.2) in order to determine $\{u\}^{2\Delta t}$. This step is repeated until the solution reaches the final time.

The forward difference technique in Eq. (5.6.1) has the local truncation error $O(\Delta t^2)$ and the global truncation error $O(\Delta t)$ where O denotes the order of error. The forward difference technique is conditionally stable so that a proper size of time step Δt should be used to have a stable solution.

The next technique is the backward difference technique. For this technique, Eq. (5.5.7) can be rewritten at time $t + \Delta t$.

$$[M]\{\dot{u}\}^{t+\Delta t} + [K]\{u\}^{t+\Delta t} = \{F\}^{t+\Delta t} \tag{5.6.3}$$

The time derivative in the backward difference is

$$\{\dot{u}\}^{t+\Delta t} = \frac{\{u\}^{t+\Delta t} - \{u\}^t}{\Delta t} \tag{5.6.4}$$

Use of Eq. (5.6.4) with Eq. (5.6.3) results in

$$([M] + \Delta t[K])\{u\}^{t+\Delta t} = \Delta t\{F\}^{t+\Delta t} + [M]\{u\}^t \tag{5.6.5}$$

The solution procedure is similar to that for the forward difference technique. The local and global truncation errors are also the same as those for the forward difference technique. However, the backward difference technique is unconditionally stable. Therefore, any size of Δt can be used without worrying about stability. However, the time step size is, of course, important for accuracy because of the truncation error.

The other technque is the Crank-Nicolson method. For this technique, we write Eq. (5.5.7) at time $t + \frac{\Delta t}{2}$ instead of t. Then,

$$[M]\{\dot{u}\}^{t+\frac{\Delta t}{2}} + [K]\{u\}^{t+\frac{\Delta t}{2}} = \{F\}^{t+\frac{\Delta t}{2}} \qquad (5.6.6)$$

The time derivative term is expressed using the central difference technique as

$$\{\dot{u}\}^{t+\frac{\Delta t}{2}} = \frac{\{u\}^{t+\Delta t} - \{u\}^t}{\Delta t} \qquad (5.6.7)$$

On the other hand, the other terms are computed as average by

$$\{u\}^{t+\frac{\Delta t}{2}} = \frac{1}{2}(\{u\}^t + \{u\}^{t+\Delta t}) \qquad (5.6.8)$$

and

$$\{F\}^{t+\frac{\Delta t}{2}} = \frac{1}{2}(\{F\}^t + \{F\}^{t+\Delta t}) \qquad (5.6.9)$$

Substitution of Eqs. (5.6.7) through (5.6.9) into Eq. (5.6.6) yields

$$(2[M] + \Delta t[K])\{u\}^{t+\Delta t} = \Delta t(\{F\}^t + \{F\}^{t+\Delta t}) + (2[M] - \Delta t[K])\{u\}^t \qquad (5.6.10)$$

The Crank-Nicolson method is also unconditionally stable and the global truncation error is $O(\Delta t^2)$ so that it is one order higher than the other two techniques.

♣ **Example 5.6.1** Let us solve the following set of ordinary differential equations using the backward difference method.

$$[M]\{\dot{u}\} + [K]\{u\} = \{F\} \qquad (5.6.11)$$

where

$$[M] = \frac{1}{6}\begin{bmatrix} 2 & 1 & 0 & 0 \\ 1 & 4 & 1 & 0 \\ 0 & 1 & 4 & 1 \\ 0 & 0 & 1 & 2 \end{bmatrix} \qquad (5.6.12)$$

$$[K] = \begin{bmatrix} 1 & -1 & 0 & 0 \\ -1 & 2 & -1 & 0 \\ 0 & -1 & 2 & -1 \\ 0 & 0 & -1 & 1 \end{bmatrix} \qquad (5.6.13)$$

and
$$\{F\} = \{\, F_1 \ \ 0 \ \ 0 \ \ F_4 \,\}^T \tag{5.6.14}$$

Here, F_1 and F_4 are unknown while $u_1 = 100$ and $u_4 = 100$ are known as boundary conditions. In addition, the initial condition states $\{u\}^0 = 0$. Substitution of Eqs. (5.6.12) through (5.6.14) into Eq. (5.6.5) yields

$$\begin{bmatrix} \frac{1}{3}+\Delta t & \frac{1}{6}-\Delta t & 0 & 0 \\ \frac{1}{6}-\Delta t & \frac{2}{3}+2\Delta t & \frac{1}{6}-\Delta t & 0 \\ 0 & \frac{1}{6}-\Delta t & \frac{2}{3}+2\Delta t & \frac{1}{6}-\Delta t \\ 0 & 0 & \frac{1}{6}-\Delta t & \frac{1}{3}+\Delta t \end{bmatrix} \begin{Bmatrix} u_1^{t+\Delta t} \\ u_2^{t+\Delta t} \\ u_3^{t+\Delta t} \\ u_4^{t+\Delta t} \end{Bmatrix} =$$

$$\begin{Bmatrix} \frac{1}{3}u_1^t + \frac{1}{6}u_2^t + F_1 \Delta t \\ \frac{1}{6}u_1^t + \frac{2}{3}u_2^t + \frac{1}{6}u_3^t \\ \frac{1}{6}u_2^t + \frac{2}{3}u_3^t + \frac{1}{6}u_4^t \\ \frac{1}{6}u_3^t + \frac{1}{3}u_4^t + F_4 \Delta t \end{Bmatrix} \tag{5.6.15}$$

Applying the boundary conditions with $\Delta t = 1$ as the time step size, Eq. (5.6.15) becomes

$$\begin{bmatrix} 1 & 0 & 0 & 0 \\ -\frac{5}{6} & \frac{8}{3} & -\frac{5}{6} & 0 \\ 0 & -\frac{5}{6} & \frac{8}{3} & -\frac{5}{6} \\ 0 & 0 & 0 & 1 \end{bmatrix} \begin{Bmatrix} u_1^{t+\Delta t} \\ u_2^{t+\Delta t} \\ u_3^{t+\Delta t} \\ u_4^{t+\Delta t} \end{Bmatrix} = \begin{Bmatrix} 100 \\ \frac{1}{6}u_1^t + \frac{2}{3}u_2^t + \frac{1}{6}u_3^t \\ \frac{1}{6}u_2^t + \frac{2}{3}u_3^t + \frac{1}{6}u_4^t \\ 100 \end{Bmatrix} \tag{5.6.16}$$

Let $t = 0$ in Eq. (5.6.16) and use the initial condition to find the solution at $t = 1$. The solution is

$$\{\, u_1^1 \ \ u_2^1 \ \ u_3^1 \ \ u_4^1 \,\} = \{\, 100 \ \ 45.5 \ \ 45.5 \ \ 100 \,\} \tag{5.6.17}$$

where superscript 1 denotes the solution at time $t = 1$. To continue the solution, let $t = 1$ in Eq. (5.6.16) and use the previous solution in Eq. (5.6.17). Then, the solution at time $t = 2$ is

$$\{\, u_1^2 \ \ u_2^2 \ \ u_3^2 \ \ u_4^2 \,\} = \{\, 100 \ \ 75.4 \ \ 75.4 \ \ 100 \,\} \tag{5.6.18}$$

This process continues until the final time. As expected, the solution approaches the steady state of uniform value of 100. ‡

♣ **Example 5.6.2**   Let us solve Example 5.6.1 using the Crank-Nicolson method. Applying Eqs. (5.6.12) through (5.6.14) to Eq. (5.6.10) yields

$$\begin{bmatrix} \frac{2}{3}+\Delta t & \frac{1}{3}-\Delta t & 0 & 0 \\ \frac{1}{3}-\Delta t & \frac{4}{3}+2\Delta t & \frac{1}{3}-\Delta t & 0 \\ 0 & \frac{1}{3}-\Delta t & \frac{4}{3}+2\Delta t & \frac{1}{3}-\Delta t \\ 0 & 0 & \frac{1}{3}-\Delta t & \frac{2}{3}+\Delta t \end{bmatrix} \begin{Bmatrix} u_1^{t+\Delta t} \\ u_2^{t+\Delta t} \\ u_3^{t+\Delta t} \\ u_4^{t+\Delta t} \end{Bmatrix} =$$

$$\begin{bmatrix} \frac{2}{3}-\Delta t & \frac{1}{3}+\Delta t & 0 & 0 \\ \frac{1}{3}+\Delta t & \frac{4}{3}-2\Delta t & \frac{1}{3}+\Delta t & 0 \\ 0 & \frac{1}{3}+\Delta t & \frac{4}{3}-2\Delta t & \frac{1}{3}+\Delta t \\ 0 & 0 & \frac{1}{3}+\Delta t & \frac{2}{3}-\Delta t \end{bmatrix} \begin{Bmatrix} u_1^t \\ u_2^t \\ u_3^t \\ u_4^t \end{Bmatrix} +$$

$$\begin{Bmatrix} \Delta t (F_1^t + F_1^{t+\Delta t}) \\ 0 \\ 0 \\ \Delta t (F_4^t + F_4^{t+\Delta t}) \end{Bmatrix} \tag{5.6.19}$$

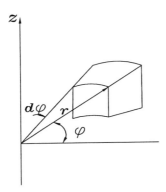

Figure 5.7.1 Cylindrical Coordinate

Applying the boundary conditions $u_1 = 100$ and $u_4 = 100$ with $\Delta t = 1$ results in the following matrix equation.

$$\begin{bmatrix} 1 & 0 & 0 & 0 \\ -\frac{2}{3} & \frac{10}{3} & -\frac{2}{3} & 0 \\ 0 & -\frac{2}{3} & \frac{10}{3} & -\frac{2}{3} \\ 0 & 0 & 0 & 1 \end{bmatrix} \begin{Bmatrix} u_1^{t+\Delta t} \\ u_2^{t+\Delta t} \\ u_3^{t+\Delta t} \\ u_4^{t+\Delta t} \end{Bmatrix} = \begin{Bmatrix} 100 \\ \frac{4}{3}u_1^t - \frac{2}{3}u_2^t + \frac{4}{3}u_3^t \\ \frac{4}{3}u_2^t - \frac{2}{3}u_3^t + \frac{4}{3}u_4^t \\ 100 \end{Bmatrix} \qquad (5.6.20)$$

Applying the initial condition with $t = 0$ yields the solution at time $t = 1$.

$$\{ u_1^1 \ \ u_2^1 \ \ u_3^1 \ \ u_4^1 \} = \{ 100 \ \ 25 \ \ 25 \ \ 100 \} \qquad (5.6.21)$$

The solution at the next step, i.e. $t = 2$, becomes

$$\{ u_1^2 \ \ u_2^2 \ \ u_3^2 \ \ u_4^2 \} = \{ 100 \ \ 81.3 \ \ 81.3 \ \ 100 \} \qquad (5.6.22)$$

using the solution in Eq. (5.6.21) and $t = 2$ for Eq. (5.6.20). ‡

5.7 Axisymmetric Analysis

Laplace's equation in the cylindrical coordinate system is written as below:

$$\frac{\partial^2 u}{\partial r^2} + \frac{1}{r}\frac{\partial u}{\partial r} + \frac{1}{r^2}\frac{\partial^2 u}{\partial \phi^2} + \frac{\partial^2 u}{\partial z^2} = 0 \qquad (5.7.1)$$

where r, ϕ, and z are the radial, circumferential, and axial axes, respectively, as shown in Fig. 5.7.1. For the axisymmetric problem, variable u is independent of the circumferential axis ϕ. This is the case where the domain is axisymmetric and all the described loading and/or boundary conditions are also axisymmetric. Therefore, the governing equation is simplified to

$$\frac{\partial^2 u}{\partial r^2} + \frac{1}{r}\frac{\partial u}{\partial r} + \frac{\partial^2 u}{\partial z^2} = 0 \qquad (5.7.2)$$

for the axisymmetric analysis.

Let us apply the weighted residual method. The integral becomes

$$\int_\Omega w\left(\frac{\partial^2 u}{\partial r^2} + \frac{1}{r}\frac{\partial u}{\partial r} + \frac{\partial^2 u}{\partial z^2}\right)d\Omega \qquad (5.7.3)$$

The first two terms in Eq. (5.7.3) can be rewritten as

$$\int_\Omega w\left\{\frac{1}{r}\frac{\partial}{\partial r}\left(r\frac{\partial u}{\partial r}\right) + \frac{\partial^2 u}{\partial z^2}\right\}d\Omega \qquad (5.7.4)$$

Now, the domain integral can be expressed as

$$\int_\Omega f(r,z)d\Omega = \int_\phi \int_r \int_z f(r,z)d\phi dr dz$$

$$= 2\pi \int_r \int_z rf(r,z)dr dz \qquad (5.7.5)$$

where $f(r,z)$ is any function which is independent of ϕ. Applying Eq. (5.7.5) to Eq. (5.7.4) gives

$$2\pi \int_r \int_z w\left\{\frac{\partial}{\partial r}\left(r\frac{\partial u}{\partial r}\right) + r\frac{\partial^2 u}{\partial z^2}\right\}d\Omega \qquad (5.7.6)$$

The weak formulation of Eq. (5.6.7) using integration by parts becomes

$$-2\pi \int_r \int_z r\left(\frac{\partial w}{\partial r}\frac{\partial u}{\partial r} + \frac{\partial w}{\partial z}\frac{\partial u}{\partial z}\right)dz dr + \int_\Gamma rw\frac{\partial u}{\partial n}d\Gamma \qquad (5.7.7)$$

where the boundary integral is on the rz-plane and n is also the outward normal unit vector to the boundary.

Equation (5.7.7) is now expressed in terms of the radial and axial axes, i.e. r and z. As a result, we need a finite element discretization in the rz-plane like a two-dimensional analysis. The same kinds of shape functions can be used for both two-dimensional and axisymmetric analyses of *Laplace's* equation. However, there is one difference between the two formulations. The axisymmetric analysis contains r within the integral while the two-dimensional analysis does not include r. Let us use the linear triangular element for the axisymmetric analysis. The element matrix for the triangular element can be written as

$$[K^e] = 2\pi \int_r \int_z r\left(\begin{Bmatrix}\frac{\partial H_1}{\partial r}\\ \frac{\partial H_2}{\partial r}\\ \frac{\partial H_3}{\partial r}\end{Bmatrix}\begin{Bmatrix}\frac{\partial H_1}{\partial r} & \frac{\partial H_2}{\partial r} & \frac{\partial H_3}{\partial r}\end{Bmatrix}\right.$$

$$\left.+ \begin{Bmatrix}\frac{\partial H_1}{\partial z}\\ \frac{\partial H_2}{\partial z}\\ \frac{\partial H_3}{\partial z}\end{Bmatrix}\begin{Bmatrix}\frac{\partial H_1}{\partial z} & \frac{\partial H_2}{\partial z} & \frac{\partial H_3}{\partial z}\end{Bmatrix}\right)dr dz \qquad (5.7.8)$$

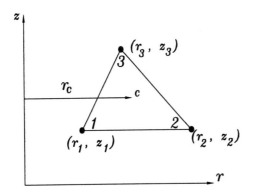

Figure 5.7.2 Triangular Axisymmetric Element

where H_i is given in Eqs. (5.2.7) through (5.2.9), replacing x and y by r and z. The element is shown in Fig. 5.7.2. As discussed in Sec. 5.2, $\frac{\partial H_i}{\partial r}$ and $\frac{\partial H_i}{\partial z}$ are independent of r and z. We also know that

$$\int_r \int_z r \, dr \, dz = A r_c \qquad (5.7.9)$$

where A is the area of the triangular element as defined in Eq. (5.2.5), and $r_c = \frac{1}{3}(r_1 + r_2 + r_3)$ is the r coordinate value of the centroid of the triangle as seen in Fig. 5.7.2. Consequently, the element matrix for the linear triangular axisymmetric element is

$$[K^e] = 2\pi r_c \begin{bmatrix} k_{11} & k_{12} & k_{13} \\ k_{21} & k_{22} & k_{23} \\ k_{31} & k_{32} & k_{33} \end{bmatrix} \qquad (5.7.10)$$

in which k_{ij} is the same as given in Eqs. (5.2.15) through (5.2.23) except that x_i and y_i are replaced by r_i and z_i for the axisymmetric analysis.

The flux at the boundary is also handled in a similar way as the two-dimensional analysis. However, the boundary integral for the axisymmetric analysis also contains r. If there is a uniform flux on the boundary of a linear triangular element for the axisymmetric problem as shown in Fig. 5.7.3, the equivalent nodal flux vector becomes $\pi q l \{(2r_i + r_j)/3 \; (r_i + 2r_j)/3\}^T$ where r_i and r_j are r coordinate values of the two boundary nodes i and j, q is the value of uniform flux per unit area, and l is the side length of the element as seen in Fig. 5.7.3.

5.8 Three-Dimensional Analysis

For the three-dimensional analysis of *Poisson's* equation, Eq. (5.1.18) can be extended directly into

$$I = -\int_\Omega \left(\frac{\partial w}{\partial x}\frac{\partial u}{\partial x} + \frac{\partial w}{\partial y}\frac{\partial u}{\partial y} + \frac{\partial w}{\partial z}\frac{\partial u}{\partial z} \right) d\Omega - \int_\Omega w g(x,y,z) d\Omega + \int_{\Gamma_n} w \frac{\partial u}{\partial n} d\Gamma \qquad (5.8.1)$$

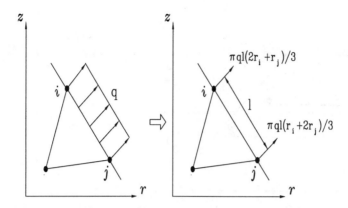

Figure 5.7.3 Flux on Axisymmetric Element

After discretization of the domain into finite elements, we can compute element matrices and vectors as before. For further explanation, we use a tetrahedral element as shown in Fig. 5.8.1. The variable interpolation for this element is assumed

$$u = \{X\}^T\{C\} \tag{5.8.2}$$

where

$$\{C\} = \{\begin{matrix} c_1 & c_2 & c_3 & c_4 \end{matrix}\}^T \tag{5.8.3}$$

and

$$\{X\} = \{\begin{matrix} 1 & x & y & z \end{matrix}\}^T \tag{5.8.4}$$

That is, the interpolation function is assumed to be linear in terms of every axis. Evaluation of the variable at the nodal points gives

$$\begin{Bmatrix} u_1 \\ u_2 \\ u_3 \\ u_4 \end{Bmatrix} = \begin{bmatrix} 1 & x_1 & y_1 & z_1 \\ 1 & x_2 & y_2 & z_2 \\ 1 & x_3 & y_3 & z_3 \\ 1 & x_4 & y_4 & z_4 \end{bmatrix} \begin{Bmatrix} c_1 \\ c_2 \\ c_3 \\ c_4 \end{Bmatrix} \tag{5.8.5}$$

or Eq. (5.8.5) can be put in the following way

$$\{u\} = [\bar{X}]\{C\} \tag{5.8.6}$$

in which

$$\{u\} = \{\begin{matrix} u_1 & u_2 & u_3 & u_4 \end{matrix}\}^T \tag{5.8.7}$$

After taking the inverse of matrix $[\bar{X}]$ and pre-multiplying it to both sides of Eq. (5.8.6), we substitute the resulting expression into Eq. (5.8.2). Then we obtain

$$u = \{X\}^T[\bar{X}]^{-1}\{u\} = \{H\}^T\{u\} \tag{5.8.8}$$

where

$$\{H\}^T = \{\begin{matrix} H_1 & H_2 & H_3 & H_4 \end{matrix}\} = \{X\}^T[\bar{X}]^{-1} \tag{5.8.9}$$

Section 5.8   Three-Dimensional Analysis

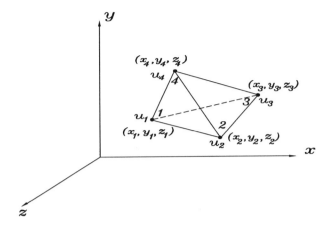

Figure 5.8.1 Tetrahedral Element

are the shape functions for the tetrahedral element with four nodes.

Substitution of the shape functions into Eq. (5.8.1) with element discretization results in an element matrix which is

$$[K^e] = \int_{\Omega^e} \left(\begin{Bmatrix} \frac{\partial H_1}{\partial x} \\ \frac{\partial H_2}{\partial x} \\ \frac{\partial H_3}{\partial x} \\ \frac{\partial H_4}{\partial x} \end{Bmatrix} \begin{Bmatrix} \frac{\partial H_1}{\partial x} & \frac{\partial H_2}{\partial x} & \frac{\partial H_3}{\partial x} & \frac{\partial H_4}{\partial x} \end{Bmatrix} + \begin{Bmatrix} \frac{\partial H_1}{\partial y} \\ \frac{\partial H_2}{\partial y} \\ \frac{\partial H_3}{\partial y} \\ \frac{\partial H_4}{\partial y} \end{Bmatrix} \begin{Bmatrix} \frac{\partial H_1}{\partial y} & \frac{\partial H_2}{\partial y} & \frac{\partial H_3}{\partial y} & \frac{\partial H_4}{\partial y} \end{Bmatrix} \right.$$

$$\left. + \begin{Bmatrix} \frac{\partial H_1}{\partial z} \\ \frac{\partial H_2}{\partial z} \\ \frac{\partial H_3}{\partial z} \\ \frac{\partial H_4}{\partial z} \end{Bmatrix} \begin{Bmatrix} \frac{\partial H_1}{\partial z} & \frac{\partial H_2}{\partial z} & \frac{\partial H_3}{\partial z} & \frac{\partial H_4}{\partial z} \end{Bmatrix} \right) d\Omega \qquad (5.8.10)$$

From Eq. (5.8.9), let

$$[\bar{X}]^{-1} = \begin{bmatrix} a_{11} & a_{12} & a_{13} & a_{14} \\ a_{21} & a_{22} & a_{23} & a_{24} \\ a_{31} & a_{32} & a_{33} & a_{34} \\ a_{41} & a_{42} & a_{43} & a_{44} \end{bmatrix} \qquad (5.8.11)$$

Then, the shape functions can be expressed as

$$H_1(x,y,z) = a_{11} + a_{21}x + a_{31}y + a_{41}z \qquad (5.8.12)$$

$$H_2(x,y,z) = a_{12} + a_{22}x + a_{32}y + a_{42}z \qquad (5.8.13)$$

$$H_3(x,y,z) = a_{13} + a_{23}x + a_{33}y + a_{43}z \qquad (5.8.14)$$

$$H_4(x,y,z) = a_{14} + a_{24}x + a_{34}y + a_{44}z \qquad (5.8.15)$$

Inserting Eqs. (5.8.12) through (5.8.15) into Eq. (5.8.10) yields

$$[K^e] = V \begin{bmatrix} k_{11} & k_{12} & k_{13} & k_{14} \\ k_{21} & k_{22} & k_{23} & k_{24} \\ k_{31} & k_{32} & k_{33} & k_{34} \\ k_{41} & k_{42} & k_{43} & k_{44} \end{bmatrix} \qquad (5.8.16)$$

in which

$$k_{11} = (a_{21})^2 + (a_{31})^2 + (a_{41})^2 \tag{5.8.17}$$

$$k_{12} = a_{21}a_{22} + a_{31}a_{32} + a_{41}a_{42} \tag{5.8.18}$$

$$k_{13} = a_{21}a_{23} + a_{31}a_{33} + a_{41}a_{43} \tag{5.8.19}$$

$$k_{14} = a_{21}a_{24} + a_{31}a_{34} + a_{41}a_{44} \tag{5.8.20}$$

$$k_{21} = k_{12} \tag{5.8.21}$$

$$k_{22} = (a_{22})^2 + (a_{32})^2 + (a_{42})^2 \tag{5.8.22}$$

$$k_{23} = a_{22}a_{23} + a_{32}a_{33} + a_{42}a_{43} \tag{5.8.23}$$

$$k_{24} = a_{22}a_{24} + a_{32}a_{34} + a_{42}a_{44} \tag{5.8.24}$$

$$k_{31} = k_{13} \tag{5.8.25}$$

$$k_{32} = k_{23} \tag{5.8.26}$$

$$k_{33} = (a_{23})^2 + (a_{33})^2 + (a_{43})^2 \tag{5.8.27}$$

$$k_{34} = a_{23}a_{24} + a_{33}a_{34} + a_{43}a_{44} \tag{5.8.28}$$

$$k_{41} = k_{14} \tag{5.8.29}$$

$$k_{42} = k_{24} \tag{5.8.30}$$

$$k_{43} = k_{34} \tag{5.8.31}$$

$$k_{44} = (a_{24})^2 + (a_{34})^2 + (a_{44})^2 \tag{5.8.32}$$

Furthermore, V is the element volume.

The flux boundary condition on the three-dimensional analysis can be treated in the following way. For a uniform flux on one side of the tetrahedral element, the flux column vector becomes approximately $\frac{A_s q}{3} \{1\ 1\ 1\}^T$ for the nodes on the element boundary as shown in Fig. 5.8.2. Here, A_s is the surface area of the element side on which a uniform flux of q is applied.

The element matrix for the transient term (i.e. $\frac{\partial u}{\partial t}$) is

$$[M^e] = \frac{V}{20} \begin{bmatrix} 2 & 1 & 1 & 1 \\ 1 & 2 & 1 & 1 \\ 1 & 1 & 2 & 1 \\ 1 & 1 & 1 & 2 \end{bmatrix} \tag{5.8.33}$$

Section 5.9   MATLAB Application to 2-D Steady State Analysis   109

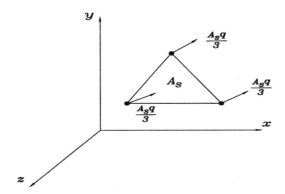

Figure 5.8.2 Triangular Boundary With Constant Flux

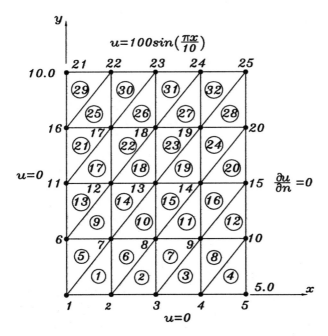

Figure 5.9.1 Mesh With Linear Triangular Elements

for the tetrahedral element.

5.9 MATLAB Application to 2-D Steady State Analysis

This section shows some examples for two-dimensional steady state problems using MATLAB programs. Both linear triangular and bilinear rectangular elements are used.

♣ **Example 5.9.1** We want to solve the two-dimensional *Laplace* equation for the following given conditions. The domain and the finite element discretization is shown in Fig. 5.9.1.

$$\frac{\partial^2 u}{\partial x^2} + \frac{\partial^2 u}{\partial y^2} = 0 \qquad (5.9.1)$$

for $0 < x < 5$ and $0 < y < 10$. The boundary conditions are $u(x,0) = 0$ for $0 < x < 5$, $u(0,y) = 0$ for $0 < y < 10$, $u(x,10) = 100 sin(\pi x/10)$ for $0 < x < 5$, and $\frac{\partial u(5,y)}{\partial x} = 0$ for $0 < y < 10$. The MATLAB main program along with function programs are listed below. Some function programs listed in previous chapters are not listed here. Appendix A lists all the function files.

```
%----------------------------------------------------------
% EX5.9.1.m
% to solve the two-dimensional Laplace equation given as
% u,xx + u,yy =0, 0 < x < 5, 0 < y < 10
% u(x,0) = 0, u(x,10) = 100sin(pi*x/10),
% u(0,y) = 0, u,x(5,y) = 0
% using linear triangular elements
%(see Fig. 5.9.1 for the finite element mesh)
%
% Variable descriptions
% k = element matrix
% f = element vector
% kk = system matrix
% ff = system vector
% gcoord = coordinate values of each node
% nodes = nodal connectivity of each element
% index = a vector containing system dofs associated with each element
% bcdof = a vector containing dofs associated with boundary conditions
% bcval = a vector containing boundary condition values associated with
%           the dofs in bcdof
%----------------------------------------------------------
%
%----------------------------------------
% input data for control parameters
%----------------------------------------
nel=32;                        % number of elements
nnel=3;                        % number of nodes per element
ndof=1;                        % number of dofs per node
nnode=25;                      % total number of nodes in system
sdof=nnode*ndof;               % total system dofs
%
%----------------------------------------------------
% input data for nodal coordinate values
% gcoord(i,j) where i-> node no. and j-> x or y
%----------------------------------------------------
gcoord(1,1)=0.0; gcoord(1,2)=0.0;
```

gcoord(2,1)=1.25; gcoord(2,2)=0.0;
gcoord(3,1)=2.5; gcoord(3,2)=0.0;
gcoord(4,1)=3.75; gcoord(4,2)=0.0;
gcoord(5,1)=5.0; gcoord(5,2)=0.0;
gcoord(6,1)=0.0; gcoord(6,2)=2.5;
gcoord(7,1)=1.25; gcoord(7,2)=2.5;
gcoord(8,1)=2.5; gcoord(8,2)=2.5;
gcoord(9,1)=3.75; gcoord(9,2)=2.5;
gcoord(10,1)=5.0; gcoord(10,2)=2.5;
gcoord(11,1)=0.0; gcoord(11,2)=5.0;
gcoord(12,1)=1.25; gcoord(12,2)=5.0;
gcoord(13,1)=2.5; gcoord(13,2)=5.0;
gcoord(14,1)=3.75; gcoord(14,2)=5.0;
gcoord(15,1)=5.0; gcoord(15,2)=5.0;
gcoord(16,1)=0.0; gcoord(16,2)=7.5;
gcoord(17,1)=1.25; gcoord(17,2)=7.5;
gcoord(18,1)=2.5; gcoord(18,2)=7.5;
gcoord(19,1)=3.75; gcoord(19,2)=7.5;
gcoord(20,1)=5.0; gcoord(20,2)=7.5;
gcoord(21,1)=0.0; gcoord(21,2)=10.;
gcoord(22,1)=1.25; gcoord(22,2)=10.;
gcoord(23,1)=2.5; gcoord(23,2)=10.;
gcoord(24,1)=3.75; gcoord(24,2)=10.;
gcoord(25,1)=5.0; gcoord(25,2)=10.;
%
%————————————————————
% input data for nodal connectivity for each element
% nodes(i,j) where i-> element no. and j-> connected nodes
%————————————————————
nodes(1,1)=1; nodes(1,2)=2; nodes(1,3)=7;
nodes(2,1)=2; nodes(2,2)=3; nodes(2,3)=8;
nodes(3,1)=3; nodes(3,2)=4; nodes(3,3)=9;
nodes(4,1)=4; nodes(4,2)=5; nodes(4,3)=10;
nodes(5,1)=1; nodes(5,2)=7; nodes(5,3)=6;
nodes(6,1)=2; nodes(6,2)=8; nodes(6,3)=7;
nodes(7,1)=3; nodes(7,2)=9; nodes(7,3)=8;
nodes(8,1)=4; nodes(8,2)=10; nodes(8,3)=9;
nodes(9,1)=6; nodes(9,2)=7; nodes(9,3)=12;
nodes(10,1)=7; nodes(10,2)=8; nodes(10,3)=13;
nodes(11,1)=8; nodes(11,2)=9; nodes(11,3)=14;
nodes(12,1)=9; nodes(12,2)=10; nodes(12,3)=15;
nodes(13,1)=6; nodes(13,2)=12; nodes(13,3)=11;
nodes(14,1)=7; nodes(14,2)=13; nodes(14,3)=12;
nodes(15,1)=8; nodes(15,2)=14; nodes(15,3)=13;
nodes(16,1)=9; nodes(16,2)=15; nodes(16,3)=14;
nodes(17,1)=11; nodes(17,2)=12; nodes(17,3)=17;
nodes(18,1)=12; nodes(18,2)=13; nodes(18,3)=18;
nodes(19,1)=13; nodes(19,2)=14; nodes(19,3)=19;

112   Laplace's and Poisson's Equations   Chapter 5

```
nodes(20,1)=14; nodes(20,2)=15; nodes(20,3)=20;
nodes(21,1)=11; nodes(21,2)=17; nodes(21,3)=16;
nodes(22,1)=12; nodes(22,2)=18; nodes(22,3)=17;
nodes(23,1)=13; nodes(23,2)=19; nodes(23,3)=18;
nodes(24,1)=14; nodes(24,2)=20; nodes(24,3)=19;
nodes(25,1)=16; nodes(25,2)=17; nodes(25,3)=22;
nodes(26,1)=17; nodes(26,2)=18; nodes(26,3)=23;
nodes(27,1)=18; nodes(27,2)=19; nodes(27,3)=24;
nodes(28,1)=19; nodes(28,2)=20; nodes(28,3)=25;
nodes(29,1)=16; nodes(29,2)=22; nodes(29,3)=21;
nodes(30,1)=17; nodes(30,2)=23; nodes(30,3)=22;
nodes(31,1)=18; nodes(31,2)=24; nodes(31,3)=23;
nodes(32,1)=19; nodes(32,2)=25; nodes(32,3)=24;
%
%———————————————————————
% input data for boundary conditions
%———————————————————————
bcdof(1)=1;                    % first node is constrained
bcval(1)=0;                    % whose described value is 0
bcdof(2)=2;                    % second node is constrained
bcval(2)=0;                    % whose described value is 0
bcdof(3)=3;                    % third node is constrained
bcval(3)=0;                    % whose described value is 0
bcdof(4)=4;                    % 4th node is constrained
bcval(4)=0;                    % whose described value is 0
bcdof(5)=5;                    % 5th node is constrained
bcval(5)=0;                    % whose described value is 0
bcdof(6)=6;                    % 6th node is constrained
bcval(6)=0;                    % whose described value is 0
bcdof(7)=11;                   % 11th node is constrained
bcval(7)=0;                    % whose described value is 0
bcdof(8)=16;                   % 16th node is constrained
bcval(8)=0;                    % whose described value is 0
bcdof(9)=21;                   % 21st node is constrained
bcval(9)=0;                    % whose described value is 0
bcdof(10)=22;                  % 22nd node is constrained
bcval(10)=38.2683;             % whose described value is 38.2683
bcdof(11)=23;                  % 23rd node is constrained
bcval(11)=70.7107;             % whose described value is 70.7107
bcdof(12)=24;                  % 24th node is constrained
bcval(12)=92.3880;             % whose described value is 92.3880
bcdof(13)=25;                  % 25th node is constrained
bcval(13)=100;                 % whose described value is 100
%
%———————————————————————
% initialization of matrices and vectors
%———————————————————————
ff=zeros(sdof,1);              % initialization of system force vector
```

```
kk=zeros(sdof,sdof);                    % initialization of system matrix
index=zeros(nnel*ndof,1);               % initialization of index vector
%
%----------------------------------------------------------------
% computation of element matrices and vectors and their assembly
%----------------------------------------------------------------
for iel=1:nel                           % loop for the total number of elements
%
nd(1)=nodes(iel,1);                     % 1st connected node for (iel)-th element
nd(2)=nodes(iel,2);                     % 2nd connected node for (iel)-th element
nd(3)=nodes(iel,3);                     % 3rd connected node for (iel)-th element
x1=gcoord(nd(1),1); y1=gcoord(nd(1),2); % coord values of 1st node
x2=gcoord(nd(2),1); y2=gcoord(nd(2),2); % coord values of 2nd node
x3=gcoord(nd(3),1); y3=gcoord(nd(3),2); % coord values of 3rd node
%
index=feeldof(nd,nnel,ndof);            % extract system dofs for the element
%
k=felp2dt3(x1,y1,x2,y2,x3,y3);          % compute element matrix
%
kk=feasmbl1(kk,k,index);                % assemble element matrices
%
end
%
%----------------------------------------
% apply boundary conditions
%----------------------------------------
[kk,ff]=feaplyc2(kk,ff,bcdof,bcval);
%
%----------------------------------------
% solve the matrix equation
%----------------------------------------
fsol=kk\ff;
%
%----------------------------------------
% analytical solution
%----------------------------------------
for i=1:nnode
x=gcoord(i,1); y=gcoord(i,2);
esol(i)=100*sinh(0.31415927*y)*sin(0.31415927*x)/sinh(3.1415927);
end
%
%----------------------------------------
% print both exact and fem solutions
%----------------------------------------
num=1:1:sdof;
store=[num' fsol esol']
%
%----------------------------------------------------------------
```

```
function [kk]=feasmbl1(kk,k,index)
%-----------------------------------------------------------------
% Purpose:
% Assembly of element matrices into the system matrix
%
% Synopsis:
% [kk]=feasmbl1(kk,k,index)
%
% Variable Description:
% kk - system matrix
% k - element matrix
% index - dof vector associated with an element
%-----------------------------------------------------------------
%
edof = length(index);
for i=1:edof
ii=index(i);
for j=1:edof
jj=index(j);
kk(ii,jj)=kk(ii,jj)+k(i,j);
end
end
%-----------------------------------------------------------------
```

```
function [index]=feeldof(nd,nnel,ndof)
%-----------------------------------------------------------------
% Purpose:
% Compute system dofs associated with each element
%
% Synopsis:
% [index]=feeldof(nd,nnel,ndof)
%
% Variable Description:
% index - system dof vector associated with element $iel$
% nd - element node numbers whose system dofs are to be determined
% nnel - number of nodes per element
% ndof - number of dofs per node
%-----------------------------------------------------------------
%
edof = nnel*ndof;
k=0;
for i=1:nnel
start = (nd(i)-1)*ndof;
for j=1:ndof
k=k+1;
index(k)=start+j;
```

```
          end
       end
%————————————————————————————————————————————

function [k]=felp2dt3(x1,y1,x2,y2,x3,y3)
%————————————————————————————————————————————
% Purpose:
% element matrix for two-dimensional Laplace's equation
% using three-node linear triangular element
%
% Synopsis:
% [k]=felp2dt3(x1,y1,x2,y2,x3,y3)
%
% Variable Description:
% k - element stiffness matrix (size of 3x3)
% x1, y1 - x and y coordinate values of the first node of element
% x2, y2 - x and y coordinate values of the second node of element
% x3, y3 - x and y coordinate values of the third node of element
%————————————————————————————————————————————
%
% element matrix
%
A=0.5*(x2*y3+x1*y2+x3*y1-x2*y1-x1*y3-x3*y2);
%                                            % area of the triangle
k(1,1)=((x3-x2)*(x3-x2)+(y2-y3)*(y2-y3))/(4*A);
k(1,2)=((x3-x2)*(x1-x3)+(y2-y3)*(y3-y1))/(4*A);
k(1,3)=((x3-x2)*(x2-x1)+(y2-y3)*(y1-y2))/(4*A);
k(2,1)=k(1,2);
k(2,2)=((x1-x3)*(x1-x3)+(y3-y1)*(y3-y1))/(4*A);
k(2,3)=((x1-x3)*(x2-x1)+(y3-y1)*(y1-y2))/(4*A);
k(3,1)=k(1,3);
k(3,2)=k(2,3);
k(3,3)=((x2-x1)*(x2-x1)+(y1-y2)*(y1-y2))/(4*A);
%————————————————————————————————————————————
```

The finite element and analytical solutions are compared below:

```
store =
dof #     fem sol    exact
1.0000    0.0000     0.0000      % at x=0.00 and y=0.0
2.0000    0.0000     0.0000      % at x=1.25 and y=0.0
3.0000    0.0000     0.0000      % at x=2.50 and y=0.0
4.0000    0.0000     0.0000      % at x=3.75 and y=0.0
5.0000    0.0000     0.0000      % at x=5.00 and y=0.0
6.0000    0.0000     0.0000      % at x=0.00 and y=2.5
7.0000    3.0516     2.8785      % at x=1.25 and y=2.5
8.0000    5.6386     5.3187      % at x=2.50 and y=2.5
```

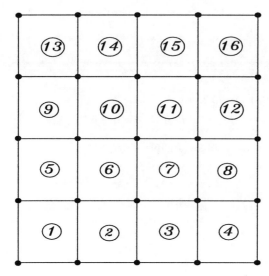

Figure 5.9.2 Mesh With Bilinear Elements

9.0000	7.3672	6.9492	% at x=3.75 and y=2.5
10.000	7.9742	7.5218	% at x=5.00 and y=2.5
11.000	0.0000	0.0000	% at x=0.00 and y=5.0
12.000	7.9615	7.6257	% at x=1.25 and y=5.0
13.000	14.711	14.090	% at x=2.50 and y=5.0
14.000	19.221	18.410	% at x=3.75 and y=5.0
15.000	20.804	19.927	% at x=5.00 and y=5.0
16.000	0.0000	0.0000	% at x=0.00 and y=7.5
17.000	17.720	17.324	% at x=1.25 and y=7.5
18.000	32.742	32.010	% at x=2.50 and y=7.5
19.000	42.779	41.823	% at x=3.75 and y=7.5
20.000	46.304	45.269	% at x=5.00 and y=7.5
21.000	0.0000	0.0000	% at x=0.00 and y=10.
22.000	38.268	38.268	% at x=1.25 and y=10.
23.000	70.711	70.711	% at x=2.50 and y=10.
24.000	92.388	92.388	% at x=3.75 and y=10.
25.000	100.00	100.00	% at x=5.00 and y=10.

‡

♣ **Example 5.9.2** We solve the same problem as given in Example 5.9.1 using bilinear rectangular elements. The mesh is shown in Fig. 5.9.2 and the MATLAB programs are listed below.

%————————————————————————————
% EX5.9.2.m

Section 5.9   MATLAB Application to 2-D Steady State Analysis   117

```
% to solve the two-dimensional Laplace equation given as
% u,xx + u,yy =0, 0 < x < 5, 0 < y < 10
% u(x,0) = 0, u(x,10) = 100sin(pi*x/10),
% u(0,y) = 0, u,x(5,y) = 0
% using bilinear rectangular elements
%(see Fig. 5.9.2 for the finite element mesh)
%
% Variable descriptions
% k = element matrix
% f = element vector
% kk = system matrix
% ff = system vector
% gcoord = coordinate values of each node
% nodes = nodal connectivity of each element
% index = a vector containing system dofs associated with each element
% bcdof = a vector containing dofs associated with boundary conditions
% bcval = a vector containing boundary condition values associated with
%                the dofs in bcdof
%------------------------------------------------
%
%------------------------------------------------
% input data for control parameters
%------------------------------------------------
nel=16;                               % number of elements
nnel=4;                               % number of nodes per element
ndof=1;                               % number of dofs per node
nnode=25;                             % total number of nodes in system
sdof=nnode*ndof;                      % total system dofs
%
%------------------------------------------------
% input data for nodal coordinate values
% gcoord(i,j) where i-> node no. and j-> x or y
%------------------------------------------------
gcoord(1,1)=0.0; gcoord(1,2)=0.0;
gcoord(2,1)=1.25; gcoord(2,2)=0.0;
gcoord(3,1)=2.5; gcoord(3,2)=0.0;
gcoord(4,1)=3.75; gcoord(4,2)=0.0;
gcoord(5,1)=5.0; gcoord(5,2)=0.0;
gcoord(6,1)=0.0; gcoord(6,2)=2.5;
gcoord(7,1)=1.25; gcoord(7,2)=2.5;
gcoord(8,1)=2.5; gcoord(8,2)=2.5;
gcoord(9,1)=3.75; gcoord(9,2)=2.5;
gcoord(10,1)=5.0; gcoord(10,2)=2.5;
gcoord(11,1)=0.0; gcoord(11,2)=5.0;
gcoord(12,1)=1.25; gcoord(12,2)=5.0;
gcoord(13,1)=2.5; gcoord(13,2)=5.0;
gcoord(14,1)=3.75; gcoord(14,2)=5.0;
gcoord(15,1)=5.0; gcoord(15,2)=5.0;
```

```
gcoord(16,1)=0.0; gcoord(16,2)=7.5;
gcoord(17,1)=1.25; gcoord(17,2)=7.5;
gcoord(18,1)=2.5; gcoord(18,2)=7.5;
gcoord(19,1)=3.75; gcoord(19,2)=7.5;
gcoord(20,1)=5.0; gcoord(20,2)=7.5;
gcoord(21,1)=0.0; gcoord(21,2)=10.;
gcoord(22,1)=1.25; gcoord(22,2)=10.;
gcoord(23,1)=2.5; gcoord(23,2)=10.;
gcoord(24,1)=3.75; gcoord(24,2)=10.;
gcoord(25,1)=5.0; gcoord(25,2)=10.;
%
%----------------------------------------
% input data for nodal connectivity for each element
% nodes(i,j) where i-> element no. and j-> connected nodes
%----------------------------------------
nodes(1,1)=1; nodes(1,2)=2; nodes(1,3)=7; nodes(1,4)=6;
nodes(2,1)=2; nodes(2,2)=3; nodes(2,3)=8; nodes(2,4)=7;
nodes(3,1)=3; nodes(3,2)=4; nodes(3,3)=9; nodes(3,4)=8;
nodes(4,1)=4; nodes(4,2)=5; nodes(4,3)=10; nodes(4,4)=9;
nodes(5,1)=6; nodes(5,2)=7; nodes(5,3)=12; nodes(5,4)=11;
nodes(6,1)=7; nodes(6,2)=8; nodes(6,3)=13; nodes(6,4)=12;
nodes(7,1)=8; nodes(7,2)=9; nodes(7,3)=14; nodes(7,4)=13;
nodes(8,1)=9; nodes(8,2)=10; nodes(8,3)=15; nodes(8,4)=14;
nodes(9,1)=11; nodes(9,2)=12; nodes(9,3)=17; nodes(9,4)=16;
nodes(10,1)=12; nodes(10,2)=13; nodes(10,3)=18; nodes(10,4)=17;
nodes(11,1)=13; nodes(11,2)=14; nodes(11,3)=19; nodes(11,4)=18;
nodes(12,1)=14; nodes(12,2)=15; nodes(12,3)=20; nodes(12,4)=19;
nodes(13,1)=16; nodes(13,2)=17; nodes(13,3)=22; nodes(13,4)=21;
nodes(14,1)=17; nodes(14,2)=18; nodes(14,3)=23; nodes(14,4)=22;
nodes(15,1)=18; nodes(15,2)=19; nodes(15,3)=24; nodes(15,4)=23;
nodes(16,1)=19; nodes(16,2)=20; nodes(16,3)=25; nodes(16,4)=24;
%
%----------------------------------------
% input data for boundary conditions
%----------------------------------------
bcdof(1)=1;           % first node is constrained
bcval(1)=0;           % whose described value is 0
bcdof(2)=2;           % second node is constrained
bcval(2)=0;           % whose described value is 0
bcdof(3)=3;           % third node is constrained
bcval(3)=0;           % whose described value is 0
bcdof(4)=4;           % 4th node is constrained
bcval(4)=0;           % whose described value is 0
bcdof(5)=5;           % 5th node is constrained
bcval(5)=0;           % whose described value is 0
bcdof(6)=6;           % 6th node is constrained
bcval(6)=0;           % whose described value is 0
bcdof(7)=11;          % 11th node is constrained
```

```matlab
bcval(7)=0;                         % whose described value is 0
bcdof(8)=16;                        % 16th node is constrained
bcval(8)=0;                         % whose described value is 0
bcdof(9)=21;                        % 21st node is constrained
bcval(9)=0;                         % whose described value is 0
bcdof(10)=22;                       % 22nd node is constrained
bcval(10)=38.2683;                  % whose described value is 38.2683
bcdof(11)=23;                       % 23rd node is constrained
bcval(11)=70.7107;                  % whose described value is 70.7107
bcdof(12)=24;                       % 24th node is constrained
bcval(12)=92.3880;                  % whose described value is 92.3880
bcdof(13)=25;                       % 25th node is constrained
bcval(13)=100;                      % whose described value is 100
%
%------------------------------------
% initialization of matrices and vectors
%------------------------------------
%
ff=zeros(sdof,1);                   % initialization of system force vector
kk=zeros(sdof,sdof);                % initialization of system matrix
index=zeros(nnel*ndof,1);           % initialization of index vector
%
%---------------------------------------------
% computation of element matrices and vectors and their assembly
%---------------------------------------------
for iel=1:nel                       % loop for the total number of elements
%
for i=1:nnel                        % loop for number of nodes per element
nd(i)=nodes(iel,i);                 % extract connected node for (iel)-th element
x(i)=gcoord(nd(i),1);               % extract x value of the node
y(i)=gcoord(nd(i),2);               % extract y value of the node
end
%
xleng = x(2)-x(1);                  % length of the element in x-axis
yleng = y(4)-y(1);                  % length of the element in y-axis
index=feeldof(nd,nnel,ndof);        % extract system dofs for the element
%
k=felp2dr4(xleng,yleng);            % compute element matrix
%
kk=feasmbl1(kk,k,index);            % assemble element matrices
%
end
%
%------------------------------
% apply boundary conditions
%------------------------------
[kk,ff]=feaplyc2(kk,ff,bcdof,bcval);
%
```

```
%------------------------------
% solve the matrix equation
%------------------------------
fsol=kk\ff;
%
%------------------------------
% analytical solution
%------------------------------
for i=1:nnode
x=gcoord(i,1); y=gcoord(i,2);
esol(i)=100*sinh(0.31415927*y)*sin(0.31415927*x)/sinh(3.1415927);
end
%
%------------------------------
% print both exact and fem solutions
%------------------------------
num=1:1:sdof;
store=[num' fsol esol']
%
```


```
function [k]=felp2dr4(xleng,yleng)
%------------------------------
% Purpose:
% element matrix for two-dimensional Laplace's equation
% using four-node bilinear rectangular element
%
% Synopsis:
% [k]=felp2dr4(xleng,yleng)
%
% Variable Description:
% k - element stiffness matrix (size 4x4)
% xleng - element size in the x-axis
% yleng - element size in the y-axis
%------------------------------
%
% element matrix
%
k(1,1)=(xleng*xleng+yleng*yleng)/(3*xleng*yleng);
k(1,2)=(xleng*xleng-2*yleng*yleng)/(6*xleng*yleng);
k(1,3)=-0.5*k(1,1);
k(1,4)=(yleng*yleng-2*xleng*xleng)/(6*xleng*yleng);
k(2,1)=k(1,2); k(2,2)=k(1,1); k(2,3)=k(1,4); k(2,4)=k(1,3);
k(3,1)=k(1,3); k(3,2)=k(2,3); k(3,3)=k(1,1); k(3,4)=k(1,2);
k(4,1)=k(1,4); k(4,2)=k(2,4); k(4,3)=k(3,4); k(4,4)=k(1,1);
%------------------------------
```

The finite element solution is shown below. The same number of nodes were used for this case as that in the previous example. By comparing the two finite element solutions using either linear triangular elements or bilinear rectangular elements, we see that the rectangular elements produced a more accurate solution in the present example.

```
store =
dof #    fem sol    exact
1.0000   0.0000     0.0000     % at x=0.00 and y=0.0
2.0000   0.0000     0.0000     % at x=1.25 and y=0.0
3.0000   0.0000     0.0000     % at x=2.50 and y=0.0
4.0000   0.0000     0.0000     % at x=3.75 and y=0.0
5.0000   0.0000     0.0000     % at x=5.00 and y=0.0
6.0000   0.0000     0.0000     % at x=0.00 and y=2.5
7.0000   2.6888     2.8785     % at x=1.25 and y=2.5
8.0000   4.9683     5.3187     % at x=2.50 and y=2.5
9.0000   6.4914     6.9492     % at x=3.75 and y=2.5
10.000   7.0263     7.5218     % at x=5.00 and y=2.5
11.000   0.0000     0.0000     % at x=0.00 and y=5.0
12.000   7.2530     7.6257     % at x=1.25 and y=5.0
13.000   13.402     14.090     % at x=2.50 and y=5.0
14.000   17.510     18.410     % at x=3.75 and y=5.0
15.000   18.953     19.927     % at x=5.00 and y=5.0
16.000   0.0000     0.0000     % at x=0.00 and y=7.5
17.000   16.876     17.324     % at x=1.25 and y=7.5
18.000   31.182     32.010     % at x=2.50 and y=7.5
19.000   40.742     41.823     % at x=3.75 and y=7.5
20.000   44.098     45.269     % at x=5.00 and y=7.5
21.000   0.0000     0.0000     % at x=0.00 and y=10.
22.000   38.268     38.268     % at x=1.25 and y=10.
23.000   70.711     70.711     % at x=2.50 and y=10.
24.000   92.388     92.388     % at x=3.75 and y=10.
25.000   100.00     100.00     % at x=5.00 and y=10.
```

‡

5.10 MATLAB Application to Axisymmetric Analysis

This section shows an example of an axisymmetric steady state problem using MATLAB programs. Linear triangular elements are used.

♣ **Example 5.10.1** An axisymmetric *Laplace* equation is solved using linear triangular elements. The governing equation is given in Eq. (5.7.2) for a cylinder whose inside and outside radii are 4 and 6, and whose height is 1. The finite element mesh used for the present analysis is shown in Fig. 5.10.1. The boundary conditions are $u = 100$ at the inside of the cylinder and $\frac{\partial u}{\partial r} = 20$ at

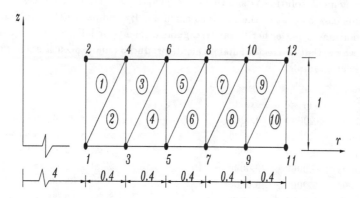

Figure 5.10.1 An Infinite Cylinder Modeled With Symmetric Boundaries

the outside of the cylinder. Both top and bottom surfaces of the cylinder have $\frac{\partial u}{\partial z} = 0$, i.e. they are insulated. Ten triangular elements with 12 nodes are used and the MATLAB programs are provided below. As seen in the main program, the constant flux at the outside surface is converted into the nodal flux at the outside surface. Each node takes half of the total flux over the element which is $2\pi \bar{r} q l = 240\pi$ where $\bar{r}=6$, $q=20$ and $l=1$, as explained in Sec. 5.7.

```
%----------------------------------------------------------------
% EX5.10.1.m
% to solve the axisymmetric Laplace equation given as
% u,rr + (u,r)/r + u,zz =0, 4 < r < 6, 0 < z < 1
% u(4,z) = 100, u,r(6,z) = 20
% u,z(r,0) = 0, u,z(r,1) = 0
% using linear triangular elements
%(see Fig. 5.10.1 for the finite element mesh)
%
% Variable descriptions
% k = element matrix
% f = element vector
% kk = system matrix
% ff = system vector
% gcoord = coordinate values of each node
% nodes = nodal connectivity of each element
% index = a vector containing system dofs associated with each element
% bcdof = a vector containing dofs associated with boundary conditions
% bcval = a vector containing boundary condition values associated with
%                the dofs in bcdof
%----------------------------------------------------------------
%
%----------------------------------------
% input data for control parameters
%----------------------------------------
```

```
nel=10;                                    % number of elements
nnel=3;                                    % number of nodes per element
ndof=1;                                    % number of dofs per node
nnode=12;                                  % total number of nodes in system
sdof=nnode*ndof;                           % total system dofs
%
%----------------------------------------
% input data for nodal coordinate values
% gcoord(i,j) where i-> node no. and j-> x or y
%----------------------------------------
gcoord(1,1)=4.0; gcoord(1,2)=0.0; gcoord(2,1)=4.0; gcoord(2,2)=1.0;
gcoord(3,1)=4.4; gcoord(3,2)=0.0; gcoord(4,1)=4.4; gcoord(4,2)=1.0;
gcoord(5,1)=4.8; gcoord(5,2)=0.0; gcoord(6,1)=4.8; gcoord(6,2)=1.0;
gcoord(7,1)=5.2; gcoord(7,2)=0.0; gcoord(8,1)=5.2; gcoord(8,2)=1.0;
gcoord(9,1)=5.6; gcoord(9,2)=0.0; gcoord(10,1)=5.6; gcoord(10,2)=1.0;
gcoord(11,1)=6.0; gcoord(11,2)=0.0; gcoord(12,1)=6.0; gcoord(12,2)=1.0;
%
%----------------------------------------
% input data for nodal connectivity for each element
% nodes(i,j) where i-> element no. and j-> connected nodes
%----------------------------------------
nodes(1,1)=1; nodes(1,2)=4; nodes(1,3)=2;
nodes(2,1)=1; nodes(2,2)=3; nodes(2,3)=4;
nodes(3,1)=3; nodes(3,2)=6; nodes(3,3)=4;
nodes(4,1)=3; nodes(4,2)=5; nodes(4,3)=6;
nodes(5,1)=5; nodes(5,2)=8; nodes(5,3)=6;
nodes(6,1)=5; nodes(6,2)=7; nodes(6,3)=8;
nodes(7,1)=7; nodes(7,2)=10; nodes(7,3)=8;
nodes(8,1)=7; nodes(8,2)=9; nodes(8,3)=10;
nodes(9,1)=9; nodes(9,2)=12; nodes(9,3)=10;
nodes(10,1)=9; nodes(10,2)=11; nodes(10,3)=12;
%
%----------------------------------------
% input data for boundary conditions
%----------------------------------------
bcdof(1)=1;                                % first node is constrained
bcval(1)=100;                              % whose described value is 100
bcdof(2)=2;                                % second node is constrained
bcval(2)=100;                              % whose described value is 100
%
%----------------------------------------
% initialization of matrices and vectors
%----------------------------------------
ff=zeros(sdof,1);                          % initialization of system force vector
kk=zeros(sdof,sdof);                       % initialization of system matrix
index=zeros(nnel*ndof,1);                  % initialization of index vector
%
pi=4*atan(1);                              % define pi
```

```
ff(11)=120*pi;              % nodal flux at the outside boundary
ff(12)=120*pi;              % nodal flux at the outside boundary
%
%------------------------------------------------
% computation of element matrices and vectors and their assembly
%------------------------------------------------
for iel=1:nel               % loop for the total number of elements
%
nd(1)=nodes(iel,1);         % 1st connected node for (iel)-th element
nd(2)=nodes(iel,2);         % 2nd connected node for (iel)-th element
nd(3)=nodes(iel,3);         % 3rd connected node for (iel)-th element
r1=gcoord(nd(1),1); z1=gcoord(nd(1),2);   % coordinate of 1st node
r2=gcoord(nd(2),1); z2=gcoord(nd(2),2);   % coordinate of 2nd node
r3=gcoord(nd(3),1); z3=gcoord(nd(3),2);   % coordinate of 3rd node
%
index=feeldof(nd,nnel,ndof);   % extract system dofs for the element
%
k=felpaxt3(r1,z1,r2,z2,r3,z3);      % compute element matrix
%
kk=feasmbl1(kk,k,index);            % assemble element matrices
%
end
%
%------------------------------------
% apply boundary conditions
%------------------------------------
[kk,ff]=feaplyc2(kk,ff,bcdof,bcval);
%
%------------------------------------
% solve the matrix equation
%------------------------------------
fsol=kk\ff;
%
%------------------------------
% analytical solution
%------------------------------
for i=1:nnode
r=gcoord(i,1); z=gcoord(i,2);
esol(i)=100-6*20*log(4)+6*20*log(r);
end
%
%----------------------------------------
% print both exact and fem solutions
%----------------------------------------
num=1:1:sdof;
store=[num' fsol esol']
%
%------------------------------------------------------
```

```
function [k]=felpaxt3(r1,z1,r2,z2,r3,z3)
%------------------------------------------------
% Purpose:
% element matrix for axisymmetric Laplace equation
% using three-node linear triangular element
%
% Synopsis:
% [k]=felpaxt3(r1,z1,r2,z2,r3,z3)
%
% Variable Description:
% k - element stiffness matrix (size 3x3)
% r1, z1 - r and z coordinate values of the first node of element
% r2, z2 - r and z coordinate values of the second node of element
% r3, z3 - r and z coordinate values of the third node of element
%------------------------------------------------
%
% element matrix
%
A=0.5*(r2*z3+r1*z2+r3*z1-r2*z1-r1*z3-r3*z2);    % area of the triangle
rc=(r1+r2+r3)/3;              % r coordinate value of the element centroid
twopirc=8*atan(1)*rc;
k(1,1)=((r3-r2)*(r3-r2)+(z2-z3)*(z2-z3))/(4*A);
k(1,2)=((r3-r2)*(r1-r3)+(z2-z3)*(z3-z1))/(4*A);
k(1,3)=((r3-r2)*(r2-r1)+(z2-z3)*(z1-z2))/(4*A);
k(2,1)=k(1,2);
k(2,2)=((r1-r3)*(r1-r3)+(z3-z1)*(z3-z1))/(4*A);
k(2,3)=((r1-r3)*(r2-r1)+(z3-z1)*(z1-z2))/(4*A);
k(3,1)=k(1,3);
k(3,2)=k(2,3);
k(3,3)=((r2-r1)*(r2-r1)+(z1-z2)*(z1-z2))/(4*A);
k=twopirc*k;
%
%------------------------------------------------
```

The results are

```
store =
dof #      fem sol    exact
1.0000     100.000    100.000        % at r=4.0 and z=0.0
2.0000     100.000    100.000        % at r=4.0 and z=1.0
3.0000     111.413    111.437        % at r=4.4 and z=0.0
4.0000     111.444    111.437        % at r=4.4 and z=1.0
5.0000     121.839    121.879        % at r=4.8 and z=0.0
6.0000     121.889    121.879        % at r=4.8 and z=1.0
7.0000     131.427    131.484        % at r=5.2 and z=0.0
8.0000     131.501    131.484        % at r=5.2 and z=1.0
9.0000     140.290    140.377        % at r=5.6 and z=0.0
10.000     140.417    140.377        % at r=5.6 and z=1.0
11.000     148.510    148.656        % at r=6.0 and z=0.0
```

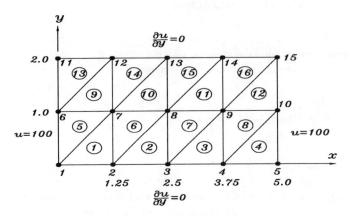

Figure 5.11.1 Mesh With Triangular Element

    12.000   148.749   148.656   % at r=6.0 and z=1.0

‡

5.11 MATLAB Application to Transient Analysis

Examples are given for some transient analyses using MATLAB programs. Forward difference, backward difference, and Crank-Nicolson techniques are used.

♣ **Example 5.11.1** The transient Laplace equation as described below is solved using the forward difference technique for time integration. The differential equation is

$$\frac{\partial u}{\partial t} = \frac{\partial^2 u}{\partial x^2} + \frac{\partial^2 u}{\partial y^2} \qquad (5.11.1)$$

over a rectangular domain defined by $0 < x < 5$ and $0 < y < 2$. The whole domain has the initial value of $u = 0$, and suddenly the left and right boundaries (i.e. edges with x=0 and x=5) are maintained at $u = 100$. On the other hand, the top and bottom boundaries (i.e. edges with y=0 and y=2) are insulated (i.e. $\frac{\partial u}{\partial y} = 0$). We want to find the solution as a function of time. A finite element mesh of the domain is shown in Fig. 5.11.1 using 16 linear triangular elements. The finite element analysis program is shown below.

```
%—————————————————————————————————
% EX5.11.1.m
% to solve the transient two-dimensional Laplace's equation
% u,t = u,xx + u,yy , 0 < x < 5, 0 < y < 2
% boundary conditions:
% u(0,y,t) = 100, u(5,y,t) = 100,
```

```
% u,y(x,0,t) = 0, u,y(x,2,t) = 0
% initial condition:
% u(x,y,0) = 0 over the domain
% using linear triangular elements and forward difference method
%(see Fig. 5.11.1 for the finite element mesh)
%
% Variable descriptions
% k = element matrix for time-independent term (u,xx + u,yy)
% m = element matrix for time-dependent term (u,t)
% f = element vector
% kk = system matrix of k
% mm = system matrix of m
% ff = system vector
% fn = effective system vector
% fsol = solution vector
% sol = time history solution of selected nodes
% gcoord = coordinate values of each node
% nodes = nodal connectivity of each element
% index = a vector containing system dofs associated with each element
% bcdof = a vector containing dofs associated with boundary conditions
% bcval = a vector containing boundary condition values associated with
%             the dofs in bcdof
%----------------------------------------------------------------
clear
%----------------------------------------------------------------
% input data for control parameters
%----------------------------------------------------------------
nel=16;                             % number of elements
nnel=3;                             % number of nodes per element
ndof=1;                             % number of dofs per node
nnode=15;                           % total number of nodes in system
sdof=nnode*ndof;                    % total system dofs
deltt=0.1;                          % time step size for transient analysis
stime=0.0;                          % initial time
ftime=10;                           % termination time
ntime=fix((ftime-stime)/deltt);     % number of time increment
%
%----------------------------------------------------------------
% input data for nodal coordinate values
% gcoord(i,j) where i->node no. and j->x or y
%----------------------------------------------------------------
gcoord(1,1)=0.0;  gcoord(1,2)=0.0;
gcoord(2,1)=1.25; gcoord(2,2)=0.0;
gcoord(3,1)=2.5;  gcoord(3,2)=0.0;
gcoord(4,1)=3.75; gcoord(4,2)=0.0;
gcoord(5,1)=5.0;  gcoord(5,2)=0.0;
gcoord(6,1)=0.0;  gcoord(6,2)=1.0;
gcoord(7,1)=1.25; gcoord(7,2)=1.0;
```

```
gcoord(8,1)=2.5;  gcoord(8,2)=1.0;
gcoord(9,1)=3.75; gcoord(9,2)=1.0;
gcoord(10,1)=5.0; gcoord(10,2)=1.0;
gcoord(11,1)=0.0; gcoord(11,2)=2.0;
gcoord(12,1)=1.25; gcoord(12,2)=2.0;
gcoord(13,1)=2.5; gcoord(13,2)=2.0;
gcoord(14,1)=3.75; gcoord(14,2)=2.0;
gcoord(15,1)=5.0; gcoord(15,2)=2.0;
%
%----------------------------------------
% input data for nodal connectivity for each element
% nodes(i,j) where i-> element no. and j-> connected nodes
%----------------------------------------
nodes(1,1)=1;  nodes(1,2)=2;  nodes(1,3)=7;
nodes(2,1)=2;  nodes(2,2)=3;  nodes(2,3)=8;
nodes(3,1)=3;  nodes(3,2)=4;  nodes(3,3)=9;
nodes(4,1)=4;  nodes(4,2)=5;  nodes(4,3)=10;
nodes(5,1)=1;  nodes(5,2)=7;  nodes(5,3)=6;
nodes(6,1)=2;  nodes(6,2)=8;  nodes(6,3)=7;
nodes(7,1)=3;  nodes(7,2)=9;  nodes(7,3)=8;
nodes(8,1)=4;  nodes(8,2)=10; nodes(8,3)=9;
nodes(9,1)=6;  nodes(9,2)=7;  nodes(9,3)=12;
nodes(10,1)=7; nodes(10,2)=8; nodes(10,3)=13;
nodes(11,1)=8; nodes(11,2)=9; nodes(11,3)=14;
nodes(12,1)=9; nodes(12,2)=10; nodes(12,3)=15;
nodes(13,1)=6; nodes(13,2)=12; nodes(13,3)=11;
nodes(14,1)=7; nodes(14,2)=13; nodes(14,3)=12;
nodes(15,1)=8; nodes(15,2)=14; nodes(15,3)=13;
nodes(16,1)=9; nodes(16,2)=15; nodes(16,3)=14;
%
%----------------------------------------
% input data for boundary conditions
%----------------------------------------
bcdof(1)=1;              % 1st node is constrained
bcval(1)=100;            % whose described value is 100
bcdof(2)=5;              % 5th node is constrained
bcval(2)=100;            % whose described value is 100
bcdof(3)=6;              % 6th node is constrained
bcval(3)=100;            % whose described value is 100
bcdof(4)=10;             % 10th node is constrained
bcval(4)=100;            % whose described value is 100
bcdof(5)=11;             % 11th node is constrained
bcval(5)=100;            % whose described value is 100
bcdof(6)=15;             % 15th node is constrained
bcval(6)=100;            % whose described value is 100
%
%----------------------------------------
% initialization of matrices and vectors
```

```
%----------------------------------------
ff=zeros(sdof,1);                   % initialization of system vector
fn=zeros(sdof,1);                   % initialization of effective system vector
fsol=zeros(sdof,1);                 % solution vector
sol=zeros(2,ntime+1);               % vector containing time history solution
kk=zeros(sdof,sdof);                % initialization of system matrix
mm=zeros(sdof,sdof);                % initialization of system matrix
index=zeros(nnel*ndof,1);           % initialization of index vector
%
%----------------------------------------
% computation of element matrices and vectors and their assembly
%----------------------------------------
for iel=1:nel                       % loop for the total number of elements
%
nd(1)=nodes(iel,1);                 % 1st connected node for (iel)-th element
nd(2)=nodes(iel,2);                 % 2nd connected node for (iel)-th element
nd(3)=nodes(iel,3);                 % 3rd connected node for (iel)-th element
x1=gcoord(nd(1),1); y1=gcoord(nd(1),2);   % coord values of 1st node
x2=gcoord(nd(2),1); y2=gcoord(nd(2),2);   % coord values of 2nd node
x3=gcoord(nd(3),1); y3=gcoord(nd(3),2);   % coord values of 3rd node
%
index=feeldof(nd,nnel,ndof);        % extract system dofs for the element
%
k=felp2dt3(x1,y1,x2,y2,x3,y3);      % compute element matrix
m=felpt2t3(x1,y1,x2,y2,x3,y3);      % compute element matrix
%
kk=feasmbl1(kk,k,index);            % assemble element matrices
mm=feasmbl1(mm,m,index);            % assemble element matrices
%
end
%
%----------------------------------------
% loop for time integration
%----------------------------------------
for in=1:sdof
fsol(in)=0.0;                       % initial condition
end
%
sol(1,1)=fsol(8);                   % store time history solution for node no. 8
sol(2,1)=fsol(9);                   % store time history solution for node no. 9
%
for it=1:ntime                      % start loop for time integration
%
fn=deltt*ff+(mm-deltt*kk)*fsol;     % compute effective column vector
%
[mm,fn]=feaplyc2(mm,fn,bcdof,bcval);   % apply boundary condition
%
fsol=mm\fn;                         % solve the matrix equation
```

```
%
sol(1,it+1)=fsol(8);          % store time history solution for node no. 8
sol(2,it+1)=fsol(9);          % store time history solution for node no. 9
%
end
%
%----------------------------------------
% plot the solution at nodes 8 and 9
%----------------------------------------
time=0:deltt:ntime*deltt;
plot(time,sol(1,:),'*',time,sol(2,:),'-');
xlabel('Time')
ylabel('Solution at nodes')
%
%----------------------------------------

function [m]=felpt2t3(x1,y1,x2,y2,x3,y3)
%----------------------------------------
% Purpose:
% element matrix for transient term of two-dimensional
% Laplace's equation using linear triangular element
%
% Synopsis:
% [m]=felpt2t3(x1,y1,x2,y2,x3,y3)
%
% Variable Description:
% m - element stiffness matrix (size of 3x3)
% x1, y1 - x and y coordinate values of the first node of element
% x2, y2 - x and y coordinate values of the second node of element
% x3, y3 - x and y coordinate values of the third node of element
%----------------------------------------
%
% element matrix
%
A=0.5*(x2*y3+x1*y2+x3*y1-x2*y1-x1*y3-x3*y2);
                                            % area of the triangle
%
m = (A/12)* [ 2 1 1;
1 2 1;
1 1 2 ];
%----------------------------------------
```

The finite element solutions are plotted in Fig. 5.11.2 and Fig. 5.11.3. The time history of nodes 8 and 9 in Fig. 5.11.1 is plotted in both figures. While $\Delta t = 0.1$ was used for Fig. 5.11.2, $\Delta t = 0.12$ was used for Fig. 5.11.3. As noticed, the finite element solution is unstable when $\Delta t = 0.12$ is used because the forward difference technique is conditionally stable.

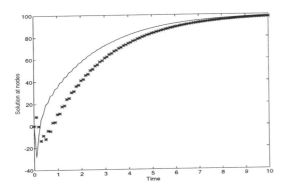

Figure 5.11.2 Finite Element Solution With $\Delta t=0.1$

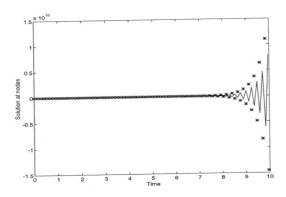

Figure 5.11.3 Finite Element Solution With $\Delta t=0.12$

♣ **Example 5.11.2**   The same example as Example 5.11.1 is solved using bilinear rectangular elements. The mesh is the same as that shown in Fig. 5.11.1 except that 8 rectangular elements are used instead of 16 triangular elements.

```
%————————————————————————————————————
% EX5.11.2.m
% to solve the transient two-dimensional Laplace's equation
% u,t = u,xx + u,yy , 0 < x < 5, 0 < y < 2
% boundary conditions:
% u(0,y,t) = 100, u(5,y,t) = 100,
% u,y(x,0,t) = 0, u,y(x,2,t) = 0
% initial condition:
% u(x,y,0) = 0 over the domain
% using bilinear rectangular elements and forward difference method
%(see Fig. 5.11.1 for the finite element mesh except for
```

```
%   8 rectangular elements instead of 16 triangular elements)
%
%  Variable descriptions
%  k = element matrix for time-independent term (u,xx + u,yy)
%  m = element matrix for time-dependent term (u,t)
%  f = element vector
%  kk = system matrix of k
%  mm = system matrix of m
%  ff = system vector
%  fn = effective system vector
%  fsol = solution vector
%  sol = time history solution of selected nodes
%  gcoord = coordinate values of each node
%  nodes = nodal connectivity of each element
%  index = a vector containing system dofs associated with each element
%  bcdof = a vector containing dofs associated with boundary conditions
%  bcval = a vector containing boundary condition values associated with
%          the dofs in bcdof
%----------------------------------------------------------------
clear
%----------------------------------------------------------------
% input data for control parameters
%----------------------------------------------------------------
nel=8;                          % number of elements
nnel=4;                         % number of nodes per element
ndof=1;                         % number of dofs per node
nnode=15;                       % total number of nodes in system
sdof=nnode*ndof;                % total system dofs
deltt=0.1;                      % time step size for transient analysis
stime=0.0;                      % initial time
ftime=10;                       % termination time
ntime=fix((ftime-stime)/deltt); % number of time increment
%
%----------------------------------------------------------------
% input data for nodal coordinate values
% gcoord(i,j) where i->node no. and j->x or y
%----------------------------------------------------------------
gcoord(1,1)=0.0;  gcoord(1,2)=0.0;
gcoord(2,1)=1.25; gcoord(2,2)=0.0;
gcoord(3,1)=2.5;  gcoord(3,2)=0.0;
gcoord(4,1)=3.75; gcoord(4,2)=0.0;
gcoord(5,1)=5.0;  gcoord(5,2)=0.0;
gcoord(6,1)=0.0;  gcoord(6,2)=1.0;
gcoord(7,1)=1.25; gcoord(7,2)=1.0;
gcoord(8,1)=2.5;  gcoord(8,2)=1.0;
gcoord(9,1)=3.75; gcoord(9,2)=1.0;
gcoord(10,1)=5.0; gcoord(10,2)=1.0;
gcoord(11,1)=0.0; gcoord(11,2)=2.0;
```

```
gcoord(12,1)=1.25; gcoord(12,2)=2.0;
gcoord(13,1)=2.5;  gcoord(13,2)=2.0;
gcoord(14,1)=3.75; gcoord(14,2)=2.0;
gcoord(15,1)=5.0;  gcoord(15,2)=2.0;
%
%----------------------------------------------
% input data for nodal connectivity for each element
% nodes(i,j) where i-> element no. and j-> connected nodes
%----------------------------------------------
nodes(1,1)=1; nodes(1,2)=2; nodes(1,3)=7;  nodes(1,4)=6;
nodes(2,1)=2; nodes(2,2)=3; nodes(2,3)=8;  nodes(2,4)=7;
nodes(3,1)=3; nodes(3,2)=4; nodes(3,3)=9;  nodes(3,4)=8;
nodes(4,1)=4; nodes(4,2)=5; nodes(4,3)=10; nodes(4,4)=9;
nodes(5,1)=6; nodes(5,2)=7; nodes(5,3)=12; nodes(5,4)=11;
nodes(6,1)=7; nodes(6,2)=8; nodes(6,3)=13; nodes(6,4)=12;
nodes(7,1)=8; nodes(7,2)=9; nodes(7,3)=14; nodes(7,4)=13;
nodes(8,1)=9; nodes(8,2)=10; nodes(8,3)=15; nodes(8,4)=14;
%
%----------------------------------------------
% input data for boundary conditions
%----------------------------------------------
bcdof(1)=1;          % 1st node is constrained
bcval(1)=100;        % whose described value is 100
bcdof(2)=5;          % 5th node is constrained
bcval(2)=100;        % whose described value is 100
bcdof(3)=6;          % 6th node is constrained
bcval(3)=100;        % whose described value is 100
bcdof(4)=10;         % 10th node is constrained
bcval(4)=100;        % whose described value is 100
bcdof(5)=11;         % 11th node is constrained
bcval(5)=100;        % whose described value is 100
bcdof(6)=15;         % 15th node is constrained
bcval(6)=100;        % whose described value is 100
%
%----------------------------------------------
% initialization of matrices and vectors
%----------------------------------------------
ff=zeros(sdof,1);            % initialization of system vector
fn=zeros(sdof,1);            % initialization of effective system vector
fsol=zeros(sdof,1);          % solution vector
sol=zeros(2,ntime+1);        % vector containing time history solution
kk=zeros(sdof,sdof);         % initialization of system matrix
mm=zeros(sdof,sdof);         % initialization of system matrix
index=zeros(nnel*ndof,1);    % initialization of index vector
%
%----------------------------------------------
% computation of element matrices and vectors and their assembly
%----------------------------------------------
```

```
for iel=1:nel                           % loop for the total number of elements
%
nd(1)=nodes(iel,1);                     % 1st connected node for (iel)-th element
nd(2)=nodes(iel,2);                     % 2nd connected node for (iel)-th element
nd(3)=nodes(iel,3);                     % 3rd connected node for (iel)-th element
nd(4)=nodes(iel,3);                     % 4th connected node for (iel)-th element
x1=gcoord(nd(1),1); y1=gcoord(nd(1),2); % coord values of 1st node
x2=gcoord(nd(2),1); y2=gcoord(nd(2),2); % coord values of 2nd node
x3=gcoord(nd(3),1); y3=gcoord(nd(3),2); % coord values of 3rd node
x4=gcoord(nd(4),1); y4=gcoord(nd(4),2); % coord values of 4th node
xleng=x2-x1;                            % element size in x-axis
yleng=y4-y1;                            % element size in y-axis
%
index=feeldof(nd,nnel,ndof);%
k=felp2dr4(xleng,yleng);                % time-independent element matrix
m=felpt2r4(xleng,yleng);                % transient element matrix
%
kk=feasmbl1(kk,k,index);                % assemble element matrices
mm=feasmbl1(mm,m,index);                % assemble element matrices
%
end
%
%--------------------------
% loop for time integration
%--------------------------
%
for in=1:sdof
fsol(in)=0.0;                           % initial condition
end
%
sol(1,1)=fsol(8);                       % store time history solution for node no. 8
sol(2,1)=fsol(9);                       % store time history solution for node no. 9
%
for it=1:ntime                          % start loop for time integration
%
fn=deltt*ff+(mm-deltt*kk)*fsol;         % compute effective column vector
%
[mm,fn]=feaplyc2(mm,fn,bcdof,bcval);    % apply boundary condition
%
fsol=mm\fn;                             % solve the matrix equation
%
sol(1,it+1)=fsol(8);                    % store time history solution for node no. 8
sol(2,it+1)=fsol(9);                    % store time history solution for node no. 9
%
end
%
%--------------------------
% analytical solution at node 8
```

```
%------------------------------
pi=4*atan(1);
esol=zeros(1,ntime+1);
xx=2.5; xl=5;
ii=0;
for ti=0:deltt:ntime*deltt;
ii=ii+1;
for i=1:100
esol(ii)=esol(ii)+(1/i)*exp(-i*i*pi*pi*ti/(xl*xl))*sin(i*pi*xx/xl);
end
end
esol=100-(100*4/pi)*esol;
%
%------------------------------
% plot fem and exact solutions at node 8
%------------------------------
time=0:deltt:ntime*deltt;
plot(time,sol(1,:),'*',time,esol,'-');
xlabel('Time')
ylabel('Solution at nodes')
%
%------------------------------

function [m]=felpt2r4(xleng,yleng)
%------------------------------
% Purpose:
% element matrix of transient term for two-dimensional Laplace's
% equation using four-node bilinear rectangular element
%
% Synopsis:
% [m]=felpt2r4(xleng,yleng)
%
% Variable Description:
% m - element stiffness matrix (size 4x4)
% xleng - element size in the x-axis
% yleng - element size in the y-axis
%------------------------------
%
% element matrix
%
m=(xleng*yleng/36)*[4 2 1 2;
 2 4 2 1;
 1 2 4 2;
 2 1 2 4];
%------------------------------
```

The finite element solution obtained using rectangular elements is comparable to that obtained using triangular elements. Figure 5.11.4 compares the finite

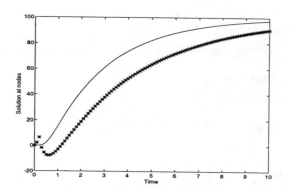

Figure 5.11.4 Analytical and Finite Element Results at Node 8

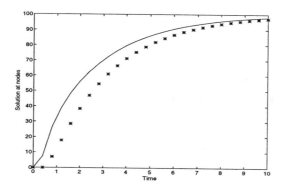

Figure 5.11.5 Time History of Nodes 8 and 9

element solution to the exact solution at node 8 (see Fig. 5.11.1). The finite element solution approaches the steady state solution more slowly than the exact solution. This is due to the very crude mesh in the x-direction.

‡

♣ **Example 5.11.3** The present example solves the same problem as that in Example 5.11.1 using the backward difference technique for time integration. Because this technique is unconditionally stable, we use a time step size $\Delta t = 0.4$ which exceeds the critical time step size for the forward difference technique. Figure 5.11.5 shows the time history of nodes 8 and 9.

%————————————————————————
% EX5.11.3.m

% to solve the transient two-dimensional Laplace's equation
% u,t = u,xx + u,yy , 0 < x < 5, 0 < y < 2
% boundary conditions:
% u(0,y,t) = 100, u(5,y,t) = 100,
% u,y(x,0,t) = 0, u,y(x,2,t) = 0
% initial condition:
% u(x,y,0) = 0 over the domain
% using linear triangular elements and backward difference method
%(see Fig. 5.11.1 for the finite element mesh)
%
% Variable descriptions
% k = element matrix for time-independent term (u,xx + u,yy)
% m = element matrix for time-dependent term (u,t)
% f = element vector
% kk = system matrix of k
% mm = system matrix of m
% ff = system vector
% fn = effective system vector
% fsol = solution vector
% sol = time history solution of selected nodes
% gcoord = coordinate values of each node
% nodes = nodal connectivity of each element
% index = a vector containing system dofs associated with each element
% bcdof = a vector containing dofs associated with boundary conditions
% bcval = a vector containing boundary condition values associated with
%   the dofs in *bcdof*
%————————————————————————————————
clear
%————————————————————————
% input data for control parameters
%————————————————————————
nel=16;   % number of elements
nnel=3;   % number of nodes per element
ndof=1;   % number of dofs per node
nnode=15;   % total number of nodes in system
sdof=nnode*ndof;   % total system dofs
deltt=0.4;   % time step size for transient analysis
stime=0.0;   % initial time
ftime=10;   % termination time
ntime=fix((ftime-stime)/deltt);   % number of time increment
%
%———————————————————————————
% input data for nodal coordinate values
% gcoord(i,j) where i->node no. and j->x or y
%———————————————————————————
gcoord(1,1)=0.0; gcoord(1,2)=0.0;
gcoord(2,1)=1.25; gcoord(2,2)=0.0;
gcoord(3,1)=2.5; gcoord(3,2)=0.0;

```
gcoord(4,1)=3.75; gcoord(4,2)=0.0;
gcoord(5,1)=5.0; gcoord(5,2)=0.0;
gcoord(6,1)=0.0; gcoord(6,2)=1.0;
gcoord(7,1)=1.25; gcoord(7,2)=1.0;
gcoord(8,1)=2.5; gcoord(8,2)=1.0;
gcoord(9,1)=3.75; gcoord(9,2)=1.0;
gcoord(10,1)=5.0; gcoord(10,2)=1.0;
gcoord(11,1)=0.0; gcoord(11,2)=2.0;
gcoord(12,1)=1.25; gcoord(12,2)=2.0;
gcoord(13,1)=2.5; gcoord(13,2)=2.0;
gcoord(14,1)=3.75; gcoord(14,2)=2.0;
gcoord(15,1)=5.0; gcoord(15,2)=2.0;
%
%---------------------------------------------
% input data for nodal connectivity for each element
% nodes(i,j) where i-> element no. and j-> connected nodes
%---------------------------------------------
nodes(1,1)=1; nodes(1,2)=2; nodes(1,3)=7;
nodes(2,1)=2; nodes(2,2)=3; nodes(2,3)=8;
nodes(3,1)=3; nodes(3,2)=4; nodes(3,3)=9;
nodes(4,1)=4; nodes(4,2)=5; nodes(4,3)=10;
nodes(5,1)=1; nodes(5,2)=7; nodes(5,3)=6;
nodes(6,1)=2; nodes(6,2)=8; nodes(6,3)=7;
nodes(7,1)=3; nodes(7,2)=9; nodes(7,3)=8;
nodes(8,1)=4; nodes(8,2)=10; nodes(8,3)=9;
nodes(9,1)=6; nodes(9,2)=7; nodes(9,3)=12;
nodes(10,1)=7; nodes(10,2)=8; nodes(10,3)=13;
nodes(11,1)=8; nodes(11,2)=9; nodes(11,3)=14;
nodes(12,1)=9; nodes(12,2)=10; nodes(12,3)=15;
nodes(13,1)=6; nodes(13,2)=12; nodes(13,3)=11;
nodes(14,1)=7; nodes(14,2)=13; nodes(14,3)=12;
nodes(15,1)=8; nodes(15,2)=14; nodes(15,3)=13;
nodes(16,1)=9; nodes(16,2)=15; nodes(16,3)=14;
%
%---------------------------------------------
% input data for boundary conditions
%---------------------------------------------
bcdof(1)=1;                    % 1st node is constrained
bcval(1)=100;                  % whose described value is 100
bcdof(2)=5;                    % 5th node is constrained
bcval(2)=100;                  % whose described value is 100
bcdof(3)=6;                    % 6th node is constrained
bcval(3)=100;                  % whose described value is 100
bcdof(4)=10;                   % 10th node is constrained
bcval(4)=100;                  % whose described value is 100
bcdof(5)=11;                   % 11th node is constrained
bcval(5)=100;                  % whose described value is 100
bcdof(6)=15;                   % 15th node is constrained
```

```
bcval(6)=100;                           % whose described value is 100
%
%------------------------------------
% initialization of matrices and vectors
%------------------------------------
ff=zeros(sdof,1);                       % initialization of system vector
fn=zeros(sdof,1);                       % initialization of effective system vector
fsol=zeros(sdof,1);                     % solution vector
sol=zeros(2,ntime+1);                   % vector containing time history solution
kk=zeros(sdof,sdof);                    % initialization of system matrix
mm=zeros(sdof,sdof);                    % initialization of system matrix
index=zeros(nnel*ndof,1);               % initialization of index vector
%
%--------------------------------------------------
% computation of element matrices and vectors and their assembly
%--------------------------------------------------
for iel=1:nel                           % loop for the total number of elements
%
nd(1)=nodes(iel,1);                     % 1st connected node for (iel)-th element
nd(2)=nodes(iel,2);                     % 2nd connected node for (iel)-th element
nd(3)=nodes(iel,3);                     % 3rd connected node for (iel)-th element
x1=gcoord(nd(1),1); y1=gcoord(nd(1),2); % coord values of 1st node
x2=gcoord(nd(2),1); y2=gcoord(nd(2),2); % coord values of 2nd node
x3=gcoord(nd(3),1); y3=gcoord(nd(3),2); % coord values of 3rd node
%
index=feeldof(nd,nnel,ndof);            % extract system dofs for the element
%
k=felp2dt3(x1,y1,x2,y2,x3,y3);          % compute element matrix
m=felpt2t3(x1,y1,x2,y2,x3,y3);          % compute element matrix
%
kk=feasmbl1(kk,k,index);                % assemble element matrices
mm=feasmbl1(mm,m,index);                % assemble element matrices
%
end
%
%------------------------------
% loop for time integration
%------------------------------
for in=1:sdof
fsol(in)=0.0;                           % initial condition
end
%
sol(1,1)=fsol(8);                       % sol contains time history solution of node 8
sol(2,1)=fsol(9);                       % sol contains time history solution of node 9
%
kk=mm+deltt*kk;
%
for it=1:ntime
```

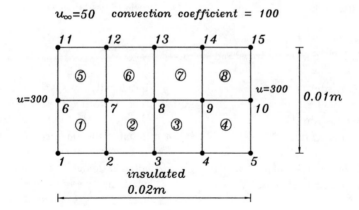

Figure 5.11.6 Finite Element Mesh

```
%
fn=deltt*ff+mm*fsol;                    % compute effective column vector
%
[kk,fn]=feaplyc2(kk,fn,bcdof,bcval);    % apply boundary condition
%
fsol=kk\fn;                             % solve the matrix equation
%
sol(1,it+1)=fsol(8);    % sol contains time history solution of node 8
sol(2,it+1)=fsol(9);    % sol contains time history solution of node 9
%
end
%
%-----------------------------------
% plot the solution at nodes 8 and 9
%-----------------------------------
time=0:deltt:ntime*deltt;
plot(time,sol(1,:),'*',time,sol(2,:),'-');
xlabel('Time')
ylabel('Solution at nodes')
%
%------------------------------------------------------
```

‡

♣ **Example 5.11.4** A plate of size 0.02 m by 0.01 m, whose heat conduction coefficient is $k = 0.3$ W/mC, is initially at a temperature of 300 C. While its left and right sides are maintained at the same temperature of 300 C, the bottom side is insulated and the top side is subjected to heat convection with convection coefficient of $h_c = 100$ W/m^2 C and an ambient temperature of 50 C. The material has also density ρ=1600 Kg/m^2 and specific heat c=0.8 J/KgC. The

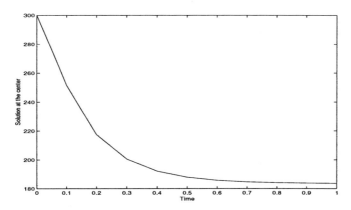

Figure 5.11.7 Time History Plot

MATLAB program using the backward difference technique is listed below and the mesh is shown in Fig. 5.11.6. The time history of the solution of node 8 is given in Fig. 5.11.7.

```
%----------------------------------------------------------------
% EX5.11.4.m
% to solve the transient two-dimensional Laplace's equation
% a*u,t = u,xx + u,yy , 0 < x < 0.02, 0 < y < 0.01
% boundary conditions:
% u(0,y,t) = 300, u(0.02,y,t) = 300,
% u,y(x,0,t) = 0, u,y(x,0.01,t) = -h_c/k*(u-50)
% initial condition:
% u(x,y,0) = 0 over the domain
% using bilinear rectangular elements and forward difference method
%(see Fig. 5.11.6 for the finite element mesh)
%
% Variable descriptions
% k = element matrix for time-independent term (u,xx + u,yy)
% m = element matrix for time-dependent term (u,t)
% f = element vector
% kk = system matrix of k
% mm = system matrix of m
% ff = system vector
% fn = effective system vector
% fsol = solution vector
% sol = time history solution vector of selected nodes
% gcoord = coordinate values of each node
% nodes = nodal connectivity of each element
% index = a vector containing system dofs associated with each element
% bcdof = a vector containing dofs associated with boundary conditions
% bcval = a vector containing boundary condition values associated with
%                the dofs in bcdof
```

```
% k1 = element matrix due to Cauchy-type flux
% f1 = element vector due to flux boundary condition
% index1 = index for nodal dofs with flux
%----------------------------------------------------------------
clear
%--------------------------------------
% input data for control parameters
%--------------------------------------
nel=8;                                          % number of elements
nnel=4;                                         % number of nodes per element
ndof=1;                                         % number of dofs per node
nnode=15;                                       % total number of nodes in system
sdof=nnode*ndof;                                % total system dofs
deltt=0.1;                                      % time step size for transient analysis
stime=0.0;                                      % initial time
ftime=1.0;                                      % termination time
ntime=fix((ftime-stime)/deltt);                 % number of time increment
a=4266.7;                                       % coefficient for the transient term
nf=4;                                           % number of element boundaries with flux
nnels=2;                                        % number of nodes per side of each element
%
%--------------------------------------
% input data for nodal coordinate values
% gcoord(i,j) where i-> node no. and j-> x or y
%--------------------------------------
gcoord(1,1)=0.0; gcoord(1,2)=0.0;
gcoord(2,1)=0.005; gcoord(2,2)=0.0;
gcoord(3,1)=0.010; gcoord(3,2)=0.0;
gcoord(4,1)=0.015; gcoord(4,2)=0.0;
gcoord(5,1)=0.020; gcoord(5,2)=0.0;
gcoord(6,1)=0.0; gcoord(6,2)=0.005;
gcoord(7,1)=0.005; gcoord(7,2)=0.005;
gcoord(8,1)=0.010; gcoord(8,2)=0.005;
gcoord(9,1)=0.015; gcoord(9,2)=0.005;
gcoord(10,1)=0.020; gcoord(10,2)=0.005;
gcoord(11,1)=0.0; gcoord(11,2)=0.01;
gcoord(12,1)=0.005; gcoord(12,2)=0.01;
gcoord(13,1)=0.010; gcoord(13,2)=0.01;
gcoord(14,1)=0.015; gcoord(14,2)=0.01;
gcoord(15,1)=0.020; gcoord(15,2)=0.01;
%
%--------------------------------------
% input data for nodal connectivity for each element
% nodes(i,j) where i-> element no. and j-> connected nodes
%--------------------------------------
nodes(1,1)=1; nodes(1,2)=2; nodes(1,3)=7; nodes(1,4)=6;
nodes(2,1)=2; nodes(2,2)=3; nodes(2,3)=8; nodes(2,4)=7;
nodes(3,1)=3; nodes(3,2)=4; nodes(3,3)=9; nodes(3,4)=8;
```

Section 5.11 MATLAB Application to Transient Analysis

```
nodes(4,1)=4; nodes(4,2)=5; nodes(4,3)=10; nodes(4,4)=9;
nodes(5,1)=6; nodes(5,2)=7; nodes(5,3)=12; nodes(5,4)=11;
nodes(6,1)=7; nodes(6,2)=8; nodes(6,3)=13; nodes(6,4)=12;
nodes(7,1)=8; nodes(7,2)=9; nodes(7,3)=14; nodes(7,4)=13;
nodes(8,1)=9; nodes(8,2)=10; nodes(8,3)=15; nodes(8,4)=14;
%
%----------------------------------------------------
% input data for boundary conditions
%----------------------------------------------------
bcdof(1)=1;                      % 1st node is constrained
bcval(1)=300;                    % whose described value is 300
bcdof(2)=5;                      % 5th node is constrained
bcval(2)=300;                    % whose described value is 300
bcdof(3)=6;                      % 6th node is constrained
bcval(3)=300;                    % whose described value is 300
bcdof(4)=10;                     % 10th node is constrained
bcval(4)=300;                    % whose described value is 300
bcdof(5)=11;                     % 11th node is constrained
bcval(5)=300;                    % whose described value is 300
bcdof(6)=15;                     % 15th node is constrained
bcval(6)=300;                    % whose described value is 300
%
%----------------------------------------------------
% input for flux boundary conditions
% nflx(i,j) where i-> element no. and j-> two side nodes
%----------------------------------------------------
nflx(1,1)=11; nflx(1,2)=12;      % nodes on 1st element side with flux
nflx(2,1)=12; nflx(2,2)=13;      % nodes on 2nd element side with flux
nflx(3,1)=13; nflx(3,2)=14;      % nodes on 3rd element side with flux
nflx(4,1)=14; nflx(4,2)=15;      % nodes on 4th element side with flux
%
b=333.3; c=50; %
%----------------------------------------------------
% initialization of matrices and vectors
%----------------------------------------------------
ff=zeros(sdof,1);                       % system vector
fn=zeros(sdof,1);                       % effective system vector
fsol=zeros(sdof,1);                     % solution vector
sol=zeros(1,ntime+1);      % time history solution of a selected node
kk=zeros(sdof,sdof);                    % of system matrix
mm=zeros(sdof,sdof);                    % system matrix
index=zeros(nnel*ndof,1);               % index vector
f1=zeros(nnels*ndof,1);                 % element flux vector
k1=zeros(nnels*ndof,nnels*ndof);        % flux matrix
index1=zeros(nnels*ndof,1);             % flux index vector
%
%----------------------------------------------------
% computation of element matrices and vectors and their assembly
```

```
%----------------------------------------------------------------
for iel=1:nel                   % loop for the total number of elements
%
nd(1)=nodes(iel,1);             % 1st connected node for (iel)-th element
nd(2)=nodes(iel,2);             % 2nd connected node for (iel)-th element
nd(3)=nodes(iel,3);             % 3rd connected node for (iel)-th element
nd(4)=nodes(iel,4);             % 4th connected node for (iel)-th element
x1=gcoord(nd(1),1); y1=gcoord(nd(1),2);   % coord values of 1st node
x2=gcoord(nd(2),1); y2=gcoord(nd(2),2);   % coord values of 2nd node
x3=gcoord(nd(3),1); y3=gcoord(nd(3),2);   % coord values of 3rd node
x4=gcoord(nd(4),1); y4=gcoord(nd(4),2);   % coord values of 4th node
xleng=x2-x1;                    % element size in x-axis
yleng=y4-y1;                    % element size in y-axis
%
index=feeldof(nd,nnel,ndof);    % extract system dofs for the element
%
k=felp2dr4(xleng,yleng);        % compute element matrix
m=a*felpt2r4(xleng,yleng);      % compute element matrix
%
kk=feasmbl1(kk,k,index);        % assemble element matrices
mm=feasmbl1(mm,m,index);        % assemble element matrices
%
end
%
%----------------------------------------------------------------
% additional computation due to flux boundary condition
%----------------------------------------------------------------
for ifx=1:nf
%
nds(1)=nflx(ifx,1);             % node with flux BC for (ifx)-th element
nds(2)=nflx(ifx,2);             % node with flux BC for (ifx)-th element
x1=gcoord(nds(1),1); y1=gcoord(nds(1),2);   % nodal coordinate
x2=gcoord(nds(2),1); y2=gcoord(nds(2),2);   % nodal coordinate
eleng=sqrt((x2-x1)*(x2-x1)+(y2-y1)*(y2-y1));  % element side length
%
index1=feeldof(nds,nnels,ndof); % find related system dofs
%
k1=b*feflxl2(eleng);            % compute element matrix due to flux
f1=b*c*fefl1(0,eleng);          % compute element vector due to flux
%
[kk,ff]=feasmbl2(kk,ff,k1,f1,index1);       % assembly
%
end
%
%----------------------------------------
% loop for time integration
%----------------------------------------
for in=1:sdof
```

Section 5.11   MATLAB Application to Transient Analysis   145

```
        fsol(in)=300.0;                         % initial condition
        end
        %
        sol(1)=fsol(8);            % sol contains time history solution at node 8
        %
        kk=mm+deltt*kk;                         % effective system matrix
        %
        for it=1:ntime
        %
        fn=deltt*ff+mm*fsol;            % compute effective column vector
        %
        [kk,fn]=feaplyc2(kk,fn,bcdof,bcval);    % apply boundary condition
        %
        fsol=kk\fn;                             % solve the matrix equation
        %
        sol(it+1)=fsol(8);         % sol contains time history solution at node 8
        %
        end
        %
        %---------------------------
        % plot the solution at node 8
        %---------------------------
        time=0:deltt:ntime*deltt;
        plot(time,sol);
        xlabel('Time')
        ylabel('Solution at the center')
        %
        %----------------------------------------------------------------



        function [k]=feflxl2(eleng)
        %------------------------------------------------------------------
        % Purpose:
        %   element matrix for Cauchy-type boundary such as du/dn=a(u-b)
        %   using linear element where a and b are known constants.
        %
        % Synopsis:
        %   [k]=feflxl2(eleng)
        %
        % Variable Description:
        %   k - element vector (size 2x2)
        %   eleng - length of element side with given flux
        %------------------------------------------------------------------
        %
        % element matrix
        %
        k=(eleng/6)*[ 2  1;
```

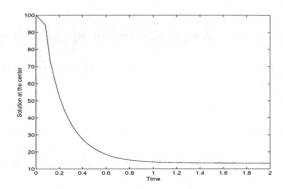

Figure 5.11.8 Transient Finite Element Solution

```
        1 2];
      %
      %————————————————————————————————————
‡
```

♣ **Example 5.11.5** We use the Crank-Nicolson technique to solve the following problem.

$$0.04\frac{\partial u}{\partial t} = \frac{\partial^2 u}{\partial x^2} + \frac{\partial^2 u}{\partial y^2} \qquad (5.11.2)$$

The problem domain is the same as that shown in Fig. 5.9.1 and the boundary conditions are the same as those described in Example 5.9.1. The initial condition is 100. The transient solution is plotted in Fig. 5.11.8. The steady state solution is that obtained in Example 5.9.1. As a result, the present solution must approach the steady state solution. The program is given below.

```
%————————————————————————————————————
% EX5.11.5.m
% to solve the two-dimensional Laplace's equation given as
% 0.04*u,t = u,xx + u,yy , 0 < x < 5, 0 < y < 10
% u(x,0) = 0, u(x,10) = 100sin(pi*x/10),
% u(0,y) = 0, u,x(5,y) = 0
% using linear triangular elements
% (see Fig. 5.9.1 for the finite element mesh)
%
% Variable descriptions
% k = element matrix
% f = element vector
% kk = system matrix
% ff = system vector
```

```
% fn = effective system vector
% kn = effective system matrix
% fsol = solution vector
% sol = time history solution of selected nodes
% gcoord = coordinate values of each node
% nodes = nodal connectivity of each element
% index = a vector containing system dofs associated with each element
% bcdof = a vector containing dofs associated with boundary conditions
% bcval = a vector containing boundary condition values associated with
%              the dofs in bcdof
%----------------------------------------------------------------------
clear
%----------------------------------------
% input data for control parameters
%----------------------------------------
nel=16;                              % number of elements
nnel=4;                              % number of nodes per element
ndof=1;                              % number of dofs per node
nnode=25;                            % total number of nodes in system
sdof=nnode*ndof;                     % total system dofs
deltt=0.04;                          % time step size for transient analysis
stime=0.0;                           % initial time
ftime=2;                             % termination time
ntime=fix((ftime-stime)/deltt);      % number of time increment
a=0.04;                              % coefficient
%
%----------------------------------------------------
% input data for nodal coordinate values
% gcoord(i,j) where i-> node no. and j-> x or y
%----------------------------------------------------
gcoord(1,1)=0.0; gcoord(1,2)=0.0;
gcoord(2,1)=1.25; gcoord(2,2)=0.0;
gcoord(3,1)=2.5; gcoord(3,2)=0.0;
gcoord(4,1)=3.75; gcoord(4,2)=0.0;
gcoord(5,1)=5.0; gcoord(5,2)=0.0;
gcoord(6,1)=0.0; gcoord(6,2)=2.5;
gcoord(7,1)=1.25; gcoord(7,2)=2.5;
gcoord(8,1)=2.5; gcoord(8,2)=2.5;
gcoord(9,1)=3.75; gcoord(9,2)=2.5;
gcoord(10,1)=5.0; gcoord(10,2)=2.5;
gcoord(11,1)=0.0; gcoord(11,2)=5.0;
gcoord(12,1)=1.25; gcoord(12,2)=5.0;
gcoord(13,1)=2.5; gcoord(13,2)=5.0;
gcoord(14,1)=3.75; gcoord(14,2)=5.0;
gcoord(15,1)=5.0; gcoord(15,2)=5.0;
gcoord(16,1)=0.0; gcoord(16,2)=7.5;
gcoord(17,1)=1.25; gcoord(17,2)=7.5;
gcoord(18,1)=2.5; gcoord(18,2)=7.5;
```

```
gcoord(19,1)=3.75; gcoord(19,2)=7.5;
gcoord(20,1)=5.0; gcoord(20,2)=7.5;
gcoord(21,1)=0.0; gcoord(21,2)=10.;
gcoord(22,1)=1.25; gcoord(22,2)=10.;
gcoord(23,1)=2.5; gcoord(23,2)=10.;
gcoord(24,1)=3.75; gcoord(24,2)=10.;
gcoord(25,1)=5.0; gcoord(25,2)=10.;
%
%─────────────────────────────────────
% input data for nodal connectivity for each element
% nodes(i,j) where i-> element no. and j-> connected nodes
%─────────────────────────────────────
nodes(1,1)=1; nodes(1,2)=2; nodes(1,3)=7; nodes(1,4)=6;
nodes(2,1)=2; nodes(2,2)=3; nodes(2,3)=8; nodes(2,4)=7;
nodes(3,1)=3; nodes(3,2)=4; nodes(3,3)=9; nodes(3,4)=8;
nodes(4,1)=4; nodes(4,2)=5; nodes(4,3)=10; nodes(4,4)=9;
nodes(5,1)=6; nodes(5,2)=7; nodes(5,3)=12; nodes(5,4)=11;
nodes(6,1)=7; nodes(6,2)=8; nodes(6,3)=13; nodes(6,4)=12;
nodes(7,1)=8; nodes(7,2)=9; nodes(7,3)=14; nodes(7,4)=13;
nodes(8,1)=9; nodes(8,2)=10; nodes(8,3)=15; nodes(8,4)=14;
nodes(9,1)=11; nodes(9,2)=12; nodes(9,3)=17; nodes(9,4)=16;
nodes(10,1)=12; nodes(10,2)=13; nodes(10,3)=18; nodes(10,4)=17;
nodes(11,1)=13; nodes(11,2)=14; nodes(11,3)=19; nodes(11,4)=18;
nodes(12,1)=14; nodes(12,2)=15; nodes(12,3)=20; nodes(12,4)=19;
nodes(13,1)=16; nodes(13,2)=17; nodes(13,3)=22; nodes(13,4)=21;
nodes(14,1)=17; nodes(14,2)=18; nodes(14,3)=23; nodes(14,4)=22;
nodes(15,1)=18; nodes(15,2)=19; nodes(15,3)=24; nodes(15,4)=23;
nodes(16,1)=19; nodes(16,2)=20; nodes(16,3)=25; nodes(16,4)=24;
%
%─────────────────────────────────────
% input data for boundary conditions
%─────────────────────────────────────
bcdof(1)=1;           % 1st node is constrained
bcval(1)=0;           % whose described value is 0
bcdof(2)=2;           % 2nd node is constrained
bcval(2)=0;           % whose described value is 0
bcdof(3)=3;           % 3rd node is constrained
bcval(3)=0;           % whose described value is 0
bcdof(4)=4;           % 4th node is constrained
bcval(4)=0;           % whose described value is 0
bcdof(5)=5;           % 5th node is constrained
bcval(5)=0;           % whose described value is 0
bcdof(6)=6;           % 6th node is constrained
bcval(6)=0;           % whose described value is 0
bcdof(7)=11;          % 11th node is constrained
bcval(7)=0;           % whose described value is 0
bcdof(8)=16;          % 16th node is constrained
bcval(8)=0;           % whose described value is 0
```

```
bcdof(9)=21;                            % 21st node is constrained
bcval(9)=0;                             % whose described value is 0
bcdof(10)=22;                           % 22nd node is constrained
bcval(10)=38.2683;                      % whose described value is 38.2683
bcdof(11)=23;                           % 23rd node is constrained
bcval(11)=70.7107;                      % whose described value is 70.7107
bcdof(12)=24;                           % 24th node is constrained
bcval(12)=92.3880;                      % whose described value is 92.3880
bcdof(13)=25;                           % 25th node is constrained
bcval(13)=100;                          % whose described value is 100
%
%------------------------------------
% initialization of matrices and vectors
%------------------------------------
ff=zeros(sdof,1);                       % system vector
fn=zeros(sdof,1);                       % effective system vector
fsol=zeros(sdof,1);                     % solution vector
sol=zeros(1,ntime+1);                   % time history solution
kk=zeros(sdof,sdof);                    % initialization of system matrix
mm=zeros(sdof,sdof);                    % initialization of system matrix
kn=zeros(sdof,sdof);                    % effective system matrix
index=zeros(nnel*ndof,1);               % initialization of index vector
%
%--------------------------------------------
% computation of element matrices and vectors and their assembly
%--------------------------------------------
for iel=1:nel                           % loop for the total number of elements
%
nd(1)=nodes(iel,1);                     % 1st connected node for (iel)-th element
nd(2)=nodes(iel,2);                     % 2nd connected node for (iel)-th element
nd(3)=nodes(iel,3);                     % 3rd connected node for (iel)-th element
nd(4)=nodes(iel,4);                     % 4th connected node for (iel)-th element
x1=gcoord(nd(1),1); y1=gcoord(nd(1),2); % coord values of 1st node
x2=gcoord(nd(2),1); y2=gcoord(nd(2),2); % coord values of 2nd node
x3=gcoord(nd(3),1); y3=gcoord(nd(3),2); % coord values of 3rd node
x4=gcoord(nd(4),1); y4=gcoord(nd(4),2); % coord values of 4th node
xleng=x2-x1;                            % element size in x-axis
yleng=y4-y1;                            % element size in y-axis
%
index=feeldof(nd,nnel,ndof);            % extract system dofs for the element
%
k=felp2dr4(xleng,yleng);                % compute element matrix
m=a*felpt2r4(xleng,yleng);              % compute element matrix
%
kk=feasmbl1(kk,k,index);                % assemble element matrices
mm=feasmbl1(mm,m,index);                % assemble element matrices
%
end
```

```
%
%---------------------------
% loop for time integration
%---------------------------
for in=1:sdof
fsol(in)=100.0;                              % initial condition
end
%
sol(1)=fsol(13);         % sol contains time history solution at node 13
%
kn=2*mm+deltt*kk;                    % compute effective system matrix
%
for it=1:ntime
%
fn=deltt*ff+(2*mm-deltt*kk)*fsol;            % compute effective vector
%
[kn,fn]=feaplyc2(kn,fn,bcdof,bcval);         % apply boundary condition
%
fsol=kn\fn;                                  % solve the matrix equation
%
sol(it+1)=fsol(13);           % sol contains time history at node 13
%
end
%
%---------------------------
% plot the solution at node 13
%---------------------------
time=0:deltt:ntime*deltt;
plot(time,sol);
xlabel('Time')
ylabel('Solution at the center')
%
%---------------------------------------------------------------
```

‡

5.12 MATLAB Application to 3-D Steady State Analysis

♣ **Example 5.12.1** A pyramid shape of three-dimensional domain as seen in Fig. 5.12.1 is analyzed for the *Laplace* equation. The bottom face of the pyramid has specified nodal variables as given in the figure while the side faces have no flux (i.e. $\frac{\partial u}{\partial n} = 0$). Four tetrahedral elements are used for the present three-dimensional analysis. The MATLAB program is also listed.

Section 5.12 MATLAB Application to 3-D Steady State Analysis

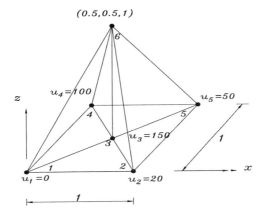

Figure 5.12.1 A Pyramid With Four Tetrahedral Elements

```
%----------------------------------------------------------------
% EX5.12.1.m
% to solve the three-dimensional Laplace equation
% for a pyramid shape of domain
% using four-node tetrahedral elements.
% Bottom face has essential boundary condition and the
% side faces are insulated.
%(see Fig. 5.12.1 for the finite element mesh)
%
% Variable descriptions
% k = element matrix
% f = element vector
% kk = system matrix
% ff = system vector
% gcoord = coordinate values of each node
% nodes = nodal connectivity of each element
% index = a vector containing system dofs associated with each element
% bcdof = a vector containing dofs associated with boundary conditions
% bcval = a vector containing boundary condition values associated with
%             the dofs in bcdof
%----------------------------------------------------------------
%
%----------------------------------------------------------------
% input data for control parameters
%----------------------------------------------------------------
nel=4;                          % number of elements
nnel=4;                         % number of nodes per element
ndof=1;                         % number of dofs per node
nnode=6;                        % total number of nodes in system
sdof=nnode*ndof;                % total system dofs
%
```

```
%----------------------------------------
% input data for nodal coordinate values
% gcoord(i,j) where i-> node no. and j-> x or y
%----------------------------------------
gcoord(1,1)=0.0; gcoord(1,2)=0.0; gcoord(1,3)=0.0;
gcoord(2,1)=1.0; gcoord(2,2)=0.0; gcoord(2,3)=0.0;
gcoord(3,1)=0.5; gcoord(3,2)=0.5; gcoord(3,3)=0.0;
gcoord(4,1)=0.0; gcoord(4,2)=1.0; gcoord(4,3)=0.0;
gcoord(5,1)=1.0; gcoord(5,2)=1.0; gcoord(5,3)=0.0;
gcoord(6,1)=0.5; gcoord(6,2)=0.5; gcoord(6,3)=1.0;
%
%----------------------------------------
% input data for nodal connectivity for each element
% nodes(i,j) where i-> element no. and j-> connected nodes
%----------------------------------------
nodes(1,1)=4; nodes(1,2)=1; nodes(1,3)=3; nodes(1,4)=6;
nodes(2,1)=1; nodes(2,2)=2; nodes(2,3)=3; nodes(2,4)=6;
nodes(3,1)=2; nodes(3,2)=5; nodes(3,3)=3; nodes(3,4)=6;
nodes(4,1)=5; nodes(4,2)=4; nodes(4,3)=3; nodes(4,4)=6;
%
%----------------------------------------
% input data for boundary conditions
%----------------------------------------
bcdof(1)=1;                     % 1st node is constrained
bcval(1)=0;                     % whose described value is 0
bcdof(2)=2;                     % 2nd node is constrained
bcval(2)=20;                    % whose described value is 20
bcdof(3)=3;                     % 3rd node is constrained
bcval(3)=150;                   % whose described value is 150
bcdof(4)=4;                     % 4th node is constrained
bcval(4)=100;                   % whose described value is 100
bcdof(5)=5;                     % 5th node is constrained
bcval(5)=50;                    % whose described value is 50
%
%----------------------------------------
% initialization of matrices and vectors
%----------------------------------------
ff=zeros(sdof,1);               % system vector
kk=zeros(sdof,sdof);            % system matrix
index=zeros(nnel*ndof,1);       % index vector
%
%----------------------------------------
% computation of element matrices and vectors and their assembly
%----------------------------------------
for iel=1:nel                   % loop for the total number of elements
%
nd(1)=nodes(iel,1);             % 1st connected node for (iel)-th element
nd(2)=nodes(iel,2);             % 2nd connected node for (iel)-th element
```

```
nd(3)=nodes(iel,3);              % 3rd connected node for (iel)-th element
nd(4)=nodes(iel,4);              % 4th connected node for (iel)-th element
x(1)=gcoord(nd(1),1); y(1)=gcoord(nd(1),2);
z(1)=gcoord(nd(1),3);            % coordinate of 1st node
x(2)=gcoord(nd(2),1); y(2)=gcoord(nd(2),2);
z(2)=gcoord(nd(2),3);            % coordinate of 2nd node
x(3)=gcoord(nd(3),1); y(3)=gcoord(nd(3),2);
z(3)=gcoord(nd(3),3);            % coordinate of 3rd node
x(4)=gcoord(nd(4),1); y(4)=gcoord(nd(4),2);
z(4)=gcoord(nd(4),3);            % coordinate of 4th node
%
index=feeldof(nd,nnel,ndof);     % extract system dofs for the element
%
k=felp3dt4(x,y,z);               % compute element matrix
%
kk=feasmbl1(kk,k,index);         % assemble element matrices
%
end
%
%----------------------------
% apply boundary conditions
%----------------------------
[kk,ff]=feaplyc2(kk,ff,bcdof,bcval);
%
%----------------------------
% solve the matrix equation
%----------------------------
fsol=kk\ff;
%
%----------------------------
% print both exact and fem solutions
%----------------------------
num=1:1:sdof;
store=[num' fsol]
%
%----------------------------------------------------------------


function [k]=felp3dt4(x,y,z)
%----------------------------------------------------------------
% Purpose:
% element matrix for three-dimensional Laplace's equation
% using four-node tetrahedral element
%
% Synopsis:
% [k]=felp3dt4(x,y,z)
%
```

```
% Variable Description:
% k - element matrix (size 4x4)
% x - x coordinate values of the four nodes
% y - y coordinate values of the four nodes
% z - z coordinate values of the four nodes
%------------------------------------------------------------
%
xbar= [ 1 x(1) y(1) z(1);
1 x(2) y(2) z(2);
1 x(3) y(3) z(3);
1 x(4) y(4) z(4) ];
xinv = inv(xbar);
vol = (1/6)*det(xbar); %
% element matrix
%
k(1,1)=xinv(2,1)*xinv(2,1)+xinv(3,1)*xinv(3,1)+xinv(4,1)*xinv(4,1);
k(1,2)=xinv(2,1)*xinv(2,2)+xinv(3,1)*xinv(3,2)+xinv(4,1)*xinv(4,2);
k(1,3)=xinv(2,1)*xinv(2,3)+xinv(3,1)*xinv(3,3)+xinv(4,1)*xinv(4,3);
k(1,4)=xinv(2,1)*xinv(2,4)+xinv(3,1)*xinv(3,4)+xinv(4,1)*xinv(4,4);
k(2,1)=k(1,2);
k(2,2)=xinv(2,2)*xinv(2,2)+xinv(3,2)*xinv(3,2)+xinv(4,2)*xinv(4,2);
k(2,3)=xinv(2,2)*xinv(2,3)+xinv(3,2)*xinv(3,3)+xinv(4,2)*xinv(4,3);
k(2,4)=xinv(2,2)*xinv(2,4)+xinv(3,2)*xinv(3,4)+xinv(4,2)*xinv(4,4);
k(3,1)=k(1,3);
k(3,2)=k(2,3);
k(3,3)=xinv(2,3)*xinv(2,3)+xinv(3,3)*xinv(3,3)+xinv(4,3)*xinv(4,3);
k(3,4)=xinv(2,3)*xinv(2,4)+xinv(3,3)*xinv(3,4)+xinv(4,3)*xinv(4,4);
k(4,1)=k(1,4);
k(4,2)=k(2,4);
k(4,3)=k(3,4);
k(4,4)=xinv(2,4)*xinv(2,4)+xinv(3,4)*xinv(3,4)+xinv(4,4)*xinv(4,4);
k=vol*k;
%
%------------------------------------------------------------
```

The finite element solution is

```
store =
node no.    fem sol
1.0000      0.00000
2.0000      20.0000
3.0000      150.000
4.0000      100.000
5.0000      50.0000
6.0000      150.000
```

Problems

5.1 Repeat Example 5.1.1 to derive Eq. (5.1.14).

5.2 A square domain is modeled using either one bilinear element or two linear triangular elements as shown in Fig. P5.2. Compute the system matrix for the Laplace equation for each discretization.

5.3 Flux through an element boundary is shown in Fig. P5.3. Determine the equivalent nodal fluxes.

5.4 A uniformly distributed flux is given on a side of a biquadratic element as shown in Fig P5.4. Compute the boundary integral to determine the equivalent nodal flux. The interpolation functions for the boundary nodes are

$$H_1(x) = \frac{1}{2}(x-1)(x-2)$$

$$H_2(x) = x(2-x)$$

$$H_3(x) = \frac{1}{2}x(x-1)$$

5.5 A linear triangular element has three vertices (x_1, y_1), (x_2, y_2), and (x_3, y_3). Evaluate Eq. (5.2.26) for the element vector if a concentrated source of magnitude Q is located at (x_s, y_s) which lies within the element.

5.6 Explain how to incorporate the boundary condition given at the edge of Fig. P5.6 into the finite element equation for the Laplace equation.

5.7 Apply the Galerkin method and the Crank-Nicolson method to solve the following parabolic partial differential equation.

$$\frac{\partial u}{\partial t} - \frac{1}{10}\frac{\partial^2 u}{\partial x^2} = 10 \quad 0 < x < 3$$

Initially u is 50 all over the domain and the domain is subjected to boundary conditions u=100 at the left end and $\frac{\partial u}{\partial x} = 100e^{-t}$ at the right end. Using $\Delta t=1$, find the nodal solution at time t=1. The domain is discretized into two linear elements. As a result, the three nodal points are located at $x_1=0$, $x_2=1$, and $x_3=3$, respectively.

5.8 Redo Prob. 5.7 using the backward difference method.

5.9 For a thermally orthotropic material, the two-dimensional heat conduction equation is

$$\frac{\partial}{\partial x}\left(k_x \frac{\partial u}{\partial x}\right) + \frac{\partial}{\partial y}\left(k_y \frac{\partial u}{\partial y}\right) + Q = 0$$

where k_x and k_y are heat conduction coefficients along the orthotropic axes x and y, respectively. Q is heat generation per unit volume. Develop the element matrix equation using linear triangular elements.

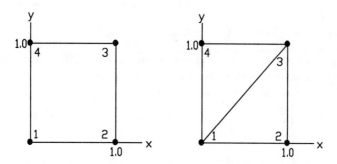

Figure P5.2 Problem 5.2

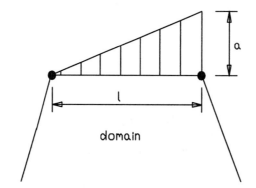

Figure P5.3 Problem 5.3

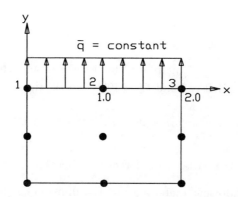

Figure P5.4 Problem 5.4

5.10 Repeat Example 5.9.1 for various mesh patterns shown in Fig. P5.10 using the computer program. Compare the solutions at the center of the domain.

5.11 A domain is normalized such that $0 < x < 1$ and $0 < y < 1$. Solve Laplace's

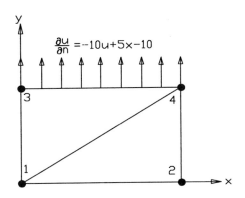

Figure P5.6 Problem 5.6

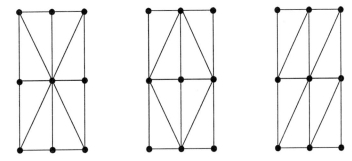

Figure P5.10 Problem 5.10

equation over the domain using the provided program and 16 bilinear elements (4 elements in the x- and y-axis, respectively). The boundary conditions are $u=0$ at $x=0$, $u=100$ at $x=1$, $u=0$ at $y=0$, and $u=200$ at $y=1$.

5.12 Solve the Laplace equation for the domain shown in Fig. P5.12. The boundary condition is also shown in the figure.

5.13 Redo Prob. 5.11 using the transient analysis assuming initial condition $u=0$ all over the domain. Use the forward difference time integration technique.

5.14 Redo Prob. 5.11 using the transient analysis assuming initial condition $u=0$ all over the domain. Use the backward difference time integration technique.

5.15 Redo Prob. 5.11 using the transient analysis assuming initial condition $u=0$ all over the domain. Use the Crank-Nicolson time integration technique.

5.16 Redo Prob. 5.12 using the transient analysis with initial condition $u=400$. Use the Crank-Nicolson technique for time integration.

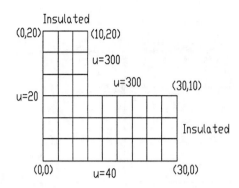

Figure P5.12 Problem 5.12

CHAPTER SIX

ISOPARAMETRIC ELEMENTS

6.0 Chapter Overview

The concept of isoparametric elements for 1-D, 2-D, and 3-D problems is introduced in this chapter because these elements are very important to application to complex shapes of domains. A numerical integration technique called Gauss-Legendre quadrature is also discussed along with the isoparametric elements. MATLAB example programs for Gauss-Legendre quadrature and the Laplace equation are given at the end of the chapter.

6.1 One-Dimensional Elements

Isoparametric elements [10] use mathematical mapping from one coordinate system into the other coordinate system. The former coordinate system is called the *natural* coordinate system while the latter is called the *physical* coordinate system. The problem domain is provided in the *physical* coordinate system denoted xyz-axes in the following discussion. On the other hand, element shape functions are defined in terms of the *natural* coordinate system denoted $\xi\eta\zeta$-axes. As a result, mapping is needed between the two coordinate systems.

We consider a linear one-dimensional isoparametric element to discuss the basic characteristics of isoparametric elements. Multi-dimensional isoparametric elements will be discussed in the subsequent sections. Shape functions for the isoparametric element are given in terms of the *natural* coordinate system as seen in Fig. 6.1.1. The two nodes are located at $\xi_1 = -1.0$ and $\xi_2 = 1.0$. These nodal positions are arbitrary but the proposed selection is very useful for numerical integration because the element in the *natural* coordinate system is normalized between -1 and 1. The shape functions can be written as

$$H_1(\xi) = \frac{1}{2}(1-\xi) \qquad (6.1.1)$$

and
$$H_2(\xi) = \frac{1}{2}(1+\xi) \qquad (6.1.2)$$

The physical linear element may be located at any position in the *physical* coordinate system as shown in Fig. 6.1.2. The element has two nodal coordinate values x_1 and x_2, with corresponding nodal variables u_1 and u_2.

Any point between $\xi_1 = -1.0$ and $\xi_2 = 1.0$ in the *natural* coordinate system can be mapped onto a point between x_1 and x_2 in the *physical* coordinate system using the shape functions defined in Eqs. (6.1.1) and (6.1.2).

$$x = H_1(\xi)x_1 + H_2(\xi)x_2 \qquad (6.1.3)$$

The same shape functions are also used to interpolate the variable u within the element.

$$u = H_1(\xi)u_1 + H_2(\xi)u_2 \qquad (6.1.4)$$

If the same shape functions are used for the geometric mapping as well as nodal variable interpolation, such as Eqs. (6.1.3) and (6.1.4), the element is called the *isoparametric* element.

In order to compute $\frac{du}{dx}$, which is necessary in most cases to compute element matrices, we use the chain rule such that

$$\begin{aligned}\frac{du}{dx} &= \frac{dH_1(\xi)}{dx}u_1 + \frac{dH_2(\xi)}{dx}u_2 \\ &= \frac{dH_1(\xi)}{d\xi}\frac{d\xi}{dx}u_1 + \frac{dH_2(\xi)}{d\xi}\frac{d\xi}{dx}u_2\end{aligned} \qquad (6.1.5)$$

where the expression requires $\frac{d\xi}{dx}$, which is the inverse of $\frac{dx}{d\xi}$. The latter can be computed from Eq. (6.1.3).

$$\frac{dx}{d\xi} = \frac{dH_1(\xi)}{d\xi}x_1 + \frac{dH_2(\xi)}{d\xi}x_2 = \frac{1}{2}(x_2 - x_1) \qquad (6.1.6)$$

Substituting Eq. (6.1.6) into Eq. (6.1.5) yields

$$\frac{du}{dx} = -\frac{1}{x_2 - x_1}u_1 + \frac{1}{x_2 - x_1}u_2 \qquad (6.1.7)$$

As a result, derivatives of shape functions with respect to the *physical* coordinate system are

$$\frac{dH_1(\xi)}{dx} = -\frac{1}{x_2 - x_1} = -\frac{1}{h_i} \qquad (6.1.8)$$

$$\frac{dH_2(\xi)}{dx} = \frac{1}{x_2 - x_1} = \frac{1}{h_i} \qquad (6.1.9)$$

in which $h_i = (x_2 - x_1)$ is the element size in the *physical* coordinate system. These derivative values are identical to those obtained directly from the linear shape

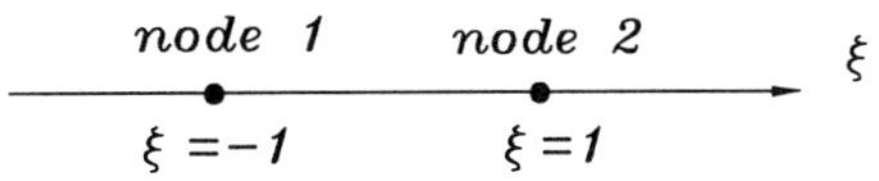

Figure 6.1.1 Linear Element in the Natural Coordinate

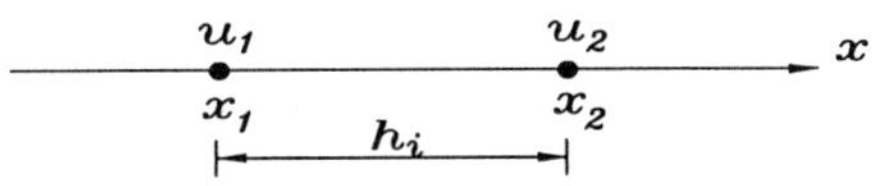

Figure 6.1.2 Linear Element in the Physical Coordinate

functions expressed in terms of the *physical* coordinate system like Eqs. (2.4.7) and (2.4.8).

Let us compute the following integral using the linear isoparametric element.

$$\int_{x_1}^{x_2} \left(\frac{dw}{dx} \frac{du}{dx} + wu \right) dx \qquad (6.1.10)$$

The integration is in terms of the *physical* coordinate system while the integrand is expressed in terms of the *natural* coordinate system because isoparametric shape functions are used for the trial and test functions u and w. Hence, we want to write the integral in terms of the *natural* coordinate system. To this end, we obtain

$$\int_{-1}^{1} \left(\frac{dw}{dx} \frac{du}{dx} + wu \right) J d\xi \qquad (6.1.11)$$

where $J = \frac{dx}{d\xi}$ is called the *Jacobian*.

Substitution of the isoparametric shape functions for both u and w results in

$$\int_{-1}^{1} \left(\frac{1}{h_i^2} \begin{bmatrix} 1 & -1 \\ -1 & 1 \end{bmatrix} + \frac{1}{4} \begin{bmatrix} (1-\xi)^2 & (1-\xi^2) \\ (1-\xi^2) & (1+\xi)^2 \end{bmatrix} \right) \frac{h_i}{2} d\xi \begin{Bmatrix} u_1 \\ u_2 \end{Bmatrix}$$
$$= \begin{bmatrix} \frac{1}{h_i} + \frac{h_i}{3} & -\frac{1}{h_i} + \frac{h_i}{6} \\ -\frac{1}{h_i} + \frac{h_i}{6} & \frac{1}{h_i} + \frac{h_i}{3} \end{bmatrix} \begin{Bmatrix} u_1 \\ u_2 \end{Bmatrix} \qquad (6.1.12)$$

This expression is the same as that obtained from the conventional linear element.

At this point, the isoparametric element does not seem to have an advantage over the conventional element because the isoparametric element requires more procedures such as mapping and the chain rule. The major advantage of isoparametric elements comes when analytical integration to compute element matrices and column vectors is either very complicated or almost impossible. This is the case when either element shapes in the physical domain are not regular such as in multi-dimensional problems or differential equations are quite complex. Therefore, a numerical integration technique is needed. Because each isoparametric element is defined in terms of the normalized

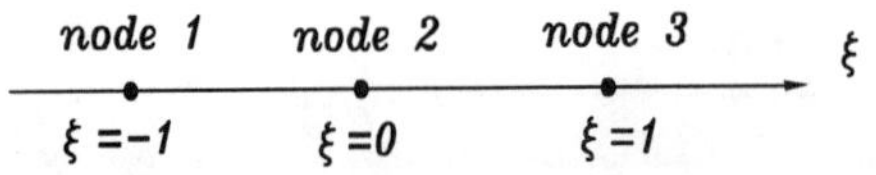

```
    node 1    node 2    node 3         ξ
      •         •         •         ───▶
    ξ =-1     ξ =0      ξ =1
```

Figure 6.1.3 Quadratic Isoparametric Element

domain such as $\xi_1 = -1$ and $\xi_2 = 1$, it is much easier to apply any numerical integration technique. The application of numerical integration technique is discussed later in this chapter.

♣ **Example 6.1.1** Let us consider a quadratic one-dimensional isoparametric element as seen in Fig 6.1.3. Shape functions for this element are

$$H_1(\xi) = \frac{(\xi^2 - \xi)}{2} \tag{6.1.13}$$

$$H_2(\xi) = 1 - \xi^2 \tag{6.1.14}$$

and

$$H_3(\xi) = \frac{(\xi^2 + \xi)}{2} \tag{6.1.15}$$

The variable u can be interpolated using these shape functions.

$$u = H_1(\xi)u_1 + H_2(\xi)u_2 + H_3(\xi)u_3 \tag{6.1.16}$$

Geometric mapping from the *natural* coordinate to the *physical* coordinate is

$$x = H_1(\xi)x_1 + H_2(\xi)x_2 + H_3(\xi)x_3 \tag{6.1.17}$$

The *Jacobian* becomes

$$J = \frac{dx}{d\xi} = \sum_{i=1}^{3} \frac{dH_i(\xi)}{d\xi} x_i = (\xi - 0.5)x_1 - 2\xi x_2 + (\xi + 0.5)x_3 \tag{6.1.18}$$

If the mid-node x_2 is located between the two end-nodes x_1 and x_3 (i.e. $x_2 = \frac{(x_1+x_3)}{2}$), the *Jacobian* becomes $\frac{h_i}{2}$ in which $h_i = x_3 - x_1$ is the element length.

Derivatives of the shape functions, Eqs. (6.1.13) through (6.1.15), are

$$\frac{dH_1(\xi)}{dx} = \frac{1}{J}\frac{dH_1}{d\xi} = \frac{1}{h_i}(2\xi - 1) \tag{6.1.19}$$

$$\frac{dH_2(\xi)}{dx} = \frac{1}{J}\frac{dH_2}{d\xi} = -\frac{4\xi}{h_i} \tag{6.1.20}$$

Section 6.2 — Quadrilateral Elements

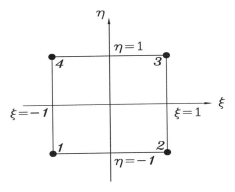

Figure 6.2.1 Bilinear Element in the Natural Coordinate

$$\frac{dH_3(\xi)}{dx} = \frac{1}{J}\frac{dH_3}{d\xi} = \frac{1}{h_i}(2\xi + 1) \qquad (6.1.21)$$

‡

6.2 Quadrilateral Elements

The shape functions for the bilinear isoparametric element are given below:

$$H_1(\xi, \eta) = \frac{1}{4}(1-\xi)(1-\eta) \qquad (6.2.1)$$

$$H_2(\xi, \eta) = \frac{1}{4}(1+\xi)(1-\eta) \qquad (6.2.2)$$

$$H_3(\xi, \eta) = \frac{1}{4}(1+\xi)(1+\eta) \qquad (6.2.3)$$

$$H_4(\xi, \eta) = \frac{1}{4}(1-\xi)(1+\eta) \qquad (6.2.4)$$

for the nodes shown in Fig. 6.2.1. These shape functions are defined in terms of the normalized *natural* domain (i.e. $-1 \leq \xi \leq 1$ and $-1 \leq \eta \leq 1$).

While the element shape is a square in the *natural* coordinate system, it can be mapped into a general quadrilateral shape with distortion as seen in Fig. 6.2.2. When this mapping is undertaken, the relative positions of nodal points should be consistent between the two elements in the *natural* and *physical* domains. In other words, the second node is next to the first node in the counter-clockwise direction and similarly for the rest of the nodes. Then, a point (ξ, η) within the *natural* element is mapped into a point (x, y) within the *physical* element using the shape functions given in Eqs. (6.2.1) through (6.2.4) as shown below:

$$x = \sum_{i=1}^{4} H_i(\xi, \eta) x_i \qquad (6.2.5)$$

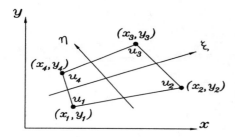

Figure 6.2.2 Bilinear Element in the Physical Coordinate

$$y = \sum_{i=1}^{4} H_i(\xi, \eta) y_i \qquad (6.2.6)$$

in which x_i and y_i are the coordinate values of the ith node. Similarly, any physical variable can be interpolated using the same shape functions:

$$u = \sum_{i=1}^{4} H_i(\xi, \eta) u_i \qquad (6.2.7)$$

in which u_i is the nodal variable at node i.

Let us apply this bilinear isoparametric element to the *Laplace* equation discussed in Chapter 5. Then, we need to compute $\frac{\partial H_i(\xi,\eta)}{\partial x}$ and $\frac{\partial H_i(\xi,\eta)}{\partial y}$, respectively. In order to compute these derivatives, we use the chain rule again.

$$\frac{\partial}{\partial \xi} = \frac{\partial}{\partial x}\frac{\partial x}{\partial \xi} + \frac{\partial}{\partial y}\frac{\partial y}{\partial \xi} \qquad (6.2.8)$$

$$\frac{\partial}{\partial \eta} = \frac{\partial}{\partial x}\frac{\partial x}{\partial \eta} + \frac{\partial}{\partial y}\frac{\partial y}{\partial \eta} \qquad (6.2.9)$$

Rewriting these in the matrix form provides

$$\left\{ \begin{array}{c} \frac{\partial}{\partial \xi} \\ \frac{\partial}{\partial \eta} \end{array} \right\} = \left[\begin{array}{cc} \frac{\partial x}{\partial \xi} & \frac{\partial y}{\partial \xi} \\ \frac{\partial x}{\partial \eta} & \frac{\partial y}{\partial \eta} \end{array} \right] \left\{ \begin{array}{c} \frac{\partial}{\partial x} \\ \frac{\partial}{\partial y} \end{array} \right\} \qquad (6.2.10)$$

Here, the derivative shown in the left-hand column vector is called the local derivative while that in the right-hand column vector is called the global derivative. Furthermore, the square matrix in this equation is called the *Jacobian* matrix for the two-dimensional domain and denoted as

$$[J] = \begin{bmatrix} J_{11} & J_{12} \\ J_{21} & J_{22} \end{bmatrix} = \begin{bmatrix} \frac{\partial x}{\partial \xi} & \frac{\partial y}{\partial \xi} \\ \frac{\partial x}{\partial \eta} & \frac{\partial y}{\partial \eta} \end{bmatrix} \qquad (6.2.11)$$

The *Jacobian* matrix can be easily extended for the three-dimensional domain. The inverse of the *Jacobian* matrix is denoted by

$$[R] = [J]^{-1} = \begin{bmatrix} R_{11} & R_{12} \\ R_{21} & R_{22} \end{bmatrix} \quad (6.2.12)$$

Then, Eq. (6.2.10) can be rewritten as

$$\left\{ \begin{array}{c} \frac{\partial}{\partial x} \\ \frac{\partial}{\partial y} \end{array} \right\} = \begin{bmatrix} R_{11} & R_{12} \\ R_{21} & R_{22} \end{bmatrix} \left\{ \begin{array}{c} \frac{\partial}{\partial \xi} \\ \frac{\partial}{\partial \eta} \end{array} \right\} \quad (6.2.13)$$

As a result, the derivatives of shape functions with respect to x and y can be obtained from the above equation.

$$\left\{ \begin{array}{c} \frac{\partial H_i}{\partial x} \\ \frac{\partial H_i}{\partial y} \end{array} \right\} = \begin{bmatrix} R_{11} & R_{12} \\ R_{21} & R_{22} \end{bmatrix} \left\{ \begin{array}{c} \frac{\partial H_i}{\partial \xi} \\ \frac{\partial H_i}{\partial \eta} \end{array} \right\} \quad (6.2.14)$$

The components in the *Jacobian* matrix are computed as shown below:

$$J_{11} = \frac{\partial x}{\partial \xi} = \sum_{i=1}^{4} \frac{\partial H_i(\xi, \eta)}{\partial \xi} x_i \quad (6.2.15)$$

$$J_{12} = \frac{\partial y}{\partial \xi} = \sum_{i=1}^{4} \frac{\partial H_i(\xi, \eta)}{\partial \xi} y_i \quad (6.2.16)$$

$$J_{21} = \frac{\partial x}{\partial \eta} = \sum_{i=1}^{4} \frac{\partial H_i(\xi, \eta)}{\partial \eta} x_i \quad (6.2.17)$$

$$J_{22} = \frac{\partial y}{\partial \eta} = \sum_{i=1}^{4} \frac{\partial H_i(\xi, \eta)}{\partial \eta} y_i \quad (6.2.18)$$

Substitution of bilinear shape functions, Eqs. (6.2.1) through (6.2.4), into the above expressions yields

$$J_{11} = -\frac{1}{4}(1-\eta)x_1 + \frac{1}{4}(1-\eta)x_2 + \frac{1}{4}(1+\eta)x_3 - \frac{1}{4}(1+\eta)x_4 \quad (6.2.19)$$

$$J_{12} = -\frac{1}{4}(1-\eta)y_1 + \frac{1}{4}(1-\eta)y_2 + \frac{1}{4}(1+\eta)y_3 - \frac{1}{4}(1+\eta)y_4 \quad (6.2.20)$$

$$J_{21} = -\frac{1}{4}(1-\xi)x_1 - \frac{1}{4}(1+\xi)x_2 + \frac{1}{4}(1+\xi)x_3 + \frac{1}{4}(1-\xi)x_4 \quad (6.2.21)$$

$$J_{22} = -\frac{1}{4}(1-\xi)y_1 - \frac{1}{4}(1+\xi)y_2 + \frac{1}{4}(1+\xi)y_3 + \frac{1}{4}(1-\xi)y_4 \quad (6.2.22)$$

These components are in general a function of ξ and η. However, they may be constant for a special case as shown in the following example. Once the *Jacobian* matrix is

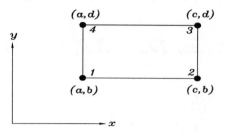

Figure 6.2.3 Rectangular Element

computed from Eqs. (6.2.19) through (6.2.22), global derivatives of shape functions are computed as

$$\frac{\partial H_i(\xi,\eta)}{\partial x} = R_{11}\frac{\partial H_i(\xi,\eta)}{\partial \xi} + R_{12}\frac{\partial H_i(\xi,\eta)}{\partial \eta} \qquad (6.2.23)$$

$$\frac{\partial H_i(\xi,\eta)}{\partial y} = R_{21}\frac{\partial H_i(\xi,\eta)}{\partial \xi} + R_{22}\frac{\partial H_i(\xi,\eta)}{\partial \eta} \qquad (6.2.24)$$

♣ **Example 6.2.1** Let us compute the *Jacobian* matrix for the physical element shown in Fig. 6.2.3. Substituting the nodal coordinate values into Eqs. (6.2.19) through (6.2.22) yields the following matrix.

$$[J] = \begin{bmatrix} \frac{c-a}{2} & 0 \\ 0 & \frac{d-b}{2} \end{bmatrix} \qquad (6.2.25)$$

As seen in this example, the *Jacobian* matrix becomes a diagonal matrix (i.e. all off-diagonal components vanish) when the element in the *physical* domain is a rectangular shape. In addition, the diagonal components are constant and not a function of ξ and η.

The inverse of the *Jacobian* matrix becomes

$$[R] = \begin{bmatrix} \frac{2}{c-a} & 0 \\ 0 & \frac{2}{d-b} \end{bmatrix} \qquad (6.2.26)$$

The global derivatives of shape functions become

$$\frac{\partial H_1}{\partial x} = -\frac{1-\eta}{2(c-a)} \qquad (6.2.27)$$

$$\frac{\partial H_2}{\partial x} = \frac{1-\eta}{2(c-a)} \qquad (6.2.28)$$

Section 6.2   Quadrilateral Elements   167

$$\frac{\partial H_3}{\partial x} = \frac{1+\eta}{2(c-a)} \quad (6.2.29)$$

$$\frac{\partial H_4}{\partial x} = -\frac{1+\eta}{2(c-a)} \quad (6.2.30)$$

$$\frac{\partial H_1}{\partial y} = -\frac{1-\xi}{2(d-b)} \quad (6.2.31)$$

$$\frac{\partial H_2}{\partial y} = -\frac{1+\xi}{2(d-b)} \quad (6.2.32)$$

$$\frac{\partial H_3}{\partial y} = \frac{1+\xi}{2(d-b)} \quad (6.2.33)$$

$$\frac{\partial H_4}{\partial y} = \frac{1-\xi}{2(d-b)} \quad (6.2.34)$$

‡

♣ **Example 6.2.2**   Let us compute the following integral using the same element as given in Example 6.2.1.

$$\int_{\Omega^e} \left[\left(\frac{\partial H_1}{\partial x}\right)^2 + \left(\frac{\partial H_1}{\partial y}\right)^2 \right] dx\, dy \quad (6.2.35)$$

Substitution of Eqs. (6.2.27) and (6.2.31) into Eq. (6.2.35) results in

$$\int_b^d \int_a^c \left[\frac{1}{4(c-a)^2}(1-\eta)^2 + \frac{1}{4(d-b)^2}(1-\xi)^2 \right] dx\, dy \quad (6.2.36)$$

The lower and upper limits of the integrals can be changed using

$$dx\, dy = |J|\, d\xi\, d\eta \quad (6.2.37)$$

where $|J|$ is the determinant of the *Jacobian* matrix and is equal to $\frac{(c-a)(d-b)}{4}$ for the present element. That is, $|J|$ is a constant value for a rectangular shape of the *physical* element. Then, we obtain

$$\int_{-1}^1 \int_{-1}^1 \left[\frac{1}{4(c-a)^2}(1-\eta)^2 + \frac{1}{4(d-b)^2}(1-\xi)^2 \right] \frac{(c-a)(d-b)}{4} d\xi\, d\eta \quad (6.2.38)$$

Integration of Eq. (6.2.37) finally yields $\frac{(c-a)^2+(d-b)^2}{3(c-a)(d-b)}$.   ‡

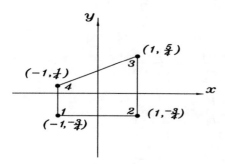

Figure 6.2.4 Element of Trapezoidal Shape

♣ **Example 6.2.3** Find the *Jacobian* matrix for the quadrilateral element shown in Fig. 6.2.4. Equations (6.2.19) through (6.2.22) along with the nodal coordinate values as specified in the figure yield

$$[J] = \begin{bmatrix} 1 & \frac{1}{4}(1+\eta) \\ 0 & \frac{1}{4}(3+\xi) \end{bmatrix} \quad (6.2.39)$$

The determinant of the *Jacobian* matrix is $|J| = \frac{3+\xi}{4}$ which is always positive for $-1 \leq \xi \leq 1$. Inverting the matrix gives

$$[R] = \begin{bmatrix} 1 & -\frac{1+\eta}{3+\xi} \\ 0 & \frac{4}{3+\xi} \end{bmatrix} \quad (6.2.40)$$

This matrix is used to compute the global derivatives of the shape functions.

In order to compute the integration as given in Eq. (6.2.35) for the present element, we first compute

$$\frac{\partial H_1}{\partial x} = -\frac{1-\eta}{4} + \frac{(1-\xi)(1+\eta)}{4(3+\xi)} \quad (6.2.41)$$

$$\frac{\partial H_1}{\partial y} = \frac{\xi - 1}{3+\xi} \quad (6.2.42)$$

The expression for the integral becomes

$$\int_{-1}^{1}\int_{-1}^{1}\left[\left\{-\frac{1-\eta}{4} + \frac{(1-\xi)(1+\eta)}{4(3+\xi)}\right\}^2 + \left\{\frac{\xi-1}{3+\xi}\right\}^2\right]\frac{(3+\xi)}{4}d\xi d\eta \quad (6.2.43)$$

This integral can be conducted analytically. However, if the shape of the *physical* element has more severe distortion, the integral becomes more complicated and may be beyond the analytical computation. Even if analytical integration may be possible, performing the analytical integration for every element of different

Section 6.2 Quadrilateral Elements

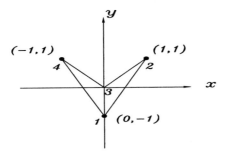

Figure 6.2.5 Element of Quadrilateral Shape

shapes is not practically possible. Therefore, the numerical integration technique is used along with the isoparametric element. ‡

♣ **Example 6.2.4** The *physical* element has a severe distortion as seen in Fig. 6.2.5. The corresponding *Jacobian* matrix is

$$[J] = \begin{bmatrix} \frac{1}{2} & \frac{1-3\eta}{4} \\ -\frac{1}{2} & \frac{1-3\xi}{4} \end{bmatrix} \qquad (6.2.44)$$

and its determinant is $\frac{1}{8}(2 - 3\xi - 3\eta)$. This determinant can be zero or negative for $-1 \leq \xi \leq 1$ and $-1 \leq \eta \leq 1$. Hence, this shape of element should be avoided in discretizing the *physical* domain. ‡

Some other popular quadrilateral isoparametric elements are eight-node and nine-node elements as shown in Figs. 6.2.6 and 6.2.7. Their shape functions are given below.

Eight-node element:

$$H_1 = \frac{1}{4}(1-\xi)(1-\eta)(-1-\xi-\eta) \qquad (6.2.45)$$

$$H_2 = \frac{1}{4}(1+\xi)(1-\eta)(-1+\xi-\eta) \qquad (6.2.46)$$

$$H_3 = \frac{1}{4}(1+\xi)(1+\eta)(-1+\xi+\eta) \qquad (6.2.47)$$

$$H_4 = \frac{1}{4}(1-\xi)(1+\eta)(-1-\xi+\eta) \qquad (6.2.48)$$

$$H_5 = \frac{1}{2}(1-\xi^2)(1-\eta) \qquad (6.2.49)$$

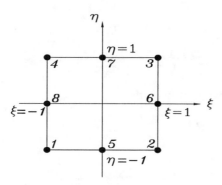

Figure 6.2.6 Eight-Node Isoparametric Element

$$H_6 = \frac{1}{2}(1+\xi)(1-\eta^2) \tag{6.2.50}$$

$$H_7 = \frac{1}{2}(1-\xi^2)(1+\eta) \tag{6.2.51}$$

$$H_8 = \frac{1}{2}(1-\xi)(1-\eta^2) \tag{6.2.52}$$

Nine-node element:

$$H_1 = \frac{1}{4}(\xi^2-\xi)(\eta^2-\eta) \tag{6.2.53}$$

$$H_2 = \frac{1}{4}(\xi^2+\xi)(\eta^2-\eta) \tag{6.2.54}$$

$$H_3 = \frac{1}{4}(\xi^2+\xi)(\eta^2+\eta) \tag{6.2.55}$$

$$H_4 = \frac{1}{4}(\xi^2-\xi)(\eta^2+\eta) \tag{6.2.56}$$

$$H_5 = \frac{1}{2}(1-\xi^2)(\eta^2-\eta) \tag{6.2.57}$$

$$H_6 = \frac{1}{2}(\xi^2+\xi)(1-\eta^2) \tag{6.2.58}$$

$$H_7 = \frac{1}{2}(1-\xi^2)(\eta^2+\eta) \tag{6.2.59}$$

$$H_8 = \frac{1}{2}(\xi^2-\xi)(1-\eta^2) \tag{6.2.60}$$

$$H_9 = (1-\xi^2)(1-\eta^2) \tag{6.2.61}$$

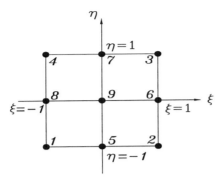

Figure 6.2.7 Nine-Node Isoparametric Element

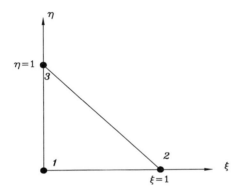

Figure 6.3.1 Three-Node Triangular Element in the Natural Coordinate

6.3 Triangular Elements

Like quadrilateral isoparametric elements, triagular isoparametric elements can be defined. Shape functions of the linear triangular element are in terms of the *natural* coordinate system

$$H_1 = 1 - \xi - \eta \tag{6.3.1}$$

$$H_2 = \xi \tag{6.3.2}$$

$$H_3 = \eta \tag{6.3.3}$$

for the nodes shown in Fig. 6.3.1. The quadratic triangular element has the following shape functions with reference to Fig. 6.3.2.

$$H_1 = (1 - \xi - \eta)(1 - 2\xi - 2\eta) \tag{6.3.4}$$

$$H_2 = \xi(2\xi - 1) \tag{6.3.5}$$

$$H_3 = \eta(2\eta - 1) \tag{6.3.6}$$

$$H_4 = 4\xi(1 - \xi - \eta) \tag{6.3.7}$$

$$H_5 = 4\xi\eta \tag{6.3.8}$$

$$H_6 = 4\eta(1 - \xi - \eta) \tag{6.3.9}$$

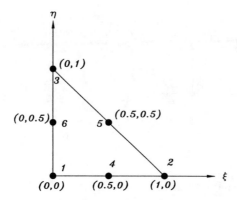

Figure 6.3.2 Six-Node Triangular Element in the Natural Coordinate

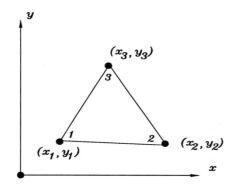

Figure 6.3.3 Three-Node Triangular Element in the Physical Coordinate

♣ **Example 6.3.1** Consider an element as shown in Fig. 6.3.3. The *Jacobian* matrix for the element is

$$[J] = \begin{bmatrix} x_2 - x_1 & y_2 - y_1 \\ x_3 - x_1 & y_3 - y_1 \end{bmatrix} \qquad (6.3.10)$$

and its determinant is $|J| = (x_2 - x_1)(y_3 - y_1) - (x_3 - x_1)(y_2 - y_1)$ which equals twice the triangular area in the *physical* domain. ‡

6.4 Gauss Quadrature

An integral is defined as

$$\int_a^b f(x)dx = \lim_{n \to \infty} \sum_{i=1}^{n} f(x_i)dx_i \qquad (6.4.1)$$

Section 6.4   Gauss Quadrature   173

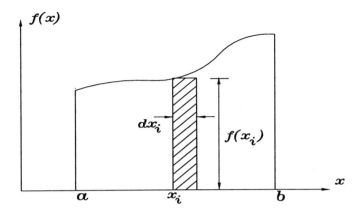

Figure 6.4.1 Integration

This is shown in Fig. 6.4.1. In numerical integration, we take a finite number of calculations. Therefore, Eq. (6.4.1) is approximated as

$$\int_a^b f(x)dx \approx \sum_{i=1}^{N} f(x_i)\Delta x_i \qquad (6.4.2)$$

where N is a finite number. Rewriting this expression in a general way gives

$$\int_a^b f(x)dx \approx \sum_{i=1}^{M} f(x_i)W_i \qquad (6.4.3)$$

in which M is the number of integration points, x_i is the integration point (or sampling point), and W_i is called the weighting coefficient. The weighting coefficient can be interpreted as the width of the rectangular strip whose height is $f(x_i)$ by comparing Eqs. (6.4.2) and (6.4.3). Any numerical integration may be expressed in this form. In order to derive standard values for the integration points and weighting coefficients, the integration domain is normalized such that $-1 \leq x \leq 1$. Of course, there are other ways for normalization.

♣ **Example 6.4.1** Let us find the proper integration points and weighting coefficients for the two-point trapezoidal rule. The trapezoidal rule gives

$$\int_{-1}^{1} g(\xi)d\xi = \bigl(g(-1) + g(1)\bigr) \qquad (6.4.4)$$

Comparing Eq. (6.4.4) to Eq. (6.4.3) indicates that the integration points for this case are $x_1 = -1$ and $x_2 = 1$ while the weighting coefficients are $W_1 = 1$ and $W_2 = 1$. ‡

Table 6.4.1 Sampling points and weights in Gauss-Legendre numerical integration

n	Int.Point	Weight
1	0.00000 00000 00000	2.00000 00000 00000
2	±0.57735 02691 89626	1.00000 00000 00000
3	±0.77459 66692 41483	0.55555 55555 55556
	0.00000 00000 00000	0.88888 88888 88889
4	±0.86113 63115 94053	0.34785 48451 37454
	±0.33998 10435 84856	0.65214 51548 62546
5	±0.90617 98459 38664	0.23692 68850 56189
	±0.53846 93101 05683	0.47862 86704 99366
	0.0000 00000 00000	0.56888 88888 88889
6	±0.93246 95142 03152	0.17132 44923 79170
	±0.66120 93864 66265	0.36076 15730 48139
	±0.23861 91860 83197	0.46791 39345 72691

♣ **Example 6.4.2** Repeat Example 6.4.1 using Simpson's $\frac{1}{3}$ rule with three-point integration. This integration results in

$$\int_{-1}^{1} g(\xi)d\xi = \frac{1}{3}\big(g(-1) + 4g(0) + g(1)\big) \tag{6.4.5}$$

Therefore, we obtain $x_1 = -1$, $x_2 = 0$, $x_3 = 1$, $W_1 = \frac{1}{3}$, $W_2 = \frac{4}{3}$, and $W_3 = \frac{1}{3}$. ‡

Gauss-Legendre quadrature is very useful for integration of polynomial functions. *It can integrate a polynomial function of order $2n - 1$ using the n-point quadrature exactly.* Integration points and weighting coefficients for Gauss-Legendre quadrature are provided in Table 6.4.1. Similarly, Table 6.4.2 gives integration points and weighting coefficients for the triangular domain shown in Fig. 6.3.1 and Fig. 6.3.2. If the integrand is not a polynomial expression, Gauss-Legendre quadrature gives an approximate result. In this case, an optimal number of integration points should be selected in consideration of accuracy and computational cost. The next example shows how to determine the integration points and weighting coefficients for the Gauss-Legendre quadrature.

Gauss Quadrature

Table 6.4.2 Numerical integrations over triangular domains

Int. order	ξ-coordinate	η-coordinate	Weight
3-points	0.16666 66666 667	0.16666 66666 667	0.16666 66666 667
	0.66666 66666 667	0.16666 66666 667	0.16666 66666 667
	0.16666 66666 667	0.66666 66666 667	0.16666 66666 667
7-points	0.10128 65073 235	0.10128 65073 235	0.06296 95902 724
	0.79742 69853 531	0.10128 65073 235	0.06296 95902 724
	0.10128 65073 235	0.79742 69853 531	0.06296 95902 724
	0.47014 20641 051	0.05971 58717 898	0.06619 70763 942
	0.47014 20641 051	0.47014 20641 051	0.06619 70763 942
	0.05971 58717 898	0.47014 20641 051	0.06619 70763 942
	0.33333 33333 333	0.33333 33333 333	0.11250 00000 000
13-points	0.06513 01029 022	0.06513 01029 022	0.02667 36178 044
	0.86973 97941 956	0.06513 01029 022	0.02667 36178 044
	0.06513 01029 022	0.86973 97941 956	0.02667 36178 044
	0.31286 54960 049	0.04869 03154 253	0.03855 68804 452
	0.63844 41885 698	0.31286 54960 049	0.03855 68804 452
	0.04869 03154 253	0.63844 41885 698	0.03855 68804 452
	0.63844 41885 698	0.04869 03154 253	0.03855 68804 452
	0.31286 54960 049	0.63844 41885 698	0.03855 68804 452
	0.04869 03154 253	0.04869 03154 253	0.03855 68804 452
	0.26034 59660 790	0.26034 59660 790	0.08780 76287 166
	0.47930 80678 419	0.26034 59660 790	0.08780 76287 166
	0.26034 59660 790	0.47930 80678 419	0.08780 76287 166
	0.33333 33333 333	0.33333 33333 333	-0.07478 50222 338

♣ **Example 6.4.3** This example shows a way to compute the sampling points and weighting coefficients for Gauss-Legendre quadrature. Let us integrate a cubic polynomial as shown in Fig. 6.4.2. In Gauss-Legendre quadrature, we want to make the integration of the cubic polynomial the same as that of a

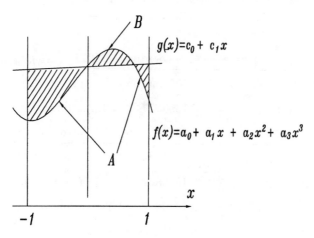

Figure 6.4.2 Two-Point Gauss-Legendre Quadrature

linear function. In other words, the two different hatched areas in Fig. 6.4.2 are the same (Area(A)=Area(B)). Then we can write

$$\int_{-1}^{1} f(x)dx = \int_{-1}^{1} g(x)dx = \sum_{s=1}^{2} W_s f(x_s) \quad (6.4.6)$$

where

$$f(x) = a_0 + a_1 x + a_2 x^2 + a_3 x^3 \quad (6.4.7)$$

$$g(x) = c_0 + c_1 x \quad (6.4.8)$$

and W_s and x_s are the weighting coefficient and sampling point for the two-point Gauss-Legendre quadrature because the two-point rule integrates a cubic polynomial exactly.

Let us rewite the cubic polynomial in the following way.

$$f(x) = c_0 + c_1 x + (x - x_1)(x - x_2)(b_0 + b_1 x) \quad (6.4.9)$$

In this expression, x_1 and x_2 are fixed constants to be determined later. However, there are still four general constants c_0, c_1, b_0, and b_1 to be determined to make Eq. (6.4.9) the same as Eq. (6.4.7) for arbitrary constants a_i. Substitution of Eq. (6.4.9) into Eq. (6.4.6) states

$$\int_{-1}^{1} (x - x_1)(x - x_2)(b_0 + b_1 x)dx = 0 \quad (6.4.10)$$

Equation (6.4.10) must be truly independent of b_0 and b_1 because the integration rule holds for a general cubic polynomial. Therefore,

$$\int_{-1}^{1} (x - x_1)(x - x_2)dx = 0 \quad (6.4.11)$$

and
$$\int_{-1}^{1} x(x-x_1)(x-x_2)dx = 0 \tag{6.4.12}$$

These two equations determine $x_1 = -\frac{1}{\sqrt{3}}$ and $x_2 = \frac{1}{\sqrt{3}}$. These are two sampling points for the two-point Gauss-Legendre quadrature. In order to find the corresponding weighting coefficients, we integrate

$$I = \int_{-1}^{1} (c_0 + c_1 x)dx = 2c_0 \tag{6.4.13}$$

From Eq. (6.4.6), this integration is equal to

$$I = \sum_{s=1}^{2} W_s f(x_s) = W_1(c_0 + c_1 x_1) + W_2(c_0 + c_1 x_2)$$
$$= c_0(W_1 + W_2) - \frac{1}{\sqrt{3}} c_1(W_1 - W_2) \tag{6.4.14}$$

Equating Eq. (6.4.13) and Eq. (6.4.14) yields two weighting coefficients $W_1 = 1$ and $W_2 = 1$. ‡

♣ **Example 6.4.4**   Perform the following integration:

$$\int_{-1}^{1} (1 + 2\xi + 3\xi^2)d\xi \tag{6.4.15}$$

Because the order of polynomial is 2, $2n - 1 = 2$. From this, we get $n = 1.5$. The number of integration points should be an integer. So, we use the two-point quadrature rule. From Table 6.4.1, the two integration points are $-\frac{1}{\sqrt{3}}$ and $\frac{1}{\sqrt{3}}$ with weighting coefficient 1 for each point, respectively. The numerical integration becomes

$$\int_{-1}^{1} (1 + 2\xi + 3\xi^2)d\xi =$$
$$(1)\{1 + 2(-\frac{1}{\sqrt{3}}) + 3(-\frac{1}{\sqrt{3}})^2\} + (1)\{1 + 2(\frac{1}{\sqrt{3}}) + 3(\frac{1}{\sqrt{3}})^2\} = 4 \tag{6.4.16}$$

This is the exact solution. If we use the three-point quadrature rule to integrate Eq. (6.4.15) (i.e. $\xi_1 = -\frac{\sqrt{15}}{5}$, $\xi_2 = 0$, $\xi_1 = \frac{\sqrt{15}}{5}$, $W_1 = \frac{5}{9}$, $W_2 = \frac{8}{9}$, and $W_3 = \frac{5}{9}$), we also obtain the same exact solution. Therefore, the quadrature rule using two or a higher number of integration points will yield the exact solution for this problem. ‡

The quadrature rule can be extended for multi-dimensional integration. For example, numerical integration in the normalized two-dimensional domain can be conducted in the following way.

$$\int_{-1}^{1}\int_{-1}^{1} g(\xi,\eta)d\xi d\eta$$

$$= \int_{-1}^{1} \sum_{i=1}^{M_1} W_i g(\xi_i,\eta)d\eta$$

$$= \sum_{j=1}^{M_2} \bar{W}_j \sum_{i=1}^{M_1} W_i g(\xi_i,\eta_j)$$

$$= \sum_{i=1}^{M_1}\sum_{j=1}^{M_2} W_i \bar{W}_j g(\xi_i,\eta_j) \qquad (6.4.17)$$

in which M_1 and M_2 are the number of integration points in the ξ and η axes, respectively. In addition, (ξ_i,η_j) are the integration points and W_i and $\bar{W}_j$ are weighting coefficients. Table 6.4.1 can be used for these values. Similarly, numerical integration in three-dimension becomes

$$\int_{-1}^{1}\int_{-1}^{1}\int_{-1}^{1} g(\xi,\eta,\zeta)d\xi d\eta d\zeta = \sum_{i=1}^{M_1}\sum_{j=1}^{M_2}\sum_{k=1}^{M_3} W_i \bar{W}_j \tilde{W}_k g(\xi_i,\eta_j,\zeta_k) \qquad (6.4.18)$$

♣ **Example 6.4.5** Integrate the following expression:

$$\int_{-1}^{1}\int_{-1}^{1} 9\xi^2\eta^2 d\xi d\eta \qquad (6.4.19)$$

The integrand is of the second order in terms of ξ and η, respectively. That is, $2M_1 - 1 = 2M_2 - 1 = 2$ for both axes. Therefore, we use two-point quadrature in both ξ and η directions. The integration points are $\xi_1 = \eta_1 = -\frac{1}{\sqrt{3}}$ and $\xi_2 = \eta_2 = \frac{1}{\sqrt{3}}$. The weighting coefficients are $W_1 = \bar{W}_1 = 1$ and $W_2 = \bar{W}_2 = 1$. Applying these values to Eq. (6.4.17) results in 4. ‡

♣ **Example 6.4.6** We want to integrate

$$\int_{-1}^{1}\int_{-1}^{1} 15\xi^2\eta^4 d\xi d\eta \qquad (6.4.20)$$

The integrand is of the second order in terms of ξ and fourth order in terms of η. Therefore $M_1 = 2$ and $M_2 = 3$. Using the two-point quadrature in the

ξ direction and three-point quadrature in the η direction from Table 6.4.1, we obtain the solution of 4. ‡

6.5 MATLAB Application to Gauss Quadrature

This section shows three MATLAB examples for numerical integration of one-, two-, or three-dimensional functions using Gauss-Legendre quadrature. The domain of integration is normalized between -1 and 1 for every axis.

♣ **Example 6.5.1**   We want to integrate

$$f(x) = 1 + x^2 - 3x^3 + 4x^5 \qquad (6.5.1)$$

over the domain $-1 < x < 1$ using Gauss-Legendre quadrature. Because the highest order of the polynomial is 5, we need the 3-point quadrature rule for exact integration from $2n - 1 = 5$. The numerical result is $\frac{8}{3}$. The MATLAB program is shown below.

```
%----------------------------------------------------
% Example 6.5.1
% Gauss-Legendre quadrature of a function in 1-dimension
%
% Problem description
% Integrate f(x)=1+x^2-3x^3+4x^5 between x=-1 and x=1
%
% Variable descriptions
% point1 = integration (or sampling) points
% weight1 = weighting coefficients
% ngl = number of integration points
%----------------------------------------------------
%
ngl=3;                                  % (2*ngl-1)=5
%
[point1,weight1]=feglqd1(ngl);    % extract sampling points and weights
%
%----------------------------------------
% summation for numerical integration
%----------------------------------------
%
value=0.0;
%
for int=1:ngl
x=point1(int);
wt=weight1(int);
func=1+x^2-3*x^3+4*x^5;    % evaluate function at sampling point
```

```
value=value+func*wt;
end
%
value                                          % print the solution
%
%————————————————————————————



function [point1,weight1]=feglqd1(ngl)
%————————————————————————————
% Purpose:
% determine the integration points and weighting coefficients
% of Gauss-Legendre quadrature for one-dimensional integration
%
% Synopsis:
% [point1,weight1]=feglqd1(ngl)
%
% Variable Description:
% ngl - number of integration points
% point1 - vector containing integration points
% weight1 - vector containing weighting coefficients
%————————————————————————————
%
% initialization
%
point1=zeros(ngl,1);
weight1=zeros(ngl,1);
%
% find corresponding integration points and weights
%
if ngl==1                                      % 1-point quadrature rule
point1(1)=0.0;
weight1(1)=2.0;
%
elseif ngl==2                                  % 2-point quadrature rule
point1(1)=-0.577350269189626;
point1(2)=-point1(1);
weight1(1)=1.0;
weight1(2)=weight1(1);
%
elseif ngl==3                                  % 3-point quadrature rule
point1(1)=-0.774596669241483;
point1(2)=0.0;
point1(3)=-point1(1);
weight1(1)=0.555555555555556;
weight1(2)=0.888888888888889;
weight1(3)=weight1(1);
```

```
%
elseif ngl==4                              % 4-point quadrature rule
point1(1)=-0.861136311594053;
point1(2)=-0.339981043584856;
point1(3)=-point1(2);
point1(4)=-point1(1);
weight1(1)=0.347854845137454;
weight1(2)=0.652145154862546;
weight1(3)=weight1(2);
weight1(4)=weight1(1);
%
else                                       % 5-point quadrature rule
point1(1)=-0.906179845938664;
point1(2)=-0.538469310105683;
point1(3)=0.0;
point1(4)=-point1(2);
point1(5)=-point1(1);
weight1(1)=0.236926885056189;
weight1(2)=0.478628670499366;
weight1(3)=0.568888888888889;
weight1(4)=weight1(2);
weight1(5)=weight1(1);
%
end
%
%————————————————————————————————
```

♣ **Example 6.5.2** Use Gauss-Legendre quadrature for integration of

$$f(x,y) = 1 + 4xy - 3x^2y^2 + x^4y^6 \qquad (6.5.2)$$

over the domain $-1 < x < 1$ and $-1 < y < 1$. We use 3-point quadrature rule along the x-axis and 4-point quadrature rule along the y-axis. The result is 2.7810.

```
%————————————————————————————————
% Example 6.5.2
% Gauss-Legendre quadrature of a function in 2-dimension
%
% Problem description
% Integrate f(x,y)=1+4xy-3x²y²+x⁴y⁶ over -1<x<1 and -1<y<1
%
% Variable descriptions
% point2 = integration (or sampling) points
% weight2 = weighting coefficients
```

```
% nglx = number of integration points along x-axis
% ngly = number of integration points along y-axis
%------------------------------------------------
%
nglx=3;                                          % (2*nglx-1)=4
ngly=4;                                          % (2*ngly-1)=6
%
[point2,weight2]=feglqd2(nglx,ngly);             % sampling points and weights
%
%-----------------------------------
% summation for numerical integration
%-----------------------------------
%
value=0.0;
%
for intx=1:nglx
x=point2(intx,1);                                % sampling point in x-axis
wtx=weight2(intx,1);                             % weight in x-axis
for inty=1:ngly
y=point2(inty,2);                                % sampling point in y-axis
wty=weight2(inty,2) ;                            % weight in y-axis
func=1+4*x*y-3*x^2*y^2+x^4*y^6;                  % evaluate function
value=value+func*wtx*wty;
end
end
%
value %
%------------------------------------------------



function [point2,weight2]=feglqd2(nglx,ngly)
%------------------------------------------------
% Purpose:
% determine the integration points and weighting coefficients
% of Gauss-Legendre quadrature for two-dimensional integration
%
% Synopsis:
% [point2,weight2]=feglqd2(nglx,ngly)
%
% Variable Description:
% nglx - number of integration points in the x-axis
% ngly - number of integration points in the y-axis
% point2 - vector containing integration points
% weight2 - vector containing weighting coefficients
%------------------------------------------------
%
% determine the largest one between nglx and ngly
```

```
%
if nglx > ngly
ngl=nglx;
else
ngl=ngly;
end
%
% initialization
%
point2=zeros(ngl,2);
weight2=zeros(ngl,2);
%
% find corresponding integration points and weights
%
[pointx,weightx]=feglqd1(nglx); [pointy,weighty]=feglqd1(ngly); %
% quadrature for two-dimension
%
for intx=1:nglx point2(intx,1)=pointx(intx);
weight2(intx,1)=weightx(intx);
end
%
for inty=1:ngly point2(inty,2)=pointy(inty);
weight2(inty,2)=weighty(inty);
end
%
%————————————————————————————
```

‡

♣ **Example 6.5.3** The following three-dimensional function is integrated using Gauss-Legendre quadrature.

$$f(x,y,z) = 1 + 4x^2y^2 - 3x^2z^4 + y^4z^6 \tag{6.5.3}$$

over the normalized domain $-1 < x < 1$, $-1 < y < 1$, and $-1 < z < 1$. The integrated value is 10.1841.

```
%————————————————————————————
% Example 6.5.3
% Gauss-Legendre quadrature of a function in 3-dimension
%
% Problem description
%   Integrate f(x,y,z)=1+4x^2y^2-3x^2z^4+y^4z^6 over -1<(x,y,z)<1
%
% Variable descriptions
%   point3 = integration (or sampling) points
%   weight3 = weighting coefficients
```

```
% nglx = number of integration points along x-axis
% ngly = number of integration points along y-axis
% nglz = number of integration points along z-axis
%------------------------------------------------
%
nglx=2;                                    % (2*nglx-1)=2
ngly=3;                                    % (2*ngly-1)=4
nglz=4;                                    % (2*nglz-1)=6
%
[point3,weight3]=feglqd3(nglx,ngly,nglz);  % sampling point & weight
%
%-----------------------------------
% summation for numerical integration
%-----------------------------------
%
value=0.0;
%
for intx=1:nglx
x=point3(intx,1);                          % sampling point in x-axis
wtx=weight3(intx,1);                       % weight in x-axis
for inty=1:ngly
y=point3(inty,2);                          % sampling point in y-axis
wty=weight3(inty,2) ;                      % weight in y-axis
for intz=1:nglz
z=point3(intz,3);                          % sampling point in z-axis
wtz=weight3(intz,3) ;                      % weight in z-axis
func=1+4*x^2*y^2-3*x^2*z^2*4+y^4*z^6;      % evaluate function
value=value+func*wtx*wty*wtz;
end
end
end
%
value                                      % print the solution
%
%------------------------------------------------


function [point3,weight3]=feglqd3(nglx,ngly,nglz)
%------------------------------------------------
% Purpose:
% determine the integration points and weighting coefficients
% of Gauss-Legendre quadrature for three-dimensional integration
%
% Synopsis:
% [point3,weight3]=feglqd3(nglx,ngly,nglz)
%
% Variable Description:
```

Section 6.5 MATLAB Application to Gauss Quadrature 185

```
%  nglx - number of integration points in the x-axis
%  ngly - number of integration points in the y-axis
%  nglz - number of integration points in the z-axis
%  point3 - vector containing integration points
%  weight3 - vector containing weighting coefficients
%-------------------------------------------------------
%
% determine the largest one between nglx and ngly
%
if nglx > ngly
if nglx > nglz
ngl=nglx;
else
ngl=nglz;
end
else
if ngly > nglz
ngl=ngly;
else
ngl=nglz;
end
end
%
% initialization
%
point3=zeros(ngl,3);
weight3=zeros(ngl,3);
%
% find corresponding integration points and weights
%
[pointx,weightx]=feglqd1(nglx);        % quadrature rule for x-axis
[pointy,weighty]=feglqd1(ngly);        % quadrature rule for y-axis
[pointz,weightz]=feglqd1(nglz);        % quadrature rule for z-axis
%
% quadrature for two-dimension
%
for intx=1:nglx                        % quadrature in x-axis
point3(intx,1)=pointx(intx);
weight3(intx,1)=weightx(intx);
end
%
for inty=1:ngly                        % quadrature in y-axis
point3(inty,2)=pointy(inty);
weight3(inty,2)=weighty(inty);
end
%
for intz=1:nglz                        % quadrature in z-axis
point3(intz,3)=pointz(intz);
```

```
weight3(intz,3)=weightz(intz);
end
%
%————————————————————————————————————————
```

‡

6.6 MATLAB Application to Laplace's Equation

Isoparametric elements are used to solve *Laplace*'s equation which was discussed in Chapter 5.

♣ **Example 6.6.1**   This example shows how to compute the element matrix for Laplace's equation. The element matrix is expressed as

$$K_{ij}^e = \int_{\Omega^e} \left\{ \frac{\partial H_i}{\partial x}\frac{\partial H_j}{\partial x} + \frac{\partial H_i}{\partial y}\frac{\partial H_j}{\partial y} \right\} d\Omega \qquad (6.6.1)$$

The element domain is shown in Fig. 6.2.4. The MATLAB program is listed below to evaluate the element matrix.

```
%————————————————————————————————————————
% Example 6.6.1
% Compute element matrix for two-dimensional Laplace's equation
%
% Problem description
% Determine the element matrix for Laplace's equation using
% isoparametric four-node quadrilateral element and Gauss-Legendre
% quadrature for a single element shown in Fig. 6.2.4.
%
% Variable descriptions
% k - element matrix
% point2 - integration (or sampling) points
% weight2 - weighting coefficients
% nglx - number of integration points along x-axis
% ngly - number of integration points along y-axis
% xcoord - x coordinate values of nodes
% ycoord - y coordinate values of nodes
% jacob2 - Jacobian matrix
% shape - four-node quadrilateral shape functions
% dhdr - derivatives of shape functions w.r.t. natural coord. r
% dhds - derivatives of shape functions w.r.t. natural coord. s
% dhdx - derivatives of shape functions w.r.t. physical coord. x
% dhdy - derivatives of shape functions w.r.t. physical coord. y
%————————————————————————————————————————
%
```

```
nnel=4;                                    % number of nodes per element
ndof=1;                                    % degrees of freedom per node
edof=nnel*ndof;                            % degrees of freedom per element
%
nglx=2; ngly=2;                            % use 2x2 integration rule
%
xcoord=[-1 1 1 -1];                        % x coordinate values
ycoord=[-0.75 -0.75 1.25 0.25];            % y coordinate values
%
[point2,weight2]=feglqd2(nglx,ngly);       % sampling points & weights
%
%---------------------
% numerical integration
%---------------------
k=zeros(edof,edof);                        % initialization to zero
%
for intx=1:nglx
x=point2(intx,1);                          % sampling point in x-axis
wtx=weight2(intx,1);                       % weight in x-axis
for inty=1:ngly
y=point2(inty,2);                          % sampling point in y-axis
wty=weight2(inty,2) ;                      % weight in y-axis
%
[shape,dhdr,dhds]=feisoq4(x,y);            % compute shape functions and
                                           % derivatives at sampling point
%
jacob2=fejacob2(nnel,dhdr,dhds,xcoord,ycoord);   % compute Jacobian
detjacob=det(jacob2);                      % determinant of Jacobian
invjacob=inv(jacob2);                      % inverse of Jacobian matrix
%
[dhdx,dhdy]=federiv2(nnel,dhdr,dhds,invjacob);   % derivatives w.r.t.
                                           % physical coordinate
%
%---------------------
% element matrix loop
%---------------------
for i=1:edof
for j=1:edof
k(i,j)=k(i,j)+(dhdx(i)*dhdx(j)+dhdy(i)*dhdy(j))*wtx*wty*detjacob;
end
end
%
end
end                                        % end of numerical integration loop
%
k                                          % print the element matrix
%
%---------------------------------------------------------------
```

```
function [dhdx,dhdy]=federiv2(nnel,dhdr,dhds,invjacob)
%------------------------------------------------------------
% Purpose:
% determine derivatives of 2-D isoparametric shape functions with
% respect to physical coordinate system
%
% Synopsis:
% [dhdx,dhdy]=federiv2(nnel,dhdr,dhds,invjacob)
%
% Variable Description:
% dhdx - derivative of shape function w.r.t. physical coordinate x
% dhdy - derivative of shape function w.r.t. physical coordinate y
% nnel - number of nodes per element
% dhdr - derivative of shape functions w.r.t. natural coordinate r
% dhds - derivative of shape functions w.r.t. natural coordinate s
% invjacob - inverse of 2-D Jacobian matrix
%------------------------------------------------------------
%
for i=1:nnel
dhdx(i)=invjacob(1,1)*dhdr(i)+invjacob(1,2)*dhds(i);
dhdy(i)=invjacob(2,1)*dhdr(i)+invjacob(2,2)*dhds(i);
end
%
%------------------------------------------------------------
```

```
function [shapeq4,dhdrq4,dhdsq4]=feisoq4(rvalue,svalue)
%------------------------------------------------------------
% Purpose:
% compute isoparametric four-node quadrilateral shape functions
% and their derivatives at the selected (integration) point
% in terms of the natural coordinate
%
% Synopsis:
% [shapeq4,dhdrq4,dhdsq4]=feisoq4(rvalue,svalue)
%
% Variable Description:
% shapeq4 - shape functions for four-node element
% dhdrq4 - derivatives of the shape functions w.r.t. r
% dhdsq4 - derivatives of the shape functions w.r.t. s
% rvalue - r coordinate value of the selected point
% svalue - s coordinate value of the selected point
%
% Notes:
% 1st node at (-1,-1), 2nd node at (1,-1)
% 3rd node at (1,1), 4th node at (-1,1)
%------------------------------------------------------------
```

```
%
% shape functions
%
shapeq4(1)=0.25*(1-rvalue)*(1-svalue);
shapeq4(2)=0.25*(1+rvalue)*(1-svalue);
shapeq4(3)=0.25*(1+rvalue)*(1+svalue);
shapeq4(4)=0.25*(1-rvalue)*(1+svalue);
%
% derivatives
%
dhdrq4(1)=-0.25*(1-svalue);
dhdrq4(2)=0.25*(1-svalue);
dhdrq4(3)=0.25*(1+svalue);
dhdrq4(4)=-0.25*(1+svalue);
%
dhdsq4(1)=-0.25*(1-rvalue);
dhdsq4(2)=-0.25*(1+rvalue);
dhdsq4(3)=0.25*(1+rvalue);
dhdsq4(4)=0.25*(1-rvalue);
%
%----------------------------------------------


function [jacob2]=fejacob2(nnel,dhdr,dhds,xcoord,ycoord)
%----------------------------------------------
% Purpose:
% determine the Jacobian for two-dimensional mapping
%
% Synopsis:
% [jacob2]=fejacob2(nnel,dhdr,dhds,xcoord,ycoord)
%
% Variable Description:
% jacob2 - Jacobian for one-dimension
% nnel - number of nodes per element
% dhdr - derivative of shape functions w.r.t. natural coordinate r
% dhds - derivative of shape functions w.r.t. natural coordinate s
% xcoord - x axis coordinate values of nodes
% ycoord - y axis coordinate values of nodes
%----------------------------------------------
%
jacob2=zeros(2,2);
%
for i=1:nnel
jacob2(1,1)=jacob2(1,1)+dhdr(i)*xcoord(i);
jacob2(1,2)=jacob2(1,2)+dhdr(i)*ycoord(i);
jacob2(2,1)=jacob2(2,1)+dhds(i)*xcoord(i);
jacob2(2,2)=jacob2(2,2)+dhds(i)*ycoord(i);
```

```
end
%
%----------------------------------------------------------------
```

The computed element matrix is

$$[K] = \begin{bmatrix} 0.7500 & 0.0000 & -0.2500 & -0.5000 \\ 0.0000 & 0.7500 & -0.2500 & -0.5000 \\ -0.2500 & -0.2500 & 0.5000 & 0.0000 \\ -0.5000 & -0.5000 & 0.0000 & 1.0000 \end{bmatrix} \qquad (6.6.2)$$

‡

♣ **Example 6.6.2** Repeat Example 5.9.2 using isoparametric elements. Four-node quadrilateral elements are used. The finite element solution is the same as that obtained in Example 5.9.2. As a result, the solution is not repeated here. The MATLAB program is shown below.

```
%----------------------------------------------------------------
% Example 6.6.2
% to solve the two-dimensional Laplace's equation given as
% u,xx + u,yy =0, 0 < x < 5, 0 < y < 10
% u(x,0) = 0, u(x,10) = 100sin(pi*x/10),
% u(0,y) = 0, u,x(5,y) = 0
% using isoparametric four-node quadrilateral elements
% (see Fig. 5.9.2 for the finite element mesh)
%
% Variable descriptions
% k = element matrix
% f = element vector
% kk = system matrix
% ff = system vector
% gcoord = coordinate values of each node
% nodes = nodal connectivity of each element
% index = a vector containing system dofs associated with each element
% bcdof = a vector containing dofs associated with boundary conditions
% bcval = a vector containing boundary condition values associated with
%             the dofs in bcdof
% point2 - integration (or sampling) points
% weight2 - weighting coefficients
% nglx - number of integration points along x-axis
% ngly - number of integration points along y-axis
% xcoord - x coordinate values of nodes
% ycoord - y coordinate values of nodes
% jacob2 - Jacobian matrix
% shape - four-node quadrilateral shape functions
```

```
% dhdr - derivatives of shape functions w.r.t. natural coord. r
% dhds - derivatives of shape functions w.r.t. natural coord. s
% dhdx - derivatives of shape functions w.r.t. physical coord. x
% dhdy - derivatives of shape functions w.r.t. physical coord. y
%----------------------------------------------------------------
clear
%----------------------------------------
% input data for control parameters
%----------------------------------------
nel=16;                              % number of elements
nnel=4;                              % number of nodes per element
ndof=1;                              % number of dofs per node
nnode=25;                            % total number of nodes in system
nglx=2; ngly=2;                      % use 2x2 integration rule
sdof=nnode*ndof;                     % total system dofs
edof=nnel*ndof;                      % dofs per element
%
%-----------------------------------------------
% input data for nodal coordinate values
% gcoord(i,j) where i-> node no. and j-> x or y
%-----------------------------------------------
gcoord(1,1)=0.0; gcoord(1,2)=0.0;
gcoord(2,1)=1.25; gcoord(2,2)=0.0;
gcoord(3,1)=2.5; gcoord(3,2)=0.0;
gcoord(4,1)=3.75; gcoord(4,2)=0.0;
gcoord(5,1)=5.0; gcoord(5,2)=0.0;
gcoord(6,1)=0.0; gcoord(6,2)=2.5;
gcoord(7,1)=1.25; gcoord(7,2)=2.5;
gcoord(8,1)=2.5; gcoord(8,2)=2.5;
gcoord(9,1)=3.75; gcoord(9,2)=2.5;
gcoord(10,1)=5.0; gcoord(10,2)=2.5;
gcoord(11,1)=0.0; gcoord(11,2)=5.0;
gcoord(12,1)=1.25; gcoord(12,2)=5.0;
gcoord(13,1)=2.5; gcoord(13,2)=5.0;
gcoord(14,1)=3.75; gcoord(14,2)=5.0;
gcoord(15,1)=5.0; gcoord(15,2)=5.0;
gcoord(16,1)=0.0; gcoord(16,2)=7.5;
gcoord(17,1)=1.25; gcoord(17,2)=7.5;
gcoord(18,1)=2.5; gcoord(18,2)=7.5;
gcoord(19,1)=3.75; gcoord(19,2)=7.5;
gcoord(20,1)=5.0; gcoord(20,2)=7.5;
gcoord(21,1)=0.0; gcoord(21,2)=10.;
gcoord(22,1)=1.25; gcoord(22,2)=10.;
gcoord(23,1)=2.5; gcoord(23,2)=10.;
gcoord(24,1)=3.75; gcoord(24,2)=10.;
gcoord(25,1)=5.0; gcoord(25,2)=10.;
%
%-----------------------------------------------
```

% input data for nodal connectivity for each element
% nodes(i,j) where i-> element no. and j-> connected nodes
%————————————————————————
nodes(1,1)=1; nodes(1,2)=2; nodes(1,3)=7; nodes(1,4)=6;
nodes(2,1)=2; nodes(2,2)=3; nodes(2,3)=8; nodes(2,4)=7;
nodes(3,1)=3; nodes(3,2)=4; nodes(3,3)=9; nodes(3,4)=8;
nodes(4,1)=4; nodes(4,2)=5; nodes(4,3)=10; nodes(4,4)=9;
nodes(5,1)=6; nodes(5,2)=7; nodes(5,3)=12; nodes(5,4)=11;
nodes(6,1)=7; nodes(6,2)=8; nodes(6,3)=13; nodes(6,4)=12;
nodes(7,1)=8; nodes(7,2)=9; nodes(7,3)=14; nodes(7,4)=13;
nodes(8,1)=9; nodes(8,2)=10; nodes(8,3)=15; nodes(8,4)=14;
nodes(9,1)=11; nodes(9,2)=12; nodes(9,3)=17; nodes(9,4)=16;
nodes(10,1)=12; nodes(10,2)=13; nodes(10,3)=18; nodes(10,4)=17;
nodes(11,1)=13; nodes(11,2)=14; nodes(11,3)=19; nodes(11,4)=18;
nodes(12,1)=14; nodes(12,2)=15; nodes(12,3)=20; nodes(12,4)=19;
nodes(13,1)=16; nodes(13,2)=17; nodes(13,3)=22; nodes(13,4)=21;
nodes(14,1)=17; nodes(14,2)=18; nodes(14,3)=23; nodes(14,4)=22;
nodes(15,1)=18; nodes(15,2)=19; nodes(15,3)=24; nodes(15,4)=23;
nodes(16,1)=19; nodes(16,2)=20; nodes(16,3)=25; nodes(16,4)=24;
%
%————————————————————————-
% input data for boundary conditions
%————————————————————————-
%
bcdof(1)=1;   % first node is constrained
bcval(1)=0;   % whose described value is 0
bcdof(2)=2;   % second node is constrained
bcval(2)=0;   % whose described value is 0
bcdof(3)=3;   % third node is constrained
bcval(3)=0;   % whose described value is 0
bcdof(4)=4;   % 4th node is constrained
bcval(4)=0;   % whose described value is 0
bcdof(5)=5;   % 5th node is constrained
bcval(5)=0;   % whose described value is 0
bcdof(6)=6;   % 6th node is constrained
bcval(6)=0;   % whose described value is 0
bcdof(7)=11;   % 11th node is constrained
bcval(7)=0;   % whose described value is 0
bcdof(8)=16;   % 16th node is constrained
bcval(8)=0;   % whose described value is 0
bcdof(9)=21;   % 21st node is constrained
bcval(9)=0;   % whose described value is 0
bcdof(10)=22;   % 22nd node is constrained
bcval(10)=38.2683;   % whose described value is 38.2683
bcdof(11)=23;   % 23rd node is constrained
bcval(11)=70.7107;   % whose described value is 70.7107
bcdof(12)=24;   % 24th node is constrained
bcval(12)=92.3880;   % whose described value is 92.3880

```
bcdof(13)=25;                              % 25th node is constrained
bcval(13)=100;                             % whose described value is 100
%
%---------------------------------
% initialization of matrices and vectors
%---------------------------------
ff=zeros(sdof,1);                          % initialization of system force vector
kk=zeros(sdof,sdof);                       % initialization of system matrix
index=zeros(nnel*ndof,1);                  % initialization of index vector
%
%---------------------------------
% loop for computation and assembly of element matrices
%---------------------------------
[point2,weight2]=feglqd2(nglx,ngly);       % sampling points & weights
%
for iel=1:nel                              % loop for the total number of elements
%
for i=1:nnel
nd(i)=nodes(iel,i);                        % extract connected node for (iel)-th element
xcoord(i)=gcoord(nd(i),1);                 % extract x value of the node
ycoord(i)=gcoord(nd(i),2);                 % extract y value of the node
end
%
k=zeros(edof,edof);                        % initialization of element matrix to zero
%
%---------------------------------
% numerical integration
%---------------------------------
for intx=1:nglx
x=point2(intx,1);                          % sampling point in x-axis
wtx=weight2(intx,1);                       % weight in x-axis
for inty=1:ngly
y=point2(inty,2);                          % sampling point in y-axis
wty=weight2(inty,2) ;                      % weight in y-axis
%
[shape,dhdr,dhds]=feisoq4(x,y);            % compute shape functions and
                                           % derivatives at sampling point
%
jacob2=fejacob2(nnel,dhdr,dhds,xcoord,ycoord);    % compute Jacobian
%
detjacob=det(jacob2);                      % determinant of Jacobian
invjacob=inv(jacob2);                      % inverse of Jacobian matrix
%
[dhdx,dhdy]=federiv2(nnel,dhdr,dhds,invjacob);    % derivatives w.r.t.
                                                  % physical coordinate
%
%---------------------------------
% compute element matrix
```

```
%------------------------
for i=1:edof
for j=1:edof
k(i,j)=k(i,j)+(dhdx(i)*dhdx(j)+dhdy(i)*dhdy(j))*wtx*wty*detjacob;
end
end
%
end
end                                    % end of numerical integration loop
%
index=feeldof(nd,nnel,ndof);           % extract system dofs for the element
%
%------------------------
% assemble element matrices
%------------------------
%
kk=feasmbl1(kk,k,index);
%
end %
%------------------------
% apply boundary conditions
%------------------------
[kk,ff]=feaplyc2(kk,ff,bcdof,bcval);
%
%------------------------
% solve the matrix equation
%------------------------
fsol=kk\ff;
%
%------------------------
% analytical solution
%------------------------
for i=1:nnode
x=gcoord(i,1); y=gcoord(i,2);
esol(i)=100*sinh(0.31415927*y)*sin(0.31415927*x)/sinh(3.1415927);
end
%
%------------------------
% print both exact and fem solutions
%------------------------
num=1:1:sdof;
store=[num' fsol esol']
%
%------------------------
```

Problems

6.1 Compute the following integral using the quadratic isoparametric element:

$$K_{11} = \int_2^6 \left[\left(\frac{dH_1}{dx} \right)^2 + (H_1)^2 \right] dx$$

The shape functions are given in Eqs. (6.1.13) through (6.1.15) and the element has nodes $x_1=2$, $x_2=4$, and $x_3=6$ in the physical coordinate system.

6.2 Consider one-dimensional isoparametric shape functions as given in Eqs. (6.1.13) through (6.1.15). The isoparametric element is mapped into a physical domain with nodal points located at $x_1 = 0$, $x_2 = a$, and $x_3 = 4$, where $a = 1.5$, $a = 1$, or $a = 0.5$. Compute the Jacobian J and its inverse for these cases.

6.3 Compute the Jacobian matrix for the following bilinear element shown in Fig. P6.3 and evaluate

$$K_{12} = \int_\Omega \left(\frac{\partial H_1}{\partial x} \frac{\partial H_2}{\partial x} + \frac{\partial H_1}{\partial y} \frac{\partial H_2}{\partial y} \right) d\Omega$$

using the isoparametric element and 3×3 Gauss-Legendre quadrature. The shape functions are provided in Eqs. (6.2.1) through (6.2.4).

6.4 For the linear triangular isoparametric element shown in Fig. P6.4, (a) compute the Jacobian matrix and (b) find $\frac{\partial H_1}{\partial x}$ in which H_1 is given in Eq. (6.3.1).

6.5 Evaluate the Jacobian matrix for the four-node element shown in Fig. P6.5 using the bilinear isoparametric element.

6.6 The Gauss-Legendre quadrature rule is used to evaluate the integral

$$\int_{-1}^1 \int_{-1}^1 H_1(\xi, \eta) H_3(\xi, \eta) |J| d\xi d\eta$$

in which H_1 and H_2 are quadratic shape functions of ξ and η, respectively. What order of integration is necessary for exact integration of the integral if the element has no distortion (i.e. a rectangular shape of element in the physical domain)?

6.7 Two different isoparametric elements are used together as shown in Fig. P6.7. There is an interelement boundary between $(x, y)=(1,1)$ and $(x, y)=(2,2)$. Show that the variable is continuous across the interface boundary. In other words, show that variable interpolation from the quadrilateral element is the same as that from the triangular element at the interface.

6.8 Consider the two elements shown in Fig. P6.7 again. For the elements, we use the following interpolation for each element. For the triangular element, we use

$$u = a_0 + a_1 x + a_2 y$$

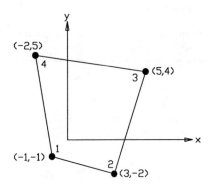

Figure P6.3 Problem 6.3

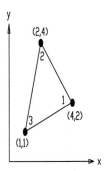

Figure P6.4 Problem 6.4

and for the quadrilateral element we use

$$u = b_0 + b_1 x + b_2 y + b_3 xy$$

Is u compatible at the element interface of the two elements?

6.9 Two kinds of quadrilateral isoparametric elements are utilized together to mesh a domain as seen in Fig. P6.9. One is a bilinear element and the other is a biquadratic element. Is there compatibility between the element interface?

6.10 Solve Prob. 5.11 using isoparametric elements and computer programs provided in this chapter.

6.11 Solve Prob. 5.12 using isoparametric elements and MATLAB programs.

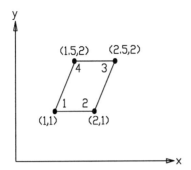

Figure P6.5 Problem 6.5

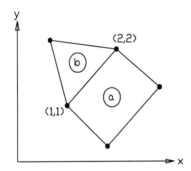

Figure P6.7 Problem 6.7

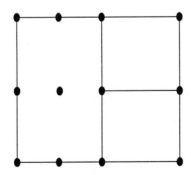

Figure P6.9 Problem 6.9

CHAPTER SEVEN

TRUSS STRUCTURES

7.0 Chapter Overview

A truss structure is an assembly of axial members using pin joints. This chapter presents the finite element matrix equations for 2-D or 3-D truss structures. Because each member has a different orientation from the reference axis, transformation of matrices from a local axis to the global reference axis is discussed here. Both static and dynamic problems are considered. MATLAB programs for static, eigenvalue, and transient analyses of truss structures are appended at the end.

7.1 One-Dimensional Truss

The one-dimensional truss is also called a rod or an axial bar which was described in Chapter 4. The governing equation to describe the motion of a rod is derived below. Let E, A and ρ indicate the elastic modulus, cross-sectional area, and density of the rod, respectively. Applying Newton's second law to the free body diagram shown in Fig. 7.1.1 gives

$$\rho A dx \frac{\partial^2 u}{\partial t^2} = \left(P + \frac{\partial P}{\partial x}dx\right) - P \qquad (7.1.1)$$

where u is the axial displacement along the rod direction, and x and t are the spatial and temporal axes, respectively. Hooke's law states

$$\frac{P}{A} = E\epsilon \qquad (7.1.2)$$

The strain-displacement relation is

$$\epsilon = \frac{\partial u}{\partial x} \qquad (7.1.3)$$

Substituting Eq. (7.1.3) into Eq. (7.1.2) and the result into (7.1.1) yields

$$\rho A \frac{\partial^2 u}{\partial t^2} = \frac{\partial}{\partial x}\left(AE \frac{\partial u}{\partial x}\right) \tag{7.1.4}$$

In Eq. (7.1.4), ρ, A, and E may vary as a function of x.

The weak formulation for Eq. (7.1.4) is

$$I = \int_0^L \left(\rho A w \frac{\partial^2 u}{\partial t^2} + AE \frac{\partial w}{\partial x}\frac{\partial u}{\partial x}\right) dx - \left[AEw \frac{\partial u}{\partial x}\right]_0^L \tag{7.1.5}$$

in which w is the test function. The first term is the inertia term and the second term is the stiffness term. Discretization of the domain into a number of elements breaks the global integral in Eq. (7.1.5) into element integrals over the element domains.

Use of Galerkin's method and linear shape functions for a rod element whose length is l results in the following stiffness matrix:

$$[K^e] = \int_0^l AE \left\{ \begin{array}{c} \frac{dH_1}{dx} \\ \frac{dH_2}{dx} \end{array} \right\} \left\{ \begin{array}{cc} \frac{dH_1}{dx} & \frac{dH_2}{dx} \end{array} \right\} dx \tag{7.1.6}$$

for the element nodal degrees of freedom $\{u_1 \; u_2\}$ as shown in Fig. 7.1.2. Superscript e denotes element. Here H_i is the linear shape function which is

$$H_1 = \frac{l - x}{l} \tag{7.1.7}$$

$$H_2 = \frac{x}{l} \tag{7.1.8}$$

Substitution of these shape functions into Eq. (7.1.6) results in the element stiffness matrix for a rod.

$$[K^e] = \frac{AE}{l} \begin{bmatrix} 1 & -1 \\ -1 & 1 \end{bmatrix} \tag{7.1.9}$$

This is the same as given in Chapter 4.

The element mass matrix for a rod is obtained from the first term in Eq. (7.1.5) using the linear shape functions.

$$[M^e] = \frac{\rho Al}{6} \begin{bmatrix} 2 & 1 \\ 1 & 2 \end{bmatrix} \tag{7.1.10}$$

for the constant density and cross-sectional area. This is called the *consistent* mass matrix for a rod. The *lumped* mass matrix is

$$[M^e] = \frac{\rho Al}{2} \begin{bmatrix} 1 & 0 \\ 0 & 1 \end{bmatrix} \tag{7.1.11}$$

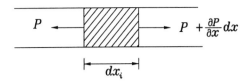

Figure 7.1.1 Free Body Diagram for Axial Member

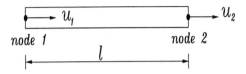

Figure 7.1.2 Two-Node Axial Bar

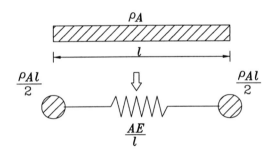

Figure 7.1.3 Equivalent Spring-Mass System

This is obtained by lumping the distributed mass within the element into the concentrated masses at the two nodal points as seen in Fig. 7.1.3.

7.2 Plane Truss

The truss is a structure which consists of axial members connected by pin joints. Therefore, each member of the truss structure supports the external load through its axial force and it does not undergo the bending deformation. The stiffness matrix for a truss member shown in Fig. 7.2.1 is given in Eq. (7.1.9). However, the matrix size becomes 4×4 because the nodal degrees of freedom of the truss element are expressed as

$$\{d^e\} = \{\, u_1 \quad v_1 \quad u_2 \quad v_2 \,\}^T \tag{7.2.1}$$

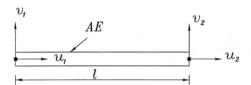

Figure 7.2.1 Two-Dimensional Truss Element

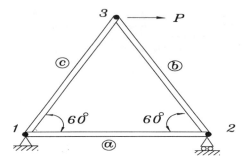

Figure 7.2.2 Triangular Truss

Here superscript e denotes the element level. The corresponding stiffness matrix is

$$[K^e] = \begin{bmatrix} k & 0 & -k & 0 \\ 0 & 0 & 0 & 0 \\ -k & 0 & k & 0 \\ 0 & 0 & 0 & 0 \end{bmatrix} \quad (7.2.2)$$

in which

$$k = \frac{AE}{l} \quad (7.2.3)$$

For a uniform member, A and E are the area of the cross-section and the elastic modulus, respectively. In addition, l is the length of the member. The second and fourth columns and rows of the stiffness matrix associated with the transverse displacement v are null since the truss member has axial deformation only.

The plane truss structure consists of axial members in different orientations. For example, Fig. 7.2.2 shows that members a, b, and c lie in three different directions. In order to assemble the stiffness matrices related to these truss members, we need to have the element degrees of freedom given in terms of the common reference axes. In other words, the element nodal displacements are expressed in terms of the fixed global coordinate system.

Figure 7.2.3 shows a plane truss element oriented in an arbitrary angle β with respect to the horizontal axis $\bar{x}$. The figure shows two sets of nodal displacements. One set has nodal displacements along and perpendicular to the element axis (i.e. u and v) while the other set has the displacements in terms of the global reference axes (i.e. $\bar{u}$ and $\bar{v}$). Because the element stiffness matrix Eq. (7.2.2) is expressed in terms

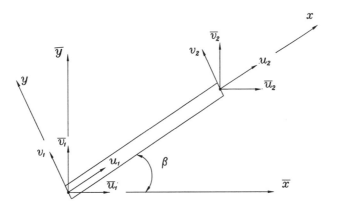

Figure 7.2.3 Generalized Two-Dimensional Truss Element

of u and v, it should be transformed such that the stiffness matrix is expressed in terms of $\bar{u}$ and $\bar{v}$.

To this end, we find the relationship between xy- and $\bar{x}\bar{y}$-coordinate systems. This relationship is called the *coordinate transformation* and the same relationship holds for the two sets of nodal displacements (i.e. u, v and $\bar{u}$, $\bar{v}$). The relationship is

$$\begin{Bmatrix} u_1 \\ v_1 \\ u_2 \\ v_2 \end{Bmatrix} = \begin{bmatrix} c & s & 0 & 0 \\ -s & c & 0 & 0 \\ 0 & 0 & c & s \\ 0 & 0 & -s & c \end{bmatrix} \begin{Bmatrix} \bar{u}_1 \\ \bar{v}_1 \\ \bar{u}_2 \\ \bar{v}_2 \end{Bmatrix} \qquad (7.2.4)$$

where $c = cos\beta$ and $s = sin\beta$. Let us rewrite Eq. (7.2.4) as

$$\{d^e\} = [T]\{\bar{d}^e\} \qquad (7.2.5)$$

To transform the element stiffness matrix from the xy-coordinate system to $\bar{x}\bar{y}$-coordinate system, consider the concept of strain energy. The strain energy is expressed as

$$U = \frac{1}{2}\{d^e\}^T[K^e]\{d^e\} \qquad (7.2.6)$$

in terms of the xy-coordinate system. If we substitute Eq. (7.2.5) into Eq. (7.2.6), we obtain

$$U = \frac{1}{2}\{\bar{d}^e\}^T[T]^T[K^e][T]\{\bar{d}^e\} \qquad (7.2.7)$$

The strain energy is now expressed in terms of the $\bar{x}\bar{y}$-coordinate system.

$$U = \frac{1}{2}\{\bar{d}^e\}^T[\bar{K}^e]\{\bar{d}^e\} \qquad (7.2.8)$$

in which $[\bar{K}^e]$ is the transformed element stiffness matrix in terms of the $\bar{x}\bar{y}$-coordinate system. The strain energy in Eq. (7.2.8) should be the same as that in Eq. (7.2.7) because strain energy is independent of the coordinate system. Equating Eq. (7.2.7) to Eq. (7.2.8) shows that

$$[\bar{K}^e] = [T]^T[K^e][T] \tag{7.2.9}$$

Substitution of Eqs. (7.2.2) and (7.2.4) into Eq. (7.2.9) results in the transformed stiffness matrix

$$[\bar{K}^e] = \frac{AE}{l} \begin{bmatrix} c^2 & cs & -c^2 & -cs \\ cs & s^2 & -cs & -s^2 \\ -c^2 & -cs & c^2 & cs \\ -cs & -s^2 & cs & s^2 \end{bmatrix} \tag{7.2.10}$$

for the nodal degrees of freedom

$$\{\bar{u}_1 \quad \bar{v}_1 \quad \bar{u}_2 \quad \bar{v}_2\} \tag{7.2.11}$$

This element stiffness matrix can be assembled into the global matrix as usual for the shared nodal points.

♣ **Example 7.2.1** Let us compute element stiffness matrices for the truss structure shown in Fig. 7.2.2. The structure has three elements (a, b, and c) and three nodes (1, 2, and 3). Let each truss member have the same material and geometric properties. In the following derivation, the superimposed (¯) is omitted for simplicity.

Member a: Let $i=1$ and $j=2$. Then $\beta=0$ so that $c = \cos\beta=1$ and $s = \sin\beta=0$. The element stiffness matrix is

$$[K^a] = \frac{A^a E^a}{l^a} \begin{bmatrix} 1 & 0 & -1 & 0 \\ 0 & 0 & 0 & 0 \\ -1 & 0 & 1 & 0 \\ 0 & 0 & 0 & 0 \end{bmatrix} \tag{7.2.12}$$

Member b: Let $i=2$ and $j=3$. Then $\beta = \frac{2\pi}{3}$ so that $c = \cos\beta=-0.5$ and $s = \sin\beta=\frac{\sqrt{3}}{2}$. The element stiffness matrix is

$$[K^b] = \frac{A^b E^b}{l^b} \begin{bmatrix} 0.250 & -0.433 & -0.250 & 0.433 \\ -0.433 & 0.750 & 0.433 & -0.750 \\ -0.250 & 0.433 & 0.250 & -0.433 \\ 0.433 & -0.750 & -0.433 & 0.750 \end{bmatrix} \tag{7.2.13}$$

Member c: Let $i=1$ and $j=3$. Then $\beta = \frac{\pi}{3}$ so that $c = \cos\beta=0.5$ and $s = \sin\beta=\frac{\sqrt{3}}{2}$. The element stiffness matrix is

$$[K^c] = \frac{A^c E^c}{l^c} \begin{bmatrix} 0.250 & 0.433 & -0.250 & -0.433 \\ 0.433 & 0.750 & -0.433 & -0.750 \\ -0.250 & -0.433 & 0.250 & 0.433 \\ -0.433 & -0.750 & 0.433 & 0.750 \end{bmatrix} \tag{7.2.14}$$

The element mass matrix for the plane truss member can be calculated using the same coordinate transformation. Using the kinetic energy expression, similar to the strain energy expression for derivation of the element stiffness matrix, we can write

$$[\bar{M}^e] = [T]^T [M^e][T] \tag{7.2.15}$$

Carrying out this matrix multiplication using the *consistent* mass matrix gives

$$[\bar{M}^e] = \frac{\rho A l}{6} \begin{bmatrix} 2c^2 & 2cs & c^2 & cs \\ 2cs & 2s^2 & cs & s^2 \\ c^2 & cs & 2c^2 & 2cs \\ cs & s^2 & 2cs & 2s^2 \end{bmatrix} \tag{7.2.16}$$

The *lumped* mass matrix can be obtained similarly and shown below:

$$[\bar{M}^e] = \frac{\rho A l}{2} \begin{bmatrix} c^2 & cs & 0 & 0 \\ cs & s^2 & 0 & 0 \\ 0 & 0 & c^2 & cs \\ 0 & 0 & cs & s^2 \end{bmatrix} \tag{7.2.17}$$

7.3 Space Truss

Development of the element stiffness matrix for the space truss member is similar to that for the plane truss member. The element stiffness matrix in terms of the global Cartesian coordinate system is obtained in the same way as given in Eq. (7.2.9). However, the sizes of both the transformation matrix and the element stiffness matrix in terms of the body coordinate system are 6 × 6 for the space truss member. Here, the body coordinate system denotes the coordinate system, one of whose axes lies along the member direction. The stiffness matrix in terms of the body coordinate system is

$$[K^e] = \frac{AE}{l} \begin{bmatrix} 1 & 0 & 0 & -1 & 0 & 0 \\ 0 & 0 & 0 & 0 & 0 & 0 \\ 0 & 0 & 0 & 0 & 0 & 0 \\ -1 & 0 & 0 & 1 & 0 & 0 \\ 0 & 0 & 0 & 0 & 0 & 0 \\ 0 & 0 & 0 & 0 & 0 & 0 \end{bmatrix} \tag{7.3.1}$$

for the nodal degrees of freedom of

$$\{d^e\} = \{ u_1 \quad v_1 \quad w_1 \quad u_2 \quad v_2 \quad w_2 \} \tag{7.3.2}$$

where u is the displacement along the x-axis as shown in Fig. 7.3.1.

The transformation matrix between the two coordinate systems is given below:

$$[T] = \begin{bmatrix} \xi_1 & \xi_2 & \xi_3 & 0 & 0 & 0 \\ \eta_1 & \eta_2 & \eta_3 & 0 & 0 & 0 \\ \zeta_1 & \zeta_2 & \zeta_3 & 0 & 0 & 0 \\ 0 & 0 & 0 & \xi_1 & \xi_2 & \xi_3 \\ 0 & 0 & 0 & \eta_1 & \eta_2 & \eta_3 \\ 0 & 0 & 0 & \zeta_1 & \zeta_2 & \zeta_3 \end{bmatrix} \quad (7.3.3)$$

where $\{\xi_1\ \eta_1\ \zeta_1\}$ is the *direction cosines* of the $\bar{x}$-axis with respect to the xyz-coordinate system. Similarly, $\{\xi_2\ \eta_2\ \zeta_2\}$ and $\{\xi_3\ \eta_3\ \zeta_3\}$ are the *direction cosines* of the $\bar{y}$- and $\bar{z}$-axis with respect to the xyz-coordinate system, respectively. Conducting matrix manipulation yields

$$[\bar{K}^e] = \frac{AE}{l} \begin{bmatrix} \xi_1^2 & \xi_1\xi_2 & \xi_1\xi_3 & -\xi_1^2 & -\xi_1\xi_2 & -\xi_1\xi_3 \\ \xi_1\xi_2 & \xi_2^2 & \xi_2\xi_3 & -\xi_1\xi_2 & -\xi_2^2 & -\xi_2\xi_3 \\ \xi_1\xi_3 & \xi_2\xi_3 & \xi_3^2 & -\xi_1\xi_3 & -\xi_2\xi_3 & -\xi_3^2 \\ -\xi_1^2 & -\xi_1\xi_2 & -\xi_1\xi_3 & \xi_1^2 & \xi_1\xi_2 & \xi_1\xi_3 \\ -\xi_1\xi_2 & -\xi_2^2 & -\xi_2\xi_3 & \xi_1\xi_2 & \xi_2^2 & \xi_2\xi_3 \\ -\xi_1\xi_3 & -\xi_2\xi_3 & -\xi_3^2 & \xi_1\xi_3 & \xi_2\xi_3 & \xi_3^2 \end{bmatrix} \quad (7.3.4)$$

The corresponding element degrees of freedom are

$$\{\bar{d}^e\} = \{\bar{u}_1 \quad \bar{v}_1 \quad \bar{w}_1 \quad \bar{u}_2 \quad \bar{v}_2 \quad \bar{w}_2\} \quad (7.3.5)$$

The *consistent* mass matrix for the space truss element is

$$[\bar{M}^e] = \frac{\rho A l}{6} \begin{bmatrix} 2\xi_1^2 & 2\xi_1\xi_2 & 2\xi_1\xi_3 & \xi_1^2 & \xi_1\xi_2 & \xi_1\xi_3 \\ 2\xi_1\xi_2 & 2\xi_2^2 & 2\xi_2\xi_3 & \xi_1\xi_2 & \xi_2^2 & \xi_2\xi_3 \\ 2\xi_1\xi_3 & 2\xi_2\xi_3 & 2\xi_3^2 & \xi_1\xi_3 & \xi_2\xi_3 & \xi_3^2 \\ \xi_1^2 & \xi_1\xi_2 & \xi_1\xi_3 & 2\xi_1^2 & 2\xi_1\xi_2 & 2\xi_1\xi_3 \\ \xi_1\xi_2 & \xi_2^2 & \xi_2\xi_3 & 2\xi_1\xi_2 & 2\xi_2^2 & 2\xi_2\xi_3 \\ \xi_1\xi_3 & \xi_2\xi_3 & \xi_3^2 & 2\xi_1\xi_3 & 2\xi_2\xi_3 & 2\xi_3^2 \end{bmatrix} \quad (7.3.6)$$

while the *lumped* mass matrix is

$$[\bar{M}^e] = \frac{\rho A l}{2} \begin{bmatrix} \xi_1^2 & \xi_1\xi_2 & \xi_1\xi_3 & 0 & 0 & 0 \\ \xi_1\xi_2 & \xi_2^2 & \xi_2\xi_3 & 0 & 0 & 0 \\ \xi_1\xi_3 & \xi_2\xi_3 & \xi_3^2 & 0 & 0 & 0 \\ 0 & 0 & 0 & \xi_1^2 & \xi_1\xi_2 & \xi_1\xi_3 \\ 0 & 0 & 0 & \xi_1\xi_2 & \xi_2^2 & \xi_2\xi_3 \\ 0 & 0 & 0 & \xi_1\xi_3 & \xi_2\xi_3 & \xi_3^2 \end{bmatrix} \quad (7.3.7)$$

7.4 MATLAB Application to Static Analysis

The static analysis of a truss structure is to solve the following matrix equation:

$$[K]\{d\} = \{F\} \quad (7.4.1)$$

Section 7.4   MATLAB Application to Static Analysis   207

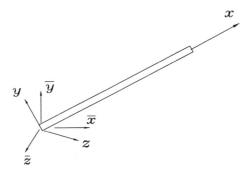

Figure 7.3.1 Generalized Three-Dimensional Truss Element

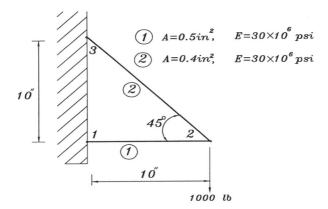

Figure 7.4.1 Truss With Two Axial Members

where the system stiffness matrix $[K]$ and the system force vector $\{F\}$ are obtained by assembling each element matrix and vector. This section shows some examples of MATLAB programs and *m-files* for static analyses of two-dimensional truss structures.

♣ **Example 7.4.1** Figure 7.4.1 shows a simple truss structure made of two members. Each member has elastic modulus of $E=30\times 10^6$ psi and cross-sectional areas are $A_1=0.4$ in^2 and $A_2=0.5$ in^2 where the subscript indicates the element number as shown in the figure. A 1000 lb force is applied at the tip in the downward direction. Find the displacements and stresses of the members. The MATLAB program and *m-files* are provided below.

```
%----------------------------------------------------
% Example 7.4.1
% to solve static 2-D truss structure
%
% Problem description
```

```
% Find the deflection and stress of the truss made of two members
% as shown in Fig. 7.4.1.
%
% Variable descriptions
% k = element stiffness matrix
% kk = system stiffness matrix
% ff = system force vector
% index = a vector containing system dofs associated with each element
% gcoord = global coordinate matrix
% disp = nodal displacement vector
% elforce = element force vector
% eldisp = element nodal displacement
% stress = stress vector for every element
% elprop = element property matrix
% nodes = nodal connectivity matrix for each element
% bcdof = a vector containing dofs associated with boundary conditions
% bcval = a vector containing boundary condition values associated with
%                the dofs in bcdof
%----------------------------------------------------------------
%
%---------------------------
% control input data
%---------------------------
nel=2;                                      % number of elements
nnel=2;                                     % number of nodes per element
ndof=2;                                     % number of dofs per node
nnode=3;                                    % total number of nodes in system
sdof=nnode*ndof;                            % total system dofs
%
%---------------------------
% nodal coordinates
%---------------------------
gcoord(1,1)=0.0; gcoord(1,2)=0.0;           % x, y-coordinate of node 1
gcoord(2,1)=10.0; gcoord(2,2)=0.0;          % x, y-coordinate of node 2
gcoord(3,1)=0.0; gcoord(3,2)=10.0;          % x, y-coordinate of node 3
%
%-------------------------------------
% material and geometric properties
%-------------------------------------
elprop(1,1)=30000000;                       % elastic modulus of 1st element
elprop(1,2)=0.4;                            % cross-section of 1st element
elprop(2,1)=30000000;                       % elastic modulus of 2nd element
elprop(2,2)=0.5;                            % cross-section of 2nd element
%
%---------------------------
% nodal connectivity
%---------------------------
nodes(1,1)=1; nodes(1,2)=2;                 % nodes associated with element 1
```

```
nodes(2,1)=2; nodes(2,2)=3;          % nodes associated with element 2
%
%------------------------
% applied constraints
%------------------------
bcdof(1)=1;                          % 1st dof (horizontal displ) is constrained
bcval(1)=0;                          % whose described value is 0
bcdof(2)=2;                          % 2nd dof (vertical displ) is constrained
bcval(2)=0;                          % whose described value is 0
bcdof(3)=5;                          % 5th dof (horizontal displ) is constrained
bcval(3)=0;                          % whose described value is 0
bcdof(4)=6;                          % 6th dof (vertical displ) is constrained
bcval(4)=0;                          % whose described value is 0
%
%------------------------
% initialization to zero
%------------------------
ff=zeros(sdof,1);                    % system force vector
kk=zeros(sdof,sdof);                 % system stiffness matrix
index=zeros(nnel*ndof,1);            % index vector
elforce=zeros(nnel*ndof,1);          % element force vector
eldisp=zeros(nnel*ndof,1);           % element nodal displacement vector
k=zeros(nnel*ndof,nnel*ndof);        % element stiffness matrix
stress=zeros(nel,1);                 % stress vector for every element
%
%------------------------
% applied nodal force
%------------------------
ff(4)=-1000;                         % 2nd node has 1000 lb in downward direction
%
%------------------------
% loop for elements
%------------------------
for iel=1:nel                        % loop for the total number of elements
%
nd(1)=nodes(iel,1);                  % 1st connected node for the (iel)-th element
nd(2)=nodes(iel,2);                  % 2nd connected node for the (iel)-th element
%
x1=gcoord(nd(1),1); y1=gcoord(nd(1),2);   % coordinate of 1st node
x2=gcoord(nd(2),1); y2=gcoord(nd(2),2);   % coordinate of 2nd node
%
leng=sqrt((x2-x1)^2+(y2-y1)^2);      % element length
%
if (x2-x1)==0;
if y2>y1;
beta=2*atan(1);                      % angle between local and global axes
else
beta=-2*atan(1);
```

```
end
else
beta=atan((y2-y1)/(x2-x1));
end
%
el=elprop(iel,1);                           % extract elastic modulus
area=elprop(iel,2);                         % extract cross-sectional area
%
index=feeldof(nd,nnel,ndof);                % extract system dofs for the element
%
k=fetruss2(el,leng,area,0,beta,1);          % compute element matrix
%
kk=feasmbl1(kk,k,index);                    % assemble into system matrix
%
end
%
%----------------------------------------
% apply constraints and solve the matrix
%----------------------------------------
[kk,ff]=feaplyc2(kk,ff,bcdof,bcval);        % apply boundary conditions
%
disp=kk\ff;          % solve matrix equation for nodal displacements
%
%----------------------------------------
% post computation for stress calculation
%----------------------------------------
for iel=1:nel                % loop for the total number of elements
%
nd(1)=nodes(iel,1);          % 1st connected node for the (iel)-th element
nd(2)=nodes(iel,2);          % 2nd connected node for the (iel)-th element
%
x1=gcoord(nd(1),1); y1=gcoord(nd(1),2);     % coordinate of 1st node
x2=gcoord(nd(2),1); y2=gcoord(nd(2),2);     % coordinate of 2nd node
%
leng=sqrt((x2-x1)^2+(y2-y1)^2);             % element length
%
if (x2-x1)==0;
if y2>y1;
beta=2*atan(1);                             % angle between local and global axes
else
beta=-2*atan(1);
end
else
beta=atan((y2-y1)/(x2-x1));
end
%
el=elprop(iel,1);                           % extract elastic modulus
area=elprop(iel,2);                         % extract cross-sectional area
```

Section 7.4   MATLAB Application to Static Analysis   211

```
%
index=feeldof(nd,nnel,ndof);         % extract system dofs for the element
%
k=fetruss2(el,leng,area,0,beta,1);   % compute element matrix
%
for i=1:(nnel*ndof)                  % extract displacements associated with
eldisp(i)=disp(index(i));            % (iel)-th element
end
%
elforce=k*eldisp;                    % element force vector
stress(iel)=sqrt(elforce(1)^2+elforce(2)^2)/area;   % stress
%
if ((x2-x1)*elforce(3)) < 0;         % check if tension or compression
stress(iel)=-stress(iel);
end
%
end
%
%------------------------------
% print fem solutions
%------------------------------
num=1:1:sdof;
displ=[num' disp]                    % print displacements
%
numm=1:1:nel;
stresses=[numm' stress]              % print stresses
%
%--------------------------------------------------------------


function [k,m]=fetruss2(el,leng,area,rho,beta,ipt)
%--------------------------------------------------------------
% Purpose:
% Stiffness and mass matrices for the 2-D truss element
% nodal dof { u_1 v_1 u_2 v_2 }
%
% Synopsis:
% [k,m]=fetruss2(el,leng,area,rho,beta,ipt)
%
% Variable Description:
% k - element stiffness matrix (size 4x4)
% m - element mass matrix (size 4x4)
% el - elastic modulus
% leng - element length
% area - area of truss cross-section
% rho - mass density (mass per unit volume)
% beta - angle between the local and global axes
```

```
%                      positive if local axis is in ccw direction from
%                      the global axis
% ipt = 1 - consistent mass matrix
% ipt = 2 - lumped mass matrix
%------------------------------------------------------------
%
% stiffness matrix
%
c=cos(beta); s=sin(beta);
k= (area*el/leng)*[ c*c     c*s     -c*c    -c*s;...
                    c*s     s*s     -c*s    -s*s;...
                   -c*c    -c*s      c*c     c*s;...
                   -c*s    -s*s      c*s     s*s];
%
% consistent mass matrix
%
if ipt==1
%
m=(rho*area*leng/6)*[ 2*c*c    2*c*s    c*c     c*s;...
                      2*c*s    2*s*s    c*s     s*s;...
                       c*c      c*s    2*c*c   2*c*s;...
                       c*s      s*s    2*c*s   2*s*s];
%
% lumped mass matrix
%
else
%
m=(rho*area*leng/2)*[ c*c     c*s     0      0;...
                      c*s     s*s     0      0;...
                       0       0     c*c    c*s;...
                       0       0     c*s    s*s];
%
end
%
%------------------------------------------------------------
```

The results from the finite element analysis are given below. The minus sign in the stress indicates compressive stress.

```
displ =
dofs       displacement
1.0000      0.0000              % horizontal displ. of node 1
2.0000      0.0000              % vertical displ. of node 1
3.0000     -0.0008              % horizontal displ. of node 2
4.0000     -0.0027              % vertical displ. of node 2
5.0000      0.0000              % horizontal displ. of node 3
6.0000      0.0000              % vertical displ. of node 3
```

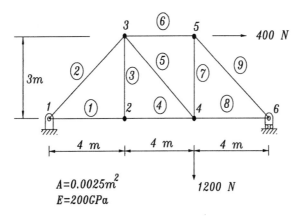

Figure 7.4.2 Truss Structure

```
        stresses =
        element    stress
        1.00000    -2500.           % compressive stress for element 1
        2.00000    2828.            % tensile stress for element 2
```

‡

♣ **Example 7.4.2**   Find the stresses of the truss structure shown in Fig. 7.4.2. All members have elastic modulus of 200 GPa and cross-sectional area of 2.5×10^{-3} m^2.

```
%------------------------------------------------------------
% Example 7.4.2
% to solve static 2-D truss structure
%
% Problem description
% Find the deflection and stress of the truss made of two members
% as shown in Fig. 7.4.2.
%
% Variable descriptions
% k = element stiffness matrix
% kk = system stiffness matrix
% ff = system force vector
% index = a vector containing system dofs associated with each element
% gcoord = global coordinate matrix
% disp = nodal displacement vector
% elforce = element force vector
% eldisp = element nodal displacement
% stress = stress vector for every element
% prop = material and geometric property matrix
```

```
% nodes = nodal connectivity matrix for each element
% bcdof = a vector containing dofs associated with boundary conditions
% bcval = a vector containing boundary condition values associated with
%         the dofs in bcdof
%---------------------------------------------------------------
%
%-------------------------
% control input data
%-------------------------
nel=9;                              % number of elements
nnel=2;                             % number of nodes per element
ndof=2;                             % number of dofs per node
nnode=6;                            % total number of nodes in system
sdof=nnode*ndof;                    % total system dofs
%
%-------------------------
% nodal coordinates
%-------------------------
gcoord(1,1)=0.0; gcoord(1,2)=0.0;
gcoord(2,1)=4.0; gcoord(2,2)=0.0;
gcoord(3,1)=4.0; gcoord(3,2)=3.0;
gcoord(4,1)=8.0; gcoord(4,2)=0.0;
gcoord(5,1)=8.0; gcoord(5,2)=3.0;
gcoord(6,1)=12.; gcoord(6,2)=0.0;
%
%---------------------------------------
% material and geometric properties
%---------------------------------------
prop(1)=200e9;                      % elastic modulus
prop(2)=0.0025;                     % cross-sectional area
%
%-------------------------
% nodal connectivity
%-------------------------
nodes(1,1)=1; nodes(1,2)=2;
nodes(2,1)=1; nodes(2,2)=3;
nodes(3,1)=2; nodes(3,2)=3;
nodes(4,1)=2; nodes(4,2)=4;
nodes(5,1)=3; nodes(5,2)=4;
nodes(6,1)=3; nodes(6,2)=5;
nodes(7,1)=4; nodes(7,2)=5;
nodes(8,1)=4; nodes(8,2)=6;
nodes(9,1)=5; nodes(9,2)=6;
%
%-------------------------
% applied constraints
%-------------------------
bcdof(1)=1;                % 1st dof (horizontal displ) is constrained
```

Section 7.4   MATLAB Application to Static Analysis

```
bcval(1)=0;                                 % whose described value is 0
bcdof(2)=2;                                 % 2nd dof (vertical displ) is constrained
bcval(2)=0;                                 % whose described value is 0
bcdof(3)=12;                                % 12th dof (vertical displ) is constrained
bcval(3)=0;                                 % whose described value is 0
%
%---------------------
% initialization to zero
%---------------------
ff=zeros(sdof,1);                           % system force vector
kk=zeros(sdof,sdof);                        % system stiffness matrix
index=zeros(nnel*ndof,1);                   % index vector
elforce=zeros(nnel*ndof,1);                 % element force vector
eldisp=zeros(nnel*ndof,1);                  % element nodal displacement vector
k=zeros(nnel*ndof,nnel*ndof);               % element stiffness matrix
stress=zeros(nel,1);                        % stress vector for every element
%
%---------------------
% applied nodal force
%---------------------
ff(8)=-600;                                 % 4th node has 600 N in downward direction
ff(9)=200;                                  % 5th node has 200 N in r.h.s. direction
%
%---------------------
% loop for elements
%---------------------
for iel=1:nel                               % loop for the total number of elements
%
nd(1)=nodes(iel,1);                         % 1st connected node for the (iel)-th element
nd(2)=nodes(iel,2);                         % 2nd connected node for the (iel)-th element
%
x1=gcoord(nd(1),1); y1=gcoord(nd(1),2);     % coordinate of 1st node
x2=gcoord(nd(2),1); y2=gcoord(nd(2),2);     % coordinate of 2nd node
%
leng=sqrt((x2-x1)^2+(y2-y1)^2);             % element length
%
if (x2-x1)==0;
if y2>y1;
beta=2*atan(1);                             % angle between local and global axes
else
beta=-2*atan(1);
end
else
beta=atan((y2-y1)/(x2-x1));
end
%
el=prop(1);                                 % extract elastic modulus
area=prop(2);                               % extract cross-sectional area
```

```
%
index=feeldof(nd,nnel,ndof);        % extract system dofs for the element
%
k=fetruss2(el,leng,area,0,beta,1);  % compute element matrix
%
kk=feasmbl1(kk,k,index);            % assemble into system matrix
%
end
%
%----------------------------------------
% apply constraints and solve the matrix
%----------------------------------------
[kk,ff]=feaplyc2(kk,ff,bcdof,bcval); % apply boundary conditions
%
disp=kk\ff;       % solve matrix equation to find nodal displacements
%
%----------------------------------------
% post computation for stress calculation
%----------------------------------------
%
for iel=1:nel                       % loop for the total number of elements
%
nd(1)=nodes(iel,1);    % 1st connected node for the (iel)-th element
nd(2)=nodes(iel,2);    % 2nd connected node for the (iel)-th element
%
x1=gcoord(nd(1),1); y1=gcoord(nd(1),2);   % coordinate of 1st node
x2=gcoord(nd(2),1); y2=gcoord(nd(2),2);   % coordinate of 2nd node
%
leng=sqrt((x2-x1)^2+(y2-y1)^2);           % element length
%
if (x2-x1)==0;
beta=2*atan(1);               % angle between local and global axes
else
beta=atan((y2-y1)/(x2-x1));
end
%
el=prop(1);                           % extract elastic modulus
area=prop(2);                         % extract cross-sectional area
%
index=feeldof(nd,nnel,ndof);          % extract system dofs for the element
%
k=fetruss2(el,leng,area,0,beta,1);    % compute element matrix
%
for i=1:(nnel*ndof)                   % extract displacements associated with
eldisp(i)=disp(index(i));             % (iel)-th element
end
%
elforce=k*eldisp;                     % element force vector
```

Section 7.4 MATLAB Application to Static Analysis

```
stress(iel)=sqrt(elforce(1)^2+elforce(2)^2)/area;        % stress
%
if ((x2-x1)*elforce(3)) < 0;              % check if tension or compression
stress(iel)=-stress(iel);
end
%
end
%
%————————————————
% print fem solutions
%————————————————
num=1:1:sdof;
displ=[num' disp]                         % print displacements
%
numm=1:1:nel;
stresses=[numm' stress]                   % print stresses
%
%————————————————————————————
```

The nodal displacements and stresses of members are shown below.

```
displ =
dofs      displacement
1.0000     0.0000e-0
2.0000     0.0000e-5
3.0000     0.3200e-5
4.0000    -1.5700e-5
5.0000     0.8650e-5
6.0000    -1.5700e-5
7.0000     0.6400e-5
8.0000    -2.2867e-5
9.0000     0.5450e-5
10.000    -2.0167e-5
11.000     1.1200e-5
12.000     0.0000e-5

stresses =
element   stress
1.0000    160000
2.0000   -100000
3.0000    000000
4.0000    160000
5.0001    100000
6.0001   -160000
7.0001    180000
8.0001    240000
9.0001   -300000
```

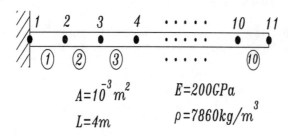

$A = 10^{-3} \, m^2$ $E = 200 GPa$
$L = 4m$ $\rho = 7860 kg/m^3$

Figure 7.5.1 Finite Element Discretization

7.5 MATLAB Application to Eigenvalue Analysis

Once the system mass and stiffness matrices are computed for the truss structure, the matrix equation becomes

$$[M]\{\ddot{u}\} + [K]\{u\} = 0 \qquad (7.5.1)$$

In order to compute natural frequencies of the structure, we assume a harmonic motion for the displacement. The resultant equation is the eigenvalue problem given as

$$([K] - \omega^2[M])\{\bar{u}\} = 0 \qquad (7.5.2)$$

where ω is the circular natural frequency and $\{\bar{u}\}$ is the vector for mode shape.

♣ **Example 7.5.1** Determine the natural frequency of a free bar using the finite element method. The bar is shown in Fig. 7.5.1 and it has elastic modulus of 200 GPa, cross-sectional area of 0.001 m², and density of 7860 Kg/m³.

```
%─────────────────────────────────────────────
% Example 7.5.1
% to solve natural frequency of 1-D bar structure
%
% Problem description
% Find the natural frequency of a bar structure
% as shown in Fig. 7.5.1.
%
% Variable descriptions
% k = element stiffness matrix
% m = element mass matrix
% kk = system stiffness matrix
% mm = system mass vector
% index = a vector containing system dofs associated with each element
% gcoord = global coordinate matrix
% prop = element property matrix
% nodes = nodal connectivity matrix for each element
```

```
% bcdof = a vector containing dofs associated with boundary conditions
% bcval = a vector containing boundary condition values associated with
%                the dofs in bcdof
%----------------------------------------------------------------
%
%----------------------------
% control input data
%----------------------------
nel=4;                                    % number of elements
nnel=2;                                   % number of nodes per element
ndof=1;                                   % number of dofs per node
nnode=5;                                  % total number of nodes in system
sdof=nnode*ndof;                          % total system dofs
%
%----------------------------
% nodal coordinates
%----------------------------
gcoord(1,1)=0.0;
gcoord(2,1)=1.0;
gcoord(3,1)=2.0;
gcoord(4,1)=3.0;
gcoord(5,1)=4.0;
%
%--------------------------------------------
% material and geometric properties
%--------------------------------------------
prop(1)=200e9;                            % elastic modulus
prop(2)=0.001;                            % cross-sectional area
prop(3)=7860;                             % density
%
%----------------------------
% nodal connectivity
%----------------------------
nodes(1,1)=1; nodes(1,2)=2;
nodes(2,1)=2; nodes(2,2)=3;
nodes(3,1)=3; nodes(3,2)=4;
nodes(4,1)=4; nodes(4,2)=5;
%
%----------------------------
% initialization to zero
%----------------------------
kk=zeros(sdof,sdof);                      % system stiffness matrix
mm=zeros(sdof,sdof);                      % system mass matrix
index=zeros(nnel*ndof,1);                 % index vector
%
%----------------------------
% loop for elements
%----------------------------
```

```
for iel=1:nel                        % loop for the total number of elements
%
nd(1)=nodes(iel,1);                  % 1st connected node for the (iel)-th element
nd(2)=nodes(iel,2);                  % 2nd connected node for the (iel)-th element
%
x1=gcoord(nd(1),1);                          % coordinate of 1st node
x2=gcoord(nd(2),1);                          % coordinate of 2nd node
%
leng=(x2-x1);                                      % element length
%
el=prop(1);                                   % extract elastic modulus
area=prop(2);                              % extract cross-sectional area
rho=prop(3);                                   % extract mass density
%
index=feeldof(nd,nnel,ndof);         % extract system dofs for the element
%
ipt=1;                                % flag for consistent mass matrix
[k,m]=fetruss1(el,leng,area,rho,ipt);                % element matrix
%
kk=feasmbl1(kk,k,index);             % assemble system stiffness matrix
mm=feasmbl1(mm,m,index);               % assemble system mass matrix
%
end
%
%--------------------
% solve for eigenvalues
%--------------------
fsol=eig(kk,mm);
fsol=sqrt(fsol);
%
%--------------------
% print fem solutions
%--------------------
num=1:1:sdof;
freqcy=[num' fsol]                            % print natural frequency
%
%--------------------------------------------------------------------


function [k,m]=fetruss1(el,leng,area,rho,ipt)
% Purpose:
% Stiffness and mass matrices for the 1-D truss element
% nodal dof { u_1 u_2 }
%
% Synopsis:
% [k,m]=fetruss1(el,leng,area,rho,ipt)
%
```

```
% Variable Description:
% k - element stiffness matrix (size 4x4)
% m - element mass matrix (size 4x4)
% el - elastic modulus
% leng - element length
% area - area of truss cross-section
% rho - mass density (mass per unit volume)
% ipt = 1 - consistent mass matrix
% ipt = 2 - lumped mass matrix
%----------------------------------------------------------------
%
% stiffness matrix
%
k= (area*el/leng)*[ 1    -1;...
                   -1     1];
%
% consistent mass matrix
%
if ipt==1
%
m=(rho*area*leng/6)*[ 2    1;...
                      1    2];
%
% lumped mass matrix
%
else
%
m=(rho*area*leng/2)*[ 1    0;...
                      0    1];
%
end
%
%----------------------------------------------------------------
```

The natural frequencies are computed from the finite element analysis and compared to the exact solution.

```
freqcy =
mode     nat. freq.
1.00     0.0000                % exact 0.0000
2.00     4060.0                % exact 3962.0
3.00     8737.0                % exact 7924.0
4.00     14198.                % exact 11895.
5.00     17474.                % exact 15847.
```

♣ **Example 7.5.2** We want to find the natural frequency of the truss structure shown in Fig. 7.4.2. Each member has density of 7860 Kg/m^3.

```
%------------------------------------------------------------
% Example 7.5.2
% to solve natural frequency of 2-D truss structure
%
% Problem description
%   Find the natural frequency of a truss structure
%   as shown in Fig. 7.4.2.
%
% Variable descriptions
%   k = element stiffness matrix
%   m = element mass matrix
%   kk = system stiffness matrix
%   mm = system mass vector
%   index = a vector containing system dofs associated with each element
%   gcoord = global coordinate matrix
%   prop = element property matrix
%   nodes = nodal connectivity matrix for each element
%   bcdof = a vector containing dofs associated with boundary conditions
%   bcval = a vector containing boundary condition values associated with
%           the dofs in 'bcdof'
%------------------------------------------------------------
%
%--------------------------
% control input data
%--------------------------
nel=9;                        % number of elements
nnel=2;                       % number of nodes per element
ndof=2;                       % number of dofs per node
nnode=6;                      % total number of nodes in system
sdof=nnode*ndof;              % total system dofs
%
%--------------------------
% nodal coordinates
%--------------------------
gcoord(1,1)=0.0;  gcoord(1,2)=0.0;
gcoord(2,1)=4.0;  gcoord(2,2)=0.0;
gcoord(3,1)=4.0;  gcoord(3,2)=3.0;
gcoord(4,1)=8.0;  gcoord(4,2)=0.0;
gcoord(5,1)=8.0;  gcoord(5,2)=3.0;
gcoord(6,1)=12.; gcoord(6,2)=0.0;
%
%--------------------------------------
% material and geometric properties
%--------------------------------------
prop(1)=200e9;                % elastic modulus
```

```
prop(2)=0.0025;                          % cross-sectional area
prop(3)=7860;                            % density
%
%----------------------
% nodal connectivity
%----------------------
nodes(1,1)=1; nodes(1,2)=2;
nodes(2,1)=1; nodes(2,2)=3;
nodes(3,1)=2; nodes(3,2)=3;
nodes(4,1)=2; nodes(4,2)=4;
nodes(5,1)=3; nodes(5,2)=4;
nodes(6,1)=3; nodes(6,2)=5;
nodes(7,1)=4; nodes(7,2)=5;
nodes(8,1)=4; nodes(8,2)=6;
nodes(9,1)=5; nodes(9,2)=6;
%
%----------------------
% applied constraints
%----------------------
bcdof(1)=1;                              % 1st dof (horizontal displ) is constrained
bcval(1)=0;                              % whose described value is 0
bcdof(2)=2;                              % 2nd dof (vertical displ) is constrained
bcval(2)=0;                              % whose described value is 0
bcdof(3)=12;                             % 12th dof (vertical displ) is constrained
bcval(3)=0;                              % whose described value is 0
%
%----------------------
% initialization to zero
%----------------------
kk=zeros(sdof,sdof);                     % system stiffness matrix
mm=zeros(sdof,sdof);                     % system mass matrix
index=zeros(nnel*ndof,1);                % index vector
%
%----------------------
% loop for elements
%----------------------
for iel=1:nel                            % loop for the total number of elements
%
nd(1)=nodes(iel,1);                      % 1st connected node for the (iel)-th element
nd(2)=nodes(iel,2);                      % 2nd connected node for the (iel)-th element
%
x1=gcoord(nd(1),1); y1=gcoord(nd(1),2);  % coordinate of 1st node
x2=gcoord(nd(2),1); y2=gcoord(nd(2),2);  % coordinate of 2nd node
%
leng=sqrt((x2-x1)^2+(y2-y1)^2);          % element length
%
if (x2-x1)==0;
if y2>y1;
```

```
        beta=2*atan(1);                   % angle between local and global axes
      else
        beta=-2*atan(1);
      end
    else
      beta=atan((y2-y1)/(x2-x1));
    end
    %
    el=prop(1);                           % extract elastic modulus
    area=prop(2);                         % extract cross-sectional area
    rho=prop(3);                          % extract mass density
    %
    index=feeldof(nd,nnel,ndof);          % extract system dofs for the element
    %
    ipt=1;                                % flag for consistent mass matrix
    [k,m]=fetruss2(el,leng,area,rho,beta,ipt);     % element matrix
    %
    kk=feasmbl1(kk,k,index);              % assemble system stiffness matrix
    mm=feasmbl1(mm,m,index);              % assemble system mass matrix
    %
    end
    %
    %----------------------------
    % apply constraints and solve
    %----------------------------
    [kk,mm]=feaplycs(kk,mm,bcdof);        % apply the boundary conditions
    %
    fsol=eig(kk,mm);
    fsol=sqrt(fsol);
    %
    %--------------------
    % print fem solutions
    %--------------------
    num=1:1:sdof;
    freqcy=[num' fsol]                    % print natural frequency
    %
    %-----------------------------------------------------------------


    function [kk,mm]=feaplycs(kk,mm,bcdof)
    %-----------------------------------------------------------------
    % Purpose:
    % Apply constraints to eigenvalue matrix equation
    %    [kk]x=lambda[mm]x
    %
    % Synopsis:
    %    [kk,mm]=feaplycs(kk,mm,bcdof)
```

```
%
% Variable Description:
% kk - system stiffness matrix before applying constraints
% mm - system mass matrix before applying constraints
% bcdof - a vector containing constrained dof
%
% Notes:
% This program does not reduce the matrix size depending on
% the number of constraints. Instead the system matrix size
% is preserved regardless of constraints. As a result, the
% matrix obtained after applying the constraints contains fictitious
% zero eigenvalues, as many as the number of constraints, in
% addition to actual eigenvalues. Users neglect the zero
% fictitious eigenvalues from the results.
%----------------------------------------
%
n=length(bcdof);
sdof=size(kk);
%
for i=1:n
c=bcdof(i);
for j=1:sdof
kk(c,j)=0;
kk(j,c)=0;
mm(c,j)=0;
mm(j,c)=0;
end
%
mm(c,c)=1;
end
%
%----------------------------------------
```

The first five natural frequencies of the truss structure are provided below.

1st frequency = 240.9 rad/s
2nd frequency = 467.9 rad/s
3rd frequency = 739.8 rad/s
4th frequency = 1243. rad/s
5th frequency = 1633. rad/s

7.6 MATLAB Application to Transient Analysis

The dynamic equation of motion for the truss structure is

$$[M]\{\ddot{d}\}^t + [K]\{d\}^t = \{F\}^t \qquad (7.6.1)$$

with prescribed initial conditions which are usually initial displacements and initial velocities. We apply the central difference technique for time integration of Eq. (7.6.1). The details of the techniques are described in Sec. 8.11 and are omitted here. In particular, the summed form of the central difference technique is used for the following examples.

♣ **Example 7.6.1**   A bar is fixed at the left end and it is subjected to a step function of magnitude 200 N (see Fig. 7.5.1). The bar has elastic modulus of 200 GPa, cross-sectional area of 0.001 m^2, and density of 7860 Kg/m^3. It is initially at rest. The MATLAB program is shown below.

```
%----------------------------------------------------------------
% Example 7.6.1
% to solve transient response of 1-D bar structure
%
% Problem description
% Find the dynamic behavior of a bar structure,
% as shown in Fig. 7.5.1, subjected to a step
% force function at the right end.
%
% Variable descriptions
% k = element stiffness matrix
% m = element mass matrix
% kk = system stiffness matrix
% mm = system mass vector
% ff = system force vector
% index = a vector containing system dofs associated with each element
% gcoord = global coordinate matrix
% prop = element property matrix
% nodes = nodal connectivity matrix for each element
% bcdof = a vector containing dofs associated with boundary conditions
% bcval = a vector containing boundary condition values associated with
%              the dofs in bcdof
%----------------------------------------------------------------
%
%---------------------------
% control input data
%---------------------------
nel=10;                    % number of elements
nnel=2;                    % number of nodes per element
ndof=1;                    % number of dofs per node
```

```
nnode=11;                              % total number of nodes in system
sdof=nnode*ndof;                       % total system dofs
dt=0.0001;                             % time step size
ti=0;                                  % initial time
tf=0.05;                               % final time
nt=fix((tf-ti)/dt);                    % number of time steps
%
%------------------------
% nodal coordinates
%------------------------
gcoord(1,1)=0.0;
gcoord(2,1)=1.0;
gcoord(3,1)=2.0;
gcoord(4,1)=3.0;
gcoord(5,1)=4.0;
gcoord(6,1)=5.0;
gcoord(7,1)=6.0;
gcoord(8,1)=7.0;
gcoord(9,1)=8.0;
gcoord(10,1)=9.0;
gcoord(11,1)=10.0;
%
%----------------------------------------
% material and geometric properties
%----------------------------------------
prop(1)=200e9;                         % elastic modulus
prop(2)=0.001;                         % cross-sectional area
prop(3)=7860;                          % density
%
%------------------------
% nodal connectivity
%------------------------
nodes(1,1)=1;  nodes(1,2)=2;
nodes(2,1)=2;  nodes(2,2)=3;
nodes(3,1)=3;  nodes(3,2)=4;
nodes(4,1)=4;  nodes(4,2)=5;
nodes(5,1)=5;  nodes(5,2)=6;
nodes(6,1)=6;  nodes(6,2)=7;
nodes(7,1)=7;  nodes(7,2)=8;
nodes(8,1)=8;  nodes(8,2)=9;
nodes(9,1)=9;  nodes(9,2)=10;
nodes(10,1)=10; nodes(10,2)=11;
%
%------------------------
% applied constraints
%------------------------
nbc=1;                                 % number of constraints
bcdof(1)=1;                            % 1st dof is constrained
```

```
%
%----------------------
% initialization to zero
%----------------------
kk=zeros(sdof,sdof);                    % system stiffness matrix
mm=zeros(sdof,sdof);                    % system mass matrix
ff=zeros(sdof,1);                       % system force vector
index=zeros(nnel*ndof,1);               % index vector
acc=zeros(sdof,nt);                     % acceleration matrix
vel=zeros(sdof,nt);                     % velocity matrix
disp=zeros(sdof,nt);                    % displacement matrix
%
%----------------------
% loop for elements
%----------------------
for iel=1:nel                           % loop for the total number of elements
%
nd(1)=nodes(iel,1);                     % 1st connected node for the (iel)-th element
nd(2)=nodes(iel,2);                     % 2nd connected node for the (iel)-th element
%
x1=gcoord(nd(1),1);                     % coordinate of 1st node
x2=gcoord(nd(2),1);                     % coordinate of 2nd node
%
leng=(x2-x1);                           % element length
%
el=prop(1);                             % extract elastic modulus
area=prop(2);                           % extract cross-sectional area
rho=prop(3);                            % extract mass density
%
index=feeldof(nd,nnel,ndof);            % extract system dofs for the element
%
ipt=1;                                  % flag for consistent mass matrix
[k,m]=fetruss1(el,leng,area,rho,ipt);   % element matrix
%
kk=feasmbl1(kk,k,index);                % assemble system stiffness matrix
mm=feasmbl1(mm,m,index);                % assemble system mass matrix
%
end
%
%----------------------
% initial condition
%----------------------
vel(:,1)=zeros(sdof,1);                 % initial zero velocity
disp(:,1)=zeros(sdof,1);                % initial zero displacement
%
ff(11)=200;                             % step force at node 11
%
%----------------------
```

Section 7.6   MATLAB Application to Transient Analysis

```
% central difference scheme for time integration
%————————————————————————
mm=inv(mm);                                    % invert the mass matrix
%
for it=1:nt
%
acc(:,it)=mm*(ff-kk*disp(:,it));               % compute acceleration
%
for i=1:nbc
ibc=bcdof(i);                                  % apply constraints
acc(ibc,it)=0;
end
%
vel(:,it+1)=vel(:,it)+acc(:,it)*dt;            % compute velocity
disp(:,it+1)=disp(:,it)+vel(:,it+1)*dt;        % compute displacement
%
end
%
acc(:,nt+1)=mm*(ff-kk*disp(:,nt+1));           % acceleration at last step
%
time=0:dt:nt*dt;
plot(time,disp(11,:))
xlabel('Time(seconds)')
ylabel('Tip displ. (m)')
%
%————————————————————————
```

The tip displacement at the right end (i.e. node 11) is plotted in Fig. 7.6.1. as a function of time. As expected, the displacement has oscillation about the static displacement.   ‡

♣ **Example 7.6.2**   Find the transient response of the truss structure as shown in Fig. 7.4.2. The structure has the same geometric and material data as those given Example 7.5.2. However, the load is applied at node 5 in the upward direction as a step function. The load magnitude is 200 N. The response of the same node where the load is applied is plotted in Fig. 7.6.2.

```
%————————————————————————
% Example 7.6.2
% to solve transient response of 2-D truss structure
%
% Problem description
% Find the dynamic behavior of a truss structure,
% as shown in Fig. 7.4.2, subjected to a step
% force function at node 5 in the upward direction.
%
```

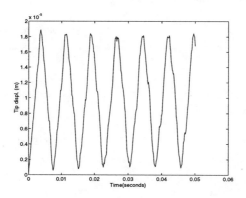

Figure 7.6.1 Time History of the Tip Displacement

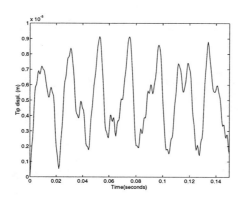

Figure 7.6.2 Time History of Node 5

```
% Variable descriptions
% k = element stiffness matrix
% m = element mass matrix
% kk = system stiffness matrix
% mm = system mass vector
% ff = system force vector
% index = a vector containing system dofs associated with each element
% gcoord = global coordinate matrix
% prop = element property matrix
% nodes = nodal connectivity matrix for each element
% bcdof = a vector containing dofs associated with boundary conditions
% bcval = a vector containing boundary condition values associated with
%             the dofs in bcdof
%----------------------------------------------------------------%
%-----------------------------
```

```
% control input data
%------------------
nel=9;                              % number of elements
nnel=2;                             % number of nodes per element
ndof=2;                             % number of dofs per node
nnode=6;                            % total number of nodes in system
sdof=nnode*ndof;                    % total system dofs
dt=0.0005;                          % time step size
ti=0; tf=0.15;                      % initial and final times
nt=fix((tf-ti)/dt);                 % number of time steps
%
%------------------
% nodal coordinates
%------------------
gcoord(1,1)=0.0; gcoord(1,2)=0.0;
gcoord(2,1)=4.0; gcoord(2,2)=0.0;
gcoord(3,1)=4.0; gcoord(3,2)=3.0;
gcoord(4,1)=8.0; gcoord(4,2)=0.0;
gcoord(5,1)=8.0; gcoord(5,2)=3.0;
gcoord(6,1)=12.; gcoord(6,2)=0.0;
%
%----------------------------------
% material and geometric properties
%----------------------------------
prop(1)=200e9;                      % elastic modulus
prop(2)=0.0025;                     % cross-sectional area
prop(3)=7860;                       % density
%
%------------------
% nodal connectivity
%------------------
nodes(1,1)=1; nodes(1,2)=2;
nodes(2,1)=1; nodes(2,2)=3;
nodes(3,1)=2; nodes(3,2)=3;
nodes(4,1)=2; nodes(4,2)=4;
nodes(5,1)=3; nodes(5,2)=4;
nodes(6,1)=3; nodes(6,2)=5;
nodes(7,1)=4; nodes(7,2)=5;
nodes(8,1)=4; nodes(8,2)=6;
nodes(9,1)=5; nodes(9,2)=6;
%
%-------------------
% applied constraints
%-------------------
nbc=3;                              % number of constraints
bcdof(1)=1;                         % 1st dof (horizontal displ) is constrained
bcval(1)=0;                         % whose described value is 0
bcdof(2)=2;                         % 2nd dof (vertical displ) is constrained
```

```
bcval(2)=0;                           % whose described value is 0
bcdof(3)=12;                          % 12th dof (horizontal displ) is constrained
bcval(3)=0;                           % whose described value is 0
%
%----------------
% initialization to zero
%----------------
kk=zeros(sdof,sdof);                  % system stiffness matrix
mm=zeros(sdof,sdof);                  % system mass matrix
ff=zeros(sdof,1);                     % system force vector
index=zeros(nnel*ndof,1);             % dofs index vector
acc=zeros(sdof,nt);                   % acceleration matrix
vel=zeros(sdof,nt);                   % velocity matrix
disp=zeros(sdof,nt);                  % displacement matrix
%
%----------------
% loop for elements
%----------------
for iel=1:nel                         % loop for the total number of elements
%
nd(1)=nodes(iel,1);                   % 1st connected node for the (iel)-th element
nd(2)=nodes(iel,2);                   % 2nd connected node for the (iel)-th element
%
x1=gcoord(nd(1),1); y1=gcoord(nd(1),2);   % coordinate of 1st node
x2=gcoord(nd(2),1); y2=gcoord(nd(2),2);   % coordinate of 2nd node
%
leng=sqrt((x2-x1)^2+(y2-y1)^2);       % element length
%
if (x2-x1)==0;
if y2>y1;
beta=2*atan(1);                       % angle between local and global axes
else
beta=-2*atan(1);
end
else
beta=atan((y2-y1)/(x2-x1));
end
%
el=prop(1);                           % extract elastic modulus
area=prop(2);                         % extract cross-sectional area
rho=prop(3);                          % extract mass density
%
index=feeldof(nd,nnel,ndof);          % extract system dofs for the element
%
ipt=1;                                % flag for consistent mass matrix
[k,m]=fetruss2(el,leng,area,rho,beta,ipt);   % element matrix
%
kk=feasmbl1(kk,k,index);              % assemble system stiffness matrix
```

```
mm=feasmbl1(mm,m,index);              % assemble system mass matrix
%
end
%
%————————————
% initial condition
%————————————
vel(:,1)=zeros(sdof,1);                % initial zero velocity
disp(:,1)=zeros(sdof,1);               % initial zero displacement
%
ff(10)=200;                            % step force at 10th dof
%
%————————————————————————
% central difference scheme for time integration
%————————————————————————
mm=inv(mm);                            % invert the mass matrix
%
for it=1:nt
%
acc(:,it)=mm*(ff-kk*disp(:,it));       % compute acceleration
%
for i=1:nbc
ibc=bcdof(i);                          % apply constraints
acc(ibc,it)=0;
end
%
vel(:,it+1)=vel(:,it)+acc(:,it)*dt;    % compute velocity
disp(:,it+1)=disp(:,it)+vel(:,it+1)*dt; % compute displacement
%
end
%
acc(:,nt+1)=mm*(ff-kk*disp(:,nt+1));   % acceleration at last step
%
time=0:dt:nt*dt;
plot(time,disp(10,:))                  % displacement plot
xlabel('Time(seconds)')
ylabel('Tip displ. (m)')
%
%————————————————————————————
```

Problems

7.1 (a) Develop the element stiffness matrix for the one-dimensional axial rod using quadratic shape functions. (b) Apply the stiffness matrix to solve an axial member whose one end is fixed and the other end is subjected to a force P. The member has elastic modulus E and cross-sectional area A, respectively. Use one quadratic element to model the axial member. (c) Compare the nodal displacement at the center of the member to the end displacement.

7.2 A telescope shape of an axial member (see Fig. P7.2) is modeled using a single linear element. Derive the element stiffness matrix.

7.3 A taper shape of an axial member (Fig. P7.3) is modeled as a single linear element. Derive the element stiffness matrix.

7.4 Develop the element mass matrix for the one-dimensional axial member using quadratic shape functions.

7.5 Develop the element mass matrix for Prob. 7.2.

7.6 Develop the element mass matrix for Prob. 7.3.

7.7 For the truss structure shown in Fig. P7.7, derive the finite element matrix equation using two elements. Find the displacements and stresses in the member.

7.8 The one-dimensional wave equation for an axial member is given as

$$\frac{\partial^2 u}{\partial t^2} - 2\frac{\partial^2 u}{\partial x^2} = 0$$

The second order equation can be rewritten as

$$\frac{\partial u}{\partial t} - v = 0 \quad \text{and} \quad \frac{\partial v}{\partial t} - 2\frac{\partial^2 u}{\partial x^2} = 0$$

using two first order equations in time. These two equations are solved using one linear finite element and the backward difference method for the time derivative. The member is initially displaced such that $u(x,0)=0.001x$. The left end of the member is held fixed all the time while the right end is released at time 0 from the initial displacement. Find the displacement u and velocity v at the right end at time $t=1$ sec. using a time step size $\Delta t=1$.

7.9 Redo Prob. 7.8 using the central finite difference method for the time derivative. Find the critical time step size for stability.

7.10 Solve the truss structure shown in Fig. P7.10 using the computer programs. Compare the finite element solution to the analytical solution obtained from statics.

7.11 Obtain the natural frequencies of the structures in Fig. P7.10 using the computer programs.

Figure P7.2 Problem 7.2

Figure P7.3 Problem 7.3

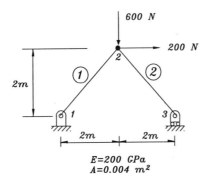

Figure P7.7 Problem 7.7

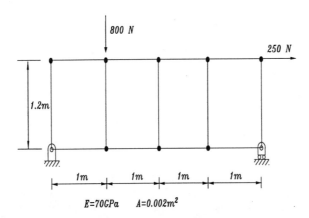

Figure P7.10 Problem 7.10

7.12 If the structure in Fig. P7.10 is initially at rest and the forces are applied suddenly at time 0, determine the dynamic response of the structure using the computer programs.

CHAPTER	EIGHT

BEAM AND FRAME STRUCTURES

8.0 Chapter Overview

Beams and frames are very common structures because of their efficient load-carrying capability. This chapter presents finite element formulations for those structures. In particular, several different beam formulations are discussed because each formulation has its own merit. The presentation starts with the Euler-Bernoulli beam equation. The weighted residual technique is applied to derive mass and stiffness matrices and load vectors. Later a shear deformable beam theory is discussed. These formulations have displacements and rotations as the primary nodal variables. Other formulations including mixed elements, hybrid elements, or elements with only displacements as nodal variables like a continuum element are also explained. Static and dynamic problems are discussed here. With coordinate transformation of the nodal variables, beam elements combined with axial and torsional elements are applied to 2-D and 3-D frame structures. Several types of problems are solved using MATLAB programs. They include static, eigenvalue, transient, modal, and frequency response analyses of beam and frame structures.

8.1 Euler-Bernoulli Beam

The Euler-Bernoulli equation for beam bending is

$$\bar{\rho}\frac{\partial^2 v}{\partial t^2} + \frac{\partial^2}{\partial x^2}\left(EI\frac{\partial^2 v}{\partial x^2}\right) = q(x,t) \tag{8.1.1}$$

where $v(x,t)$ is the transverse displacement of the beam, $\bar{\rho}$ is mass density per length, EI is the beam rigidity, $q(x,t)$ is the externally applied pressure loading, and t and x indicate time and the spatial axis along the beam axis. We apply one of the methods of weighted residual, Galerkin's method, to the beam equation Eq. (8.1.1) to develop the finite element formulation and the corresponding matrix equations.

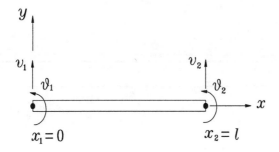

Figure 8.1.1 Two-Noded Beam Element

The averaged weighted residual of Eq. (8.1.1) is

$$I = \int_0^L \left(\bar{\rho}\frac{\partial^2 v}{\partial t^2} + \frac{\partial^2}{\partial x^2}\left(EI\frac{\partial^2 v}{\partial x^2}\right) - q \right) w \, dx = 0 \qquad (8.1.2)$$

where L is the length of the beam and w is a test function. The weak formulation of Eq. (8.1.2) is obtained from integrations by parts twice for the second term of the equation. In addition, discretization of the beam into a number of finite elements gives

$$I = \sum_{i=1}^n \left[\int_{\Omega^e} \bar{\rho}\frac{\partial^2 v}{\partial t^2} w \, dx + \int_{\Omega^e} EI \frac{\partial^2 v}{\partial x^2}\frac{\partial^2 w}{\partial x^2} dx - \int_{\Omega^e} q w \, dx \right] + \left[-Vw - M\frac{\partial w}{\partial x} \right]_0^L = 0 \qquad (8.1.3)$$

where $V = -EI(\partial^3 v/\partial x^3)$ is the shear force, $M = EI(\partial^2 v/\partial x^2)$ is the bending moment, Ω^e is an element domain and n is the number of elements for the beam.

We consider shape functions for spatial interpolation of the transverse deflection, v, in terms of nodal variables. Interpolation in terms of the time domain will be discussed later. To this end, we consider an element which has two nodes, one at each end, as shown in Fig. 8.1.1. The deformation of a beam must have continuous slope as well as continuous deflection at any two neighboring beam elements. To satisfy this continuity requirement each node has both deflection, v_i, and slope, θ_i, as nodal variables. In this case, any two neighboring beam elements have common deflection and slope at the shared nodal point. This satisfies the continuity of both deflection and slope. The Euler-Bernoulli beam equation is based on the assumption that the plane normal to the neutral axis before deformation remains normal to the neutral axis after deformation (see Fig. 8.1.2). This assumption denotes $\theta = dv/dx$ (i.e. slope is the first derivative of deflection in terms of x). Because there are four nodal variables for the beam element, we assume a cubic polynomial function for $v(x)$:

$$v(x) = c_0 + c_1 x + c_2 x^2 + c_3 x^3 \qquad (8.1.4)$$

From the assumption for the Euler-Bernoulli beam, slope is computed from Eq. (8.1.4)

$$\theta(x) = c_1 + 2c_2 x + 3c_3 x^2 \qquad (8.1.5)$$

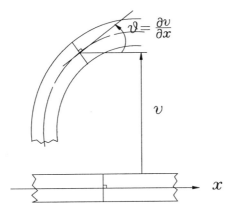

Figure 8.1.2 Euler-Bernoulli Beam

Evaluation of deflection and slope at both nodes yields

$$v(0) = c_0 = v_1$$
$$\theta(0) = c_1 = \theta_1$$
$$v(l) = c_0 + c_1 l + c_2 l^2 + c_3 l^3 = v_2 \quad (8.1.6)$$
$$\theta(l) = c_1 + 2c_2 l + 3c_3 l^2 = \theta_2$$

Solving Eq. (8.1.6) for c_i in terms of the nodal variables v_i and θ_i and substituting the results into Eq. (8.1.4) give

$$v(x) = H_1(x)v_1 + H_2(x)\theta_1 + H_3(x)v_2 + H_4(x)\theta_2 \quad (8.1.7)$$

where

$$H_1(x) = 1 - \frac{3x^2}{l^2} + \frac{2x^3}{l^3}$$
$$H_2(x) = x - \frac{2x^2}{l} + \frac{x^3}{l^2}$$
$$H_3(x) = \frac{3x^2}{l^2} - \frac{2x^3}{l^3} \quad (8.1.8)$$
$$H_4(x) = -\frac{x^2}{l} + \frac{x^3}{l^2}$$

The functions $H_i(x)$ are called *Hermitian* shape functions and shown in Fig. 8.1.3. The *Hermitian* shape functions are of C^1-type which means they make both v and $\partial v/\partial x$ continuous between two neighboring elements. In general, C^n-type continuity means the shape functions have continuity up to the nth order derivative between two neighboring elements.

Application of *Hermitian* shape functions and Galerkin's method to the second term of Eq. (8.1.3) results in the stiffness matrix of the beam element. That is,

$$[K^e] = \int_0^l [B]^T EI [B] dx \quad (8.1.9)$$

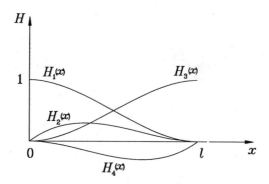

Figure 8.1.3 Hermitian Beam Element

where
$$[B] = \{H_1'' \ H_2'' \ H_3'' \ H_4''\} \tag{8.1.10}$$

and the corresponding element nodal degrees of freedom vector is $\{d^e\} = \{v_1 \ \theta_1 \ v_2 \ \theta_2\}^T$. In Eq. (8.1.10), double prime denotes the second derivative of the function and l in Eq. (8.1.9) is the length of a beam element. Assuming the beam rigidity EI is constant within the element, the element stiffness matrix is

$$[K^e] = \frac{EI}{l^3} \begin{bmatrix} 12 & 6l & -12 & 6l \\ 6l & 4l^2 & -6l & 2l^2 \\ -12 & -6l & 12 & -6l \\ 6l & 2l^2 & -6l & 4l^2 \end{bmatrix} \tag{8.1.11}$$

In case the beam rigidity is not constant within a beam element, the integral in Eq. (8.1.9) must be evaluated including EI as a function of x. If the beam element is relatively short, for example in a refined mesh, the average value of EI for the element may be used with Eq. (8.1.11) for a simple and reasonable approximation.

The third term in Eq. (8.1.3) results in the element force vector. For a generally distributed pressure loading, we need to compute

$$\{F^e\} = \int_0^l q(x) \begin{Bmatrix} H_1 \\ H_2 \\ H_3 \\ H_4 \end{Bmatrix} dx \tag{8.1.12}$$

in which $\{F^e\}$ is the element force vector. If we have a uniform pressure load q_0 within the element, the element force vector becomes

$$\{F^e\} = q_0 \int_0^l \begin{Bmatrix} H_1 \\ H_2 \\ H_3 \\ H_4 \end{Bmatrix} dx = \frac{q_0}{12} \begin{Bmatrix} 6l \\ l^2 \\ 6l \\ -l^2 \end{Bmatrix} \tag{8.1.13}$$

Another common load type is a concentrated force within a beam element as shown in Fig. 8.1.4. In this case, the element force vector is

$$\{F^e\} = \int_0^l P_0 \delta(x - x_0) \begin{Bmatrix} H_1 \\ H_2 \\ H_3 \\ H_4 \end{Bmatrix} dx = P_0 \begin{Bmatrix} H_1(x_0) \\ H_2(x_0) \\ H_3(x_0) \\ H_4(x_0) \end{Bmatrix} \tag{8.1.14}$$

where P_0 is the concentrated force applied at $x = x_0$ and $\delta(x - x_0)$ is the Dirac delta function. Element force vectors for some other cases are summarized in Fig. 8.1.4.

The last term in Eq. (8.1.3) represents the boundary conditions of shear force and bending moment at the two boundary points, $x = 0$ and $x = L$, of the beam. If these boundary conditions are known, the known shear force and/or bending moment are included in the system force vector at the two boundary nodes. Otherwise, they remain as unknowns. However, deflection and/or slope are known as geometric boundary conditions for this case. For static bending analyses of beams, the first term in Eq. (8.1.3), which is the inertia force term, is neglected. As a result, assembling the element stiffness matrices and vectors results in the system matrix equation given below:

$$[K]\{d\} = \{F\} \tag{8.1.15}$$

Given boundary conditions are applied to Eq. (8.1.15) and the matrix equations are solved for the unknown nodal variables, deflections, and slopes. An example is given in Example 8.1.1.

♣ **Example 8.1.1** Consider a cantilever beam subjected to a tip load as shown in Fig. 8.1.5. Let us use one *Hermitian* beam element to solve the tip deflection. In this case, the element stiffness matrix is the same as the system stiffness matrix. The resultant element matrix equation is:

$$\frac{EI}{L^3} \begin{bmatrix} 12 & 6L & -12 & 6L \\ 6L & 4L^2 & -6L & 2L^2 \\ -12 & -6L & 12 & -6L \\ 6L & 2L^2 & -6L & 4L^2 \end{bmatrix} \begin{Bmatrix} v_1 \\ \theta_1 \\ v_2 \\ \theta_2 \end{Bmatrix} = \begin{Bmatrix} V_1 \\ M_1 \\ -P \\ 0 \end{Bmatrix} \tag{8.1.16}$$

In this equation, V_1 and M_1 are unknown reactions at the clamped support. The minus sign indicates the tip force is applied to the opposite direction to the deflection. The boundary conditions prescribed in terms of nodal variables are $v_1 = 0$ and $\theta_1 = 0$. Applying these conditions to Eq. (8.1.16) as described in the previous chapter and solving the resultant equation yield

$$v_2 = -\frac{PL^3}{3EI} \tag{8.1.17}$$

which is the exact solution for the cantilever beam. The reason the finite element analysis with one element results in the exact solution is the following. The *Hermitian* shape functions are based on a general cubic polynomial as seen in Eq. (8.1.4). The exact solution for the cantilever beam with a tip force is also

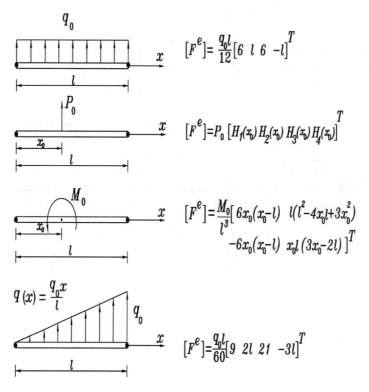

Figure 8.1.4 Element Force Vectors for Various Pressure Loads

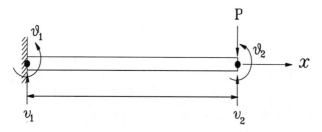

Figure 8.1.5 Cantilever Beam

a cubic function. As a result, the *Hermitian* shape functions can result in the exact solution. ‡

For dynamic analyses of beams, the inertia force needs to be included. In this case, the transverse deflection is a function of x and t. The deflection is interpolated within a beam element as given below:

$$v(x,t) = H_1(x)v_1(t) + H_2(x)\theta_1(t) + H_3(x)v_2(t) + H_4(x)\theta_2(t) \qquad (8.1.18)$$

Equation (8.1.18) states that the shape functions are used to interpolate the deflection in terms of the spatial domain and the nodal variables are functions of time. The first term in Eq. (8.1.3) becomes

$$\int_0^l \rho[H]^T[H]dx\{\ddot{d}^e\} \tag{8.1.19}$$

where

$$[H] = [H_1 \ H_2 \ H_3 \ H_4] \tag{8.1.20}$$

and the superimposed dot denotes temporal derivative. From Eq. (8.1.19) and $\bar{\rho} = \rho A$, the element mass matrix becomes

$$[M^e] = \int_0^l \rho A[H]^T[H]dx$$

$$= \frac{\rho A l}{420} \begin{bmatrix} 156 & 22l & 54 & -13l \\ 22l & 4l^2 & 13l & -3l^2 \\ 54 & 13l & 156 & -22l \\ -13l & -3l^2 & -22l & 4l^2 \end{bmatrix} \tag{8.1.21}$$

where ρ is the mass density per volume and Eq. (8.1.21) is called the *consistent* mass matrix. Archer [11,12] is credited for the first development of the *consistent* mass matrix. Adding the components in the mass matrix, which are associated with only the displacement nodal variables (i.e. v_1 and v_2), yields $\rho A l$, total mass of the beam element. The beam element conserves the mass in terms of its translational degrees of freedom.

In the dynamic analysis, the system mass matrix is usually required to be inverted. From this aspect, a diagonalized mass matrix has a computational advantage. One such matrix is

$$[M^e] = \frac{\rho A l}{2} \begin{bmatrix} 1 & 0 & 0 & 0 \\ 0 & 0 & 0 & 0 \\ 0 & 0 & 1 & 0 \\ 0 & 0 & 0 & 0 \end{bmatrix} \tag{8.1.22}$$

This matrix is called the *lumped* mass matrix, which was developed earlier than the *consistent* mass matrix. This matrix has half of the element mass at each translational nodal degree of freedom. Both mass matrices conserve the mass associated with their translational degrees of freedom.

Another way to develop a diagonalized mass matrix from the *consistent* mass matrix is summarized below[13].

1. Add the diagonal components of the *consistent* mass matrix associated with the translational degrees of freedom, i.e. the first and third diagonal components for the present beam element. The sum is called α.
2. Divide the diagonal components by α and also multiply them by the element total mass.
3. Set all off-diagonal components to zero.

Applying this procedure to Eq. (8.1.21) results in

$$\frac{\rho A l}{78} \begin{bmatrix} 39 & 0 & 0 & 0 \\ 0 & l^2 & 0 & 0 \\ 0 & 0 & 39 & 0 \\ 0 & 0 & 0 & l^2 \end{bmatrix} \tag{8.1.23}$$

This matrix is called the *diagonal* mass matrix and also conserves the mass for the translational degrees of freedom. Another technique to develop a diagonalized mass matrix is discussed in Refs. [14,15] using numerical quadrature points located only at the nodes.

The element stiffness matrix does not change for the dynamic analysis because the shape functions are the same for both static and dynamic analyses. However, the force term may vary as function of time. The force vector is for the dynamic analysis

$$\{F^e(t)\} = \int_0^l q(x,t)[H]^T dx \tag{8.1.24}$$

Thus, Eq. (8.1.24) is in general different from Eq. (8.1.12). As a result, the matrix equation for a dynamic beam analysis is, after assembly of element matrices and vectors,

$$[M]\{\ddot{d}\} + [K]\{d\} = \{F(t)\} \tag{8.1.25}$$

For free vibration of a beam, the eigenvalue problem is as shown in Ref. [16],

$$([K] - \omega^2[M])\{\bar{d}\} = 0 \tag{8.1.26}$$

where ω is the angular natural frequency in radians per second and $\{\bar{d}\}$ is the mode shape. Example problems for static, dynamic, and eigenvalue analyses of beams are provided at later sections using the MATLAB program.

8.2 Timoshenko Beam

The Timoshenko beam theory includes the effect of transverse shear deformation. As a result, a plane normal to the beam axis before deformation does not remain normal to the beam axis any longer after deformation. Figure 8.2.1 shows the deformation in contrast to that in Fig. 8.1.2. While Galerkin's method was used to derive the finite element matrix equation for the Euler-Bernoulli beam equation, the energy method is used for the present formulation for the Timoshenko beam.

Let u and v be the axial and transverse displacements of a beam, respectively. Because of the transverse shear deformation, the slope of the beam θ is different from dv/dx. Instead, the slope equals $(dv/dx) - \gamma$ where γ is the transverse shear strain. As a result, the displacement field in the Timoshenko beam can be written as

$$u(x,y) = -y\theta(x) \tag{8.2.1}$$

Timoshenko Beam

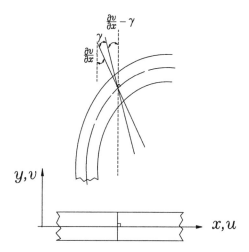

Figure 8.2.1 Timoshenko Beam

$$v(x) = v \tag{8.2.2}$$

where the x-axis is located along the neutral axis of the beam and the beam is not subjected to an axial load such that the neutral axis does not have the axial strain. A beam subjected to both axial and transverse loads is considered in a next section for frame structures. From Eq. (8.2.1) and Eq. (8.2.2), the axial and shear strains are

$$\epsilon = -y\frac{d\theta}{dx} \tag{8.2.3}$$

$$\gamma = -\theta + \frac{dv}{dx} \tag{8.2.4}$$

As explained in the previous chapter on the energy method, the element stiffness matrix can be obtained from the strain energy expression for an element. The strain energy for an element of length l is

$$U = \frac{b}{2}\int_0^l \int_{-h/2}^{h/2} \epsilon^T E \epsilon \, dy \, dx + \frac{b\mu}{2}\int_0^l \int_{-h/2}^{h/2} \gamma^T G \gamma \, dy \, dx \tag{8.2.5}$$

in which the first term is the bending strain energy and the second term is the shear strain energy. Moreover, b and h are the width and height of the beams respectively, and μ is the correction factor for shear energy whose value is normally $\frac{5}{6}$.

First, substituting Eq. (8.2.3) and Eq. (8.2.4) into Eq. (8.2.5) and taking integration with respect to y gives

$$U = \frac{1}{2}\int_0^l \left(\frac{d\theta}{dx}\right)^T EI \left(\frac{d\theta}{dx}\right) dx + \frac{\mu}{2}\int_0^l \left(-\theta + \frac{dv}{dx}\right)^T GA \left(-\theta + \frac{dv}{dx}\right) dx \tag{8.2.6}$$

where I and A are the moment of inertia and area of the beam cross-section.

In order to derive the element stiffness matrix for the Timoshenko beam, the variables v and θ need to be interpolated within each element. As seen in Eq. (8.2.6), v and θ are independent variables. That is, we can interpolate them independently using proper shape functions. This results in satisfaction of inter-element compatibility, i.e. continuity of both the transverse displacement v and slope θ between two neighboring elements. As a result, any kind of C^0 shape functions can be used for the present beam element. Shape functions of order C^0 are much easier to construct than shape functions of order C^1. It is especially very difficult to construct shape functions of order C^1 for two-dimensional and three-dimensional analyses such as the classical plate theory.

We use the simple linear shape functions for both variables. That is,

$$v = [H_1 \ H_2] \begin{Bmatrix} v_1 \\ v_2 \end{Bmatrix} \tag{8.2.7}$$

$$\theta = [H_1 \ H_2] \begin{Bmatrix} \theta_1 \\ \theta_2 \end{Bmatrix} \tag{8.2.8}$$

where H_1 and H_2 are linear shape functions. The linear element looks like that in Fig. 8.1.1, but the shape functions used are totally different from those for the *Hermitian* beam element. Using Eq. (8.2.7) and Eq. (8.2.8) along with the strain energy expression Eq. (8.2.6) yields the following element stiffness matrix for the Timoshenko beam:

$$[K^e] = [K_b^e] + [K_s^e] \tag{8.2.9}$$

where

$$[K_b^e] = \frac{EI}{l} \begin{bmatrix} 0 & 0 & 0 & 0 \\ 0 & 1 & 0 & -1 \\ 0 & 0 & 0 & 0 \\ 0 & -1 & 0 & 1 \end{bmatrix} \tag{8.2.10}$$

$$[K_s^e] = \frac{\mu GA}{4l} \begin{bmatrix} 4 & 2l & -4 & 2l \\ 2l & l^2 & -2l & l^2 \\ -4 & -2l & 4 & -2l \\ 2l & l^2 & -2l & l^2 \end{bmatrix} \tag{8.2.11}$$

One thing to be noted here is that the bending stiffness term, Eq. (8.2.10), is obtained using the exact integration of the bending strain energy but the shear stiffness term, Eq. (8.2.11), is obtained using the reduced integration technique [17,18]. For the present calculation, the one-point Gauss quadrature rule is used as shown in the example given below. The major reason is if the beam thickness becomes so small compared to its length, the shear energy dominates over the bending energy. As seen in Eq. (8.2.10) and Eq. (8.2.11), the bending stiffness is proportional to $\frac{h^3}{l}$ while the transverse shear stiffness is proportional to hl, where h and l are the thickness and length of a beam element, respectively. Hence, as $\frac{h}{l}$ becomes smaller for a very *thin beam*, the bending term becomes negligible compared to the shear term. This is not correct in the physical sense. As the beam becomes thinner, the bending strain energy is more significant than the shear energy. This phenomenon is called *shear locking*.

In order to avoid *shear locking*, the shear strain energy is under-integrated. Because of the under-integration the present beam stiffness matrix is rank deficient. That is, it contains some fictitious rigid body modes (i.e. zero energy modes). Example 8.2.1 shows the computation of the shear stiffness term.

♣ **Example 8.2.1** We use the linear isoparametric element to integrate the shear energy term in Eq. (8.2.6) to produce the shear stiffness matrix Eq. (8.2.11). Using the concept of isoparametric mapping explained in Chapter 6, the shear stiffness term becomes

$$[K_s^e] = \mu G A \int_{-1}^{1} \begin{bmatrix} -1/l \\ -(1-r)/2 \\ 1/l \\ -(1+r)/2 \end{bmatrix} \begin{bmatrix} -\dfrac{1}{l} & -\dfrac{1-r}{2} & \dfrac{1}{l} & -\dfrac{1+r}{2} \end{bmatrix} \dfrac{l}{2} dr \qquad (8.2.12)$$

The expression is a quadratic polynomial in terms of r so that the two-point Gauss quadrature will evaluate the integration exactly. For under-integration of one order less, we use the one-point Gauss quadrature rule. The integration point is 0 and the weight is 2. Applying this to Eq. (8.2.12) results in Eq. (8.2.11). ‡

These kinds of beam elements can be derived for any order of shape functions higher than one. That is, beam elements can have three or more nodes per element depending on the order of shape functions. For each case, the shear stiffness matrix should be under-integrated consistently. The order of integration for the shear stiffness matrix is one less than what is required for exact integration.

The *consistent* mass matrix for the Timoshenko beam is computed from

$$[M^e] = \int_0^l \rho A [N]^T [N] dx \qquad (8.2.13)$$

where

$$[N] = \begin{bmatrix} \dfrac{l-x}{l} & 0 & \dfrac{x}{l} & 0 \end{bmatrix} \qquad (8.2.14)$$

for a linear beam element. This equation results in

$$[M^e] = \dfrac{\rho A l}{6} \begin{bmatrix} 2 & 0 & 1 & 0 \\ 0 & 0 & 0 & 0 \\ 1 & 0 & 2 & 0 \\ 0 & 0 & 0 & 0 \end{bmatrix} \qquad (8.2.15)$$

The same *lumped* and *diagonal* mass matrices as given in Eq. (8.1.22) can be used for the present beam element.

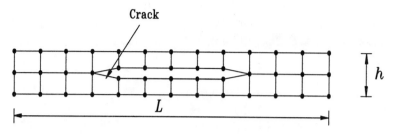

Figure 8.3.1 A Beam with an Imbedded Crack ($L \gg h$)

8.3 Beam Elements with Only Displacement Degrees of Freedom

In this section, we develop a family of beam elements which have only displacements as nodal degrees of freedom and no slope as nodal degrees of freedom [19]. In this aspect, these beam elements are similar to plane stress elements as given in Chapter 9. Therefore, when a beam needs to be discretized along its thickness direction as well as along its axial direction, these beam elements can be easily applied for the mesh. Figure 8.3.1 shows one example which uses stacked beam elements along the beam thickness. If there are multiple embedded cracks in a beam like interlaminar delamination in a laminated composite beam, it may require more beam elements through its thickness. If we plan to use beam elements which have displacements and rotations as nodal degrees of freedom for this application, we need special care at the interface of the neighboring top and bottom beam elements. As seen in Fig. 8.3.2, complicated constraint equations should be applied to maintain the continuous deformation across the interface.

Let us derive a linear beam element with displacement degrees of freedom only, which is the simplest element of this family. The element has six degrees of freedom which are axial and lateral displacements. There are axial displacements at the four corner points and lateral displacements at the two ends of the element as seen in Fig. 8.3.3.

In Fig. 8.3.3 and the subsequent formulation, u represents the axial displacements at the corner points and v represents the lateral displacements at the ends. The subscripts '1' and '2' refer to the left and right ends while the superscripts 't' and 'b' indicate the top and bottom sides of the element, respectively.

The displacement field of the element is

$$\bar{u} = \begin{Bmatrix} u(x,y) \\ v(x) \end{Bmatrix} = [N]\{d^e\} \tag{8.3.1}$$

where $[N]$ is the matrix of shape functions and $\{d^e\}$ is a vector of nodal displacements. The axial displacement is assumed to vary linearly along both axial and lateral directions. It can be written as

$$u(x,y) = \sum_{i=1}^{2} N_i(x) \left[H_1(y) u_i^b + H_2(y) u_i^t \right] \tag{8.3.2}$$

$$= N_1(x) H_1(y) u_1^b + N_1(x) H_2(y) u_1^t + N_2(x) H_1(y) u_2^b + N_2(x) H_2(y) u_2^t$$

Section 8.3 Beam Elements with Only Displacement Degrees of Freedom   249

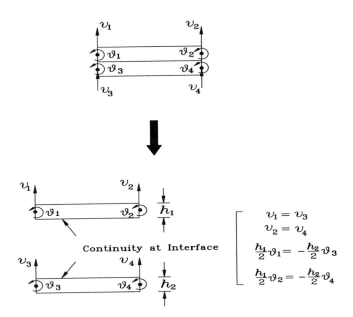

Figure 8.3.2 Continuity Requirements at the Interface for Conventional Beam Elements

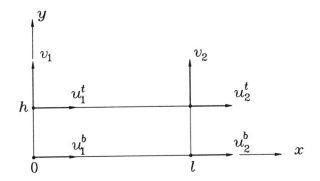

Figure 8.3.3 Four-Noded Beam Element with Six Degrees of Freedom

The lateral displacement, which is assumed to be constant through the thickness of the element, varies linearly along the axial direction and can be written as

$$v(x) = \sum_{i=1}^{2} N_i(x) v_i$$
$$= N_1(x)v_1 + N_2(x)v_2 \qquad (8.3.3)$$

Here, N_i and H_i are the linear shape functions in the axial and lateral directions. The beam element may use a higher order shape function for the axial direction if there are more nodal points in the axial direction, and the linear shape function for lateral direction. However, in this study, the linear shape function is used for both N_i and

H_i for simplicity. That is,

$$N_1(x) = 1 - \frac{x}{l}$$
$$N_2(x) = \frac{x}{l}$$
$$N_3(y) = 1 - \frac{y}{h}$$
$$N_4(y) = \frac{y}{h}$$
(8.3.4)

where l and h are the length and height of the beam element, respectively, if the beam is a rectangular shape.

For simplicity, notations N_1, N_2, H_1 and H_2 will be used instead of $N_1(x)$, $N_2(x)$, $H_1(y)$, and $H_2(y)$ in the following derivation. Axial normal strain can be written as

$$\epsilon_x = \frac{\partial u}{\partial x} = \frac{\partial N_1}{\partial x} H_1 u_1^b + \frac{\partial N_1}{\partial x} H_2 u_1^t + \frac{\partial N_2}{\partial x} H_1 u_2^b + \frac{\partial N_2}{\partial x} H_2 u_2^t \qquad (8.3.5)$$

and the shear strain is

$$\gamma_{xy} = \frac{\partial u}{\partial y} + \frac{\partial v}{\partial x} \qquad (8.3.6)$$
$$= \frac{\partial H_1}{\partial y} N_1 u_1^b + \frac{\partial H_2}{\partial y} N_1 u_1^t + \frac{\partial H_1}{\partial y} N_2 u_2^b + \frac{\partial H_2}{\partial y} N_2 u_2^t + \frac{\partial N_1}{\partial x} v_1 + \frac{\partial N_2}{\partial x} v_2$$

The element stiffness matrix can be obtained by minimizing the total strain energy which contains both bending and the transverse shear energy. This minimization yields the following element stiffness matrix

$$[K^e] = [K_b^e] + [K_s^e] \qquad (8.3.7)$$

where the subscripts 'b' and 's' indicate bending and the transverse shear, respectively.

The bending and transverse shear stiffness matrices are given as

$$[K_b^e] = \int_0^l \int_0^h \{B_b\}^T E \{B_b\} \, dy dx \qquad (8.3.8)$$

$$[K_s^e] = \int_0^l \int_0^h \{B_s\}^T G \{B_s\} \, dy dx \qquad (8.3.9)$$

where E and G are the elastic and shear modulii of the beam and the vectors $\{B_b\}$ and $\{B_s\}$ are derived below.

The strain-displacement relationship for the axial strain and shear strain can be written as from Eqs. (8.3.5) and (8.3.6)

$$\epsilon_x = \{B_b\} \{d^e\} \qquad (8.3.10)$$

and

$$\gamma_{xy} = \{B_s\} \{d^e\} \qquad (8.3.11)$$

Section 8.3 Beam Elements with Only Displacement Degrees of Freedom

where

$$\{B_b\} = \left\{ \frac{\partial N_1}{\partial x} H_1 \quad \frac{\partial N_1}{\partial x} H_2 \quad 0 \quad \frac{\partial N_2}{\partial x} H_1 \quad \frac{\partial N_2}{\partial x} H_2 \quad 0 \right\}$$

$$\{B_s\} = \left\{ N_1 \frac{\partial H_1}{\partial y} \quad N_1 \frac{\partial H_2}{\partial y} \quad \frac{\partial N_1}{\partial x} \quad N_2 \frac{\partial H_1}{\partial y} \quad N_2 \frac{\partial H_2}{\partial y} \quad \frac{\partial N_2}{\partial x} \right\} \quad (8.3.12)$$

$$\{d^e\} = \{u_1^b \ u_1^t \ v_1 \ u_2^b \ u_2^t \ v_2\}^T$$

The bending stiffness matrix can be obtained by carrying out the integration in Eq. (8.3.8) which will result in

$$[K_b^e] = \frac{Eh}{6l} \begin{bmatrix} 2 & 1 & 0 & -2 & -1 & 0 \\ 1 & 2 & 0 & -1 & -2 & 0 \\ 0 & 0 & 0 & 0 & 0 & 0 \\ -2 & -1 & 0 & 2 & 1 & 0 \\ -1 & -2 & 0 & 1 & 2 & 0 \\ 0 & 0 & 0 & 0 & 0 & 0 \end{bmatrix} \quad (8.3.13)$$

For Eq. (8.3.9) the reduced integration technique is used along the x-axis to prevent *shear locking* which occurs when the ratio of beam length to beam thickness is large. That is, one-point Gauss quadrature is used for integration in the x-direction. This integration yields the transverse shear stiffness matrix of the form

$$[K_s^e] = \frac{1}{4lh} \begin{bmatrix} Gl^2 & -Gl^2 & 2Glh & Gl^2 & -Gl^2 & -2Glh \\ -Gl^2 & Gl^2 & -2Glh & -Gl^2 & Gl^2 & 2Glh \\ 2Glh & -2Glh & 4Gh^2 & 2Glh & -2Glh & -4Gh^2 \\ Gl^2 & -Gl^2 & 2Glh & Gl^2 & -Gl^2 & -2Glh \\ -Gl^2 & Gl^2 & -2Glh & -Gl^2 & Gl^2 & 2Glh \\ -2Glh & 2Glh & -4Gh^2 & -2Glh & 2Glh & 4Gh^2 \end{bmatrix} \quad (8.3.14)$$

The element stiffness matrix, which is obtained by adding the bending and transverse stiffness matrices, can be expressed in the following form

$$[K^e] = \begin{bmatrix} a_1 + 2a_3 & -a_1 + a_3 & a_4 & a_1 - 2a_3 & -a_1 - a_3 & -a_4 \\ -a_1 + a_3 & a_1 + 2a_3 & -a_4 & -a_1 - a_3 & a_1 - 2a_3 & a_4 \\ a_4 & -a_4 & a_2 & a_4 & -a_4 & -a_2 \\ a_1 - 2a_3 & -a_1 - a_3 & a_4 & a_1 + 2a_3 & -a_1 + a_3 & -a_4 \\ -a_1 - a_3 & a_1 - 2a_3 & -a_4 & -a_1 + a_3 & a_1 + 2a_3 & a_4 \\ -a_4 & a_4 & -a_2 & -a_4 & a_4 & a_2 \end{bmatrix} \quad (8.3.15)$$

where each symbol denotes

$$a_1 = \frac{Gl}{4h} \quad a_2 = \frac{Gh}{l} \quad a_3 = \frac{Eh}{6l} \quad a_4 = \frac{G}{2} \quad (8.3.16)$$

The mass matrix can be derived similarly as shown in previous sections. The *lumped* mass matrix for the linear element is

$$[M^e] = \frac{\rho ab}{4} \begin{bmatrix} 1 & 0 & 0 & 0 & 0 & 0 \\ 0 & 1 & 0 & 0 & 0 & 0 \\ 0 & 0 & 2 & 0 & 0 & 0 \\ 0 & 0 & 0 & 1 & 0 & 0 \\ 0 & 0 & 0 & 0 & 1 & 0 \\ 0 & 0 & 0 & 0 & 0 & 2 \end{bmatrix} \quad (8.3.17)$$

Here ρab is the element mass, and the element is assumed to have a unit width. Otherwise, the matrix is multiplied by the beam width.

8.4 Mixed Beam Element

This section develops *mixed* beam elements [20,21]. The elements have the transverse deflection and bending moment as primary degrees of freedom. In the finite element method, the primary variables are more accurate than secondary variables which are usually obtained from derivatives of the primary variables. When the transverse deflection and slope are the primary variables, bending stress is a secondary variable which is related to the derivative of the primary variables. On the other hand, the *mixed* beam elements have the bending moment as a primary variable and the bending stress is computed directly from the bending moment without taking any derivative. As a result, there is no loss in accuracy in computing the bending stress in the *mixed* beam elements. The bending stress is usually one of the most important solutions needed in the beam analysis.

In order to derive the *mixed* elements, we consider the governing equations shown below:

$$\frac{M}{EI} - \frac{d^2v}{dx^2} = 0 \tag{8.4.1}$$

$$\frac{d^2M}{dx^2} = q \tag{8.4.2}$$

Galerkin's method is applied to Eqs. (8.4.1) and (8.4.2) using the same shape functions for both v and M. Then, for an element of length l, we obtain

$$\frac{1}{EI}\int_0^l [N]^T[N]\,dx\,[M] + \int_0^l \left[\frac{dN}{dx}\right]^T \left[\frac{dN}{dx}\right] dx\,[v] = \left[[N]^T\theta\right]_0^l \tag{8.4.3}$$

$$\int_0^l \left[\frac{dN}{dx}\right]^T \left[\frac{dN}{dx}\right] dx\,[M] = -\int_0^l [N]^T q\,dx + \left[[N]^T V\right]_0^l \tag{8.4.4}$$

in which $[N]$ is the vector of shape functions, and θ and V are the slope and shear force, respectively. The element stiffness matrix from these equations becomes

$$[K^e] = \begin{bmatrix} K_{11} & K_{12} \\ K_{21} & 0 \end{bmatrix} \tag{8.4.5}$$

where

$$[K_{11}] = \frac{1}{EI}\int_0^l [N]^T[N]\,dx \tag{8.4.6}$$

$$[K_{12}] = [K_{21}]^T = \int_0^l \left[\frac{dN}{dx}\right]^T \left[\frac{dN}{dx}\right] dx \tag{8.4.7}$$

Figure 8.4.1 Linear Mixed Beam Element

For the linear *mixed* beam element shown in Fig. 8.4.1, Eq. (8.4.5) becomes

$$[K^e]\{d^e\} = \{f^e\} \qquad (8.4.8)$$

where

$$[K^e] = \frac{1}{6EIl}\begin{bmatrix} 2l & l & 6EI & -6EI \\ l & 2l & -6EI & 6EI \\ 6EI & -6EI & 0 & 0 \\ -6EI & 6EI & 0 & 0 \end{bmatrix} \qquad (8.4.9)$$

$$\{d^e\} = \{M_1 \ M_2 \ v_1 \ v_2\}^T \qquad (8.4.10)$$

$$\{f^e\} = \{\theta_1 \ \theta_2 \ V_1 - Q_1 \ V_2 - Q_2\}^T \qquad (8.4.11)$$

and Q_i is the equivalent pressure load applied to the nodal points.

Boundary conditions are applied to the element in the following way. For a simply supported node, both displacement v and bending moment M are set to zero while only displacement v and θ are set to zero at a clamped node. If a node is free without any applied moment, moment M is zero at the node. Example 8.4.1 shows the application of these boundary conditions. Any higher order shape functions may be introduced to Eq. (8.4.3) and Eq. (8.4.4) to obtain a stiffness matrix for a higher order *mixed* beam element.

♣ **Example 8.4.1** Figure 8.4.2 shows a beam whose half is modeled using two linear *mixed* beam elements. The left end is either simply supported or clamped and the right end is symmetric. Assembly of two beam elements gives the following system matrix equation.

$$\frac{1}{6EI}\begin{bmatrix} 2 & 1 & 0 & 6EI & -6EI & 0 \\ 1 & 4 & 1 & -6EI & 12EI & -6EI \\ 0 & 1 & 2 & 0 & -6EI & 6EI \\ 6EI & -6EI & 0 & 0 & 0 & 0 \\ -6EI & 12EI & -6EI & 0 & 0 & 0 \\ 0 & -6EI & 6EI & 0 & 0 & 0 \end{bmatrix}\begin{bmatrix} M_1 \\ M_2 \\ M_3 \\ v_1 \\ v_2 \\ v_3 \end{bmatrix} = \begin{bmatrix} \theta_1 \\ 0 \\ 0 \\ V_1 + 0.5 \\ 1.0 \\ 0.5 \end{bmatrix}$$
$$(8.4.12)$$

Since the right end is symmetric in this problem, the slope at the right end node (i.e. the third node in Fig. 8.4.2) which corresponds to the third component in

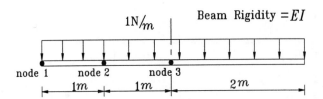

Figure 8.4.2 A Uniformly Loaded Beam Modeled Using Two Linear Mixed Beam Elements

the right-hand-side column vector in this equation is set to zero. On the other hand, the slope at the left end (i.e. the first node) is given as θ_1 while the shear force at the left end support is given as V_1. Depending on the boundary condition, the slope may or may not be known.

First of all, let us consider the simply supported left end. In this case, $v_1 = 0$ and $M_1 = 0$. However, θ_1 and V_1 are unknowns. Applying these conditions to Eq. (8.4.12) gives

$$\frac{1}{6EI}\begin{bmatrix} 4 & 1 & 12EI & -6EI \\ 1 & 2 & -6EI & 6EI \\ 12EI & -6EI & 0 & 0 \\ -6EI & 6EI & 0 & 0 \end{bmatrix}\begin{bmatrix} M_2 \\ M_3 \\ v_2 \\ v_3 \end{bmatrix} = \begin{bmatrix} 0 \\ 0 \\ 1.0 \\ 0.5 \end{bmatrix} \quad (8.4.13)$$

Solving this matrix equation provides $v_2 = -\frac{9}{4EI}$ and $v_3 = -\frac{19}{6EI}$ for deflections, and also gives bending moments of 1.5 and 2 at nodes 2 and 3. Substituting these solutions back into Eq. (8.4.12) gives the shear force at the support equal to 2. These shear force and bending moments are exact solutions.

Next, we consider the clamped left end. The corresponding boundary conditions are $v_1 = 0$ and $\theta_1 = 0$. Applying these conditions to Eq. (8.4.12) results in

$$\frac{1}{6EI}\begin{bmatrix} 2 & 1 & 0 & -6EI & 0 \\ 1 & 4 & 1 & 12EI & -6EI \\ 0 & 1 & 2 & -6EI & 6EI \\ -6EI & 12EI & -6EI & 0 & 0 \\ 0 & -6EI & 6EI & 0 & 0 \end{bmatrix}\begin{bmatrix} M_1 \\ M_2 \\ M_3 \\ v_2 \\ v_3 \end{bmatrix} = \begin{bmatrix} 0 \\ 0 \\ 0 \\ 1.0 \\ 0.5 \end{bmatrix} \quad (8.4.14)$$

The nodal deflections from this matrix equation are $v_2 = -\frac{3}{8EI}$ and $v_3 = -\frac{2}{3EI}$. The bending moments are $M_1 = -\frac{5}{4}$, $M_2 = \frac{1}{4}$ and $M_3 = \frac{3}{4}$. The shear force at the left support is 2. In this case the shear force turns out to be exact but the bending moment is not exact. ‡

The formulation provided above is based on the Euler-Bernoulli beam assumption. If the transverse shear deformation is included in the formulation, the governing equations are modified as given in Ref. [22]

$$\frac{M}{EI} - \frac{1}{\mu G h}\frac{d^2 M}{dx^2} - \frac{d^2 v}{dx^2} = 0 \quad (8.4.15)$$

$$\frac{d^2 M}{dx^2} = q \tag{8.4.2}$$

where μ is the shear correction factor, h is the beam thickness and the beam is assumed to have unit width. Application of Galerkin's method to these equations yields the element matrix shown as

$$[\hat{K}^e] = \begin{bmatrix} \hat{K}_{11} & \hat{K}_{12} \\ \hat{K}_{21} & 0 \end{bmatrix} \tag{8.4.16}$$

where

$$[\hat{K}_{11}] = \frac{1}{EI} \int_0^l [N]^T [N] \, dx + \int_0^l \frac{1}{\mu G h} \left[\frac{dN}{dx}\right]^T \left[\frac{dN}{dx}\right] dx \tag{8.4.17}$$

$$[\hat{K}_{12}] = [\hat{K}_{21}]^T = \int_0^1 \left[\frac{dN}{dx}\right]^T \left[\frac{dN}{dx}\right] dx \tag{8.4.18}$$

Again, the same shape functions are used for v and M in Eq. (8.4.17) and Eq. (8.4.18). For the linear element, Eq. (8.4.18) becomes

$$[K^e] = \frac{1}{6EIl} \begin{bmatrix} 2l^2 + a & l^2 - a & 6EI & -6EI \\ l^2 - a & 2l^2 + a & -6EI & 6EI \\ 6EI & -6EI & 0 & 0 \\ -6EI & 6EI & 0 & 0 \end{bmatrix} \tag{8.4.19}$$

where

$$a = \frac{6EI}{\mu G h} \tag{8.4.20}$$

We call the latter the *thick beam* element and the former the *thin beam* element because the effect of transverse shear deformation increases as the plate thickness-to-length ratio increases. Table 8.4.1 compares the *thin beam* and *thick beam* solutions for various loading and boundary conditions.

The *lumped* mass matrix for the linear *mixed* beam element is

$$[M^e] = \frac{\rho A l}{2} \begin{bmatrix} 0 & 0 & 0 & 0 \\ 0 & 0 & 0 & 0 \\ 0 & 0 & 1 & 0 \\ 0 & 0 & 0 & 1 \end{bmatrix} \tag{8.4.21}$$

8.5 Hybrid Beam Element

A hybrid beam element is introduced in this section. The hybrid element is based on the assumed strains within the beam element [23]. This element requires

Table 8.4.1 Thin and Thick Beam Solutions for Various Beams with Unit Width

Configuration	Maximum deflections
Cantilever with point load P at free end, length L, height h	Thin Beam : $w_{max} = \dfrac{PL^3}{3EI}$ Thick Beam: $w_{max} = \dfrac{PL^3}{3EI} + \dfrac{PLh^2}{10GI}$
Cantilever with uniform load q, length L, height h	Thin Beam : $w_{max} = \dfrac{qL^4}{8EI}$ Thick Beam: $w_{max} = \dfrac{qL^4}{8EI} + \dfrac{L^2h^2}{20GI}$
Simply supported beam with uniform load q, length L, height h	Thin Beam : $w_{max} = \dfrac{5qL^4}{384EI}$ Thick Beam: $w_{max} = \dfrac{5qL^4}{384EI} + \dfrac{qL^2h^2}{80GI}$
Simply supported beam with point load at $L/2$, height h	Thin Beam : $w_{max} = \dfrac{PL^3}{48EI}$ Thick Beam: $w_{max} = \dfrac{PL^3}{48EI} + \dfrac{PLh^2}{40GI}$
Fixed-fixed beam with uniform load q, length L	Thin Beam : $w_{max} = \dfrac{qL^4}{384EI}$ Thick Beam: $w_{max} = \dfrac{qL^4}{384EI} + \dfrac{qL^2h^2}{80GI}$
Fixed-fixed beam with point load P at $L/2$	Thin Beam : $w_{max} = \dfrac{PL^3}{192EI}$ Thick Beam: $w_{max} = \dfrac{PL^3}{192EI} + \dfrac{PLh^2}{40GI}$

(E: Elastic modulus, G: Shear modulus, I: Moment of inertia of cross section)

Section 8.5   Hybrid Beam Element

C^0 continuity. The formulation is based on a modified potential energy expression as given below for a beam with unit width

$$\Pi = \int_0^l \left(-\frac{1}{2}\epsilon_b^T D_b \epsilon_b - \frac{1}{2}\epsilon_s^T D_s \epsilon_s + \epsilon_b^T D_b L_b \{d\} + \epsilon_s^T D_s L_s \{d\} \right) dx - \int_0^l \{d\}^T \{q\} \, dx \tag{8.5.1}$$

where

$$\epsilon_b = \frac{d\theta}{dx} \tag{8.5.2}$$

$$\epsilon_s = -\theta + \frac{dv}{dx} \tag{8.5.3}$$

$$\{d\} = \{\theta \ v\}^T \tag{8.5.4}$$

and other parameters are defined below: D_b is the bending stiffness equal to EI, D_s is the shear stiffness equal to μGA, L_b is the bending strain-displacement operator, and L_s is the shear strain-displacement operator. Invoking a stationary value of the equation results in the equilibrium equation and the generalized strain-displacement relation. In order to obtain the finite element model, generalized strains and displacements are discretized as the following:

$$\epsilon_b = [B_b]\{\alpha_b\} \tag{8.5.5}$$

$$\epsilon_s = [B_s]\{\alpha_s\} \tag{8.5.6}$$

$$\{d\} = [N]\{\hat{d}\} \tag{8.5.7}$$

where generalized strains are assumed independently within each element and generalized displacements are interpolated using generalized nodal displacement $\{\hat{d}\}$. Thus, $[B_b]$ and $[B_s]$ are matrices (or vectors for the beam problem) consisting of the polynomial terms of the generalized strain parameter vectors $\{\alpha_b\}$ and $\{\alpha_s\}$, respectively. Substituting Eqs. (8.5.5) through (8.5.7) into Eq. (8.5.1) yields

$$\Pi = -\frac{1}{2}\{\alpha_b\}^T[G_b]\{\alpha_b\} - \frac{1}{2}\{\alpha_s\}^T[G_s]\{\alpha_s\} + \{\alpha_b\}^T[H_b]\{\hat{d}\}$$
$$+ \{\alpha_s\}^T[H_s]\{\hat{d}\} - \{\hat{d}\}^T\{F\} \tag{8.5.8}$$

where

$$[G_b] = \int_0^l [B_b]^T D_b [B_b] dx \tag{8.5.9}$$

$$[G_s] = \int_0^l [B_s]^T D_s [B_s] dx \tag{8.5.10}$$

$$[H_b] = \int_0^l [B_b]^T D_b L_b [N] dx \tag{8.5.11}$$

$$[H_s] = \int_0^l [B_s]^T D_s L_s [N] dx \tag{8.5.12}$$

and
$$\{F\} = \int_0^l [N]^T \{q\} dx \qquad (8.5.13)$$

Invoking stationary values of Eq. (8.5.8) with respect to α_b and α_s respectively results in
$$-[G_b]\{\alpha_b\} + [H_b]\{\hat{d}\} = 0 \qquad (8.5.14)$$
$$-[G_s]\{\alpha_s\} + [H_s]\{\hat{d}\} = 0 \qquad (8.5.15)$$

Eliminating $\{\alpha_b\}$ and $\{\alpha_s\}$ from Eq. (8.5.8), Eq. (8.5.14), and Eq. (8.5.15) gives
$$\Pi = \frac{1}{2}\{\hat{d}\}^T ([H_b]^T [G_b]^{-1} [H_b] + [H_s]^T [G_s]^{-1} [H_s])\{\hat{d}\} - \{\hat{d}\}^T \{F\} \qquad (8.5.16)$$

Equation (8.5.16) finally gives the following finite element system of equations
$$[K]\{\hat{d}\} = \{F\} \qquad (8.5.17)$$

in which
$$[K] = [H_b]^T [G_b]^{-1} [H_b] + [H_s]^T [G_s]^{-1} [H_s] \qquad (8.5.18)$$

For a linear beam element, the generalized strain vectors are assumed as
$$[B_b] = [1 \quad x] \qquad (8.5.19)$$

and
$$[B_s] = 1 \qquad (8.5.20)$$

These expressions represent that the bending strain varies linearly and the shear strain is constant within the linear beam element. Example 8.5.1 shows the derivation of the stiffness matrix for the linear beam element. The hybrid beam element can also be generalized for general higher order shape functions.

♣ **Example 8.5.1** Substituting the generalized strain vectors, Eq. (8.5.19) and Eq. (8.5.20), into Eqs. (8.5.9) through (8.5.12) gives

$$[G_b] = \frac{EI}{6} \begin{bmatrix} 6l & 3l^2 \\ 3l^2 & 2l^3 \end{bmatrix} \qquad (8.5.21)$$

$$[G_s] = \mu G A l \qquad (8.5.22)$$

$$[H_b] = \frac{EI}{2} \begin{bmatrix} 0 & -2 & 0 & 2 \\ 0 & -l & 0 & l \end{bmatrix} \qquad (8.5.23)$$

$$[H_s] = \frac{\mu G A}{2} [-2 \quad -l \quad 2 \quad -l] \qquad (8.5.24)$$

Applying these expressions into Eq. (8.5.18) yields the same element stiffness matrices as in Eqs. (8.2.9) through (8.2.11). The first term in Eq. (8.5.18) results in Eq. (8.2.10) while the second term in Eq. (8.5.18) yields that in Eq. (8.2.11). However, no reduced integration technique is used for the present stiffness matrix. ‡

8.6 Composite Beams

Laminated composite beams are made of multiple layers which have, in general, different material properties. More general cases are dealt with in the chapter for plates and shells. In this section, we consider a simple case. The laminated beam is symmetric about the midplane axis so that there is no coupling between the inplane deformation and bending deformation. For this simple case, the beam formulations developed in the previous sections are directly applicable to the laminated beam. One thing to be generalized for the laminated beam is the beam rigidity. For a symmetric laminated composite beam with unit width, the equivalent beam rigidity is computed as

$$(EI)_{eqiv} = \frac{1}{3} \sum_{i=1}^{n} E_i(y_i^3 - y_{i-1}^3) \qquad (8.6.1)$$

Here, n is the number of layers, and y_{i-1} and y_i are the y-coordinate values of the bottom and top planes of the ith layer as seen in Fig. 8.6.1. In addition E_i is the elastic modulus in the x-direction of the ith layer. This equivalent beam rigidity is substituted into the previous beam elements to compute the stiffness matrices for laminated composite beams.

However, one more important fact in laminated beam applications is the effect of transverse shear. Composite beams are not usually isotropic and their shear modulus is in general much lower than the elastic modulus. In this case, the shear deformation plays an important role [24,25]. For example, see Fig. 8.4.3. As the shear modulus G becomes much smaller than the elastic modulus E in the *thick beam* solutions, the *thick beam* solutions deviate much more from *thin beam* solutions. As a result, *thin beam* solutions may not be accurate anymore. In other words, the Euler-Bernoulli beam equation may not be suitable anymore, especially for rather thick laminated composite beams.

As a result, the beam formulations, including the effect of shear deformation, can be used for analyses of laminated composite beams. In this case, the equivalent shear modulus is computed from

$$(GA)_{eqiv} = \sum_{i=1}^{n} bG_i(y_i - y_{i-1}) \qquad (8.6.2)$$

where b is the width of the beam. The bending stress in a laminated composite beam can be determined from

$$\sigma_i = -\frac{ME_i y}{(EI)_{eqiv}} \qquad (8.6.3)$$

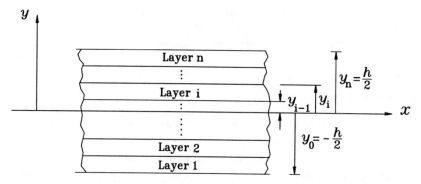

Figure 8.6.1 Laminated Beam

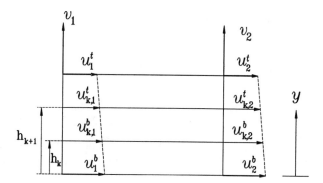

Figure 8.6.2 Laminated Beam Element

where σ_i is the bending stress in the ith layer and y is a coordinate value in the ith layer. Example 8.6.1 shows an application of beam elements with displacement degrees of freedom only developed in section 8.3 to an analysis of a laminated composite beam.

♣ **Example 8.6.1** Two different techniques can be used to model a laminated beam using beam elements with displacement degrees of freedom only. The first technique discretizes respective layers in the finite element analysis. As a result, the total number of elements is proportional to the number of layers in a laminated beam. The condensation technique can be applied to reduce the number of total degrees of freedom by eliminating internal layers' degrees of freedom. This modeling technique is computationally expensive but it can describe a general shape of inplane deformation through the beam thickness. The second technique uses one beam element through the beam thickness regardless of the number of layers. This technique is computationally efficient but it assumes a linear deformation through the beam thickness. The development of this technique is described in the following paragraphs.

Let u_k and v_k represent the axial and lateral displacements of the kth layer and let h be the beam thickness while h_k and h_{k+1} represent heights of the top and bottom sides of the kth layer measured from the bottom of the beam (see Fig. 8.6.2). The relationship between the layer displacements and the global beam displacements can be written as

$$u_{k,1}^b = c_1 u_1^b + c_2 u_1^t \tag{8.6.4}$$

$$u_{k,1}^t = c_3 u_1^b + c_4 u_1^t \tag{8.6.5}$$

$$v_{k,1} = v_1 \tag{8.6.6}$$

$$u_{k,2}^b = c_1 u_2^b + c_2 u_2^t \tag{8.6.7}$$

$$u_{k,2}^t = c_3 u_2^b + c_4 u_2^t \tag{8.6.8}$$

$$v_{k,2} = v_2 \tag{8.6.9}$$

where constants cs are

$$c_1 = \frac{h - h_k}{h} \tag{8.6.10}$$

$$c_2 = \frac{h_k}{h} \tag{8.6.11}$$

$$c_3 = \frac{h - h_{k+1}}{h} \tag{8.6.12}$$

$$c_4 = \frac{h_{k+1}}{h} \tag{8.6.13}$$

This relationship can be expressed in the matrix form

$$\begin{Bmatrix} u_{k,1}^b \\ u_{k,1}^t \\ v_{k,1} \\ u_{k,2}^b \\ u_{k,2}^t \\ v_{k,2} \end{Bmatrix} = \begin{bmatrix} c_1 & c_2 & 0 & 0 & 0 & 0 \\ c_3 & c_4 & 0 & 0 & 0 & 0 \\ 0 & 0 & 1 & 0 & 0 & 0 \\ 0 & 0 & 0 & c_1 & c_2 & 0 \\ 0 & 0 & 0 & c_3 & c_4 & 0 \\ 0 & 0 & 0 & 0 & 0 & 1 \end{bmatrix} \begin{Bmatrix} u_1^b \\ u_1^t \\ v_1 \\ u_2^b \\ u_2^t \\ v_2 \end{Bmatrix} \tag{8.6.14}$$

or in a short notation

$$\{d^k\} = [T]\{d\} \tag{8.6.15}$$

where $\{d^k\}$ and $\{d\}$ are the displacement vectors of the kth layer and the beam, respectively. $[T]$ is the transformation matrix shown in Eq. (8.6.14). Now, the stiffness and mass matrices of a laminated beam element can be expressed as

$$[K^e] = \sum_{k=1}^{n} [T]^T [K^k] [T] \tag{8.6.16}$$

$$[M^e] = \sum_{k=1}^{n} [T]^T [M^k] [T] \tag{8.6.17}$$

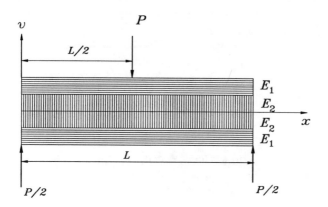

Figure 8.6.3 Simply Supported Laminated Beam with Four Layers

Here, $[K^k]$ and $[M^k]$ are the stiffness and mass matrices of the kth layer, respectively. Using this technique, a single beam element can include all the layers of a laminated beam.

Figure 8.6.3 illustrates a simply supported laminated beam with four layers. The elastic modulus E_1 is assumed to be either 20 or 100 times greater than E_2. The finite element solutions are provided in Fig. 8.6.4. Figure 8.6.5 also shows the solutions for the same beam but with eight layers. ‡

8.7 Two-Dimensional Frame Element

A frame structure is made of many beam members connected together. It may be of planar or spatial geometry. For a planar frame structure, each beam member is generally subjected to both bending and axial loads as illustrated in Fig. 8.7.1. As a result, a planar (2-D) frame element must include both axial and bending deformation. If the deformation is small, we may neglect the coupling between the two deformations. As a result, the element stiffness matrix for a 2-D frame element can be constructed by superimposing both axial and bending stiffnesses. For example, the stiffness matrix of a linear 2-D frame element is using *Hermitian* beam element

$$[K^e] = \frac{E}{l^3} \begin{bmatrix} Al^2 & 0 & 0 & -Al^2 & 0 & 0 \\ 0 & 12I & 6Il & 0 & -12I & 6Il \\ 0 & 6Il & 4Il^2 & 0 & -6Il & 2Il^2 \\ -Al^2 & 0 & 0 & Al^2 & 0 & 0 \\ 0 & -12I & -6Il & 0 & 12I & -6Il \\ 0 & 6Il & 2Il^2 & 0 & -6Il & 4Il^2 \end{bmatrix} \qquad (8.7.1)$$

for the element degrees of freedom $\{u_1 \; v_1 \; \theta_1 \; u_2 \; v_2 \; \theta_2\}$ as seen in Fig. 8.7.2. Other C^0 type beam elements may be used to develop a frame stiffness matrix.

In general, a beam member is inclined to the global coordinate system as shown in Fig. 8.7.3. In this case, the element stiffness matrix requires the planar

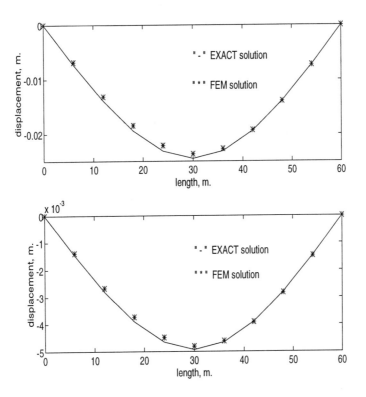

Figure 8.6.4 Static Deflections of the Laminated Composite Beam with 4 Layers: $E_1/E_2=20, 100$ for top and bottom

transformation. Figure 8.7.4 shows two coordinate systems: local and global systems. The global system is denoted by a superimposed bar on both coordinate axes and displacements. The relation between the local and global displacements is

$$\begin{Bmatrix} u_1 \\ v_1 \\ \theta_1 \\ u_2 \\ v_2 \\ \theta_2 \end{Bmatrix} = \begin{bmatrix} c & s & 0 & 0 & 0 & 0 \\ -s & c & 0 & 0 & 0 & 0 \\ 0 & 0 & 1 & 0 & 0 & 0 \\ 0 & 0 & 0 & c & s & 0 \\ 0 & 0 & 0 & -s & c & 0 \\ 0 & 0 & 0 & 0 & 0 & 1 \end{bmatrix} \begin{Bmatrix} \bar{u}_1 \\ \bar{v}_1 \\ \bar{\theta}_1 \\ \bar{u}_2 \\ \bar{v}_2 \\ \bar{\theta}_2 \end{Bmatrix} \qquad (8.7.2)$$

where $c = cos\beta$ and $s = sin\beta$. In short notation, Eq. (8.7.2) can be written as

$$\{d^e\} = [T]\{\bar{d}^e\} \qquad (8.7.3)$$

Then, the stiffness matrix for a planar frame element is expressed in terms of the global coordinate system as given below:

$$[\bar{K}^e] = [T]^T[K^e][T] \qquad (8.7.4)$$

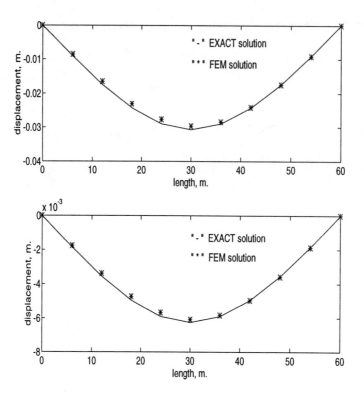

Figure 8.6.5 Static Deflections of the Laminated Composite Beam with 8 Layers: $E_1/E_2=20, 100$ for top and bottom

Carrying out the matrix multiplication gives

$$[\bar{K}^e] = \begin{bmatrix} a_3 & a_4 & a_5 & -a_3 & -a_4 & a_5 \\ a_4 & a_6 & a_7 & -a_4 & -a_6 & a_7 \\ a_5 & a_7 & a_1 & -a_5 & -a_7 & a_2 \\ -a_3 & -a_4 & -a_5 & a_3 & a_4 & -a_5 \\ -a_4 & -a_6 & -a_7 & a_4 & a_6 & -a_7 \\ a_5 & a_7 & a_2 & -a_5 & -a_7 & a_1 \end{bmatrix} \quad (8.7.5)$$

where

$$a_1 = \frac{4EI}{l} \quad (8.7.6)$$

$$a_2 = \frac{2EI}{l} \quad (8.7.7)$$

$$a_3 = \left(\frac{AE}{l}\right)c^2 + \left(\frac{12EI}{l^3}\right)s^2 \quad (8.7.8)$$

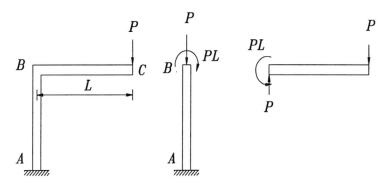

Figure 8.7.1 A Planar Frame Structure with Free Body Diagrams

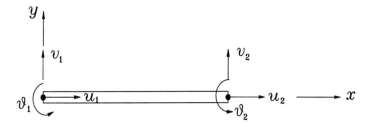

Figure 8.7.2 A Linear Frame Element

$$a_4 = \left(\frac{AE}{l}\right)cs - \left(\frac{12EI}{l^3}\right)cs \tag{8.7.9}$$

$$a_5 = -\left(\frac{6EI}{l^2}\right)s \tag{8.7.10}$$

$$a_6 = \left(\frac{AE}{l}\right)s^2 + \left(\frac{12EI}{l^3}\right)c^2 \tag{8.7.11}$$

$$a_7 = \left(\frac{6EI}{l^2}\right)c \tag{8.7.12}$$

8.8 Three-Dimensional Frame Element

A beam member in a spatial (3-D) frame is generally subjected to axial, bending, and torsional loads as illustrated in Fig. 8.8.1. If beam members have circular cross-sections, the element stiffness in a local axis is, as seen in Fig. 8.8.2,

$$[K^e] = \begin{bmatrix} K_{11}^e & K_{12}^e \\ K_{21}^e & K_{22}^e \end{bmatrix} \tag{8.8.1}$$

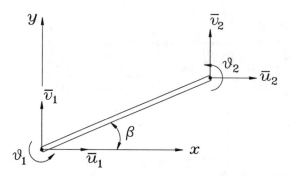

Figure 8.7.3 Inclined Frame Element

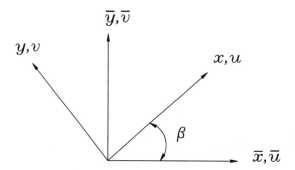

Figure 8.7.4 Coordinate Transformation

where

$$[K_{11}^e] = \begin{bmatrix} a_1 & 0 & 0 & 0 & 0 & 0 \\ 0 & b_1 & 0 & 0 & 0 & b_2 \\ 0 & 0 & c_1 & 0 & -c_2 & 0 \\ 0 & 0 & 0 & a_2 & 0 & 0 \\ 0 & 0 & -c_2 & 0 & 2c_3 & 0 \\ 0 & b_2 & 0 & 0 & 0 & 2b_3 \end{bmatrix} \qquad (8.8.2)$$

$$[K_{12}^e] = [K_{21}^e]^T = \begin{bmatrix} -a_1 & 0 & 0 & 0 & 0 & 0 \\ 0 & -b_1 & 0 & 0 & 0 & b_2 \\ 0 & 0 & -c_1 & 0 & -c_2 & 0 \\ 0 & 0 & 0 & -a_2 & 0 & 0 \\ 0 & 0 & c_2 & 0 & c_3 & 0 \\ 0 & -b_2 & 0 & 0 & 0 & b_3 \end{bmatrix} \qquad (8.8.3)$$

Section 8.8 Three-Dimensional Frame Element

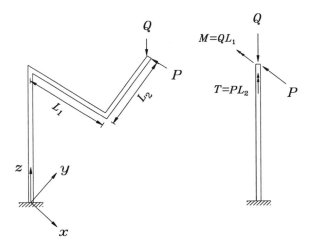

Figure 8.8.1 A Simply Supported Beam

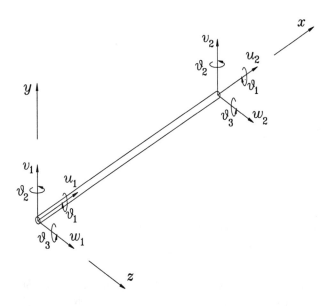

Figure 8.8.2 Spatial Frame Element

$$[K_{22}^e] = \begin{bmatrix} a_1 & 0 & 0 & 0 & 0 & 0 \\ 0 & b_1 & 0 & 0 & 0 & -b_2 \\ 0 & 0 & c_1 & 0 & c_2 & 0 \\ 0 & 0 & 0 & a_2 & 0 & 0 \\ 0 & 0 & c_2 & 0 & 2c_3 & 0 \\ 0 & -b_2 & 0 & 0 & 0 & 2b_3 \end{bmatrix} \quad (8.8.4)$$

$$a_1 = \frac{EA}{l}, \qquad a_2 = \frac{GJ}{l} \quad (8.8.5)$$

$$b_1 = \frac{12EI_z}{l^3}, \qquad b_2 = \frac{6EI_z}{l^2}, \qquad b_3 = \frac{2EI_z}{l} \qquad (8.8.6)$$

$$c_1 = \frac{12EI_y}{l^3}, \qquad c_2 = \frac{6EI_y}{l^2}, \qquad c_3 = \frac{2EI_y}{l} \qquad (8.8.7)$$

In these equations, I_y and I_z are moments of inertia of the cross-section about the $y-$ and $z-$axes and J is the polar moment of inertia. The corresponding element degrees of freedom vector is

$$\{d^e\} = \{\, u_1 \;\; v_1 \;\; w_1 \;\; \theta_1 \;\; \theta_2 \;\; \theta_3 \;\; u_2 \;\; v_2 \;\; w_2 \;\; \theta_3 \;\; \theta_4 \;\; \theta_6 \,\}^T \qquad (8.8.8)$$

in which θ_1 and θ_4 are the rotational degrees of freedom associated with the twisting moment, and θ_2, θ_3, θ_5 and θ_6 are slopes associated with bending moments. This stiffness matrix in terms of a local coordinate system needs to be transformed into that in terms of the global coordinate system in the same way as shown in Eq. (8.7.4). In this case, the transformation matrix $[T]$ is of size 12 × 12.

8.9 MATLAB Application to Static Analysis

The static analysis of a beam or a frame is to solve the following matrix equation:

$$[K]\{d\} = \{F\} \qquad (8.9.1)$$

where the system stiffness matrix $[K]$ and the system force vector $\{F\}$ are obtained by assembling each element matrix and vector. Several *m-files* are written to compute an element stiffness matrix and a mass matrix, which is used for dynamic problems in the next section, for various beam and frame elements formulated in this chapter. The names of *m-files* are given below:

febeam1.m : Hermitian beam element (see Sec. 8.1)
febeam2.m : Timoshenko beam element (see Sec. 8.2)
febeam3.m : beam element with displacement degrees of freedom (see Sec. 8.3)
febeam4.m : mixed beam element (see Sec. 8.4)
feframe2.m: 2-D frame element (see Sec. 8.7)

Detailed information regarding these *m-files* is provided in Appendix A. The following examples show computer programs for finite element analyses of beam and frame structures written in MATLAB and the *m-files* described above.

♣ **Example 8.9.1** Figure 8.9.1 shows a simply supported beam whose length is 20 in. The beam also has elastic modulus of 10×10^6 psi and its cross-section is 1 in. by 1 in. The beam is subjected to a center load of 100 lb. We use 5 *Hermitian* beam elements for one half of the beam due to symmetry to find the deflection of the beam. Figure 8.9.1 shows the finite element discretization. The constraints applied to this problem are no deflection at the left boundary support (i.e $v_1 = 0$) and zero slope at the symmetric node (i.e. $\theta_6 = 0$). Their system

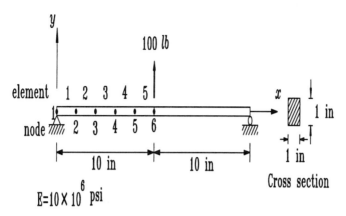

Figure 8.9.1 A Simply Supported Beam

degrees of freedom are 1 and 12, respectively, in the present mesh. Furthermore, only half of the center load is applied at the symmetric node because of symmetry. The finite element analysis program for the present static problem is listed below.

```
%————————————————————————————
% EX891.m: MATLAB program to solve a static beam deflection using
% Hermitian beam elements
%
% Variable descriptions
% k = element stiffness matrix
% kk = system stiffness matrix
% ff = system force vector
% index = a vector containing system dofs associated with each element
% bcdof = a vector containing dofs associated with boundary conditions
% bcval = a vector containing boundary condition values associated with
%         the dofs in bcdof
%————————————————————————————
nel=5;                          % number of elements
nnel=2;                         % number of nodes per element
ndof=2;                         % number of dofs per node
nnode=(nnel-1)*nel+1;           % total number of nodes in system
sdof=nnode*ndof;                % total system dofs
el=10^7;                        % elastic modulus
xi=1/12;                        % moment of inertia of cross-section
tleng=10;                       % length of half of the beam
leng=10/nel;                    % element length of equal size
area=1;                         % cross-sectional area of the beam
rho=1;          % mass density (arbitrary value for this problem because
                % it is not used for the static problem)
ipt=1;     % option for mass matrix (arbitrary value and not used here)
bcdof(1)=1;         % first dof (deflection at left end) is constrained
bcval(1)=0;                     % whose described value is 0
```

```
bcdof(2)=12;        % 12th dof (slope at the symmetric end) is constrained
bcval(2)=0;         % whose described value is 0
ff=zeros(sdof,1);   % initialization of system force vector
kk=zeros(sdof,sdof); % initialization of system matrix
index=zeros(nnel*ndof,1);  % initialization of index vector
ff(11)=50;          % because half of the load is applied due to symmetry
for iel=1:nel       % loop for the total number of elements
index=feeldof1(iel,nnel,ndof);   % extract system dofs for each element
k=febeam1(el,xi,leng,area,rho,ipt);  % compute element stiffness matrix
kk=feasmbl1(kk,k,index);  % assembly into system matrix
end
[kk,ff]=feaplyc2(kk,ff,bcdof,bcval);  % apply the boundary conditions
fsol=kk\ff;         % solve the matrix equation
%--------------------
% Analytical solution
%--------------------
e=10^7; l=20; xi=1/12; P=100;
for i = 1:nnode
x=(i-1)*2;
c=P/(48*e*xi);
k=(i-1)*ndof+1;
esol(k)=c*(3*l^2-4*x^2)*x;
esol(k+1)=c*(3*l^2-12*x^2);
end
%-----------------------------
% Print both exact and fem solutions
%-----------------------------
num=1:1:sdof;
store=[num' fsol esol']
%-----------------------------

function [k,m]=febeam1(el,xi,leng,area,rho,ipt)
%-----------------------------
% Purpose:
% Stiffness and mass matrices for Hermitian beam element
% nodal dof $v_1$ $theta_1$ $v_2$ $theta_2$
%
% Synopsis:
% [k,m]=febeam1(el,xi,leng,area,rho,ipt)
%
% Variable Description:
% k - element stiffness matrix (size 4x4)
% m - element mass matrix (size 4x4)
% el - elastic modulus
% xi - second moment of inertia of cross-section
% leng - element length
% area - area of beam cross-section
% rho - mass density (mass per unit volume)
```

```
% ipt = 1: consistent mass matrix
%     2: lumped mass matrix
%     otherwise: diagonal mass matrix
%--------------------------------------------------
%
% stiffness matrix
%
  c=el*xi/(leng^3);
  k=c*[12 6*leng -12 6*leng;...
      6*leng 4*leng^2 -6*leng 2*leng^2;...
      -12 -6*leng 12 -6*leng;...
      6*leng 2*leng^2 -6*leng 4*leng^2];
%
% consistent mass matrix
%
  if ipt==1
%
  mm=rho*area*leng/420;
  m=mm*[156 22*leng 54 -13*leng;...
      22*leng 4*leng^2 13*leng -3*leng^2;...
      54 13*leng 156 -22*leng;...
      -13*leng -3*leng^2 -22*leng 4*leng^2];
%
% lumped mass matrix
%
  elseif ipt==2
%
  m=zeros(4,4);
  mass=rho*area*leng;
  m=diag([mass/2 0 mass/2 0]);
%
% diagonal mass matrix
%
  else
%
  m=zeros(4,4);
  mass=rho*area*leng;
  m=mass*diag([1/2 leng^2/78 1/2 leng^2/78]);
%
  end
%--------------------------------------------------
```

The finite element solution obtained from this MATLAB program as well as the exact solution are

```
store =
dof #      fem sol      exact
1.0000     0.0000       0.0000              % deflection at node 1
```

2.0000	0.0030	0.0030	% slope at node 1
3.0000	0.0059	0.0059	% deflection at node 2
4.0000	0.0029	0.0029	% slope at node 2
5.0000	0.0114	0.0114	% deflection at node 3
6.0000	0.0025	0.0025	% slope at node 3
7.0000	0.0158	0.0158	% deflection at node 4
8.0000	0.0019	0.0019	% slope at node 4
9.0000	0.0189	0.0189	% deflection at node 5
10.000	0.0011	0.0011	% slope at node 5
11.000	0.0200	0.0200	% deflection at node 6
12.000	0.0000	0.0000	% slope at node 6

‡

♣ **Example 8.9.2** We want to solve Example 8.9.1 using Timoshenko beam elements. The computer program is almost the same as that given in Example 8.9.1. Instead of calling *febeam1.m* we need to call *febeam2.m* to compute the element stiffness matrix. In the beginning of the program list shown in Example 8.9.1, we add

   sh=3.8*10^6;   % shear modulus of the beam

and the following line

   k=febeam1(el,xi,leng,area,rho,ipt);

is replaced by

   k=febeam2(el,xi,leng,sh,area,rho,ipt);

The computed solution is also compared to the exact answer below.

store =			
dof #	fem sol	exact	
1.0000	0.0000	0.0000	% deflection at node 1
2.0000	0.0030	0.0030	% slope at node 1
3.0000	0.0059	0.0059	% deflection at node 2
4.0000	0.0029	0.0029	% slope at node 2
5.0000	0.0113	0.0114	% deflection at node 3
6.0000	0.0025	0.0025	% slope at node 3
7.0000	0.0158	0.0158	% deflection at node 4
8.0000	0.0019	0.0019	% slope at node 4
9.0000	0.0188	0.0189	% deflection at node 5
10.000	0.0011	0.0011	% slope at node 5
11.000	0.0200	0.0200	% deflection at node 6
12.000	0.0000	0.0000	% slope at node 6

‡

Example 8.9.3
This example again solves the same problem in Example 8.9.1 using beam elements with displacement degrees of freedom. This beam element is different from the beam elements used in previous examples. As a result, the complete program is included in the following.

```
%————————————————————————————————
% EX893.m: MATLAB program to solve a static beam deflection using
% beam elements with displacement degrees of freedom only
%
% Variable descriptions
% k = element stiffness matrix
% kk = system stiffness matrix
% ff = system force vector
% index = a vector containing system dofs associated with each element
% bcdof = a vector containing dofs associated with boundary conditions
% bcval = a vector containing boundary condition values associated with
%         the dofs in bcdof
%————————————————————————————————
nel=5;                              % number of elements
nnel=2;                             % number of nodes per element
ndof=3;                             % number of dofs per node
nnode=(nnel-1)*nel+1;               % total number of nodes in system
sdof=nnode*ndof;                    % total system dofs
el=10^7;                            % elastic modulus
sh=3.8*10^6                         % shear modulus
tleng=10;                           % length of half of the beam
leng=10/nel;                        % element length of equal size
heig=1;                             % height of the beam
width=1;                            % width of the beam
rho=1;           % mass density (arbitrary value for this problem because
                                    % it is not used for the static problem)
bcdof(1)=3;                         % deflection at left end is constrained
bcval(1)=0;                         % whose described value is 0
bcdof(2)=16;        % inplane displ. at the right end is constrained
bcval(2)=0;                         % whose described value is 0
bcdof(3)=17;        % inplane displ. at the right end is constrained
bcval(3)=0;                         % whose described value is 0
ff=zeros(sdof,1);                   % initialization of system force vector
kk=zeros(sdof,sdof);                % initialization of system matrix
index=zeros(nnel*ndof,1);           % initialization of index vector
ff(18)=50;       % because half of the load is applied due to symmetry
for iel=1:nel                       % loop for the total number of elements
index=feeldof1(iel,nnel,ndof);      % extract system dofs for each element
k=febeam3(el,sh,leng,heig,width,rho);   % compute element matrix
kk=feasmbl1(kk,k,index);            % assembly into system matrix
end
[kk,ff]=feaplyc2(kk,ff,bcdof,bcval);    % apply the boundary conditions
fsol=kk\ff;                         % solve the matrix equation
```

```
%------------------------
% Analytical solution
%------------------------
e=10^7; l=20; xi=1/12; P=100;
for i = 1:nnode
x=(i-1)*2;
c=P/(48*e*xi);
k=(i-1)*ndof+1;
esol(k+2)=c*(3*l^2-4*x^2)*x;
esol(k+1)=c*(3*l^2-12*x^2)*(-0.5);
esol(k)=c*(3*l^2-12*x^2)*(0.5);
end
%------------------------
% print both exact and fem solutions
%------------------------
num=1:1:sdof;
store=[num' fsol esol']
%------------------------

function [k,m]=febeam3(el,sh,leng,heig,width,rho)
%------------------------------------------------------
% Purpose:
% Stiffness and mass matrices for beam element with displacement
% degrees of freedom only
% nodal dof $u_1^b$ $u_1^t$ $v_1$ $u_2^b$ $u_2^t$ $v_2$
%
% Synopsis:
% [k,m]=febeam1(el,sh,leng,heig,rho,area,ipt)
%
% Variable Description:
% k - element stiffness matrix (size 6x6)
% m - element mass matrix (size 6x6)
% el - elastic modulus
% sh - shear modulus
% leng - element length
% heig - element thickness
% width - width of the beam element
% rho - mass density of the beam element (mass per unit volume)
%       lumped mass matrix only
%------------------------------------------------------
%
% stiffness matrix
%
a1=(sh*leng*width)/(4*heig);
a2=(sh*heig*width)/leng;
a3=(el*heig*width)/(6*leng);
a4=sh*width/2;
k= [ a1+2*a3 -a1+a3 a4 a1-2*a3 -a1-a3 -a4;...
```

```
        -a1+a3 a1+2*a3 -a4 -a1-a3 a1-2*a3 a4;...
        a4 -a4 a2 a4 -a4 -a2;...
        a1-2*a3 -a1-a3 a4 a1+2*a3 -a1+a3 -a4;...
        -a1-a3 a1-2*a3 -a4 -a1+a3 a1+2*a3 a4;...
        -a4 a4 -a2 -a4 a4 a2];
%
% lumped mass matrix
%
m=zeros(6,6);
mass=rho*heig*width*leng/4;
m=mass*diag([1 1 2 1 1 2]);
%—————————————————————————————————
```

The solution output is

```
store =
dof #     fem sol    exact
1.0000    0.0015     0.0015     % axial displ. at bottom side of node 1
2.0000   -0.0015    -0.0015     % axial displ. at top side of node 1
3.0000    0.0000     0.0000     % transverse displ. at node 1
4.0000    0.0014     0.0014     % axial displ. at bottom side of node 2
5.0000   -0.0014    -0.0014     % axial displ. at top side of node 2
6.0000    0.0059     0.0059     % transverse displ. at node 2
7.0000    0.0013     0.0013     % axial displ. at bottom side of node 3
8.0000   -0.0013    -0.0013     % axial displ. at top side of node 3
9.0000    0.0113     0.0114     % transverse displ. at node 3
10.000    0.0010     0.0010     % axial displ. at bottom side of node 4
11.000   -0.0010    -0.0010     % axial displ. at top side of node 4
12.000    0.0158     0.0158     % transverse displ. at node 4
13.000    0.0005     0.0005     % axial displ. at bottom side of node 5
14.000   -0.0005    -0.0005     % axial displ. at top side of node 5
15.000    0.0188     0.0189     % transverse displ. at node 5
16.000    0.0000     0.0000     % axial displ. at bottom side of node 6
17.000    0.0000     0.0000     % axial displ. at top side of node 6
18.000    0.0199     0.0200     % transverse displ. at node 6
```

‡

♣ **Example 8.9.4** Solve the same example again using the mixed beam elements. The computer program list is provided below.

```
%—————————————————————————————————
% EX894.m: MATLAB program to solve a static beam deflection
% problem using mixed beam elements
%
% Variable descriptions
% k = element stiffness matrix
```

```
% kk = system stiffness matrix
% ff = system force vector
% index = a vector containing system dofs associated with each element
% bcdof = a vector containing dofs associated with boundary conditions
% bcval = a vector containing boundary condition values associated with
% the dofs in bcdof
%----------------------------------------------------------------
nel=5;                          % number of elements
nnel=2;                         % number of nodes per element
ndof=2;                         % number of dofs per node
nnode=(nnel-1)*nel+1;           % total number of nodes in system
sdof=nnode*ndof;                % total system dofs
bcdof(1)=1;                     % bending moment at node 1 is constrained
bcval(1)=0;                     % whose described value is 0
bcdof(2)=2;                     % deflection at node 1 is constrained
bcval(2)=0;                     % whose described value is 0
ff=zeros(sdof,1);               % initialization of system force vector
kk=zeros(sdof,sdof);            % initialization of system matrix
index=zeros(nnel*ndof,1);       % initialization of index vector
ff(12)=-50;      % because a half of the load is applied due to symmetry
for iel=1:nel                   % loop for the total number of elements
index=feeldof1(iel,nnel,ndof);  % extract system dofs for each element
k=febeam4(10^7,0.083333,2,0,1,1,1);  % compute element stiffness matrix
kk=feasmbl1(kk,k,index);        % assembly into system matrix
end
[kk,ff]=feaplyc2(kk,ff,bcdof,bcval);  % apply the boundary conditions
fsol=kk\ff;                     % solve the matrix equation
%------------------------------
% analytical solution
%------------------------------
e=10^7; l=20; xi=1/12; P=100;
for i = 1:nnode
x=(i-1)*2;
c=P/(48*e*xi);
k=(i-1)*ndof+1;
esol(k+1)=c*(3*l^2-4*x^2)*x;
esol(k)=-50*x;
end
%----------------------------------------
% print both exact and fem solutions
%----------------------------------------
num=1:1:sdof;
store=[num' fsol esol']
%----------------------------------------------------------------

function [k,m]=febeam4(el,xi,leng,sh,heig,rho,ipt)
%----------------------------------------------------------------
% Purpose:
```

```
% Stiffness and mass matrices for mixed beam element
% bending moment and deflection as nodal degrees of freedom
% nodal dof M_1 v_1 M_2 v_2
%
% Synopsis:
% [k,m]=febeam4(el,xi,leng,sh,heig,rho,ipt)
%
% Variable Description:
% k - element stiffness matrix (size 4x4)
% m - element mass matrix (size 4x4)
% el - elastic modulus
% xi - second moment of inertia of cross-section
% leng - length of the beam element
% sh - shear modulus
% heig - beam thickness
% rho - mass density of the beam element (mass per unit volume)
% ipt = 1 - lumped mass matrix
%     = otherwise - diagonalized mass matrix
%--------------------------------------------------------------
%
% stiffness matrix
%
if sh == 0
%
% thin beam (no shear deformation)
%
  k= [ leng/(3*el*xi) 1/leng leng/(6*el*xi) -1/leng;...
       1/leng 0 -1/leng 0;...
       leng/(6*el*xi) -1/leng leng/(3*el*xi) 1/leng;...
       -1/leng 0 1/leng 0 ];
%
else
%
% thick beam (includes shear deformation)
%
  a=6/(5*sh*leng*heig);
  k= [ 1/(3*el*xi)+a 1/leng 1/(6*el*xi)-a -1/leng;...
       1/leng 0 -1/leng 0;...
       1/(6*el*xi)-a -1/leng 1/(3*el*xi)+a 1/leng;...
       -1/leng 0 1/leng 0 ];
%
end
%
% lumped mass matrix
%
if ipt==1
```

```
%
  m=zeros(4,4);
  mass=rho*heig*leng/2;
  m=diag([0 1 0 1]);
%
% diagonal mass matrix
%
else
%
  m=zeros(4,4);
  mass=rho*heig*leng/2;
  m=mass*diag([1 1 1 1]);
%
end
%——————————————————————————————
```

The solution from the computer program is given below.

```
store =
dof #     fem sol    exact
1.0000    0.0000     0.0000       % bending moment at node 1
2.0000    0.0000     0.0000       % deflection at node 1
3.0000   -100.00    -100.00       % bending moment at node 2
4.0000    0.0059     0.0059       % deflection at node 2
5.0000   -200.00    -200.00       % bending moment at node 3
6.0000    0.0114     0.0114       % deflection at node 3
7.0000   -300.00    -300.00       % bending moment at node 4
8.0000    0.0158     0.0158       % deflection at node 4
9.0000   -400.00    -400.00       % bending moment at node 5
10.000    0.0189     0.0189       % deflection at node 5
11.000   -500.00    -500.00       % bending moment at node 6
12.000    0.0200     0.0200       % deflection at node 6
```

‡

♣ **Example 8.9.5** Find the deflection of a frame of L-shape (see Fig. 8.9.2) which is made of two beams of lengths of 60 in. and 20 in., respectively. Both beams have cross-sections of 2 in. height by 1 in. width. The elastic modulus is 30×10^6 psi. The frame is subjected to a concentrated load of 60 lb at the end of the smaller beam and one end of the long member is fixed. Use 6 elements to find the deflection of the frame. The MATLAB program is written below using 2-D frame elements.

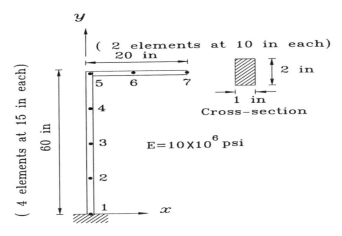

Figure 8.9.2 An L-Shaped Frame

```
%----------------------------------------------------------------
% EX895.m: MATLAB program to solve static deflection for a 2-D frame
%
% Variable descriptions
%   x and y = global x and y coordinates of each node
%   k = element stiffness matrix
%   kk = system stiffness matrix
%   ff = system force vector
%   index = a vector containing system dofs associated with each element
%   bcdof = a vector containing dofs associated with boundary conditions
%   bcval = a vector containing boundary condition values associated with
%           the dofs in bcdof
%----------------------------------------------------------------
nel=6;                          % number of elements
nnel=2;                         % number of nodes per element
ndof=3;                         % number of dofs per node
nnode=(nnel-1)*nel+1;           % total number of nodes in system
sdof=nnode*ndof;                % total system dofs
x(1)=0; y(1)=0;                 % x, y coords. of node 1 in terms of the global axis
x(2)=0; y(2)=15;                % x, y coords. of node 2 in terms of the global axis
x(3)=0; y(3)=30;                % x, y coords. of node 3 in terms of the global axis
x(4)=0; y(4)=45;                % x, y coords. of node 4 in terms of the global axis
x(5)=0; y(5)=60;                % x, y coords. of node 5 in terms of the global axis
x(6)=10; y(6)=60;               % x, y coords. of node 6 in terms of the global axis
x(7)=20; y(7)=60;               % x, y coords. of node 7 in terms of the global axis
el=30*10^6;                     % elastic modulus
area=2;                         % cross-sectional area
xi=2/3;                         % moment of inertia of cross-section
rho=1;                          % mass density per volume (dummy value for static analysis)
bcdof(1)=1;                     % transverse deflection at node 1 is constrained
```

```
bcval(1)=0;                              % whose described value is 0
bcdof(2)=2;                % axial displacement at node 1 is constrained
bcval(2)=0;                              % whose described value is 0
bcdof(3)=3;                      % slope at node 1 is constrained
bcval(3)=0;                              % whose described value is 0
ff=zeros(sdof,1);                    % initialization of system force vector
kk=zeros(sdof,sdof);                 % initialization of system matrix
index=zeros(nnel*ndof,1);            % initialization of index vector
ff(20)=-60;        % load applied at node 7 in the negative y direction
for iel=1:nel                   % loop for the total number of elements
index=feeldof1(iel,nnel,ndof);       % extract system dofs for each element
node1=iel;                      % starting node number for element 'iel'
node2=iel+1;                    % ending node number for element 'iel'
x1=x(node1); y1=y(node1);       % x and y coordinate values of 'node1'
x2=x(node2); y2=y(node2);       % x and y coordinate values of 'node2'
leng=sqrt((x2-x1)^2+(y2-y1)^2);           % length of element 'iel'
if (x2-x1)==0;    % compute the angle between the local and global axes
if y2 > y1;
beta=pi/2;
else
beta=-pi/2;
end
else
beta=atan((y2-y1)/(x2-x1));
end
k=feframe2(el,xi,leng,area,rho,beta,1);      % compute element matrix
kk=feasmbl1(kk,k,index);               % assembly into system matrix
end
[kk,ff]=feaplyc2(kk,ff,bcdof,bcval);     % apply the boundary conditions
fsol=kk\ff;                    % solve the matrix equation and print
%————————————————————————————
% Print both exact and fem solutions
%————————————————————————————
num=1:1:sdof;
store=[num' fsol]
%————————————————————————————

function [k,m]=feframe2(el,xi,leng,area,rho,beta,ipt)
%————————————————————————————
% Purpose:
% Stiffness and mass matrices for the 2-D frame element
% nodal dof u_1 v_1 theta_1 u_2 v_2 theta_2
%
% Synopsis:
% [k,m]=feframe2(el,xi,leng,area,rho,beta,ipt)
%
% Variable Description:
% k - element stiffness matrix (size 6x6)
```

```
% m - element mass matrix (size 6x6)
% el - elastic modulus
% xi - second moment of inertia of cross-section
% leng - element length
% area - area of beam cross-section
% rho - mass density (mass per unit volume)
% beta - angle between the local and global axes
%        is positive if the local axis is in the ccw direction from
%        the global axis
% ipt = 1 - consistent mass matrix
%     = 2 - lumped mass matrix
%     = 3 - diagonal mass matrix
%————————————————————————————————
%
% stiffness matrix at the local axis
%
a=el*area/leng;
c=el*xi/(leng^3);
kl=[a 0 0 -a 0 0;...
    0 12*c 6*leng*c 0 -12* c 6*leng*c;...
    0 6*leng*c 4*leng^2*c 0 -6*leng*c 2*leng^2*c;...
    -a 0 0 a 0 0;...
    0 -12*c -6*leng*c 0 12*c -6*leng*c;...
    0 6*leng*c 2*leng^2*c 0 -6*leng*c 4*leng^2*c];
%
% rotation matrix
%
  r=[ cos(beta) sin(beta) 0 0 0 0;...
     -sin(beta) cos(beta) 0 0 0 0;...
      0 0 1 0 0 0;...
      0 0 0 cos(beta) sin(beta) 0;...
      0 0 0 -sin(beta) cos(beta) 0;...
      0 0 0 0 0 1];
%
% stiffness matrix at the global axis
%
k=r'*kl*r;
% consistent mass matrix
%
if ipt==1
```

```
%
  mm=rho*area*leng/420;
  ma=rho*area*leng/6;
  ml=[2*ma 0 0 ma 0 0;...
     0 156*mm 22*leng*mm 0 54*mm -13*leng*mm;...
     0 22*leng*mm 4*leng^2*mm 0 13*leng*mm -3*leng^2*mm;...
     ma 0 0 2*ma 0 0;...
     0 54*mm 13*leng*mm 0 156*mm -22*leng*mm;...
     0 -13*leng*mm -3*leng^2*mm 0 -22*leng*mm 4*leng^2*mm];
%
% lumped mass matrix
%
elseif ipt==2
%
  ml=zeros(6,6);
  mass=rho*area*leng;
  ml=mass*diag([0.5 0.5 0 0.5 0.5 0]);
%
% diagonal mass matrix
%
else
%
  ml=zeros(6,6);
  mass=rho*area*leng;
  ml=mass*diag([0.5 0.5 leng^2/78 0.5 0.5 leng^2/78]);
%
end
%
% mass in the global system
%
m=r'*ml*r;
%----------------------------------------------------------------
```

The finite element solution is compared to the exact solution at some selected nodes as given below:

```
    store =
    dof #      fem sol     exact
    1.0000     0.0000      0.0000       % horizontal displ. at node 1
    2.0000     0.0000      0.0000       % vertical displ. at node 1
    3.0000     0.0000      0.0000       % slope at node 1
    4.0000     0.0068                   % horizontal displ. at node 2
    5.0000     0.0000                   % vertical displ. at node 2
    6.0000    -0.0009                   % slope at node 2
    7.0000     0.0270                   % horizontal displ. at node 3
    8.0000     0.0000                   % vertical displ. at node 3
      :
    13.000     0.1080      0.1080       % horizontal displ. at node 5
```

14.000	-0.0001		% vertical displ. at node 5
15.000	-0.0036	-0.0036	% slope at node 5
16.000	0.1080		% horizontal displ. at node 6
17.000	-0.0386		% vertical displ. at node 6
18.000	-0.0040		% slope at node 6
19.000	0.1080	0.1080	% horizontal displ. at node 7
20.000	-0.0801	-0.0801	% vertical displ. node 7
21.000	-0.0042	-0.0042	% slope at node 7

‡

8.10 MATLAB Application to Eigenvalue Analysis

Eigenvalue problems of a beam or a frame structure are solved using the finite element method written in MATLAB programs. The *m-files* described in the previous section compute both the element stiffness and mass matrices so that they are used in the present programs in order to compute the natural frequencies of a beam or a frame structure. To this end, we need to assemble element stiffness and mass matrices into the system stiffness and mass matrices. One *m-file* used here is

feaplycs.m : application of constraints to both mass and stiffness matrices

This *m-file* modifies the eigenvalue matrix equation with given constraints. Instead of redimensioning the matrix size because of the constraints, the original matrix size is conserved. However, the modified eigenvalue matrix equation will contain fictitious zero eigenvalues in the same number of the constraints. As a result, the user should exclude these zero eigenvalues from the computer solution. Except for these, the structure of computer programs is the same as that in examples in the last section. The following examples show the computer programs written in MATLAB to compute the natural frequencies.

♣ **Example 8.10.1**   Find the natural frequencies of a free beam of unit length. It has a cross-section 1 by 1 and it has also mass density of 1. The elastic modulus of the beam is 12. All the units are consistent. Use 4 elements to model the whole beam so that nonsymmetric mode shapes can be included. Use Hermitian beam elements and consistent mass matrices. The computer program is listed below:

```
%————————————————————————————
% EX8101.m: MATLAB program to find the natural frequencies of a free
% beam using Hermitian elements
%
% Variable descriptions
% k = element stiffness matrix
```

```
% m = element mass matrix
% kk = system stiffness matrix
% mm = system mass matrix
% index = a vector containing system dofs associated with each element
% bcdof = a vector containing dofs associated with boundary conditions
% bcval = a vector containing boundary condition values associated with
%                the dofs in bcdof
%----------------------------------------------------------------
nel=4;                              % number of elements
nnel=2;                             % number of nodes per element
ndof=2;                             % number of dofs per node
nnode=(nnel-1)*nel+1;               % total number of nodes in system
sdof=nnode*ndof;                    % total system dofs
el=12;                              % elastic modulus
xi=1/12;                            % moment of inertia of cross-section
rho=1;                              % mass density
tleng=1;                            % total length of the beam
leng=tleng/nel;                     % uniform mesh (equal size of elements)
area=1;                             % cross-sectional area
ipt=1;                              % flag for consistent mass matrix
kk=zeros(sdof,sdof);                % initialization of system stiffness matrix
mm=zeros(sdof,sdof);                % initialization of system mass matrix
index=zeros(nnel*ndof,1);           % initialization of index vector
for iel=1:nel                       % loop for the total number of elements
  index=feeldof1(iel,nnel,ndof);    % extract system dofs for each element
  [k,m]=febeam1(el,xi,leng,area,rho,ipt);   % compute element matrices
  kk=feasmbl1(kk,k,index);          % assembly of system stiffness matrix
  mm=feasmbl1(mm,m,index);          % assembly of system mass matrix
end
fsol=eig(kk,mm);                    % solve the eigenvalue problem
fsol=sqrt(fsol)                     % print circular frequencies
%----------------------------------------------------------------
```

The finite element solution is compared to the exact solution below:

mode #	fem sol	exact	
0	0.0000	0.0000	% rigid body mode
1	0.0000	0.0000	% rigid body mode
2	22.400	22.373	% first non-zero circular frequency
3	62.060	61.673	% second non-zero circular frequency
4	121.86	120.90	% third non-zero circular frequency
5	223.29	178.27	% fourth non-zero circular frequency

As seen in the comparison, the two solutions agree well for lower frequencies. However, the discrepancy becomes larger for higher natural frequencies. In order to obtain more accurate higher modes, the finite element model should have a refined mesh to represent the corresponding mode shapes properly. As a result, if we want to improve the fourth non-zero frequency in this example, we need to refine the mesh. ‡

♣ **Example 8.10.2** Find the natural frequencies of a cantilever beam whose length is 1 m long. The beam has the cross-section of 0.02 m by 0.02 m and the mass density is 1000 Kg/m^3. The elastic and shear modulii are 100 GPa and 40 GPa, respectively. Use 4 elements. The MATLAB program is shown below.

```
%─────────────────────────────────
% EX8102.m: MATLAB program to solve the natural frequencies of
% a beam using beam elements with displacement degrees of
% freedom only
%
% Variable descriptions
% k = element stiffness matrix
% kk = system stiffness matrix
% m = element mass matrix
% mm = system mass matrix
% index = a vector containing system dofs associated with each element
% bcdof = a vector containing dofs associated with boundary conditions
%─────────────────────────────────
nel=4;                              % number of elements
nnel=2;                             % number of nodes per element
ndof=3;                             % number of dofs per node
nnode=(nnel-1)*nel+1;               % total number of nodes in system
sdof=nnode*ndof;                    % total system dofs
el=100*10^9;                        % elastic modulus
sh=40*10^9;                         % shear modulus
tleng=1;                            % total beam length
leng=tleng/nel;                     % same size of beam elements
heig=0.02;                          % height (or thickness) of the beam
width=0.02;                         % width of the beam
rho=1000;                           % mass density of the beam
bcdof(1)=1;                         % bottom inplane displ. at node 1 is constrained
bcdof(2)=2;                         % top inplane displ. at node 1 is constrained
bcdof(3)=3;                         % transverse displ. at node 1 is constrained
kk=zeros(sdof,sdof);                % initialization of system stiffness matrix
mm=zeros(sdof,sdof);                % initialization of system mass matrix
index=zeros(nnel*ndof,1);           % initialization of index vector
for iel=1:nel                       % loop for the total number of elements
index=feeldof1(iel,nnel,ndof);      % extract system dofs for each element
[k,m]=febeam3(el,sh,leng,heig,width,rho);  % compute element matrices
kk=feasmbl1(kk,k,index);            % assembly of system stiffness matrix
mm=feasmbl1(mm,m,index);            % assembly of system mass matrix
end
[kk,mm]=feaplycs(kk,mm,bcdof);      % apply the boundary conditions
fsol=eig(kk,mm);                    % solve the matrix equation and print
fsol=sqrt(fsol)
%─────────────────────────────────
```

The natural frequencies obtained from the finite element program are compared to the exact solutions

mode #	fem sol	exact	
1	200.00	203.00	% first circular natural frequency
2	1260.0	1272.0	% second circular natural frequency
3	4040.0	3562.0	% third circular natural frequency

‡

♣ **Example 8.10.3** Find the natural frequencies of a frame of L-shape which is made of two beams of length of 1 m each as seen in Fig. 8.9.2. Both beams have cross-sections of 0.01 m by 0.01 m. The elastic modulus is 100 GPa. The beam has mass density of 1000 Kg/m^3. Use 10 elements.

```
%————————————————————————————————————————
% EX8103.m: MATLAB program to find the natural frequencies for a 2-D
% frame using frame elements
%
% Variable descriptions
% x and y = global x and y coordinates of each node
% k = element stiffness matrix
% kk = system stiffness matrix
% m = element mass matrix
% mm = system mass matrix
% index = a vector containing system dofs associated with each element
% bcdof = a vector containing dofs associated with boundary conditions
%————————————————————————————————————————
nel=10;                    % number of elements
nnel=2;                    % number of nodes per element
ndof=3;                    % number of dofs per node
nnode=(nnel-1)*nel+1;      % total number of nodes in system
sdof=nnode*ndof;           % total system dofs
x(1)=0; y(1)=0;        % x, y coord. of node 1 in terms of the global axis
x(2)=0; y(2)=0.2;      % x, y coord. of node 2 in terms of the global axis
x(3)=0; y(3)=0.4;      % x, y coord. of node 3 in terms of the global axis
x(4)=0; y(4)=0.6;      % x, y coord. of node 4 in terms of the global axis
x(5)=0; y(5)=0.8;      % x, y coord. of node 5 in terms of the global axis
x(6)=0; y(6)=1;        % x, y coord. of node 6 in terms of the global axis
x(7)=0.2; y(7)=1;      % x, y coord. of node 7 in terms of the global axis
x(8)=0.4; y(8)=1;      % x, y coord. of node 8 in terms of the global axis
x(9)=0.6; y(9)=1;      % x, y coord. of node 9 in terms of the global axis
x(10)=0.8; y(10)=1;    % x, y coord. of node 10 in terms of the global axis
x(11)=1; y(11)=1;      % x, y coord. of node 11 in terms of the global axis
el=100*10^9;               % elastic modulus
area=0.0001;               % cross-sectional area
xi=8.3333*10^(-10);        % moment of inertia of cross-section
rho=1000;                  % mass density per volume
bcdof(1)=1;                % transverse deflection at node 1 is constrained
bcdof(2)=2;                % axial displacement at node 1 is constrained
```

```
bcdof(3)=3;                              % slope at node 1 is constrained
kk=zeros(sdof,sdof);                     % initialization of system stiffness matrix
mm=zeros(sdof,sdof);                     % initialization of system mass matrix
index=zeros(nnel*ndof,1);                % initialization of index vector
for iel=1:nel                            % loop for the total number of elements
index=feeldof1(iel,nnel,ndof);           % extract system dofs for each element
node1=iel;                               % starting node number for element 'iel'
node2=iel+1;                             % ending node number for element 'iel'
x1=x(node1); y1=y(node1);                % x and y coordinate values of 'node1'
x2=x(node2); y2=y(node2);                % x and y coordinate values of 'node2'
leng=sqrt((x2-x1)^2+(y2-y1)^2);          % length of element 'iel'
if (x2-x1)==0;    % compute the angle between the local and global axes
beta=pi/2;
else
beta=atan((y2-y1)/(x2-x1));
end
[k,m]=feframe2(el,xi,leng,area,rho,beta,1);      % element matrix
kk=feasmbl1(kk,k,index);                 % assembly of system stiffness matrix
mm=feasmbl1(mm,m,index);                 % assembly of system mass matrix
end
[kk,mm]=feaplycs(kk,mm,bcdof);           % apply the boundary conditions
fsol=eig(kk,mm);                         % solve the matrix equation and print
fsol=sqrt(fsol)
%——————————————————————————————————————
```

The numerical solutions are

mode #	fem sol.	
1	34	% first circular natural frequency
2	92	% second circular natural frequency
3	455	% third circular natural frequency
4	667	% fourth circular natural frequency

‡

8.11 MATLAB Application to Transient Analysis

In the transient analysis of a structure, the equation of motion at time t is

$$[M]\{\ddot{d}\}^t + [C]\{\dot{d}\}^t + [K]\{d\}^t = \{F\}^t \qquad (8.11.1)$$

where $[M]$, $[C]$ and $[K]$ are the system mass, damping, and stiffness matrices and they are assumed to be independent of time. Superscript t denotes time. We will present the direct time integration scheme to solve Eq. (8.11.1). There are many integration techniques which can be applied to the matrix equation. Readers may refer to Refs. [16-18]. In this section, the central difference scheme is explained because it is one of the most popular techniques in the structural mechanics application.

There are two versions of the central difference scheme. The first method is summarized below. Detailed derivation of this technique is provided in [16].

1. Compute system matrices like $[M]$, $[C]$, and $[K]$.
2. Solve for the initial acceleration $\{\ddot{d}\}^0$ from
$$\{\ddot{d}\}^0 = [M]^{-1}\left(\{F\}^0 - [C]\{\dot{d}\}^0 - [K]\{d\}^0\right)$$
where $\{d\}^0$ and $\{\dot{d}\}^0$ are the initial displacement and velocity vectors.
3. Compute the fictitious displacement at time $-\Delta t$ from
$$\{d\}^{-\Delta t} = \{d\}^0 - (\Delta t)\{\dot{d}\}^0 + \frac{\Delta t^2}{2}\{\ddot{d}\}^0$$
4. Compute the effective mass matrix.
$$[\tilde{M}] = \frac{1}{\Delta t^2}[M] + \frac{1}{2\Delta t}[C]$$
5. Repeat 6 through 9 for each time step.
6. Compute the effective force vector.
$$\{\tilde{F}\}^t = \{F\}^t - \left([K] - \frac{2}{\Delta t^2}[M]\right)\{d\}^t - \left(\frac{1}{\Delta t^2}[M] - \frac{1}{2\Delta t}[C]\right)\{d\}^{t-\Delta t}$$
7. Find the displacement at time $t + \Delta t$ from
$$\{d\}^{t+\Delta t} = [\tilde{M}]^{-1}\{\tilde{F}\}^t$$
8. Find the acceleration at time t.
$$\{\ddot{d}\}^t = \frac{1}{\Delta t^2}\left(\{d\}^{t+\Delta t} - 2\{d\}^t + \{d\}^{t-\Delta t}\right)$$
9. Find the velocity at time t.
$$\{\dot{d}\} = \frac{1}{2\Delta t}\left(\{d\}^{t+\Delta t} - \{d\}^{t-\Delta t}\right)$$
where Δt is the time step size.

The second form of the central difference scheme, called *summed form* [19], is described below.

1. Repeat 2 through 4 for each time step.
2. Compute the acceleration.
$$\{\ddot{d}\}^t = [M]^{-1}\left(\{F\}^t - [C]\{\dot{d}\}^t - [K]\{d\}^t\right)$$
3. Compute the velocity from the acceleration.
$$\{\dot{d}\}^{t+0.5\Delta t} = \{\dot{d}\}^{t-0.5\Delta t} + \Delta t\{\ddot{d}\}^t$$
4. Compute the displacement from the velocity.
$$\{d\}^{t+\Delta t} = \{d\}^t + \Delta t\{\dot{d}\}^{t+0.5\Delta t}$$

Both central difference techniques are conditionally stable. The critical time step size for stability is

$$\Delta t_{crit} = \frac{T_{min}}{\pi} \qquad (8.11.2)$$

where T_{min} is the smallest period of the discretized system with finite degrees of freedom. Therefore, the time step size Δt must be smaller than or equal to this critical size to maintain numerical stability.

Comparison of the two central difference techniques are discussed below.

(1) The first technique computes nodal displacements, velocities, and accelerations at the same time steps while the second technique computes them at different time steps. That is, nodal displacements and accelerations are determined at the same time steps but velocities are found at the middle of the time steps. As a result, in order to compute both kinetic and strain energies, for example, at the same time step, an interpolation technique is used for the second central difference scheme to have both displacements and velocities at the same time steps.

(2) The first technique does not have to calculate nodal velocities and accelerations to march along the time if a user does not need them. On the other hand, the second technique needs to compute all of them to progress along the time.
(3) In terms of computer programming, the second method is much easier than the first method.
(4) Both techniques require initial solutions at some fictitious time steps. The first scheme needs the displacements at time $-\Delta t$ while the second requires the velocities at time $-0.5\Delta t$ to initiate the computations. The first technique has the consistent way to determine the solution at the fictitious time step but the second technique does not have that. Hence, the first technique can be used to find the velocity solution at the fictitious time step for the second technique and the procedure for the second scheme is used after that. In other words, we compute $\{d\}^{-\Delta t}$ from the first central difference technique. Then the fictitious velocity vector $\{\dot{d}\}^{-0.5\Delta t}$ is obtained from

$$\{\dot{d}\}^{-0.5\Delta t} = \frac{\{d\}^0 - \{d\}^{-\Delta t}}{\Delta t} \qquad (8.11.3)$$

Numerical experimentation was conducted in Ref. [30]. Both central difference schemes were applied to a crack propagation problem. The study showed that solutions from the both schemes were almost identical. In the study, the fictitious velocity for the second technique was assumed to be the same as the initial velocity at time 0. The solution obtained with this assumption and the second method was almost identical to the solution from the first method. The following example shows the second central difference scheme applied to a beam problem.

♣ **Example 8.11.1** Find the transient response of a cantilever beam whose length is 1 m long. The cross-sectional area of the beam is 0.004 m^2, the moment of inertia is 1.3333×10^{-8} m^4, and the mass density is 1000 Kg/m^3. The elastic modulus is 100 GPa. The beam is initially at rest and subjected to a constant tip load of 100 N at time 0. The following computer program uses the second central difference method to determine the transient response. Four *Hermitian* beam elements are used. The critical time step size for this finite element system is 1.149×10^{-4} sec. The program uses $\Delta t = 1 \times 10^{-4}$ sec. The program is listed below.

```
%————————————————————————
% EX8111.m: MATLAB program to find the transient response
% of a cantilever beam with a tip load
%
% Variable descriptions
% k = element stiffness matrix
% kk = system stiffness matrix
% m = element mass matrix
% mm = system mass matrix
% index = a vector containing system dofs associated with each element
% bcdof = a vector containing dofs associated with boundary conditions
```

```
% acc = acceleration of nodal variables
% vel = velocity of nodal variables
% disp = displacement of nodal variables
%----------------------------------------------------------------
nel=4;                              % number of elements
nnel=2;                             % number of nodes per element
ndof=2;                             % number of dofs per node
nnode=(nnel-1)*nel+1;               % total number of nodes in system
sdof=nnode*ndof;                    % total system dofs
el=100*10^9;                        % elastic modulus
tleng=1;                            % total beam length
leng=tleng/nel;                     % same size of beam elements
xi=1.3333*10^-8;                    % height (or thickness) of the beam
area=0.004;                         % cross-sectional area of the beam
rho=1000;                           % mass density of the beam
ipt=1;             % option flag for mass matrix (consistent mass matrix)
dt=0.0001;                          % time step size
ti=0;                               % initial time
tf=0.2;                             % final time
nt=fix((tf-ti)/dt);                 % number of time steps
nbc=2;                              % number of constraints
bcdof(1)=1;          % transverse displ. at node 1 is constrained
bcdof(2)=2;          % slope at node 1 is constrained
kk=zeros(sdof,sdof);                % initialization of system stiffness matrix
mm=zeros(sdof,sdof);                % initialization of system mass matrix
force=zeros(sdof,1);                % initialization of force vector
index=zeros(nnel*ndof,1);           % initialization of index vector
acc=zeros(sdof,nt);                 % initialization of acceleration matrix
vel=zeros(sdof,nt);                 % initialization of velocity matrix
disp=zeros(sdof,nt);                % initialization of displ. matrix
vel(:,1)=zeros(sdof,1);             % initial zero velocity
disp(:,1)=zeros(sdof,1);            % initial zero displacement
force(9)=100;                       % tip load of 100
for iel=1:nel                       % loop for the total number of elements
index=feeldof1(iel,nnel,ndof);      % extract system dofs for each element
[k,m]=febeam1(el,xi,leng,area,rho,ipt);   % compute element matrices
kk=feasmbl1(kk,k,index);            % assembly of system stiffness matrix
mm=feasmbl1(mm,m,index);            % assembly of system mass matrix
end
mminv=inv(mm);                      % invert the mass matrix
% central difference scheme for time integration
for it=1:nt                         % time integration loop
acc(:,it)=mminv*(force-kk*disp(:,it));    % compute acceleration
% application of constrained conditions
for i=1:nbc                         % loop for number of constraints
ibc=bcdof(i);                       % nodal dof where constraint is applied
acc(ibc,it)=0;          % acceleration at the constrained dof set to 0
end
```

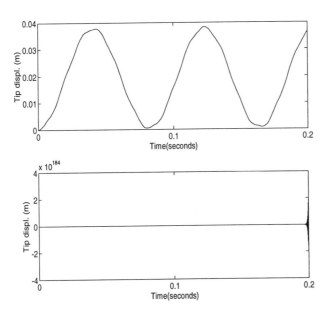

Figure 8.11.1 Time Responses using Central Difference Method

```
vel(:,it+1)=vel(:,it)+acc(:,it)*dt;                    % compute velocity
disp(:,it+1)=disp(:,it)+vel(:,it+1)*dt;                % compute displacement
end
acc(:,nt+1)=mminv*(force-kk*disp(:,nt+1));             % accel. at last time step
%————————————————
% plot of the tip deflection
%————————————————
time=0:dt:nt*dt;
plot(time,disp(9,:))
xlabel('Time(seconds)')
ylabel('Tip displ. (m)')
%————————————————————————————————
```

The plot of the tip deflection is shown in Fig. 8.11.1. The tip deflection shows an oscillation around the static deflection. Figure 8.11.1 also shows the same deflection when the time step size is 1.15×10^{-4}. Because the time step size is larger than the critical step size, the deflection becomes unstable and diverges.

‡

8.12 MATLAB Application to Modal Analysis of Undamped System

In this section, the dynamic analysis of multiple degrees of freedom systems is presented and applied to beam structure examples. The multiple degrees of freedom systems are quite different from single degrees of freedom systems in terms of mathematical formulations and associated time responses. For a multiple degrees of freedom system, we define *modes* which represent each component of overall dynamic responses. The *modes* are essential in describing the nature of motion and provide physical understanding of the dynamic behavior of the system.

The *modes* are characterized by so-called eigenvalues and eigenvectors of the system. The eigenvalues are related to usually natural frequencies and eigenvectors to mode shapes of the given system. Unfortunately, these eigenvalues and eigenvectors are limited mostly to linear systems. This limitation is not significant; in fact, the majority of the dynamic systems are represented by linear systems. There are also plenty of computer software tools available for linear system analyses.

Some key concepts are introduced in this section including a solution technique for eigenvalue/eigenvector problems, which is sometimes called the *modal analysis*. The time response of a system is obtained in a straightfoward manner once the modal analysis results are ready. We consider an undamped system here and a damped system is discussed in the subsequent section.

For a given n degree of freedom linear second order system, the governing differential equation of motion is described by the second order matrix equation as

$$[M]\{\ddot{d}\} + [K]\{d\} = \{F\} \tag{8.12.1}$$

We seek to find the natural motion of the system, i.e. response without any forcing function. The form of response or solution is assumed as

$$\{d(t)\} = \{\phi\} e^{i\omega t} \tag{8.12.2}$$

where $\{\phi\}$ is the mode shape (eigenvector) and ω is the natural frequency of the motion. In other words, the motion is assumed to be purely sinusoidal due to zero damping in the system. The general solution turns out to be a linear combination of each mode as

$$\{d(t)\} = c_1\{\phi_1\}e^{i\omega_1 t} + c_2\{\phi_2\}e^{i\omega_2 t} + \cdots + c_n\{\phi_n\}e^{i\omega_n t} \tag{8.12.3}$$

where each constant(c_i) is evaluated from initial conditions. Substituting Eq. (8.12.2) into Eq. (8.12.1) with $\{F\} = 0$ yields [31,32]

$$(-\omega^2[M] + [K])\{\phi\}e^{i\omega t} = 0 \tag{8.12.4}$$

The above equation has a nontrivial solution if $(-\omega^2[M] + [K])$ becomes singular. In other words, there exist n number of ωs which satisfy

$$|-\omega^2[M] + [K]| = |-\lambda[M] + [K]| = 0 \tag{8.12.5}$$

Section 8.12 MATLAB Application to Modal Analysis of Undamped System

where $\lambda = \omega^2$ is the eigenvalue of the system. Equation (8.12.5) produces solutions $\omega_1^2, \omega_2^2, \ldots, \omega_n^2$. Since the mass matrix is positive definite and the stiffness matrix is at least positive semidefinite, all ω_is are nonnegative. This can be easily proven from

$$\omega_i^2 [M]\{\phi_i\} = [K]\{\phi_i\} \tag{8.12.6}$$

for the ith eigenvalue and eigenvector. Let us multiply $\{\phi_i\}^T$ on both sides of the equation

$$\omega_i^2 \{\phi_i\}^T [M]\{\phi_i\} = \{\phi_i\}^T [K]\{\phi_i\} \tag{8.12.7}$$

The properties of the mass and stiffness matrices state that

$$\{\mathbf{x}\}^T [M]\{\mathbf{x}\} > 0, \qquad \{\mathbf{x}\}^T [K]\{\mathbf{x}\} \geq 0 \qquad \text{for } \{\mathbf{x}\} \neq 0$$

In addition, the eigenvectors are orthogonal to each other, which can be easily shown from

$$\omega_i^2 [M]\{\phi_i\} = [K]\{\phi_i\} \tag{8.12.8}$$
$$\omega_j^2 [M]\{\phi_j\} = [K]\{\phi_j\} \tag{8.12.9}$$

For proof, we premultiply $\{\phi_j\}^T$ on both sides of Eq. (8.12.8) and subtract the transpose of Eq. (8.12.9) which is post-multiplied by $\{\phi_i\}$. The result becomes

$$(\omega_i^2 - \omega_j^2)\{\phi_j\}^T [M]\{\phi_i\} = \{\phi_j\}^T [K]\{\phi_i\} - \{\phi_j\}^T [K]\{\phi_i\} = 0 \tag{8.12.10}$$

Therefore, if $i \neq j$

$$\{\phi_j\}^T [M]\{\phi_i\} = \{\phi_j\}^T [K]\{\phi_i\} = 0 \tag{8.12.11}$$

and the eigenvectors are orthogonal with respect to the mass and stiffness matrices. The above orthogonality property includes systems with non-repeated rigid body degrees of freedom. For multiple rigid body modes, for example a three dimensional translational motion, a special form of orthogonality exists (see Ref. [16] for the special form). Orthogonality of eigenvectors in conjunction with positive definiteness and positive semidefiniteness of mass and stiffness matrices of a vibrational system is one of the distinct features of linear dynamic systems.

The orthogonality of eigenvectors provides a useful normalization technique in the form

$$\{\phi_i\}^T [M]\{\phi_i\} = 1, \qquad \{\phi_j\}^T [K]\{\phi_i\} = \omega_i^2 \tag{8.12.12}$$

Once the eigenvectors are normalized, the following coordinate transformation is proposed

$$\{d\} = [\Phi]\{\eta\} \tag{8.12.13}$$

where

$$[\Phi] = [\phi_1, \phi_2, \ldots, \phi_n] \tag{8.12.14}$$

is called a *modal matrix* whose columns consist of normalized eigenvectors, and $\{\eta\} = [\eta_1, \eta_2, \ldots, \eta_n]^T$ is the vector of modal coordinates. Substitution of Eq. (8.12.14) into Eq. (8.12.1) yields

$$[M][\Phi]\{\ddot{\eta}\} + [K][\Phi]\{\eta\} = \{F\} \tag{8.12.15}$$

Next, we premultipy Φ^T on both sides of Eq. (8.12.15), so that

$$[\Phi]^T M[\Phi]\{\ddot{\eta}\} + [\Phi]^T [K][\Phi] \{\eta\} = \Phi^T \{F\} \qquad (8.12.16)$$

According to the orthogonality in Eq. (8.12.12), Eq. (8.12.16) can be rewritten as

$$\{\ddot{\eta}\} + diag[\omega_i^2] \{\eta\} = \Phi^T \{F\} = \{f_i\} \qquad (8.12.17)$$

In other words, the system of equations is decoupled.

$$\ddot{\eta}_i + \omega_i^2 \eta_i = f_i \qquad (8.12.18)$$

where $i = 1, 2, \ldots, n$, and f_i is the ith row of $\Phi^T \{F\}$. Equation (8.12.18) represents the modal coordinate form of equations of motion, for which each independent vibrational mode is described by a decoupled second order differential equation. The modal coordinate equations are so useful since they provide the analytical solution for each mode. Also, the input function into the ith modal coordinate (f_i) represents how much the mode is excited from the external input.

♣ **Example 8.12.1**   Consider a Euler-Bernoulli beam model with one end fixed as in Fig. 8.12.1. For simplicity, the beam is modeled by two finite elements using the *consistent* mass matrix. The numerical data for the structure are ρ (linear mass density)=0.024 kg/m, EI= 6.09 N-m^2, and L=1.27 m. A MATLAB m-file called *femodal.m* produces the following mass and stiffness matrices after applying the boundary condition

$$[M] = \begin{bmatrix} 0.0929 & 0 & 0.0161 & -0.0967 \\ 0 & 1.4881 & 0.0967 & -0.5580 \\ 0.0161 & 0.0967 & 0.0464 & -0.1637 \\ -0.0967 & -0.5580 & -0.1637 & 0.7440 \end{bmatrix}$$

and

$$[K] = 10^4 \times \begin{bmatrix} 0.0052 & 0 & -0.0026 & 0.0326 \\ 0 & 1.0880 & -0.0326 & 0.2720 \\ -0.0026 & -0.0326 & 0.0026 & -0.0326 \\ 0.0326 & 0.2720 & -0.0326 & 0.5440 \end{bmatrix}$$

The corresponding degrees of freedom for these matrices are $\{v_1 \; \theta_1 \; v_2 \; \theta_2\}$ as shown in Fig. 8.12.1. The natural frequency and modal matrix are computed from the MATLAB function file *femodal.m* as follows

$$\begin{Bmatrix} \omega_1 \\ \omega_2 \\ \omega_3 \\ \omega_4 \end{Bmatrix} = \begin{Bmatrix} 3.6692 \\ 23.1786 \\ 78.3943 \\ 227.5337 \end{Bmatrix}, \quad [\Phi] = \begin{bmatrix} -1.3594 & -2.9157 & -0.4570 & 1.9102 \\ -0.0931 & 0.0351 & 0.6871 & 0.7852 \\ -4.0039 & 4.0394 & -4.4925 & 7.5441 \\ -0.1102 & 0.3890 & -0.8665 & 2.9165 \end{bmatrix}$$

Section 8.12   MATLAB Application to Modal Analysis of Undamped System   295

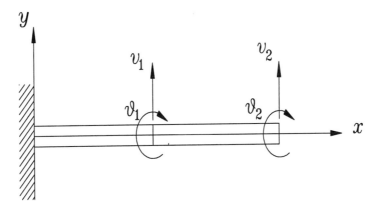

Figure 8.12.1 Two Elements Beam Model

and the modal input force matrix is

$$[\Phi]^T\{F\} = [-4.0039 \quad 4.0394 \quad -4.4925 \quad 7.5441]^T$$

where the original input influence matrix is given by

$$\{F\} = [0 \quad 0 \quad 1 \quad 0]^T$$

The MATLAB source file is provided below as a reference

```
function [Omega,Phi,ModF]=femodal(M,K,F);
%------------------------------------------------
% Purpose
%    The function subroutine femodal.m calculates modal parameters
%    for a given structural system. It calculates natural frequency and
%    eigenvector. The eigenvectors are normalized so that the modal
%    mass matrix becomes an identity matrix.
%
% Synopsis:
%    [Omega, Phi, ModF]=femodal(M,K,F)
%
% Variable Description:
%    Input parameters -
%       M, K - Mass and stiffness matrices
%       F - Input or forcing function
%    Output parameters -
%       Omega - Natural frequency(rad/sec) in ascending order
%       Phi - Modal matrix with each column corresponding to
%             the eigenvector.
%       ModF - Modal input matrices.
%------------------------------------------------
disp(' ')
disp('Please wait!! - The job is being performed.')
```

```
%----------------------------------------
% Solve the eigenvalue problem and normalize the eigenvectors
%----------------------------------------
[n,n]=size(M);[n,m]=size(F);
[V,D]=eig(K,M);
[lambda,k]=sort(diag(D)); V=V(:,k);
Factor=diag(V'*M*V);
Phi=V*inv(sqrt(diag(Factor)));
Omega=diag(sqrt(Vnorm'*K*Vnorm));
Modf=Vnorm'*F;
%----------------------------------------
```

Note that each modal coordinate or finite element degree of freedom can be taken as the output variables of *femodal.m*.

‡

In order to find out the solution to Eq. (8.12.18), the Laplace transformation technique is used.

$$\eta_i(s) = \frac{s\eta_i(0) + \dot{\eta}_i(0)}{s^2 + \omega^2} + \frac{f_i(s)}{s^2 + \omega_i^2} \tag{8.12.19}$$

where $\eta_i(0)$ and $\dot{\eta}_i(0)$ are related to the initial conditions as explained below. Taking the inverse Laplace transform of Eq. (8.12.19) yields the time domain solution

$$\eta_i(t) = \eta_i(0)cos\omega_i t + \frac{\dot{\eta}_i(0)}{\omega_i}sin\omega_i t + \int_0^t \frac{1}{\omega_i}sin\omega_i(t-\tau)f_i(\tau)d\tau \tag{8.12.20}$$

As one might have expected, the solution consists of two parts: i) excitation by the initial condition and ii) response due to the external forcing input. The convolution integral in the solution is not easy to evaluate in general, except for some special cases such as impulse and step inputs.

As shown above the initial conditions $(\eta_i(0), \dot{\eta}_i(0))$ for the modal coordinate are needed for the complete solution. This information can be directly obtained from the original transformation equation, Eq. (8.12.13). That is,

$$\{d(0)\} = [\Phi]\{\eta(0)\} \tag{8.12.21}$$

so that

$$\{\eta(0)\} = [\Phi]^T[M][\Phi]\{d(0)\} \tag{8.12.22}$$

Similarly

$$\{\dot{\eta}(0)\} = [\Phi]^T[M][\Phi]\{\dot{d}(0)\} \tag{8.12.23}$$

Now the solutions of the modal coordinates are combined to produce the solution in physical coordinates.

$$\{d(t)\} = [\Phi]\{\eta(t)\} \tag{8.12.24}$$

Section 8.12   MATLAB Application to Modal Analysis of Undamped System   297

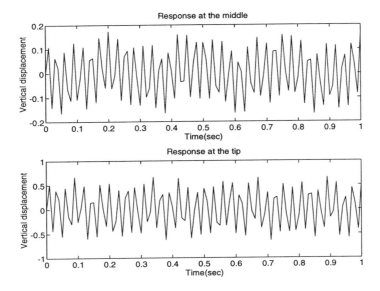

Figure 8.12.2 Impulse Responses by Modal Analysis

In other words,

$$d_i(t) = \sum_{j=1}^{n} \phi_{ij} \eta_j(t), \quad \text{for } i = 1, 2, \ldots, n \quad (8.12.25)$$

♣ **Example 8.12.2**   In this example, the same model is used as in Example 8.12.1 to demonstrate the evaluation of impulse response of the beam. The impulsive force is applied at the tip of the beam. The analytical solution, Eq. (8.12.20), is incorporated into a MATLAB *m-file feiresp.m*. The initial conditions $\{d(0)\}$ and $\{\dot{d}(0)\}$ are set to zero.

The time response results are presented in Fig. 8.12.2. Note that the response time interval is critical to show higher modes in the response. If the time interval is too large, higher modes will not show up in the response. This issue will be discussed later in frequency response analysis in Sec. 8.14.

The source MATLAB *m-file* is presented below:

```
function [eta,yim]=feiresp(M,K,F,u,t,C,q0,dq0);
%----------------------------------------------------------
%  Purpose:
%     The function subroutine feiresp.m calculates impulse response
%     for a given structural system using modal analysis. It uses modal
%     coordinate equations to evaluate modal responses analytically, then
%     convert modal coordinates into physical responses
%
```

```
% Synopsis:
%    [eta,yim]=feiresp(M,K,F,u,t,C,q0,dq0)
%
% Variable Description:
%    Input parameters -
%       M, K - Mass and stiffness matrices
%       F - Input or forcing function
%       u - Index for excitation
%       t - Time period of evaluation
%       C - Output matrix
%       q0, dq0 - Initial conditions
%    Output parameters -
%       eta - modal coordinate response
%       yim - physical coordinate response
%--------------------------------------------------------------
disp(' ')
disp('Please wait!! - The job is being performed.')
%--------------------------------------------------------------
% Solve the eigenvalue problem and normalize the eigenvectors
%--------------------------------------------------------------
[n,n]=size(M);[n,m]=size(F);
nstep=size(t');
[V,D]=eig(K,M);
[lambda,k]=sort(diag(D)); V=V(:,k);
Factor=diag(V'*M*V);
Vnorm=V*inv(sqrt(diag(Factor)));
omega=diag(sqrt(Vnorm'*K*Vnorm));
Fnorm=Vnorm'*F;
%--------------------------------------------------------------
% Find out impulse response of each modal coordinate analytically
%--------------------------------------------------------------
eta0=Vnorm'*M*q0; deta0=Vnorm'*M*dq0; eta=zeros(nstep,n);
for i=1:n phase=omega(i)*t;
eta(:,i)=eta0(i)*cos(phase')+deta0(i)*sin(phase')/omega(i)+...
sin(phase')*Fnorm(i,u);
end
% Convert into physical coordinates
yim=C*Vnorm*eta';
%--------------------------------------------------------------
```

8.13 MATLAB Application to Modal Analysis of Damped System

For an n degree of freedom system with inherent damping, the governing equation of motion can be written as

$$[M]\{\ddot{d}\} + [C]\{\dot{d}\} + [K]\{d\} = \{F\} \tag{8.13.1}$$

where $[C]$ is an n by n damping matrix. The above system is stable due to the introduction of the damping term as explained below. The damping can be classified into inherent structural damping or damping by active control. The stability of the above system can be discussed by taking the total energy(kinetic plus potential) of the system

$$U = \frac{1}{2}\{\dot{d}\}^T[M]\{\dot{d}\} + \frac{1}{2}\{d\}^T[K]\{d\} \tag{8.13.2}$$

Assuming free vibration with $\{F\}=0$, the time rate of change of U becomes

$$\frac{dU}{dt} = \{\dot{d}\}^T([M]\{\ddot{d}\} + [K]\{d\}) \tag{8.13.3}$$

Furthermore, using Eq. (8.13.1)

$$\frac{dU}{dt} = -\{\dot{d}\}^T[C]\{\dot{d}\} \tag{8.13.4}$$

Therefore, as long as the damping matrix $[C]$ satisfies

$$\{\dot{d}\}[C]\{\dot{d}\} > 0, \quad \text{for } \{\dot{d}\} \neq 0 \tag{8.13.5}$$

it follows

$$\frac{dU}{dt} < 0 \tag{8.13.6}$$

and the system is stable with respect to the equilibrium state $(\{d\}, \{\dot{d}\}) = (0,0)$. Estimating the damping matrix for a physical system is not easy in general. There are some methods of modeling the damping matrix. One of the special cases is to use so-called *proportional damping* or *Rayleigh damping* in the form

$$[C] = \alpha[M] + \beta[K] \tag{8.13.7}$$

where α and β are constants. In other words, the damping matrix is proportional to the mass and/or stiffness matrix. Proportional damping has an advantage of possessing the same characteristics as the mass or stiffness matrix. The eigenvectors obtained from the mass and stiffness matrices conserve orthogonality with respect to the damping matrix. That is,

$$\{\phi_i\}^T[C]\{\phi_j\} = 0, \quad \text{for } i \neq j \tag{8.13.8}$$

and $[\Phi]^T[C][\Phi]$ becomes a diagonal matrix. Now, the original governing equation, Eq. (8.13.1), can be rewritten in the modal coordinate form

$$\ddot{\eta}_i + 2\zeta_i\omega_i\dot{\eta}_i + \omega_i^2\eta_i = f_i \tag{8.13.9}$$

Application of the Laplace transform yields

$$\eta_i(s) = \frac{\dot{\eta}_i(0) + (s + 2\zeta_i\omega_i)\eta_i(0) + f_i(s)}{(s^2 + 2\zeta_i\omega_i s + \omega_i^2)} \tag{8.13.10}$$

The inverse Laplace transform of $\eta_i(s)$ becomes

$$\eta_i(t) = \eta_i(0)e^{-\zeta_i\omega_i t}\cos\omega_d t + (\dot{\eta}_i(0) - \frac{\zeta_i}{\sqrt{1-\zeta_i^2}}\eta_i(0))e^{-\zeta_i\omega_i t}\sin(\omega_d t)$$

$$+ \frac{1}{\omega_d}\int_0^t e^{-\zeta_i\omega_i(t-\tau)}\sin\omega_d(t-\tau)f(\tau)d\tau \tag{8.13.11}$$

where $\omega_d = \omega_i\sqrt{1-\zeta_i^2}$ is the damped natural frequency and ω_i is the undamped natural frequency. In most practical cases, the modal damping ratio ζ_i is less than unity so that the damped natural frequency is smaller than the undamped natural frequency. The modal coordinate solution in Eq.(8.13.11) can be used to produce the solution of the physical coordinates

$$\{d(t)\} = [\Phi]\{\eta(t)\} \tag{8.13.12}$$

where $[\Phi]$ is the modal matrix obtained from the damped system.

♣ **Example 8.13.1** In this example, we test the impulse response of a damped system. The system damping matrix is assumed as a proportional damping, and the mass and stiffness matrices are the same as in Example 8.12.1. The proportional constants are chosen as $\alpha=0.2$, $\beta=0.005$. The same unit impulsive force at the tip as in Example 8.12.1, is applied. A MATLAB *fediresp.m* file is written as presented below.

The simulation results are presented in Fig. 8.13.1. They show the damped responses at the two different positions of the beam.

```
function [eta,yim]=fediresp(M,K,F,u,t,C,q0,dq0,a,b);
%----------------------------------------------------------------
% Purpose:
%    The function subroutine fediresp.m calculates impulse response
%    for a damped structural system using modal analysis. It uses modal
%    coordinate equations to evaluate modal responses analytically, then
%    convert modal coordinates into physical responses
%
% Synopsis:
%    [eta,yim]=fediresp(M,K,F,u,t,C,q0,dq0,a,b)
```

Section 8.13 MATLAB Application to Modal Analysis of Damped System

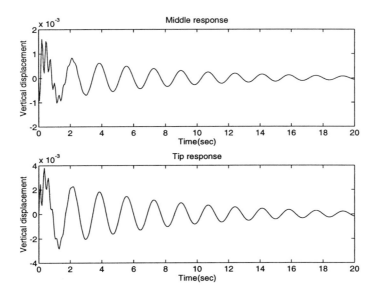

Figure 8.13.1 Impulse Responses for a Damped System

```
%
% Variable Description:
%    Input parameters : M, K - Mass and stiffness matrices
%       F - Input or forcing influence matrix
%       u - Index for excitation
%       t - Time of evaluation
%       u - Index for the excitation
%       C - Output matrix
%       q0, dq0 - Initial conditions
%       a, b - Parameters for proportional damping [C]=a[M]+b[K]
%    Output parameters : eta - modal coordinate response
%       yim - physical coordinate response
%----------------------------------------------------------------
disp(' ')
disp('Please wait!! - The job is being performed.')
%----------------------------------------------------------------
% Solve the eigenvalue problem and normalize the eigenvectors
%----------------------------------------------------------------
[n,n]=size(M);[n,m]=size(F);
nstep=size(t');
[V,D]=eig(K,M);
[lambda,k]=sort(diag(D));         % Sort the eigenvalues and eigenvectors
V=V(:,k);
Factor=diag(V'*M*V);
Vnorm=V*inv(sqrt(diag(Factor)));  % Eigenvectors are normalized
omega=diag(sqrt(Vnorm'*K*Vnorm)); % Natural frequencies
```

```
Fnorm=Vnorm'*F;
%——————————————————————————————————
% Compute modal damping matrix from the proportional damping matrix
%——————————————————————————————————
Modamp=Vnorm'*(a*M+b*K)*Vnorm;
zeta=diag((1/2)*Modamp*inv(diag(omega)))
if (max(zeta) >= 1),
disp('Warning - Maximum damping ratio is greater than or equal to 1')
disp('You have to reselect a and b ')
pause
disp('If you want to continue, type return key')
end
eta0=Vnorm'*M*q0;           % Initial conditions for modal coordinates
deta0=Vnorm'*M*dq0;         % - both displacement and velocity
eta=zeros(nstep,n);
for i=1:n                   % Responses are obtained for n modes
omegad=omega(i)*sqrt(1-zeta(i)^2);
phase=omegad*t;
tcons=zeta(i)*omega(i)*t;
eta(:,i)=exp(-tcons)'.*(eta0(i)*(cos(phase')+zeta(i)/sqrt(1-zeta(i)^2)*...
sin(phase'))+deta0(i)*sin(phase')/omegad+sin(phase')*Fnorm(i,u)...
/omegad);
end
%——————————————————————————————————
% Convert modal coordinate responses to physical coordinate responses
%——————————————————————————————————
yim=C*Vnorm*eta';
%——————————————————————————————————
```

‡

8.14 MATLAB Application to Frequency Response Analysis

The previous modal analysis of a system is mainly based upon the time domain approach. The eigenvalues and eigenvectors directly produce solutions in time domain in the form of time response functions. Modal coordinates make it possible to derive sets of decoupled equations of motion. Each individual modal coordinate solution is combined to result in the physical coordinate solution. The modal analysis provides very convenient tools for understanding behavior of multiple degrees of dynamic systems.

Sometimes, the time domain analysis is not the best choice, especially for modal testing and other applications. One supplementary approach is the frequency domain analysis. The frequency domain method has major advantages over the counterpart, i.e., time domain analysis. In fact, it is being more widely adopted in signal processing, active control system design, modal testing, etc.

Section 8.14 MATLAB Application to Frequency Response Analysis

Most of the vibrational systems can be characterized by their inherent frequency components which dictate both time and frequency responses. One key advantage of the frequency domain analysis is that one can span a whole range of frequencies, which is not possible or impractical in the time domain analysis. Conversion of time domain signals into frequency domain signals and vice versa is relatively easy due to the modern computational power. In this section, discussion on the frequency domain analysis is presented. The Fast Fourier Transform (FFT) and evaluation of the frequency response function for multiple degrees of freedom systems are presented.

Consider a general continuous time domain signal given by $x(t)$. The signal $x(t)$ can be periodic or nonperiodic. It can be represented in the following expression

$$x(t) = \frac{1}{2\pi} \int_{-\infty}^{\infty} X(\Omega) e^{i\Omega t} d\Omega \tag{8.14.1}$$

where $X(\Omega)$ is the *Fourier transform* of the time signal $x(t)$

$$X(\Omega) = \int_{-\infty}^{\infty} x(t) e^{-i\Omega t} dt \tag{8.14.2}$$

The time signal $x(t)$ is also called the *inverse Fourier transform* of $X(\Omega)$. For the existence of the *Fourier transform* the following condition should be satisfied for $x(t)$:

$$\int_{-\infty}^{\infty} |x(t)| dx \tag{8.14.3}$$

should have a finite value. The above constraint is not strict in the sense that it covers a wide range of signals of actual dynamic systems. On the other hand, introducing a new variable $f = \Omega/2\pi$, we have

$$x(t) = \int_{-\infty}^{\infty} X(f) e^{i(2\pi f)} df \tag{8.14.4}$$

and

$$X(f) = \int_{-\infty}^{\infty} x(t) e^{-i(2\pi f t)} dt \tag{8.14.5}$$

Since the *Fourier transform* involves an integral of general time varying complex variables, it is not easy to carry out the integration. Except for some special cases, a numerical integration technique is needed. One efficient algorithm to use is the *Discrete Fourier transform*. The numerical integration is conducted by a finite number of summations at discrete points.

Assume that there are N sampled values as

$$x_k = x(t_k), \quad t_k = k\Delta t, \quad k = 0, 1, 2, \ldots, N-1 \tag{8.14.6}$$

Based upon the sample data points, we assume that the time domain data project into the corresponding frequency domain data. In other words, the *Fourier transform* is defined for the N discrete frequency points

$$(f_1, f_2, \ldots, f_N) \tag{8.14.7}$$

where the frequency points should be in the range of the so-called *Nyquist critical frequency*

$$f_c = \frac{1}{2\Delta t} \qquad (8.14.8)$$

In other words, the sampling period (Δt) should be at least half of the period of a signal to sufficiently represent the signal.

$$X(f_n) = \int_{-\infty}^{\infty} x(t) e^{2\pi i f_n t} dt \sim \sum_{k=0}^{N-1} x_k e^{2\pi i f_n t_k} \Delta t \qquad (8.14.9)$$

The above equation is called the *Discrete Fourier transform*. The *Discrete Fourier transform* (DFT) has the symmetry property with respect to the input frequency (f_n). In other words,

$$X(f_n) = X(f_{N-n}), \qquad f_n = \frac{n}{N\Delta t}, \quad n = -\frac{N}{2}, ..., \frac{N}{2} \qquad (8.14.10)$$

and only half of the transform is needed to represent all frequency components. The maximum frequency range is given by

$$0 < f < \frac{1}{2\Delta t} \qquad (8.14.11)$$

An enhanced version of the DFT is called *Fast Fourier Transform* (FFT) which improves computational efficiency significantly. It turns out that the $e^{i(2\pi f_n t_k)}$ term in Eq. (8.14.9) repeats over the frequency range, and the FFT makes use of this property. The FFT algorithm is implemented in a number of computer software packages and is being used in many different areas. The algorithm is known to be highly efficient in terms of the number of numerical operations. The number of operations for FFT is $(N/2)log_2 N$ when compared to N^2 for DFT. The detailed discussion on the FFT is available in Refs. [16,33].

♣ **Example 8.14.1** The same model used in Example 8.12.1 and the impulse response results in Example 8.12.2 are used in this example to demonstrate the *FFT* technique. A MATLAB *fefft.m* file is written and the input data include both time response data and the sampling time interval. The sampling time interval is transformed into the corresponding frequency scale based upon the *Nyquist critical frequency* using Eq. (8.14.20). The number of data points in the FFT should be power of 2. Otherwise, the MATLAB built-in function *fft* fills the discrepancy with blank data. Figure 8.14.1 represents both the time domain impulse response and the corresponding FFT results. Also, provided below is the MATLAB source file for *fefft.m*.

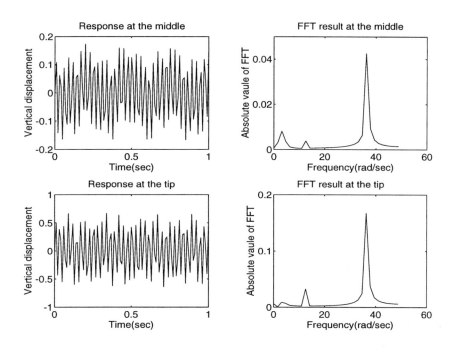

Figure 8.14.1 Impulsive Time Response and FFT

```
function [yfft,freq]=fefft(y,t)
%------------------------------------------------------------
% Purpose:
%    This function subroutine calculates Fast Fourier Transform (FFT)
%    using the time domain signal. The time domain data are provided
%    with corresponding time interval.
%
% Synopsis:
%    [yf, freq]=fefft(y,t)
%
% Variable Description:
%    Input parameters -
%       y - Time domain data n by 1
%       t - Time interval for y of n by 1 size
%    Output parameters -
%       yf - Absolute value of FFT of the time domain data y
%       freq - Frequency axis values
%
% Notes:
%    The number of data points for y should be power of 2, and
%    truncation is needed to achieve the requirement
%------------------------------------------------------------
```

```
%------------------------------------------------
% Compute number of data points and sampling time interval
%------------------------------------------------
ntime=max(size(t));
dt=(t(1,ntime)-t(1,1))/ntime;
%------------------------------------------------
% Extract data points at the power of 2. Truncate extra data points
% so that the final number of data points is in the power of two and
% also as close as possible to the given number of data points
%------------------------------------------------
N=fix(log10(ntime)/log10(2))
% Calculate FFT of the time domain data and take absolute value
yfft=fft(yN(1:2^N,:));
yfft=abs(yfft(1:2^N/2,:))*dt;
%------------------------------------------------
% Set up the frequency scale from the given sampling interval.
% Apply the Nyquist criterion to establish the maximum frequency.
%------------------------------------------------
freq0=0;
freqf= (1/dt)/2;              % Maximum or final frequency value
df=freqf/(2^N/2);             % Frequency interval
freq=0:df:freqf-df;           % Frequency axis values
%------------------------------------------------
```

‡

Problems

8.1 A 4 ft beam is subjected to a uniform load 10 lb/ft and clamped at one side and simply supported at the other side. The beam has elastic modulus of 10×10^6 psi and square cross-section of 2 in. by 2 in. Determine the system stiffness matrix and column vector using two equal size Hermitian beam elements. Find the maximum deflection.

8.2 Redo Prob. 8.1 using the linear Timoshenko beam elements.

8.3 Redo Prob. 8.1 using the linear mixed beam elements.

8.4 Redo Prob. 8.1 using the linear beam elements with displacement degrees of freedom only.

8.5 A beam is 6 in. long, 0.2 in thick and 0.1 in wide as seen in Fig. P8.5. It is subjected to a concentrated moment and a linear pressure load. (a) Construct the system stiffness matrix and system column vector using the Hermitian beam element, (b) apply the boundary conditions, and (c) determine the maximum deflection and bending stress. Use $E = 10 \times 10^6$ psi.

8.6 One beam element is loaded as seen in Fig. P8.6. Determine the element load vector using the Hermitian beam element.

8.7 A Hermitian beam element is loaded as shown in Fig. P8.7. Find the element column vector.

8.8 Redo Prob. 8.5 using the linear Timoshenko beam element.

8.9 Redo Prob. 8.6 using the linear Timoshenko beam element.

8.10 Redo Prob. 8.7 using the linear Timoshenko beam element.

8.11 Repeat Prob. 8.1 using the provided computer programs with 10 elements.

8.12 Repeat Prob. 8.2 using the provided computer programs with 10 elements.

8.13 Repeat Prob. 8.3 using the computer programs with 10 elements.

8.14 Repeat Prob. 8.4 using the computer programs with 10 elements.

8.15 Find the natural frequencies of a beam simply supported at both ends as well as at the center of the beam. The beam is 1m long, 2cm thick, and 1cm wide. It has elastic modulus 10^8 Pa and density 400Kg/m^3. Use 10 Hermitian elements with the provided computer programs.

8.16 A simply supported beam is subjected to a load at the center with a sine function. The beam is 2m long and 4cm thick and 2cm wide. The beam has also an elastic modulus of 50GPa and density 2000Kg/m^3. The applied load is $1000 sin(\pi t)$N. If the beam is initially at rest, find the motion of the center of the beam using 10 Hermitian beam elements and provided computer programs.

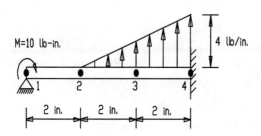

Figure P8.5 Problem 8.5

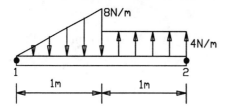

Figure P8.6 Problem 8.6

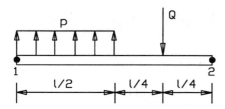

Figure P8.7 Problem 8.7

8.17 A frame structure is shown in Fig. P8.17 with the applied load. The frame is made of a circular cross-sectional beam whose diameter is 0.05m. The elastic modulus of the beam is 200GPa. Find the nodal deflection using the computer programs when the frame is subjected to a concentrated force as seen in the figure.

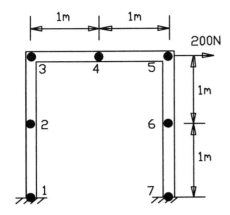

Figure P8.17 Problem 8.17

CHAPTER NINE

ELASTICITY PROBLEM

9.0 Chapter Overview

Continuum finite elements are presented in the chapter using both the weighted residual method and the energy method. Element stiffness matrices are developed for plane stress/strain, 3-D, and axisymmetric problems. Load column vectors are also discussed. In addition, mass matrices for dynamic analysis are derived and thermal stress analysis is presented. MATLAB example programs are attached for 2-D, 3-D, and axisymmetric analyses of solids.

9.1 Plane Stress and Plane Strain

First of all, we derive the basic equations for theory of elasticity. Considering the free body diagram of the infinitesimal element as shown in Fig. 9.1.1, summation of forces in the horizontal and vertical axes become

$$\sum F_x = (\sigma_x + \frac{\partial \sigma_x}{\partial x}dx)dy - \sigma_x dy + (\tau_{xy} + \frac{\partial \tau_{xy}}{\partial y}dy)dx - \tau_{xy}dx + f_x dx dy = 0 \quad (9.1.1)$$

and

$$\sum F_y = (\tau_{xy} + \frac{\partial \tau_{xy}}{\partial x}dx)dy - \tau_{xy}dy + (\sigma_y + \frac{\partial \sigma_y}{\partial y}dy)dx - \sigma_y dx + f_y dx dy = 0 \quad (9.1.2)$$

where f_x and f_y are body forces per unit area (or per unit volume assuming unit thickness perpendicular to the plane) in the x- and y-axes which are assumed to be positive when acted along the positive axes. All the stress components in Fig. 9.1.1 are shown as positive. Simplifying these expressions yields *equations of equilibrium* as given below:

$$\frac{\partial \sigma_x}{\partial x} + \frac{\partial \tau_{xy}}{\partial y} + f_x = 0 \quad (9.1.3)$$

$$\frac{\partial \tau_{xy}}{\partial x} + \frac{\partial \sigma_y}{\partial y} + f_y = 0 \qquad (9.1.4)$$

The next set of equations is the *constitutive equation*. This set of equations states the relationship between the stresses and strains. For an isotropic material, the *constitutive equation* becomes

$$\{\sigma\} = [D]\{\epsilon\} \qquad (9.1.5)$$

where $\{\sigma\} = \{\sigma_x\ \sigma_y\ \tau_{xy}\}^T$ denotes the stress and $\{\epsilon\} = \{\epsilon_x\ \epsilon_y\ \gamma_{xy}\}^T$ is the strain. The material property matrix $[D]$ becomes

$$[D] = \frac{E}{1-\nu^2} \begin{bmatrix} 1 & \nu & 0 \\ \nu & 1 & 0 \\ 0 & 0 & \frac{1-\nu}{2} \end{bmatrix} \qquad (9.1.6)$$

for the *plane stress* condition. Here, E and ν are the elastic modulus and Poisson's ratio, respectively. For the *plane strain* condition, matrix $[D]$ becomes

$$[D] = \frac{E(1-\nu)}{(1+\nu)(1-2\nu)} \begin{bmatrix} 1 & \frac{\nu}{1-\nu} & 0 \\ \frac{\nu}{1-\nu} & 1 & 0 \\ 0 & 0 & \frac{1-2\nu}{2(1-\nu)} \end{bmatrix} \qquad (9.1.7)$$

The *kinematic equations*, which relate strains to displacements, are

$$\begin{Bmatrix} \epsilon_x \\ \epsilon_y \\ \gamma_{xy} \end{Bmatrix} = \begin{Bmatrix} \frac{\partial u}{\partial x} \\ \frac{\partial v}{\partial y} \\ \frac{\partial u}{\partial y} + \frac{\partial v}{\partial x} \end{Bmatrix} \qquad (9.1.8)$$

where u and v are displacements in the x and y directions, respectively. Combining Eqs. (9.1.3), (9.1.4), (9.1.5), and (9.1.8) gives eight unknowns (three stresses, three strains, and two displacements) for eight equations (two equilibrium, three constitutive, and three kinematic equations).

Boundary conditions are either essential (or geometric) or natural (or traction) types. Essential conditions are prescribed displacements and natural boundary conditions are prescribed tractions which are expressed as

$$\Phi_x = \sigma_x n_x + \tau_{xy} n_y = \bar{\Phi}_x \qquad (9.1.9)$$

$$\Phi_y = \tau_{xy} n_x + \sigma_y n_y = \bar{\Phi}_y \qquad (9.1.10)$$

where n_x are n_y are direction cosines of the outward unit normal vector at the boundary; and $\bar{\Phi}$ is the given traction value.

In order to develop the finite element formulation for the elasticity problem, let us apply Galerkin's method. The energy method is used to derive the finite element formulation in an upcoming section. Applying the weighted residual method to Eqs. (9.1.3) and (9.1.4) and writing them together give

$$\int_\Omega \begin{Bmatrix} \omega_1 \left(\frac{\partial \sigma_x}{\partial x} + \frac{\partial \tau_{xy}}{\partial y} \right) \\ \omega_2 \left(\frac{\partial \tau_{xy}}{\partial x} + \frac{\partial \sigma_y}{\partial y} \right) \end{Bmatrix} d\Omega + \int_\Omega \begin{Bmatrix} \omega_1 f_x \\ \omega_2 f_y \end{Bmatrix} d\Omega - \int_{\Gamma_e} \begin{Bmatrix} \omega_1 \Phi_x \\ \omega_2 \Phi_y \end{Bmatrix} d\Gamma = 0 \qquad (9.1.11)$$

Section 9.1 Plane Stress and Plane Strain

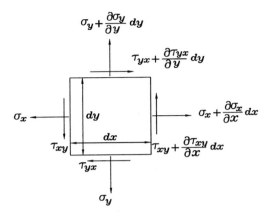

Figure 9.1.1 Free Body Diagram of Two-Dimensional Body

where Γ_e is the boundary for essential condition and ω_i ($i = 1, 2$) is the weighting function.

Applying integration by parts to the terms in the first integral in Eq. (9.1.11) yields

$$-\int_\Omega \left\{ \begin{array}{c} \frac{\partial \omega_1}{\partial x}\sigma_x + \frac{\partial \omega_1}{\partial y}\tau_{xy} \\ \frac{\partial \omega_2}{\partial x}\tau_{xy} + \frac{\partial \omega_2}{\partial y}\sigma_y \end{array} \right\} d\Omega + \int_\Omega \left\{ \begin{array}{c} \omega_1 f_x \\ \omega_2 f_y \end{array} \right\} d\Omega + \int_{\Gamma_n} \left\{ \begin{array}{c} \omega_1 \bar{\Phi}_x \\ \omega_2 \bar{\Phi}_y \end{array} \right\} d\Gamma = 0 \quad (9.1.12)$$

where Γ_n is the boundary for natural conditions and Eqs. (9.1.9) and (9.1.10) are used to come to Eq. (9.1.12). Eq. (9.1.12) can be rewritten as

$$\int_\Omega \begin{bmatrix} \frac{\partial \omega_1}{\partial x} & 0 & \frac{\partial \omega_1}{\partial y} \\ 0 & \frac{\partial \omega_2}{\partial y} & \frac{\partial \omega_2}{\partial x} \end{bmatrix} \left\{ \begin{array}{c} \sigma_x \\ \sigma_y \\ \tau_{xy} \end{array} \right\} d\Omega = \int_\Omega \left\{ \begin{array}{c} \omega_1 f_x \\ \omega_2 f_y \end{array} \right\} d\Omega + \int_{\Gamma_n} \left\{ \begin{array}{c} \omega_1 \bar{\Phi}_x \\ \omega_2 \bar{\Phi}_y \end{array} \right\} d\Gamma \quad (9.1.13)$$

Substitution of the constitutive equation into Eq. (9.1.13) results in

$$\int_\Omega \begin{bmatrix} \frac{\partial \omega_1}{\partial x} & 0 & \frac{\partial \omega_1}{\partial y} \\ 0 & \frac{\partial \omega_2}{\partial y} & \frac{\partial \omega_2}{\partial x} \end{bmatrix} [D] \left\{ \begin{array}{c} \epsilon_x \\ \epsilon_y \\ \gamma_{xy} \end{array} \right\} d\Omega = \int_\Omega \left\{ \begin{array}{c} \omega_1 f_x \\ \omega_2 f_y \end{array} \right\} d\Omega + \int_{\Gamma_n} \left\{ \begin{array}{c} \omega_1 \bar{\Phi}_x \\ \omega_2 \bar{\Phi}_y \end{array} \right\} d\Gamma \quad (9.1.14)$$

One more substitution of the kinematic equation into Eq. (9.1.14) gives

$$\int_\Omega \begin{bmatrix} \frac{\partial \omega_1}{\partial x} & 0 & \frac{\partial \omega_1}{\partial y} \\ 0 & \frac{\partial \omega_2}{\partial y} & \frac{\partial \omega_2}{\partial x} \end{bmatrix} [D] \left\{ \begin{array}{c} \frac{\partial u}{\partial x} \\ \frac{\partial v}{\partial y} \\ \frac{\partial u}{\partial y} + \frac{\partial v}{\partial x} \end{array} \right\} d\Omega = \int_\Omega \left\{ \begin{array}{c} \omega_1 f_x \\ \omega_2 f_y \end{array} \right\} d\Omega + \int_{\Gamma_n} \left\{ \begin{array}{c} \omega_1 \bar{\Phi}_x \\ \omega_2 \bar{\Phi}_y \end{array} \right\} d\Gamma$$

(9.1.15)

Let us discretize the domain using linear triangular elements as seen in Fig. 5.2.1. Then, both displacements u and v are interpolated using the same shape functions such as

$$u(x, y) = \sum_{i=1}^{3} H_i(x, y) u_i \quad (9.1.16)$$

$$v(x,y) = \sum_{i=1}^{3} H_i(x,y)v_i \qquad (9.1.17)$$

These displacements can be also expressed as

$$\begin{Bmatrix} u \\ v \end{Bmatrix} = \begin{bmatrix} H_1 & 0 & H_2 & 0 & H_3 & 0 \\ 0 & H_1 & 0 & H_2 & 0 & H_3 \end{bmatrix} \begin{Bmatrix} u_1 \\ v_1 \\ u_2 \\ v_2 \\ u_3 \\ v_3 \end{Bmatrix} = [N]\{d\} \qquad (9.1.18)$$

where $\{d\} = \{ u_1 \ v_1 \ u_2 \ v_2 \ u_3 \ v_3 \}^T$ is the nodal displacement vector. Use of this expression for strains yields

$$\begin{Bmatrix} \frac{\partial u}{\partial x} \\ \frac{\partial v}{\partial y} \\ \frac{\partial u}{\partial y} + \frac{\partial v}{\partial x} \end{Bmatrix} = \begin{bmatrix} \frac{\partial H_1}{\partial x} & 0 & \frac{\partial H_2}{\partial x} & 0 & \frac{\partial H_3}{\partial x} & 0 \\ 0 & \frac{\partial H_1}{\partial y} & 0 & \frac{\partial H_2}{\partial y} & 0 & \frac{\partial H_3}{\partial y} \\ \frac{\partial H_1}{\partial y} & \frac{\partial H_1}{\partial x} & \frac{\partial H_2}{\partial y} & \frac{\partial H_2}{\partial x} & \frac{\partial H_3}{\partial y} & \frac{\partial H_3}{\partial x} \end{bmatrix} \{d\} \qquad (9.1.19)$$

We use symbol $[B]$ to denote the matrix expression in the above equation. That is,

$$\{\epsilon\} = \begin{Bmatrix} \frac{\partial u}{\partial x} \\ \frac{\partial v}{\partial y} \\ \frac{\partial u}{\partial y} + \frac{\partial v}{\partial x} \end{Bmatrix} = [B]\{d\} \qquad (9.1.20)$$

Galerkin's method states $\omega_1 = H_i$ ($i = 1, 2, 3$) and $\omega_2 = H_i$ ($i = 1, 2, 3$). Applying these weighting functions and Eq. (9.1.20) into Eq. (9.1.15) gives for the finite element domain integral

$$\int_{\Omega^e} [B]^T [D][B] d\Omega \{d\} \qquad (9.1.21)$$

in which Ω^e denotes the element domain. As a result, the element stiffness matrix for elasticity can be expressed as

$$[K^e] = \int_{\Omega^e} [B]^T [D][B] d\Omega \qquad (9.1.22)$$

Equation (9.1.22) holds for any kind of element in any dimension.
Evaluation of the linear shape function provides

$$[B] = \frac{1}{2A} \begin{bmatrix} (y_2 - y_3) & 0 & (y_3 - y_1) & 0 & (y_1 - y_2) & 0 \\ 0 & (x_3 - x_2) & 0 & (x_1 - x_3) & 0 & (x_2 - x_1) \\ (x_3 - x_2) & (y_2 - y_3) & (x_1 - x_3) & (y_3 - y_1) & (x_2 - x_1) & (y_1 - y_2) \end{bmatrix} \qquad (9.1.23)$$

Substitution of Eq.(9.1.23) into Eq. (9.1.22) results in

$$[K^e] = \int_{\Omega^e} [B]^T[D][B]d\Omega = [B]^T[D][B]A \qquad (9.1.24)$$

since both $[B]$ and $[D]$ are constant matrices independent of x and y. Here A is the area of the element. This expression is true for both plane stress and plane strain conditions. The material property matrix $[D]$ is selected properly for plane stress (i.e. Eq. (9.1.6)) and plane strain (i.e. Eq. (9.1.7)) conditions. We assume a unit thickness for the plane stress condition because the solution is independent of the thickness direction for this case. However, if we want to include the thickness, the matrix in Eq. (9.1.24) is multiplied by the thickness. When other kinds of shape functions are used for the plane stress/strain condition, we just need to develop matrix $[B]$ as shown in Eq. (9.1.20) and put it into Eq. (9.1.22). The size of row of $[B]$ is always three for the plane stress/strain condition while the size of column equals twice the number of nodes per element because there are two degrees of freedoms per node.

9.2 Force Vector

The two right-hand-side terms in Eq. (9.1.15) are the force vector. The first term is due to body forces and the other is due to tractions. The body force term is a domain integral. As a result, the same computation can be performed to this term as the stiffness matrix. Applying Galerkin's method to this term in an element domain yields

$$\{F\} = \int_{\Omega^e} \begin{Bmatrix} \omega_1 f_x \\ \omega_2 f_y \end{Bmatrix} d\Omega = \int_{\Omega^e} \begin{bmatrix} \omega_1 & 0 \\ 0 & \omega_2 \end{bmatrix} \begin{Bmatrix} f_x \\ f_y \end{Bmatrix} d\Omega = \int_{\Omega^e} [N]^T \begin{Bmatrix} f_x \\ f_y \end{Bmatrix} d\Omega \qquad (9.2.1)$$

where $[N]$ is defined in Eq. (9.1.18).

On the other hand, the traction vector is a boundary integral. This boundary integral is very similar to what is described in Sec. 5.4. Let us consider a traction as shown in Fig. 9.2.1. The traction term can be evaluated as given below:

$$\{\Phi\} = \int_{\Gamma_n} \begin{Bmatrix} \omega_1 \bar{\Phi}_x \\ \omega_2 \bar{\Phi}_y \end{Bmatrix} d\Gamma = \int_{s_m}^{s_n} \begin{Bmatrix} \frac{s_n-s}{s_n-s_m} & 0 \\ 0 & \frac{s_n-s}{s_n-s_m} \\ \frac{s-s_m}{s_n-s_m} & 0 \\ 0 & \frac{s-s_m}{s_n-s_m} \end{Bmatrix} \begin{Bmatrix} \bar{\Phi}_x \\ \bar{\Phi}_y \end{Bmatrix} ds \qquad (9.2.2)$$

where s_m and s_n are the coordinate values along the temporary boundary axis s, and m and n are the two nodes on the element boundary where the traction is described. If the traction is constant, the traction vector becomes

$$\begin{Bmatrix} u_m \\ v_m \\ u_n \\ v_n \end{Bmatrix} = \frac{1}{2} \begin{Bmatrix} s_d \bar{\Phi} \cos\theta \\ s_d \bar{\Phi} \sin\theta \\ s_d \bar{\Phi} \cos\theta \\ s_d \bar{\Phi} \sin\theta \end{Bmatrix} \qquad (9.2.3)$$

316 Elasticity Problem Chapter 9

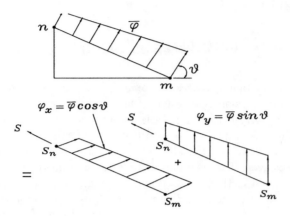

Figure 9.2.1 Boundary Traction

in which $s_d = s_n - s_m$ is the length of the boundary segment and $\bar{\Phi}$ is a constant traction value.

♣ **Example 9.2.1** Find the nodal displacements in Fig. 9.2.2. We use two linear triangular elements as seen in the figure. Each element is also shown in a separate figure indicating global node numbers and local node numbers necessary for construction of the element stiffness matrix. Using the plane stress condition, the element stiffness matrix for the first element is given below along with the associated nodal displacements:

$$[K^1]\{d^1\} = 10^9 \begin{bmatrix} 7.3 & 3.3 & -5.3 & -2.0 & -2.0 & -1.3 \\ 3.3 & 7.3 & -1.3 & -2.0 & -2.0 & -5.3 \\ -5.3 & -1.3 & 5.3 & 0.0 & 0.0 & 1.3 \\ -2.0 & -2.0 & 0.0 & 2.0 & 2.0 & 0.0 \\ -2.0 & -2.0 & 0.0 & 2.0 & 2.0 & 0.0 \\ -1.3 & -5.3 & 1.3 & 0.0 & 0.0 & 5.3 \end{bmatrix} \begin{Bmatrix} u_1 \\ v_1 \\ u_3 \\ v_3 \\ u_2 \\ v_2 \end{Bmatrix} \quad (9.2.4)$$

The stiffness matrix of the second element is the same as that given in Eq. (9.2.4) because the two elements have the same size and shape as well as the same way of local node numbering.

Assembling the element matrices into the system matrix results in

$$[K] = 10^9 \begin{bmatrix} 7.3 & 3.3 & -2.0 & -1.3 & -5.3 & -2.0 & 0.0 & 0.0 \\ 3.3 & 7.3 & -2.0 & -5.3 & -1.3 & -2.0 & 0.0 & 0.0 \\ -2.0 & -2.0 & 7.3 & 0.0 & 0.0 & 3.3 & -5.3 & -1.3 \\ -1.3 & -5.3 & 0.0 & 7.3 & 3.3 & 0.0 & -2.0 & -2.0 \\ -5.3 & -1.3 & 0.0 & 3.3 & 7.3 & 0.0 & -2.0 & -2.0 \\ -2.0 & -2.0 & 3.3 & 0.0 & 0.0 & 7.3 & -1.3 & -5.3 \\ 0.0 & 0.0 & -5.3 & -2.0 & -2.0 & -1.3 & 7.3 & 3.3 \\ 0.0 & 0.0 & -1.3 & -2.0 & -2.0 & -5.3 & 3.3 & 7.3 \end{bmatrix} \quad (9.2.5)$$

The nodal force vector is computed using Eq. (9.2.3) as below:

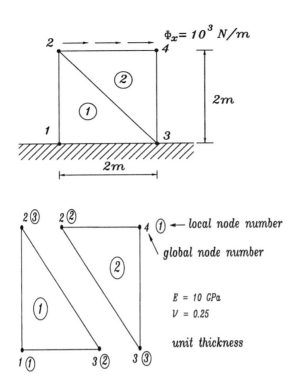

Figure 9.2.2 Square Plate With Tangential Traction

$$\{F\} = \begin{Bmatrix} F_{1x} \\ F_{1y} \\ 10^3 \\ 0 \\ F_{3x} \\ F_{3y} \\ 10^3 \\ 0 \end{Bmatrix} \quad (9.2.6)$$

where F_{1x}, F_{1y}, F_{3x} and F_{3y} are unknown forces while the essential boundary conditions state $u_1=v_1=u_3=v_3=0$. Applying these conditions to Eqs. (9.2.5) and (9.2.6) and solving for the unknown nodal displacements give u_2=6.135 × 10^{-7}, v_2=1.450 × 10^{-7}, u_4=6.975 × 10^{-7}, and v_4=-1.660 × 10^{-7}. This is a very crude mesh so that these solutions are not accurate. ‡

9.3 Energy Method

The total potential energy denoted by Π consists of two parts: internal energy U and external energy W which is equal to work done by external loads. That is,

$$\Pi = U - W \quad (9.3.1)$$

The internal energy is the strain energy caused by deformation of the body and can be written as

$$U = \frac{1}{2}\int_\Omega \{\sigma\}^T\{\epsilon\}d\Omega \tag{9.3.2}$$

where $\{\sigma\} = \{\sigma_x\ \sigma_y\ \tau_{xy}\}^T$ denotes the stress and $\{\epsilon\} = \{\epsilon_x\ \epsilon_y\ \gamma_{xy}\}^T$ is the strain. Equation (9.3.2) also holds for three-dimensional state of stresses. Use of the constitutive equation for Eq. (9.3.2) gives

$$U = \frac{1}{2}\int_\Omega \{\epsilon\}^T[D]\{\epsilon\}d\Omega \tag{9.3.3}$$

since $[D] = [D]^T$.

On the other hand, the external work can be written as

$$W = \int_\Omega \{u\ v\}\begin{Bmatrix}f_x\\f_y\end{Bmatrix}d\Omega + \int_{\Gamma_n} \{u\ v\}\begin{Bmatrix}\bar{\Phi}_x\\\bar{\Phi}_y\end{Bmatrix}d\Gamma \tag{9.3.4}$$

Substitution of Eqs. (9.3.3) and (9.3.4) into Eq. (9.3.1) and discretization of the domain into a number of finite element domains yields

$$\Pi = \sum_{e=1} \Pi^e \tag{9.3.5}$$

$$\Pi^e = \frac{1}{2}\int_{\Omega^e}\{\epsilon\}^T[D]\{\epsilon\}d\Omega - \int_{\Omega^e}\{u\ v\}\begin{Bmatrix}f_x\\f_y\end{Bmatrix}d\Omega - \int_{\Gamma^e}\{u\ v\}\begin{Bmatrix}\bar{\Phi}_x\\\bar{\Phi}_y\end{Bmatrix}d\Gamma \tag{9.3.6}$$

For each finite element, applying Eqs. (9.1.18) and (9.1.20) to Eq. (9.3.6) gives

$$\Pi^e = \frac{1}{2}\{d\}^T\int_{\Omega^e}[B]^T[D][B]d\Omega\{d\} - \{d\}^T\int_{\Omega^e}[N]^T\begin{Bmatrix}f_x\\f_y\end{Bmatrix}d\Omega$$

$$- \{d\}^T\int_{\Gamma^e}[N]^T\begin{Bmatrix}\bar{\Phi}_x\\\bar{\Phi}_y\end{Bmatrix}d\Gamma \tag{9.3.7}$$

In order to find the equilibrium solution, we apply the *principle of minimum total potential energy*. The principle states:

Of all kinematically admissible configurations, the deformation producing the minimum total potential energy is the stable equilibrium condition.

Invoking the stationary value for Eqs. (9.3.5) and (9.3.7) using this energy principle, we obtain

$$\sum_{e=1}\frac{\partial\Pi^e}{\partial\{d\}} = \sum_{e=1}([K^e]\{d\} - \{F\} - \{\bar{\Phi}\}) = 0 \tag{9.3.8}$$

where

$$[K^e] = \int_{\Omega^e}[B]^T[D][B]d\Omega \tag{9.3.9}$$

$$\{F\} = \int_{\Omega^e} [N]^T \left\{ \begin{array}{c} f_x \\ f_y \end{array} \right\} d\Omega \qquad (9.3.10)$$

$$\{\bar{\Phi}\} = \int_{\Gamma^e} [N]^T \left\{ \begin{array}{c} \bar{\Phi}_x \\ \bar{\Phi}_y \end{array} \right\} d\Gamma \qquad (9.3.11)$$

Here, Eqs. (9.3.9), (9.3.10), and (9.3.11) are the element stiffness matrix, body force vector, and surface traction vector, respectively. By comparing Eq. (9.1.22) to Eq. (9.3.9), it can be shown that Galerkin's method results in the same matrix equation as the energy method. In addition, the force terms are identical to those obtained from *Galerkin's* method. Especially, Eq. (9.3.11) looks different from Eq. (9.2.2) but they are the same when actual calculation is performed.

9.4 Three-Dimensional Solid

The governing equations for three-dimensional elasticity are given below.

Equations of equilibrium are:

$$\frac{\partial \sigma_x}{\partial x} + \frac{\partial \tau_{xy}}{\partial y} + \frac{\partial \tau_{xz}}{\partial z} + f_x = 0 \qquad (9.4.1)$$

$$\frac{\partial \tau_{xy}}{\partial x} + \frac{\partial \sigma_y}{\partial y} + \frac{\partial \tau_{yz}}{\partial z} + f_y = 0 \qquad (9.4.2)$$

$$\frac{\partial \tau_{xz}}{\partial x} + \frac{\partial \tau_{yz}}{\partial y} + \frac{\partial \sigma_z}{\partial z} + f_z = 0 \qquad (9.4.3)$$

where stresses are shown in Fig. 9.4.1 in the positive direction and f_x, f_y, and f_z are body forces per unit volume.

The constitutive equation for an isotropic material is:

$$\{\sigma\} = [D]\{\epsilon\} \qquad (9.4.4)$$

where

$$\{\sigma\} = \{ \sigma_x \quad \sigma_y \quad \sigma_z \quad \tau_{xy} \quad \tau_{yz} \quad \tau_{xz} \}^T \qquad (9.4.5)$$

$$\{\epsilon\} = \{ \epsilon_x \quad \epsilon_y \quad \epsilon_z \quad \gamma_{xy} \quad \gamma_{yz} \quad \gamma_{xz} \}^T \qquad (9.4.6)$$

$$[D] = \frac{E}{(1+\nu)(1-2\nu)} \begin{bmatrix} 1-\nu & \nu & \nu & 0 & 0 & 0 \\ \nu & 1-\nu & \nu & 0 & 0 & 0 \\ \nu & \nu & 1-\nu & 0 & 0 & 0 \\ 0 & 0 & 0 & \frac{1-2\nu}{2} & 0 & 0 \\ 0 & 0 & 0 & 0 & \frac{1-2\nu}{2} & 0 \\ 0 & 0 & 0 & 0 & 0 & \frac{1-2\nu}{2} \end{bmatrix} \qquad (9.4.7)$$

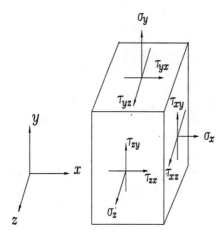

Figure 9.4.1 Three-Dimensional State of Stress

The kinematic equation for small displacements is:

$$\begin{Bmatrix} \epsilon_x \\ \epsilon_y \\ \epsilon_z \\ \gamma_{xy} \\ \gamma_{yz} \\ \gamma_{xz} \end{Bmatrix} = \begin{Bmatrix} \frac{\partial u}{\partial x} \\ \frac{\partial v}{\partial y} \\ \frac{\partial w}{\partial z} \\ \frac{\partial u}{\partial y} + \frac{\partial v}{\partial x} \\ \frac{\partial v}{\partial z} + \frac{\partial w}{\partial y} \\ \frac{\partial u}{\partial z} + \frac{\partial w}{\partial x} \end{Bmatrix} \qquad (9.4.8)$$

where u, v, and w are displacements in the x, y, and z directions, respectively.

The traction boundary condition is:

$$\Phi_x = \sigma_x n_x + \tau_{xy} n_y + \tau_{xz} n_z = \bar{\Phi}_x \qquad (9.4.9)$$

$$\Phi_y = \tau_{xy} n_x + \sigma_y n_y + \tau_{yz} n_z = \bar{\Phi}_y \qquad (9.4.10)$$

$$\Phi_z = \tau_{xz} n_x + \tau_{yz} n_y + \sigma_z n_z = \bar{\Phi}_z \qquad (9.4.11)$$

where n_x, n_y, and n_z are cosine directions of the outward unit normal vector on the traction surface and $\bar{\Phi}$ is the known value.

We want to derive the element stiffness matrix for a tetrahedron element as seen in Fig. 9.4.2. The element has four nodes. The shape functions for this element can be derived as given below: Let us assume a linear function in terms of x, y, and z.

$$u = \begin{bmatrix} 1 & x & y & z \end{bmatrix} \begin{Bmatrix} a_1 \\ a_2 \\ a_3 \\ a_4 \end{Bmatrix} = [X]\{A\} \qquad (9.4.12)$$

Section 9.4 Three-Dimensional Solid

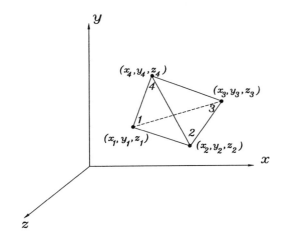

Figure 9.4.2 Tetrahedron Element

Evaluation of u at every node yields

$$\{u\} = [\bar{X}]\{A\} \tag{9.4.13}$$

where

$$[\bar{X}] = \begin{bmatrix} 1 & x_1 & y_1 & z_1 \\ 1 & x_2 & y_2 & z_2 \\ 1 & x_3 & y_3 & z_3 \\ 1 & x_4 & y_4 & z_4 \end{bmatrix} \tag{9.4.14}$$

The inverse of matrix $[\bar{X}]$ in Eq. (9.4.13) and substitution of the resulting expression into Eq. (9.4.12) gives

$$u = [X][\bar{X}]^{-1}\{u\} = [H]\{u\} \tag{9.4.15}$$

where the shape functions are

$$[H] = [\,H_1(x,y,z) \quad H_2(x,y,z) \quad H_3(x,y,z) \quad H_4(x,y,z)\,] \tag{9.4.16}$$

We use the same shape functions for the three displacements.

$$\begin{Bmatrix} u \\ v \\ w \end{Bmatrix} = \begin{bmatrix} H_1 & 0 & 0 & H_2 & 0 & 0 & H_3 & 0 & 0 & H_4 & 0 & 0 \\ 0 & H_1 & 0 & 0 & H_2 & 0 & 0 & H_3 & 0 & 0 & H_4 & 0 \\ 0 & 0 & H_1 & 0 & 0 & H_2 & 0 & 0 & H_3 & 0 & 0 & H_4 \end{bmatrix} \begin{Bmatrix} u_1 \\ v_1 \\ w_1 \\ u_2 \\ v_2 \\ w_2 \\ u_3 \\ v_3 \\ w_3 \\ u_4 \\ v_4 \\ w_4 \end{Bmatrix}$$

$$= [N]\{d\} \tag{9.4.17}$$

where $\{d\}$ is the nodal displacement vector. Substituting Eq. (9.4.17) into the three-dimensional kinematic equation Eq. (9.4.8) results in

$$\{\epsilon\} = [B]\{d\} \tag{9.4.18}$$

in which

$$[B] = \begin{bmatrix} \frac{\partial H_1}{\partial x} & 0 & 0 & \frac{\partial H_2}{\partial x} & 0 & 0 & \frac{\partial H_3}{\partial x} & 0 & 0 & \frac{\partial H_4}{\partial x} & 0 & 0 \\ 0 & \frac{\partial H_1}{\partial y} & 0 & 0 & \frac{\partial H_2}{\partial y} & 0 & 0 & \frac{\partial H_3}{\partial y} & 0 & 0 & \frac{\partial H_4}{\partial y} & 0 \\ 0 & 0 & \frac{\partial H_1}{\partial z} & 0 & 0 & \frac{\partial H_2}{\partial z} & 0 & 0 & \frac{\partial H_3}{\partial z} & 0 & 0 & \frac{\partial H_4}{\partial z} \\ \frac{\partial H_1}{\partial y} & \frac{\partial H_1}{\partial x} & 0 & \frac{\partial H_2}{\partial y} & \frac{\partial H_2}{\partial x} & 0 & \frac{\partial H_3}{\partial y} & \frac{\partial H_3}{\partial x} & 0 & \frac{\partial H_4}{\partial y} & \frac{\partial H_4}{\partial x} & 0 \\ 0 & \frac{\partial H_1}{\partial z} & \frac{\partial H_1}{\partial y} & 0 & \frac{\partial H_2}{\partial z} & \frac{\partial H_2}{\partial y} & 0 & \frac{\partial H_3}{\partial z} & \frac{\partial H_3}{\partial y} & 0 & \frac{\partial H_4}{\partial z} & \frac{\partial H_4}{\partial y} \\ \frac{\partial H_1}{\partial z} & 0 & \frac{\partial H_1}{\partial x} & \frac{\partial H_2}{\partial z} & 0 & \frac{\partial H_2}{\partial x} & \frac{\partial H_3}{\partial z} & 0 & \frac{\partial H_3}{\partial x} & \frac{\partial H_4}{\partial z} & 0 & \frac{\partial H_4}{\partial x} \end{bmatrix} \tag{9.4.19}$$

Putting matrix $[D]$ from Eq. (9.4.7) and matrix $[B]$ from Eq. (9.4.19) into Eq. (9.1.22) computes the element stiffness matrix.

$$[K^e] = [B]^T[D][B]V \tag{9.4.20}$$

where V is the volume of the tetrahedron element.

9.5 Axisymmetric Solid

When the elasticity problem degenerates from three-dimension to axisymmetry, two shearing stress components vanish. These vanishing components due to symmetry are $\tau_{r\theta}$ and $\tau_{z\theta}$ in the $r\theta z$ coordinate system where r is the radial direction, θ is the circumferential direction, and z is the axial direction. Hence, the remaining stress components are

$$\{\sigma\} = \{\, \sigma_r \quad \sigma_\theta \quad \sigma_z \quad \tau_{rz} \,\} \tag{9.5.1}$$

Similarly, the remaining strains are

$$\{\epsilon\} = \{\, \epsilon_r \quad \epsilon_\theta \quad \epsilon_z \quad \gamma_{rz} \,\} \tag{9.5.2}$$

The material property matrix $[D]$ for the axisymmetric problem is

$$[D] = \frac{E}{(1+\nu)(1-2\nu)} \begin{bmatrix} 1-\nu & \nu & \nu & 0 \\ \nu & 1-\nu & \nu & 0 \\ \nu & \nu & 1-\nu & 0 \\ 0 & 0 & 0 & \frac{1-2\nu}{2} \end{bmatrix} \tag{9.5.3}$$

The kinematic equation is

$$\begin{Bmatrix} \epsilon_r \\ \epsilon_\theta \\ \epsilon_z \\ \gamma_{rz} \end{Bmatrix} = \begin{Bmatrix} \frac{\partial u}{\partial r} \\ \frac{u}{r} \\ \frac{\partial w}{\partial z} \\ \frac{\partial u}{\partial z} + \frac{\partial w}{\partial r} \end{Bmatrix} \tag{9.5.4}$$

Section 9.5 Axisymmetric Solid

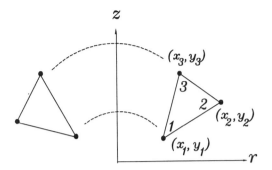

Figure 9.5.1 Free Body Diagram of Two-Dimensional Body

where u and w are the radial and axial displacements, respectively.

♣ **Example 9.5.1** Let us determine the circumferential strain ϵ_θ if a hoop of radial r is uniformly displaced along the radial direction by displacement u. Then, the deformed hoop has a uniform strain along the circumferential direction and the strain is computed as

$$\epsilon_\theta = \frac{2\pi(r+u) - 2\pi r}{2\pi r} = \frac{u}{r} \tag{9.5.5}$$

‡

In order to develop the element stiffness matrix, we use the linear triangular element again. We substitute x and y in the shape functions, Eqs. (5.2.7) through (5.2.9), with r and z for the axisymmetric problem. In addition, the axisymmetric element is a ring element as shown in Fig. 9.5.1. Substitution of the shape functions into the kinematic equation Eq. (9.5.4) gives

$$\{\epsilon\} = \begin{bmatrix} \frac{\partial H_1}{\partial r} & 0 & \frac{\partial H_2}{\partial r} & 0 & \frac{\partial H_3}{\partial r} & 0 \\ \frac{H_1}{r} & 0 & \frac{H_2}{r} & 0 & \frac{H_3}{r} & 0 \\ 0 & \frac{\partial H_1}{\partial z} & 0 & \frac{\partial H_2}{\partial z} & 0 & \frac{\partial H_3}{\partial z} \\ \frac{\partial H_1}{\partial z} & \frac{\partial H_1}{\partial r} & \frac{\partial H_2}{\partial z} & \frac{\partial H_2}{\partial r} & \frac{\partial H_3}{\partial z} & \frac{\partial H_3}{\partial r} \end{bmatrix} \{d^e\} = [B]\{d^e\} \tag{9.5.6}$$

The element matrix can be expressed as

$$[K^e] = \int_z \int_r \int_\theta [B]^T [D][B] r \, d\theta \, dr \, dz = 2\pi \int_r \int_z r[B]^T [D][B] \, dr \, dz \tag{9.5.7}$$

Because of the term $\frac{H_i}{r}$ in matrix $[B]$, the matrix is not a constant matrix like the plane stress/strain case. As a result, the integration needs to be undertaken. One simple approximation for the integration is to evaluate $[B]$ at the centroid of the element. That is, we calculate $\frac{H_i(\bar{r},\bar{z})}{\bar{r}}$ where $\bar{r} = \frac{r_1+r_2+r_3}{3}$ and $\bar{z} = \frac{z_1+z_2+z_3}{3}$. Then, the element stiffness matrix can be written as

$$[K^e] = 2\pi \bar{r} A [\bar{B}]^T [D][\bar{B}] \tag{9.5.8}$$

in which $[\bar{B}]$ is the matrix $[B]$ evaluated at the centroid of the element cross-section and A is its area.

9.6 Dynamic Analysis

While previous sections consider static problems, this section considers dynamic problems. That is, we include the inertia force in equations of equilibrium. These equations are also called equations of motion. For the two-dimensional case, these equations are

$$\rho \frac{\partial^2 u}{\partial t^2} = \frac{\partial \sigma_x}{\partial x} + \frac{\partial \tau_{xy}}{\partial y} + f_x \tag{9.6.1}$$

$$\rho \frac{\partial^2 v}{\partial t^2} = \frac{\partial \tau_{xy}}{\partial x} + \frac{\partial \sigma_y}{\partial y} + f_y \tag{9.6.2}$$

where t indicates time and ρ is the mass density. It can be easily extended to the three-dimensional case. Therefore, the finite element formulation for the dynamic problem contains one extra term compared to that for the static problem and the term is derived as follows using Galerkin's method.

$$\int_\Omega \left\{ \begin{array}{c} \rho \omega_1 \ddot{u} \\ \rho \omega_2 \ddot{v} \end{array} \right\} d\Omega = \sum_{e=1} \int_{\Omega^e} \rho \begin{bmatrix} \omega_1 & 0 \\ 0 & \omega_2 \end{bmatrix} \left\{ \begin{array}{c} \ddot{u} \\ \ddot{v} \end{array} \right\} d\Omega \tag{9.6.3}$$

in which the superimposed dot denotes temporal derivative.

Using linear triangular elements, the accelerations can be interpolated as

$$\left\{ \begin{array}{c} \ddot{u} \\ \ddot{v} \end{array} \right\} = \begin{bmatrix} H_1 & 0 & H_2 & 0 & H_3 & 0 \\ 0 & H_1 & 0 & H_2 & 0 & H_3 \end{bmatrix} \left\{ \begin{array}{c} \ddot{u}_1 \\ \ddot{v}_1 \\ \ddot{u}_2 \\ \ddot{v}_2 \\ \ddot{u}_3 \\ \ddot{v}_3 \end{array} \right\} = [N]\{\ddot{d}\} \tag{9.6.4}$$

In Eq. (9.6.4) we assume that the shape functions are functions of spatial variables only and the nodal displacements are functions of time. Hence, the temporal derivative is performed for the nodal variable. Substituting Eq. (9.6.4) into Eq. (9.6.3) for each element results in

$$\int_{\Omega^e} \rho \begin{bmatrix} \omega_1 & 0 \\ 0 & \omega_2 \end{bmatrix} \left\{ \begin{array}{c} \ddot{u} \\ \ddot{v} \end{array} \right\} d\Omega = \int_{\Omega^e} \rho [N]^T [N] d\Omega \{\ddot{d}\} \tag{9.6.5}$$

As a result, the mass matrix is defined as

$$[M^e] = \int_{\Omega^e} \rho [N]^T [N] d\Omega \qquad (9.6.6)$$

This is the *consistent* mass matrix and it becomes for the linear triangular element

$$[M^e] = \frac{\rho A}{12} \begin{bmatrix} 2 & 0 & 1 & 0 & 1 & 0 \\ 0 & 2 & 0 & 1 & 0 & 1 \\ 1 & 0 & 2 & 0 & 1 & 0 \\ 0 & 1 & 0 & 2 & 0 & 1 \\ 1 & 0 & 1 & 0 & 2 & 0 \\ 0 & 1 & 0 & 1 & 0 & 2 \end{bmatrix} \qquad (9.6.7)$$

This matrix is based on unit thickness of the element. Otherwise, it should be multiplied by the plate thickness. On the other hand, the *lumped* mass matrix for the linear triangular element is

$$[M^e] = \frac{\rho A}{3}[I] \qquad (9.6.8)$$

where $[I]$ is the identity matrix of size 6.

The *consistent* mass matrix for the bilinear element as shown in Fig. 5.3.1 is

$$[M^e] = \frac{\rho A}{36} \begin{bmatrix} 4 & 0 & 2 & 0 & 1 & 0 & 2 & 0 \\ 0 & 4 & 0 & 2 & 0 & 1 & 0 & 2 \\ 2 & 0 & 4 & 0 & 2 & 0 & 1 & 0 \\ 0 & 2 & 0 & 4 & 0 & 2 & 0 & 1 \\ 1 & 0 & 2 & 0 & 4 & 0 & 2 & 0 \\ 0 & 1 & 0 & 2 & 0 & 4 & 0 & 2 \\ 2 & 0 & 1 & 0 & 2 & 0 & 4 & 0 \\ 0 & 2 & 0 & 1 & 0 & 2 & 0 & 4 \end{bmatrix} \qquad (9.6.9)$$

This matrix is based on unit thickness of the element. Otherwise, it should be multiplied by the plate thickness. The *lumped* mass matrix for this element is

$$[M^e] = \frac{\rho A}{4}[I] \qquad (9.6.10)$$

where $[I]$ is the identity matrix of size 8.

9.7 Thermal Stress

Temperature change in a body causes thermal strain. When the thermal strain is constrained, thermal stress occurs in the body. The effect of thermal strain is implemented in the following way.

The total strain is the sum of mechanical and thermal strains. The mechanical strain is caused by an applied mechanical load such as pressure loading.

$$\{\epsilon\} = \{\epsilon^{me}\} + \{\epsilon^{th}\} \qquad (9.7.1)$$

where superscripts *me* and *th* denote mechanical and thermal strains, respectively.

For an isotropic material, temperature change results in a body expansion or shrinkage but no distortion. In other words, temperature change affects the normal strains but not shear strains. Thus, the thermal strain vector is expressed as

$$\{\epsilon^{th}\} = \{\alpha\Delta T \quad \alpha\Delta T \quad \alpha\Delta T \quad 0 \quad 0 \quad 0\}^T \tag{9.7.2}$$

in which α is the coefficient of thermal expansion and ΔT indicates the temperature change. In the vector, the first three components are the normal strains.

The total strain is expressed in terms of the nodal displacements as given in Eq. (9.4.18). The mechanical strain is related to stress through a constitutive equation. That is,

$$\{\sigma\} = [D]\{\epsilon^{me}\} \tag{9.7.3}$$

where $[D]$ is the material property matrix as given in Eq. (9.4.7).

Substituting the mechanical strain from Eq. (9.7.1) into Eq. (9.7.3) yields

$$\{\sigma\} = [D]\left(\{\epsilon\} - \{\epsilon^{th}\}\right) \tag{9.7.4}$$

Further substitution of Eq. (9.4.18) into Eq. (9.7.4) gives

$$\{\sigma\} = [D][B]\{d\} - [D]\{\epsilon^{th}\} \tag{9.7.5}$$

Applying the weighted residual technique or the energy principle to Eq. (9.7.5) will result in the following matrix equation:

$$[K]\{d\} = \{F^{th}\} + \{F^{me}\} \tag{9.7.6}$$

where $[K]$ and $\{F^{me}\}$ are the same as before, and the new thermal force term is

$$\{F^{th}\} = \int_\Omega [B]^T[D]\{\epsilon^{th}\}d\Omega \tag{9.7.7}$$

Once the nodal displacement vector is computed after applying boundary conditions, it is substituted into Eq. (9.7.5) in order to compute the stress.

If a body has any initial strain before applying an external load such as residual strain, the initial strain can be treated in the same way as the thermal strain. If there are more than one kind of initial strains, they can be added together and fomulated in the same way.

Section 9.8   MATLAB Application to 2-D Stress Analysis   327

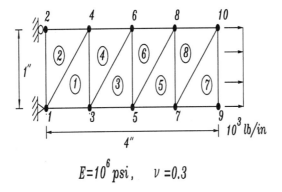

$E = 10^6 \, psi, \quad \nu = 0.3$

Figure 9.8.1 Plate Subjected to Axial Load

9.8 MATLAB Application to 2-D Stress Analysis

Two-dimensional stress analyses are performed using both conventional finite elements and isoparametric elements in the following examples.

♣ **Example 9.8.1**   A strip shown in Fig. 9.8.1 is subjected to an axial load. A MATLAB program is written to solve the problem using linear triangular elements. Eight elements are used as seen in the figure.

```
%----------------------------------------------------------------
% Example 9.8.1
% plane stress analysis of a solid using linear triangular elements
% (see Fig. 9.8.1 for the finite element mesh)
%
% Variable descriptions
% k = element matrix
% f = element vector
% kk = system matrix
% ff = system vector
% disp = system nodal displacement vector
% eldisp = element nodal displacement vector
% stress = matrix containing stresses
% strain = matrix containing strains
% gcoord = coordinate values of each node
% nodes = nodal connectivity of each element
% index = a vector containing system dofs associated with each element
% bcdof = a vector containing dofs associated with boundary conditions
% bcval = a vector containing boundary condition values associated with
%         the dofs in bcdof
%----------------------------------------------------------------
%
%----------------------------------
```

```
% input data for control parameters
%————————————————————————
nel=8;                           % number of elements
nnel=3;                          % number of nodes per element
ndof=2;                          % number of dofs per node
nnode=10;                        % total number of nodes in system
sdof=nnode*ndof;                 % total system dofs
edof=nnel*ndof;                  % degrees of freedom per element
emodule=100000.0;                % elastic modulus
poisson=0.3;                     % Poisson's ratio
%
%————————————————————————
% input data for nodal coordinate values
% gcoord(i,j) where i-> node no. and j-> x or y
%————————————————————————
gcoord=[0.0 0.0; 0.0 1.0; 1.0 0.0; 1.0 1.0; 2.0 0.0;
2.0 1.0; 3.0 0.0; 3.0 1.0; 4.0 0.0; 4.0 1.0];
%
%————————————————————————
% input data for nodal connectivity for each element
% nodes(i,j) where i-> element no. and j-> connected nodes
%————————————————————————
nodes=[1 3 4; 1 4 2; 3 5 6; 3 6 4;
5 7 8; 5 8 6;7 9 10; 7 10 8];
%
%————————————————————————
% input data for boundary conditions
%————————————————————————
bcdof=[1 2 3];                   % first three dofs are constrained
bcval=[0 0 0];                   % whose described values are 0
%
%————————————————————————
% initialization of matrices and vectors
%————————————————————————
ff=zeros(sdof,1);                % system force vector
kk=zeros(sdof,sdof);             % system matrix
disp=zeros(sdof,1);              % system displacement vector
eldisp=zeros(edof,1);            % element displacement vector
stress=zeros(nel,3);             % matrix containing stress components
strain=zeros(nel,3);             % matrix containing strain components
index=zeros(edof,1);             % index vector
kinmtx=zeros(3,edof);            % kinematic matrix
matmtx=zeros(3,3);               % constitutive matrix
%
%————————————
% force vector
%————————————
ff(17)=500;                      % force applied at node 9 in x-axis
```

```
ff(19)=500;                              % force applied at node 10 in x-axis
%
%————————————————————————
% compute element matrices and vectors, and assemble
%————————————————————————
matmtx=fematiso(1,emodule,poisson);      % constitutive matrix
%
for iel=1:nel                            % loop for the total number of elements
%
nd(1)=nodes(iel,1);                      % 1st connected node for (iel)-th element
nd(2)=nodes(iel,2);                      % 2nd connected node for (iel)-th element
nd(3)=nodes(iel,3);                      % 3rd connected node for (iel)-th element
%
x1=gcoord(nd(1),1); y1=gcoord(nd(1),2);  % coord values of 1st node
x2=gcoord(nd(2),1); y2=gcoord(nd(2),2);  % coord values of 2nd node
x3=gcoord(nd(3),1); y3=gcoord(nd(3),2);  % coord values of 3rd node
%
index=feeldof(nd,nnel,ndof);             % extract system dofs for the element
%
%————————————————————————
% find the derivatives of shape functions
%————————————————————————
area=0.5*(x1*y2+x2*y3+x3*y1-x1*y3-x2*y1-x3*y2);  % area of triangle
area2=area*2;
dhdx=(1/area2)*[(y2-y3) (y3-y1) (y1-y2)];        % derivatives w.r.t. x
dhdy=(1/area2)*[(x3-x2) (x1-x3) (x2-x1)];        % derivatives w.r.t. y
%
kinmtx2=fekine2d(nnel,dhdx,dhdy);        % kinematic matrix
%
k=kinmtx2'*matmtx*kinmtx2*area;          % element stiffness matrix
%
kk=feasmbl1(kk,k,index);                 % assemble element matrices
%
end                                      % end of loop for the total number of elements
%
%————————————————————————
% apply boundary conditions
%————————————————————————
[kk,ff]=feaplyc2(kk,ff,bcdof,bcval);
%
%————————————————————————
% solve the matrix equation
%————————————————————————
disp=kk\ff;
%
%————————————————————————
% element stress computation (post computation)
%————————————————————————
```

```
for ielp=1:nel                          % loop for the total number of elements
%
nd(1)=nodes(ielp,1);                    % 1st connected node for (iel)-th element
nd(2)=nodes(ielp,2);                    % 2nd connected node for (iel)-th element
nd(3)=nodes(ielp,3);                    % 3rd connected node for (iel)-th element
%
x1=gcoord(nd(1),1); y1=gcoord(nd(1),2); % coord values of 1st node
x2=gcoord(nd(2),1); y2=gcoord(nd(2),2); % coord values of 2nd node
x3=gcoord(nd(3),1); y3=gcoord(nd(3),2); % coord values of 3rd node
%
index=feeldof(nd,nnel,ndof);            % extract system dofs for the element
%
%----------------------------------
% extract element displacement vector
%----------------------------------
for i=1:edof
eldisp(i)=disp(index(i));
end
%
area=0.5*(x1*y2+x2*y3+x3*y1-x1*y3-x2*y1-x3*y2);   % area of triangle
area2=area*2;
dhdx=(1/area2)*[(y2-y3) (y3-y1) (y1-y2)];   % derivatives w.r.t. x
dhdy=(1/area2)*[(x3-x2) (x1-x3) (x2-x1)];   % derivatives w.r.t. y
%
kinmtx2=fekine2d(nnel,dhdx,dhdy);           % kinematic matrix
%
estrain=kinmtx2*eldisp;                     % compute strains
estress=matmtx*estrain;                     % compute stresses
%
for i=1:3
strain(ielp,i)=estrain(i);                  % store for each element
stress(ielp,i)=estress(i);                  % store for each element
end
%
end
%
%----------------------
% print fem solutions
%----------------------
num=1:1:sdof;
displace=[num' disp]                        % print nodal displacements
%
for i=1:nel
stresses=[i stress(i,:)]                    % print stresses
end
%
%----------------------------------------------------------------
```

```
function [kinmtx2]=fekine2d(nnel,dhdx,dhdy)
%-------------------------------------------------------------
% Purpose:
% determine the kinematic equation between strains and displacements
% for two-dimensional solids
%
% Synopsis:
% [kinmtx2]=fekine2d(nnel,dhdx,dhdy)
%
% Variable Description:
% nnel - number of nodes per element
% dhdx - derivatives of shape functions with respect to x
% dhdy - derivatives of shape functions with respect to y
%-------------------------------------------------------------
%
for i=1:nnel
i1=(i-1)*2+1;
i2=i1+1;
kinmtx2(1,i1)=dhdx(i);
kinmtx2(2,i2)=dhdy(i);
kinmtx2(3,i1)=dhdy(i);
kinmtx2(3,i2)=dhdx(i);
end
%
%-------------------------------------------------------------


function [matmtrx]=fematiso(iopt,elastic,poisson)
%-------------------------------------------------------------
% Purpose:
% determine the constitutive equation for isotropic material
%
% Synopsis:
% [matmtrx]=fematiso(iopt,elastic,poisson)
%
% Variable Description:
% elastic - elastic modulus
% poisson - Poisson's ratio
% iopt=1 - plane stress analysis
% iopt=2 - plane strain analysis
% iopt=3 - axisymmetric analysis
% iopt=4 - three-dimensional analysis
%-------------------------------------------------------------
%
if iopt==1                                      % plane stress
matmtrx= elastic/(1-poisson*poisson)* ...
[1 poisson 0; ...
```

```
            poisson 1 0; ...
            0 0 (1-poisson)/2];
            %
            elseif iopt==2                              % plane strain
            matmtrx= elastic/((1+poisson)*(1-2*poisson))* ...
            [(1-poisson) poisson 0;
            poisson (1-poisson) 0;
            0 0 (1-2*poisson)/2];
            %
            elseif iopt==3                              % axisymmetry
            matmtrx= elastic/((1+poisson)*(1-2*poisson))* ...
            [(1-poisson) poisson poisson 0;
            poisson (1-poisson) poisson 0;
            poisson poisson (1-poisson) 0;
            0 0 0 (1-2*poisson)/2];
            %
            else                                        % three-dimension
            matmtrx= elastic/((1+poisson)*(1-2*poisson))* ...
            [(1-poisson) poisson poisson 0 0 0;
            poisson (1-poisson) poisson 0 0 0;
            poisson poisson (1-poisson) 0 0 0;
            0 0 0 (1-2*poisson)/2 0 0;
            0 0 0 0 (1-2*poisson)/2 0;
            0 0 0 0 0 (1-2*poisson)/2];
            %
            end
            %
            %————————————————————————————————
```

The nodal displacements are listed below and they agree with the analytical solutions. On the other hand, the state of stress of each element is $\sigma_x = 1000$ and $\sigma_y = \tau_{xy} = 0$ as expected.

```
            displace =
            d.o.f.      displ.
            1.0000      0.0000              % x-displacement of node 1
            2.0000      0.0000              % y-displacement of node 1
            3.0000      0.0000              % x-displacement of node 2
            4.0000     -0.0030              % y-displacement of node 2
            5.0000      0.0100              % x-displacement of node 3
            6.0000      0.0000              % y-displacement of node 3
            7.0000      0.0100              % x-displacement of node 4
            8.0000     -0.0030              % y-displacement of node 4
            9.0000      0.0200              % x-displacement of node 5
           10.000       0.0000              % y-displacement of node 5
           11.000       0.0200              % x-displacement of node 6
           12.000      -0.0030              % y-displacement of node 6
           13.000       0.0300              % x-displacement of node 7
```

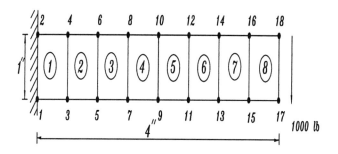

Figure 9.8.2 Cantilever Beam Subjected to a Tip Load

14.000	0.0000	% y-displacement of node 7
15.000	0.0300	% x-displacement of node 8
16.000	-0.0030	% y-displacement of node 8
17.000	0.0400	% x-displacement of node 9
18.000	0.0000	% y-displacement of node 9
19.000	0.0400	% x-displacement of node 10
20.000	-0.0030	% y-displacement of node 10

‡

♣ **Example 9.8.2** We want to analyze a short cantilever beam using two-dimensional isoparametric elements assuming plane stress condition. To this end, the beam is modeled using ten four-node quadrilateral elements as seen in Fig. 9.8.2.

```
%————————————————————————————
% Example 9.8.2
% plane stress analysis of a cantilever beam using isoparametric
% four-node elements
% (see Fig. 9.8.2 for the finite element mesh)
%
% Variable descriptions
% k = element matrix
% f = element vector
% kk = system matrix
% ff = system vector
% disp = system nodal displacement vector
% eldisp = element nodal displacement vector
% stress = matrix containing stresses
% strain = matrix containing strains
```

```
% gcoord = coordinate values of each node
% nodes = nodal connectivity of each element
% index = a vector containing system dofs associated with each element
% point2 = matrix containing sampling points
% weight2 = matrix containing weighting coefficients
% bcdof = a vector containing dofs associated with boundary conditions
% bcval = a vector containing boundary condition values associated with
%             the dofs in bcdof
%----------------------------------------------------------------
%
%----------------------------------------------
% input data for control parameters
%----------------------------------------------
nel=8;                          % number of elements
nnel=4;                         % number of nodes per element
ndof=2;                         % number of dofs per node
nnode=18;                       % total number of nodes in system
sdof=nnode*ndof;                % total system dofs
edof=nnel*ndof;                 % degrees of freedom per element
emodule=1e6;                    % elastic modulus
poisson=0.3;                    % Poisson's ratio
nglx=2; ngly=2;                 % 2x2 Gauss-Legendre quadrature
nglxy=nglx*ngly;                % number of sampling points per element
%
%------------------------------------------------
% input data for nodal coordinate values
% gcoord(i,j) where i-> node no. and j-> x or y
%------------------------------------------------
gcoord=[0.0 0.0; 0.0 1.0; 0.5 0.0; 0.5 1.0; 1.0 0.0;
1.0 1.0; 1.5 0.0; 1.5 1.0; 2.0 0.0; 2.0 1.0;
2.5 0.0; 2.5 1.0; 3.0 0.0; 3.0 1.0; 3.5 0.0;
3.5 1.0; 4.0 0.0; 4.0 1.0];
%
%---------------------------------------------------
% input data for nodal connectivity for each element
% nodes(i,j) where i-> element no. and j-> connected nodes
%---------------------------------------------------
nodes=[1 3 4 2; 3 5 6 4; 5 7 8 6; 7 9 10 8;
9 11 12 10; 11 13 14 12; 13 15 16 14; 15 17 18 16];
%
%-----------------------------------------
% input data for boundary conditions
%-----------------------------------------
bcdof=[1 2 3 4];                % first four dofs are constrained
bcval=[0 0 0 0];                % whose described values are 0
%
%-----------------------------------------
% initialization of matrices and vectors
```

Section 9.8 MATLAB Application to 2-D Stress Analysis

```
%------------------------------
ff=zeros(sdof,1);                          % system force vector
kk=zeros(sdof,sdof);                       % system matrix
disp=zeros(sdof,1);                        % system displacement vector
eldisp=zeros(edof,1);                      % element displacement vector
stress=zeros(nglxy,3);                     % matrix containing stress components
strain=zeros(nglxy,3);                     % matrix containing strain components
index=zeros(edof,1);                       % index vector
kinmtx=zeros(3,edof);                      % kinematic matrix
matmtx=zeros(3,3);                         % constitutive matrix
%
%------------------
% force vector
%------------------
ff(34)=500;                                % force applied at node 17 in y-axis
ff(36)=500;                                % force applied at node 18 in y-axis
%
%---------------------------------------------------
% compute element matrices and vectors, and assemble
%---------------------------------------------------
[point2,weight2]=feglqd2(nglx,ngly);       % sampling points & weights
matmtx=fematiso(1,emodule,poisson);        % constitutive matrix
%
for iel=1:nel                              % loop for the total number of elements
%
for i=1:nnel
nd(i)=nodes(iel,i);                        % extract nodes for (iel)-th element
xcoord(i)=gcoord(nd(i),1);                 % extract x value of the node
ycoord(i)=gcoord(nd(i),2);                 % extract y value of the node
end
%
k=zeros(edof,edof);                        % initialization of element matrix
%
%------------------------
% numerical integration
%------------------------
for intx=1:nglx
x=point2(intx,1);                          % sampling point in x-axis
wtx=weight2(intx,1);                       % weight in x-axis
for inty=1:ngly
y=point2(inty,2);                          % sampling point in y-axis
wty=weight2(inty,2) ;                      % weight in y-axis
%
[shape,dhdr,dhds]=feisoq4(x,y);            % compute shape functions and
                                           % derivatives at sampling point
%
jacob2=fejacob2(nnel,dhdr,dhds,xcoord,ycoord);   % compute Jacobian
%
```

```
            detjacob=det(jacob2);                    % determinant of Jacobian
            invjacob=inv(jacob2);                    % inverse of Jacobian matrix
            %
            [dhdx,dhdy]=federiv2(nnel,dhdr,dhds,invjacob);    % derivatives w.r.t.
                                                              % physical coordinate
            %
            kinmtx2=fekine2d(nnel,dhdx,dhdy);        % compute kinematic matrix
            %
            %------------------------
            % compute element matrix
            %------------------------
            k=k+kinmtx2'*matmtx*kinmtx2*wtx*wty*detjacob;    % element matrix
            %
            end
            end                                      % end of numerical integration loop
            %
            index=feeldof(nd,nnel,ndof);             % extract system dofs for the element
            %
            kk=feasmbl1(kk,k,index);                 % assemble element matrices
            %
            end                      % end of loop for the total number of elements
            %
            %---------------------------
            % apply boundary conditions
            %---------------------------
            [kk,ff]=feaplyc2(kk,ff,bcdof,bcval);
            %
            %---------------------------
            % solve the matrix equation
            %---------------------------
            disp=kk\ff;
            %
            num=1:1:sdof;
            displace=[num' disp]                     % print nodal displacements
            %
            %---------------------------
            % element stress computation
            %---------------------------
            for ielp=1:nel                           % loop for the total number of elements
            %
            for i=1:nnel
            nd(i)=nodes(ielp,i);                     % extract nodes for (iel)-th element
            xcoord(i)=gcoord(nd(i),1);               % extract x value of the node
            ycoord(i)=gcoord(nd(i),2):               % extract y value of the node
            end
            %
            %---------------------------
            % numerical integration
```

```
%------------------
intp=0;
for intx=1:nglx
x=point2(intx,1);                          % sampling point in x-axis
wtx=weight2(intx,1);                       % weight in x-axis
for inty=1:ngly
y=point2(inty,2);                          % sampling point in y-axis
wty=weight2(inty,2) ;                      % weight in y-axis
intp=intp+1;
%
[shape,dhdr,dhds]=feisoq4(x,y);            % compute shape functions and
                                           % derivatives at sampling point
%
jacob2=fejacob2(nnel,dhdr,dhds,xcoord,ycoord);   % compute Jacobian
%
detjacob=det(jacob2);                      % determinant of Jacobian
invjacob=inv(jacob2);                      % inverse of Jacobian matrix
%
[dhdx,dhdy]=federiv2(nnel,dhdr,dhds,invjacob);   % derivatives w.r.t.
                                           % physical coordinate
%
kinmtx2=fekine2d(nnel,dhdx,dhdy);          % kinematic matrix
%
index=feeldof(nd,nnel,ndof);        % extract system dofs for the element
%
%--------------------------------
% extract element displacement vector
%--------------------------------
for i=1:edof
eldisp(i)=disp(index(i));
end
%
kinmtx2=fekine2d(nnel,dhdx,dhdy);          % compute kinematic matrix
%
estrain=kinmtx2*eldisp;                    % compute strains
estress=matmtx*estrain;                    % compute stresses
%
for i=1:3
strain(intp,i)=estrain(i);                 % store for each sampling point
stress(intp,i)=estress(i);                 % store for each sampling point
end
%
location=[ielp,intx,inty]                  % print location for stress
stress(intp,:)                             % print stress values
%
end
end                                        % end of integration loop
%
```

```
end                    % end of loop for total number of elements
%
%————————————————————————————————————————————————————
```

As expected, the displacements in the x-axis are positive at the bottom side of the strip and negative at the top side because of bending. The tip displacement in the y-axis is 0.2238 in. while the beam bending theory gives 0.256. As a result, the mesh needs to be refined to improve the accuracy. On the other hand, the bending stress σ_x is 11950 psi at the integration point nearest the fixed edge. The point is located 0.1057 in. away from the fixed edge in the x-axis and 0.2887 in. above the midplane in the y-axis. The beam theory gives bending stress of 13820.

```
displace =
d.o.f.     displ.
1.0000     0.0000           % x-displacement of node 1
2.0000     0.0000           % y-displacement of node 1
3.0000     0.0000           % x-displacement of node 2
4.0000     0.0000           % y-displacement of node 2
5.0000     0.0094           % x-displacement of node 3
6.0000     0.0060           % y-displacement of node 3
7.0000    -0.0094           % x-displacement of node 4
8.0000     0.0060           % y-displacement of node 4
9.0000     0.0176           % x-displacement of node 5
10.000     0.0208           % y-displacement of node 5
11.000    -0.0176           % x-displacement of node 6
12.000     0.0208           % y-displacement of node 6
13.000     0.0245           % x-displacement of node 7
14.000     0.0431           % y-displacement of node 7
15.000    -0.0245           % x-displacement of node 8
16.000     0.0431           % y-displacement of node 8
17.000     0.0301           % x-displacement of node 9
18.000     0.0717           % y-displacement of node 9
19.000    -0.0301           % x-displacement of node 10
20.000     0.0717           % y-displacement of node 10
21.000     0.0345           % x-displacement of node 11
22.000     0.1534           % y-displacement of node 11
23.000    -0.0345           % x-displacement of node 12
24.000     0.1053           % y-displacement of node 12
25.000     0.0377           % x-displacement of node 13
26.000     0.1427           % y-displacement of node 13
27.000    -0.0377           % x-displacement of node 14
28.000     0.1427           % y-displacement of node 14
29.000     0.0395           % x-displacement of node 15
30.000     0.1826           % y-displacement of node 15
31.000    -0.0395           % x-displacement of node 16
32.000     0.1826           % y-displacement of node 16
33.000     0.0402           % x-displacement of node 17
```

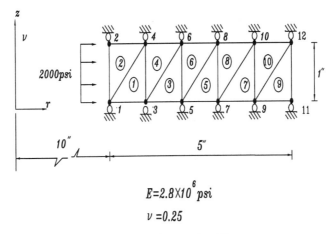

Figure 9.9.1 Axisymmetric Solid With Triangular Elements

34.000	0.2238	% y-displacement of node 17
35.000	-0.0402	% x-displacement of node 18
36.000	0.2238	% y-displacement of node 18

‡

9.9 MATLAB Application to Axisymmetric Analysis

The same axisymmetric solid is analyzed using both conventional triangular elements and isoparametric quadrilateral elements.

♣ **Example 9.9.1** A thick walled cylinder is subjected to a uniform internal pressure of 2000 psi. The cylinder has inside radius of 10 in. and outside radius of 15 in. It is made of steel whose elastic modulus is 28×10^6 psi and Poisson's ratio is 0.3. Figure 9.9.1 shows the finite element mesh using 10 triangular elements. For the present analysis, the cylinder is assumed to be constrained along the axial direction. The resultant force applied on the inside surface is $2000 \times 2\pi \times 10 = 20,000\pi$ assuming unit length along the axial direction. As a result, nodes 1 and 2 at the inside boundary have a concentrated nodal force of $10,000\pi$ along the radial direction, respectively. The value is provided to the force vector in the finite element analysis program.

```
%————————————————————————————————
% Example 9.9.1
% axisymmetric analysis of a solid subjected to an internal
% pressure using linear triangular elements
% (see Fig. 9.9.1 for the finite element mesh)
```

```
%
% Variable descriptions
% k = element matrix
% f = element vector
% kk = system matrix
% ff = system vector
% disp = system nodal displacement vector
% eldisp = element nodal displacement vector
% stress = matrix containing stresses
% strain = matrix containing strains
% gcoord = coordinate values of each node
% nodes = nodal connectivity of each element
% index = a vector containing system dofs associated with each element
% bcdof = a vector containing dofs associated with boundary conditions
% bcval = a vector containing boundary condition values associated with
%             the dofs in bcdof
%----------------------------------------------------------------
%
%----------------------------------
% input data for control parameters
%----------------------------------
%
nel=10;                         % number of elements
nnel=3;                         % number of nodes per element
ndof=2;                         % number of dofs per node
nnode=12;                       % total number of nodes in system
sdof=nnode*ndof;                % total system dofs
edof=nnel*ndof;                 % degrees of freedom per element
emodule=28e6;                   % elastic modulus
poisson=0.25;                   % Poisson's ratio
%
%-----------------------------------------------
% input data for nodal coordinate values
% gcoord(i,j) where i-> node no. and j-> x or y
%-----------------------------------------------
gcoord=[10. 0.; 10. 1.; 11. 0.; 11. 1.; 12. 0.; 12. 1.;
13. 0.; 13. 1.; 14. 0.; 14. 1.; 15. 0.; 15. 1.];
%
%-------------------------------------------------------
% input data for nodal connectivity for each element
% nodes(i,j) where i-> element no. and j-> connected nodes
%-------------------------------------------------------
nodes=[1 3 4; 1 4 2; 3 5 6; 3 6 4; 5 7 8;
5 8 6;7 9 10; 7 10 8; 9 11 12; 9 12 10];
%
%-----------------------------------
% input data for boundary conditions
%-----------------------------------
```

Section 9.9   MATLAB Application to Axisymmetric Analysis   341

```
bcdof=[2 4 6 8 10 12 14 16 18 20 22 24];      % axial motion constrained
bcval=[0 0 0 0 0 0 0 0 0 0 0 0];              % constrained values are 0
%
%------------------------------
% initialization of matrices and vectors
%------------------------------
ff=zeros(sdof,1);                             % system force vector
kk=zeros(sdof,sdof);                          % system matrix
disp=zeros(sdof,1);                           % system displacement vector
eldisp=zeros(edof,1);                         % element displacement vector
stress=zeros(nel,4);                          % matrix containing stress components
strain=zeros(nel,4);                          % matrix containing strain components
index=zeros(edof,1);                          % index vector
kinmtax=zeros(4,edof);                        % kinematic matrix
matmtx=zeros(4,4);                            % constitutive matrix
%
%------------------
% force vector
%------------------
pi=4.0*atan(1);                               % pi=3.141592
%
ff(1)=2e3*pi*2*10;                            % force applied at node 1 in x-axis
ff(3)=2e3*pi*2*10;                            % force applied at node 2 in x-axis
%
%-----------------------------------------
% compute element matrices and vectors, and assemble
%-----------------------------------------
matmtx=fematiso(3,emodule,poisson);           % constitutive matrix
%
for iel=1:nel                                 % loop for the total number of elements
%
nd(1)=nodes(iel,1);                           % 1st node for (iel)-th element
nd(2)=nodes(iel,2);                           % 2nd node for (iel)-th element
nd(3)=nodes(iel,3);                           % 3rd node for (iel)-th element
%
x1=gcoord(nd(1),1); y1=gcoord(nd(1),2);       % coord values of 1st node
x2=gcoord(nd(2),1); y2=gcoord(nd(2),2);       % coord values of 2nd node
x3=gcoord(nd(3),1); y3=gcoord(nd(3),2);       % coord values of 3rd node
%
index=feeldof(nd,nnel,ndof);                  % extract system dofs for the element
%
%------------------------------
% find the derivatives of shape functions
%------------------------------
area=0.5*(x1*y2+x2*y3+x3*y1-x1*y3-x2*y1-x3*y2);   % area of triangle
area2=area*2;
xcenter=(x1+x2+x3)/3;                         % x-centroid of triangle
ycenter=(y1+y2+y3)/3;                         % y-centroid of triangle
```

```
%
shape(1)=((x2*y3-x3*y2)+(y2-y3)*xcenter+(x3-x2)*ycenter)/area2;
shape(2)=((x3*y1-x1*y3)+(y3-y1)*xcenter+(x1-x3)*ycenter)/area2;
shape(3)=((x1*y2-x2*y1)+(y1-y2)*xcenter+(x2-x1)*ycenter)/area2;
%
dhdx=(1/area2)*[(y2-y3) (y3-y1) (y1-y2)];        % derivatives w.r.t. x
dhdy=(1/area2)*[(x3-x2) (x1-x3) (x2-x1)];        % derivatives w.r.t. y
%
kinmtax=fekineax(nnel,dhdx,dhdy,shape,xcenter);   % kinematic matrix
%
k=2*pi*xcenter*area*kinmtax'*matmtx*kinmtax;      % element matrix
%
kk=feasmbl1(kk,k,index);                          % assemble element matrices
%
end                      % end of loop for total number of elements
%
%----------------------
% apply boundary conditions
%----------------------
[kk,ff]=feaplyc2(kk,ff,bcdof,bcval);
%
%----------------------
% solve the matrix equation
%----------------------
disp=kk\ff;
%
%----------------------
% element stress computation (post-computation)
%----------------------
for ielp=1:nel                % loop for the total number of elements
%
nd(1)=nodes(ielp,1);                    % 1st node for (iel)-th element
nd(2)=nodes(ielp,2);                    % 2nd node for (iel)-th element
nd(3)=nodes(ielp,3);                    % 3rd node for (iel)-th element
%
x1=gcoord(nd(1),1); y1=gcoord(nd(1),2);    % coord values of 1st node
x2=gcoord(nd(2),1); y2=gcoord(nd(2),2);    % coord values of 2nd node
x3=gcoord(nd(3),1); y3=gcoord(nd(3),2);    % coord values of 3rd node
%
index=feeldof(nd,nnel,ndof);           % extract system dofs for the element
%
%----------------------
% extract element displacement vector
%----------------------
for i=1:edof
eldisp(i)=disp(index(i));
end
%
```

```
area=0.5*(x1*y2+x2*y3+x3*y1-x1*y3-x2*y1-x3*y2);   % area of triangle
area2=area*2;
xcenter=(x1+x2+x3)/3;                             % x-centroid of triangle
ycenter=(y1+y2+y3)/3;                             % y-centroid of triangle
%
shape(1)=((x2*y3-x3*y2)+(y2-y3)*xcenter+(x3-x2)*ycenter)/area2;
shape(2)=((x3*y1-x1*y3)+(y3-y1)*xcenter+(x1-x3)*ycenter)/area2;
shape(3)=((x1*y2-x2*y1)+(y1-y2)*xcenter+(x2-x1)*ycenter)/area2;
%
dhdx=(1/area2)*[(y2-y3) (y3-y1) (y1-y2)];         % derivatives w.r.t. x
dhdy=(1/area2)*[(x3-x2) (x1-x3) (x2-x1)];         % derivatives w.r.t. y
%
kinmtax=fekineax(nnel,dhdx,dhdy,shape,xcenter);   % kinematic matrix
%
estrain=kinmtax*eldisp;                           % compute strains
estress=matmtx*estrain;                           % compute stresses
%
for i=1:4
strain(ielp,i)=estrain(i);                        % store for each element
stress(ielp,i)=estress(i);                        % store for each element
end
%
end
%
%————————————
% print fem solutions
%————————————
num=1:1:sdof;
displace=[num' disp]                              % print nodal displacements
%
for i=1:nel
stresses=[i stress(i,:)]                          % print stresses
end
%
%————————————————————————————

function [kinmtxax]=fekineax(nnel,dhdx,dhdy,shape,radist)
%————————————————————————————
% Purpose:
% determine kinematic equations between strains and displacements
% for axisymmetric solids
%
% Synopsis:
% [kinmtxax]=fekineax(nnel,dhdx,dhdy,shape,radist)
%
% Variable Description:
```

```
% nnel - number of nodes per element
% shape - shape functions
% dhdx - derivatives of shape functions with respect to x
% dhdy - derivatives of shape functions with respect to y
% radist - radial distance of integration point or central point
%           for hoop strain component
%------------------------------------------------------------
%
for i=1:nnel
i1=(i-1)*2+1;
i2=i1+1;
kinmtxax(1,i1)=dhdx(i);
kinmtxax(2,i1)=shape(i)/radist;
kinmtxax(3,i2)=dhdy(i);
kinmtxax(4,i1)=dhdy(i);
kinmtxax(4,i2)=dhdx(i);
end
%
%------------------------------------------------------------
```

The results are

displace =

d.o.f.	displ.	
1.0000	0.0039	% radial displacement of node 1
2.0000	0.0000	
3.0000	0.0039	% radial displacement of node 2
4.0000	0.0000	
5.0000	0.0037	% radial displacement of node 3
6.0000	0.0000	
7.0000	0.0037	% radial displacement of node 4
8.0000	0.0000	
9.0000	0.0035	% radial displacement of node 5
10.000	0.0000	
11.000	0.0035	% radial displacement of node 6
12.000	0.0000	
13.000	0.0034	% radial displacement of node 7
14.000	0.0000	
15.000	0.0034	% radial displacement of node 8
16.000	0.0000	
17.000	0.0033	% radial displacement of node 9
18.000	0.0000	
19.000	0.0033	% radial displacement of node 10
20.000	0.0000	
21.000	0.0032	% radial displacement of node 11
22.000	0.0000	
23.000	0.0032	% radial displacement of node 12
24.000	0.0000	

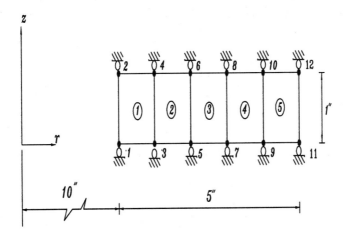

Figure 9.9.2 Axisymmetric Solid With Rectangular Elements

```
stresses =
elem    radial   hoop    axial
1.00    -3277.   9486.   1552.
2.00    -3408.   10000   1649.
3.00    -2281.   8434.   1538.
4.00    -2221.   8877.   1664.
5.00    -1457.   7637.   1545.
6.00    -1368.   7995.   1657.
7.00    -783.0   7017.   1558.
8.00    -724.0   7295.   1643.
9.00    -213.0   6529.   1579.
10.0    -242.0   6727.   1621.
```

‡

♣ **Example 9.9.2**   Example 9.9.1 is solved using isoparametric elements. The same number of nodes is used but the number of elements is 5 as seen in Fig. 9.9.2.

```
%————————————————————————————————
% Example 9.9.2
% axisymmetric analysis of a thick walled cylinder
% subjected to internal pressure using isoparametric
% four-node elements
% (see Fig. 9.9.2 for the finite element mesh)
%
% Variable descriptions
% k = element matrix
```

```
%   f = element vector
%   kk = system matrix
%   ff = system vector
%   disp = system nodal displacement vector
%   eldisp = element nodal displacement vector
%   stress = matrix containing stresses
%   strain = matrix containing strains
%   gcoord = coordinate values of each node
%   nodes = nodal connectivity of each element
%   index = a vector containing system dofs associated with each element
%   point2 = matrix containing sampling points
%   weight2 = matrix containing weighting coefficients
%   bcdof = a vector containing dofs associated with boundary conditions
%   bcval = a vector containing boundary condition values associated with
%                the dofs in bcdof
%---------------------------------------------------------------
%
%---------------------------------------------------------------
%  input data for control parameters
%---------------------------------
nel=5;                      % number of elements
nnel=4;                     % number of nodes per element
ndof=2;                     % number of dofs per node
nnode=12;                   % total number of nodes in system
sdof=nnode*ndof;            % total system dofs
edof=nnel*ndof;             % degrees of freedom per element
emodule=28.0e6;             % elastic modulus
poisson=0.25;               % Poisson's ratio
nglx=2; ngly=2;             % 2x2 Gauss-Legendre quadrature
nglxy=nglx*ngly;            % number of sampling points per element
%
%-------------------------------------
%  input data for nodal coordinate values
%  gcoord(i,j) where i-> node no. and j-> x or y
%-------------------------------------
gcoord=[10. 0.; 10. 1.; 11. 0.; 11. 1.; 12. 0.; 12. 1.;
13. 0.; 13. 1.; 14. 0.; 14. 1.; 15. 0.; 15. 1.];
%
%-------------------------------------------------
%  input data for nodal connectivity for each element
%  nodes(i,j) where i-> element no. and j-> connected nodes
%-------------------------------------------------
nodes=[1 3 4 2; 3 5 6 4; 5 7 8 6; 7 9 10 8; 9 11 12 10];
%
%-------------------------------------
%  input data for boundary conditions
%-------------------------------------
bcdof=[2 4 6 8 10 12 14 16 18 20 22 24];      % axial motion constrained
```

Section 9.9 MATLAB Application to Axisymmetric Analysis 347

```matlab
bcval=[0 0 0 0 0 0 0 0 0 0 0];          % constrained values are 0
%
%----------------------------------------
% initialization of matrices and vectors
%----------------------------------------
ff=zeros(sdof,1);                        % system force vector
kk=zeros(sdof,sdof);                     % system matrix
disp=zeros(sdof,1);                      % system displacement vector
eldisp=zeros(edof,1);                    % element displacement vector
stress=zeros(nglxy,4);                   % matrix containing stress components
strain=zeros(nglxy,4);                   % matrix containing strain components
index=zeros(edof,1);                     % index vector
kinmtx=zeros(4,edof);                    % kinematic matrix
matmtx=zeros(4,4);                       % constitutive matrix
%
%----------------------
% force vector
%----------------------
pi=4.0*atan(1.0);                        % pi=3.141592
%
ff(1)=2e3*2*pi*10;                       % force applied at node 1 in x-axis
ff(3)=2e3*2*pi*10;                       % force applied at node 2 in x-axis
%
%--------------------------------------------------
% compute element matrices and vectors, and assemble
%--------------------------------------------------
[point2,weight2]=feglqd2(nglx,ngly);     % sampling points & weights
matmtx=fematiso(3,emodule,poisson);      % constitutive matrix
%
for iel=1:nel                            % loop for the total number of elements
%
for i=1:nnel
nd(i)=nodes(iel,i);                      % extract node for (iel)-th element
xcoord(i)=gcoord(nd(i),1);               % extract x value of the node
ycoord(i)=gcoord(nd(i),2);               % extract y value of the node
end
%
k=zeros(edof,edof);                      % initialization of element matrix
%
%----------------------
% numerical integration
%----------------------
for intx=1:nglx
x=point2(intx,1);                        % sampling point in x-axis
wtx=weight2(intx,1);                     % weight in x-axis
for inty=1:ngly
y=point2(inty,2);                        % sampling point in y-axis
wty=weight2(inty,2) ;                    % weight in y-axis
```

```
%
[shape,dhdr,dhds]=feisoq4(x,y);           % compute shape functions and
                                          % derivatives at sampling point
%
jacob2=fejacob2(nnel,dhdr,dhds,xcoord,ycoord);     % compute Jacobian
%
detjacob=det(jacob2);                     % determinant of Jacobian
invjacob=inv(jacob2);                     % inverse of Jacobian matrix
%
[dhdx,dhdy]=federiv2(nnel,dhdr,dhds,invjacob);     % derivatives w.r.t.
                                                   % physical coordinate
%
xcenter=0;
for i=1:nnel                              % x-coordinate value
xcenter=xcenter+shape(i)*xcoord(i);       % of the integration point
end
%
kinmtx=fekineax(nnel,dhdx,dhdy,shape,xcenter);     % kinematic matrix
%
%------------------------
% compute element matrix
%------------------------
k=k+2*pi*xcenter*kinmtx'*matmtx*kinmtx*wtx*wty*detjacob;
                                          % element matrix
%
end
end                                       % end of numerical integration loop
%
index=feeldof(nd,nnel,ndof);              % extract system dofs for the element
%
kk=feasmbl1(kk,k,index);                  % assemble element matrices
%
end                    % end of loop for total number of elements
%
%------------------------
% apply boundary conditions
%------------------------
[kk,ff]=feaplyc2(kk,ff,bcdof,bcval);
%
%------------------------
% solve the matrix equation
%------------------------
disp=kk\ff;
%
num=1:1:sdof;
displace=[num' disp]                      % print nodal displacements
%
%------------------------
```

```
% element stress computation
%----------------------------
for ielp=1:nel                          % loop for the total number of elements
%
for i=1:nnel
nd(i)=nodes(ielp,i);                    % extract node for (iel)-th element
xcoord(i)=gcoord(nd(i),1);              % extract x value of the node
ycoord(i)=gcoord(nd(i),2);              % extract y value of the node
end
%
%----------------------------
% numerical integration
%----------------------------
intp=0;
for intx=1:nglx
x=point2(intx,1);                       % sampling point in x-axis
wtx=weight2(intx,1);                    % weight in x-axis
for inty=1:ngly
y=point2(inty,2);                       % sampling point in y-axis
wty=weight2(inty,2) ;                   % weight in y-axis
intp=intp+1;
%
[shape,dhdr,dhds]=feisoq4(x,y);         % compute shape functions and
                                        % derivatives at sampling point
%
jacob2=fejacob2(nnel,dhdr,dhds,xcoord,ycoord);   % compute Jacobian
%
detjacob=det(jacob2);                   % determinant of Jacobian
invjacob=inv(jacob2);                   % inverse of Jacobian matrix
%
[dhdx,dhdy]=federiv2(nnel,dhdr,dhds,invjacob);   % derivatives w.r.t.
                                        % physical coordinate
%
xcenter=0;
for i=1:nnel                            % x-coordinate value
xcenter=xcenter+shape(i)*xcoord(i);     % of the integration point
end
%
kinmtx=fekineax(nnel,dhdx,dhdy,shape,xcenter);   % kinematic matrix
%
index=feeldof(nd,nnel,ndof);            % extract system dofs for the element
%
%----------------------------
% extract element displacement vector
%----------------------------
for i=1:edof
eldisp(i)=disp(index(i));
end
```

```
%
estrain=kinmtx*eldisp;                    % compute strains
estress=matmtx*estrain;                   % compute stresses
%
for i=1:4
strain(intp,i)=estrain(i);                % store for each element
stress(intp,i)=estress(i);                % store for each element
end
%
end
end                                       % end of integration loop
%
for j=1:nglxy
stresses=[ielp stress(j,:)]               % print stresses
end
%
end                       % end of loop for total number of elements
%
%─────────────────────────────────────────────────────────────
```

The nodal displacements are the same as those obtained from Example 9.8.1. The stresses are also very similar but a little different. Stresses for the first two elements are listed below. They are printed for each integration point. Because 2×2 quadrature was used for numerical integration, there are four integration points for each element.

```
stresses for element 1 =
1st integration point
radial stress=-3146. hoop stress=10289 axial stress=1786
2nd integration point
radial stress=-3146. hoop stress=10289 axial stress=1786.
3rd integration point
radial stress=-3505. hoop stress=9211. axial stress=1426.
4th integration point
radial stress=-3506. hoop stress=9211. axial stress=1426.
stresses for element 2 =
1st integration point
radial stress=-2104. hoop stress=9066. axial stress=1741.
2nd integration point
radial stress=-2104. hoop stress=9066. axial stress=1741.
3rd integration point
radial stress=-2377. hoop stress=8245. axial stress=1467.
4th integration point
radial stress=-2377. hoop stress=8245. axial stress=1467.
```

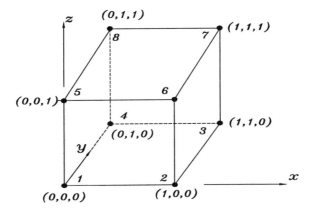

Figure 9.10.1 A Cubic Solid

9.10 MATLAB Application to 3-D Stress Analysis

♣ **Example 9.10.1**   A unit cube is subjected to a uniform load as shown in Fig. 9.10.1. We use one eight-node isoparametric solid element for the problem.

```
%----------------------------------------------------------------
% Example 9.10.1
% three-dimensional analysis of a cube using isoparametric
% eight-node elements
% (see Fig. 9.10.1 for the finite element mesh)
%
% Variable descriptions
% k = element matrix
% f = element vector
% kk = system matrix
% ff = system vector
% disp = system nodal displacement vector
% eldisp = element nodal displacement vector
% stress = matrix containing stresses
% strain = matrix containing strains
% gcoord = coordinate values of each node
% nodes = nodal connectivity of each element
% index = a vector containing system dofs associated with each element
% point3 = matrix containing sampling points
% weight3 = matrix containing weighting coefficients
% bcdof = a vector containing dofs associated with boundary conditions
% bcval = a vector containing boundary condition values associated with
%         the dofs in bcdof
%----------------------------------------------------------------
%
%----------------------------------------------
% input data for control parameters
```

```
%----------------------------------------
nel=1;                          % number of elements
nnel=8;                         % number of nodes per element
ndof=3;                         % number of dofs per node
nnode=8;                        % total number of nodes in system
sdof=nnode*ndof;                % total system dofs
edof=nnel*ndof;                 % degrees of freedom per element
emodule=1e5;                    % elastic modulus
poisson=0.3;                    % Poisson's ratio
nglx=2; ngly=2; nglz=2;         % 2x2x2 Gauss-Legendre quadrature
nglxyz=nglx*ngly*nglz;          % number of sampling points per element
%
%----------------------------------------
% input data for nodal coordinate values
% gcoord(i,j) where i-> node no. and j-> x or y
%----------------------------------------
gcoord=[0.0 0.0 0.0; 1.0 0.0 0.0; 1.0 1.0 0.0; 0.0 1.0 0.0;
0.0 0.0 1.0; 1.0 0.0 1.0; 1.0 1.0 1.0; 0.0 1.0 1.0];
%
%----------------------------------------
% input data for nodal connectivity for each element
% nodes(i,j) where i-> element no. and j-> connected nodes
%----------------------------------------
nodes=[1 2 3 4 5 6 7 8];
%
%----------------------------------------
% input data for boundary conditions
%----------------------------------------
bcdof=[1 2 3 5 6 9 12];         % constrained dofs
bcval=[0 0 0 0 0 0 0];          % constrained values
%
%----------------------------------------
% initialization of matrices and vectors
%----------------------------------------
ff=zeros(sdof,1);               % system force vector
kk=zeros(sdof,sdof);            % system matrix
disp=zeros(sdof,1);             % system displacement vector
eldisp=zeros(edof,1);           % element displacement vector
stress=zeros(nglxyz,6);         % matrix containing stress components
strain=zeros(nglxyz,6);         % matrix containing strain components
index=zeros(edof,1);            % index vector
kinmtx=zeros(6,edof);           % kinematic matrix
matmtx=zeros(6,6);              % constitutive matrix
%
%----------------------
% force vector
%----------------------
ff(15)=250;                     % force applied at node 5 in z-axis
```

Section 9.10 MATLAB Application to 3-D Stress Analysis 353

```
ff(18)=250;                              % force applied at node 6 in z-axis
ff(21)=250;                              % force applied at node 7 in z-axis
ff(24)=250;                              % force applied at node 8 in z-axis
%
%------------------------------------------
% compute element matrices and vectors, and assemble
%------------------------------------------
[point3,weight3]=feglqd3(nglx,ngly,nglz);   % sampling points & weights
matmtx=fematiso(4,emodule,poisson);         % compute constitutive matrix
%
for iel=1:nel                            % loop for the total number of elements
%
for i=1:nnel
nd(i)=nodes(iel,i);                      % extract node for (iel)-th element
xcoord(i)=gcoord(nd(i),1);               % extract x value of the node
ycoord(i)=gcoord(nd(i),2);               % extract y value of the node
zcoord(i)=gcoord(nd(i),3);               % extract z value of the node
end
%
k=zeros(edof,edof);                      % initialization of element matrix
%
%------------------------
% numerical integration
%------------------------
for intx=1:nglx
x=point3(intx,1);                        % sampling point in x-axis
wtx=weight3(intx,1);                     % weight in x-axis
for inty=1:ngly
y=point3(inty,2);                        % sampling point in y-axis
wty=weight3(inty,2) ;                    % weight in y-axis
for intz=1:nglz
z=point3(intz,3);                        % sampling point in z-axis
wtz=weight3(intz,3) ;                    % weight in z-axis
%
[shape,dhdr,dhds,dhdt]=feisos8(x,y,z);   % compute shape functions
                                         % and derivatives at sampling point
%
jacob3=fejacob3(nnel,dhdr,dhds,dhdt,xcoord,ycoord,zcoord);
                                         % compute Jacobian
%
detjacob=det(jacob3);                    % determinant of Jacobian
invjacob=inv(jacob3);                    % inverse of Jacobian matrix
%
[dhdx,dhdy,dhdz]=federiv3(nnel,dhdr,dhds,dhdt,invjacob);
                                         % derivatives w.r.t. physical coordinate
%
kinmtx=fekine3d(nnel,dhdx,dhdy,dhdz);    % kinematic matrix
%
```

```
%----------------------
% compute element matrix
%----------------------
k=k+kinmtx'*matmtx*kinmtx*wtx*wty*wtz*detjacob;        % element
%
end
end
end                                    % end of numerical integration loop
%
index=feeldof(nd,nnel,ndof);           % extract system dofs for the element
%
kk=feasmbl1(kk,k,index);               % assemble element matrices
%
end                                    % end of loop for total number of elements
%
%-------------------------
% apply boundary conditions
%-------------------------
[kk,ff]=feaplyc2(kk,ff,bcdof,bcval);
%
%-------------------------
% solve the matrix equation
%-------------------------
disp=kk\ff;
%
num=1:1:sdof;
displace=[num' disp]                   % print nodal displacements
%
%-------------------------
% element stress computation
%-------------------------
for ielp=1:nel                         % loop for the total number of elements
%
for i=1:nnel
nd(i)=nodes(ielp,i);                   % extract node for (iel)-th element
xcoord(i)=gcoord(nd(i),1);             % extract x value of the node
ycoord(i)=gcoord(nd(i),2);             % extract y value of the node
zcoord(i)=gcoord(nd(i),3);             % extract z value of the node
end
%
%-------------------------
% numerical integration
%-------------------------
intp=0;
for intx=1:nglx
x=point3(intx,1);                      % sampling point in x-axis
wtx=weight3(intx,1);                   % weight in x-axis
for inty=1:ngly
```

```
y=point3(inty,2);                           % sampling point in y-axis
wty=weight3(inty,2) ;                       % weight in y-axis
for intz=1:nglz
z=point3(intz,3);                           % sampling point in z-axis
wtz=weight3(intz,3) ;                       % weight in z-axis
intp=intp+1;
%
[shape,dhdr,dhds,dhdt]=feisos8(x,y,z);      % compute shape functions
                                            % and derivatives at sampling point
%
jacob3=fejacob3(nnel,dhdr,dhds,dhdt,xcoord,ycoord,zcoord);
                                            % compute Jacobian
%
detjacob=det(jacob3);                       % determinant of Jacobian
invjacob=inv(jacob3);                       % inverse of Jacobian matrix
%
[dhdx,dhdy,dhdz]=federiv3(nnel,dhdr,dhds,dhdt,invjacob);
                                            % derivatives w.r.t. physical coordinate
%
kinmtx=fekine3d(nnel,dhdx,dhdy,dhdz);       % compute kinematic matrix
%
index=feeldof(nd,nnel,ndof);                % extract system dofs for the element
%
%-----------------------------------
% extract element displacement vector
%-----------------------------------
for i=1:edof
eldisp(i)=disp(index(i));
end
%
estrain=kinmtx*eldisp;                      % compute strains
estress=matmtx*estrain;                     % compute stresses
%
for i=1:6
strain(intp,i)=estrain(i);                  % store for each element
stress(intp,i)=estress(i);                  % store for each element
end
%
location=[ielp,intx,inty,intz]              % print location for stress
stress(intp,:)                              % print stress values
%
end
end
end                                         % end of integration loop
%
end                                         % end of loop for total number of elements
%
%-----------------------------------
```

```
function [dhdx,dhdy,dhdz]=federiv3(nnel,dhdr,dhds,dhdt,invjacob)
%------------------------------------------------------------
% Purpose:
% determine derivatives of 3-D isoparametric shape functions with
% respect to physical coordinate system
%
% Synopsis:
% [dhdx,dhdy,dhdz]=federiv3(nnel,dhdr,dhds,dhdt,invjacob)
%
% Variable Description:
% dhdx - derivative of shape function w.r.t. physical coordinate x
% dhdy - derivative of shape function w.r.t. physical coordinate y
% dhdz - derivative of shape function w.r.t. physical coordinate z
% nnel - number of nodes per element
% dhdr - derivative of shape functions w.r.t. natural coordinate r
% dhds - derivative of shape functions w.r.t. natural coordinate s
% dhdt - derivative of shape functions w.r.t. natural coordinate t
% invjacob - inverse of 3-D Jacobian matrix
%------------------------------------------------------------
%
for i=1:nnel
dhdx(i)=invjacob(1,1)*dhdr(i)+invjacob(1,2)*dhds(i) ...
        +invjacob(1,3)*dhdt(i);
dhdy(i)=invjacob(2,1)*dhdr(i)+invjacob(2,2)*dhds(i) ...
        +invjacob(2,3)*dhdt(i);
dhdz(i)=invjacob(3,1)*dhdr(i)+invjacob(3,2)*dhds(i) ...
        +invjacob(3,3)*dhdt(i);
end
%
%------------------------------------------------------------



function [shapes8,dhdrs8,dhdss8,dhdts8]=feisos8(rvalue,svalue,tvalue)
%------------------------------------------------------------
% Purpose:
% compute isoparametric eight-node solid shape functions
% and their derivatives at the selected (integration) point
% in terms of the natural coordinate
%
% Synopsis:
% [shapes8,dhdrs8,dhdss8,dhdts8]=feisos8(rvalue,svalue,tvalue)
%
% Variable Description:
% shapes8 - shape functions for four-node element
% dhdrs8 - derivatives of the shape functions w.r.t. r
% dhdss8 - derivatives of the shape functions w.r.t. s
% dhdts8 - derivatives of the shape functions w.r.t. t
```

% rvalue - r coordinate value of the selected point
% svalue - s coordinate value of the selected point
% tvalue - t coordinate value of the selected point
%
% Notes:
% 1st node at (-1,-1,-1), 2nd node at (1,-1,-1)
% 3rd node at (1,1,-1), 4th node at (-1,1,-1)
% 5th node at (-1,-1,1), 6th node at (1,-1,1)
% 7th node at (1,1,1), 8th node at (-1,1,1)
%———————————————————————————
%
% shape functions
%
shapes8(1)=0.125*(1-rvalue)*(1-svalue)*(1-tvalue);
shapes8(2)=0.125*(1+rvalue)*(1-svalue)*(1-tvalue);
shapes8(3)=0.125*(1+rvalue)*(1+svalue)*(1-tvalue);
shapes8(4)=0.125*(1-rvalue)*(1+svalue)*(1-tvalue);
shapes8(5)=0.125*(1-rvalue)*(1-svalue)*(1+tvalue);
shapes8(6)=0.125*(1+rvalue)*(1-svalue)*(1+tvalue);
shapes8(7)=0.125*(1+rvalue)*(1+svalue)*(1+tvalue);
shapes8(8)=0.125*(1-rvalue)*(1+svalue)*(1+tvalue);
%
% derivatives
%
dhdrs8(1)=-0.125*(1-svalue)*(1-tvalue);
dhdrs8(2)=0.125*(1-svalue)*(1-tvalue);
dhdrs8(3)=0.125*(1+svalue)*(1-tvalue);
dhdrs8(4)=-0.125*(1+svalue)*(1-tvalue);
dhdrs8(5)=-0.125*(1-svalue)*(1+tvalue);
dhdrs8(6)=0.125*(1-svalue)*(1+tvalue);
dhdrs8(7)=0.125*(1+svalue)*(1+tvalue);
dhdrs8(8)=-0.125*(1+svalue)*(1+tvalue);
%
dhdss8(1)=-0.125*(1-rvalue)*(1-tvalue);
dhdss8(2)=-0.125*(1+rvalue)*(1-tvalue);
dhdss8(3)=0.125*(1+rvalue)*(1-tvalue);
dhdss8(4)=0.125*(1-rvalue)*(1-tvalue);
dhdss8(5)=-0.125*(1-rvalue)*(1+tvalue);
dhdss8(6)=-0.125*(1+rvalue)*(1+tvalue);
dhdss8(7)=0.125*(1+rvalue)*(1+tvalue);
dhdss8(8)=0.125*(1-rvalue)*(1+tvalue);
%
dhdts8(1)=-0.125*(1-rvalue)*(1-svalue);
dhdts8(2)=-0.125*(1+rvalue)*(1-svalue);
dhdts8(3)=-0.125*(1+rvalue)*(1+svalue);
dhdts8(4)=-0.125*(1-rvalue)*(1+svalue);
dhdts8(5)=0.125*(1-rvalue)*(1-svalue);
dhdts8(6)=0.125*(1+rvalue)*(1-svalue);

dhdts8(7)=0.125*(1+rvalue)*(1+svalue);
dhdts8(8)=0.125*(1-rvalue)*(1+svalue);
%
%————————————————————————

function [jacob3]=fejacob3(nnel,dhdr,dhds,dhdt,xcoord,ycoord,zcoord)
%————————————————————————
% Purpose:
% determine the Jacobian for three-dimensional mapping
%
% Synopsis:
% [jacob3]=fejacob3(nnel,dhdr,dhds,dhdt,xcoord,ycoord,zcoord)
%
% Variable Description:
% jacob3 - Jacobian for one-dimension
% nnel - number of nodes per element
% dhdr - derivative of shape functions w.r.t. natural coordinate r
% dhds - derivative of shape functions w.r.t. natural coordinate s
% dhdt - derivative of shape functions w.r.t. natural coordinate t
% xcoord - x axis coordinate values of nodes
% ycoord - y axis coordinate values of nodes
% zcoord - z axis coordinate values of nodes
%————————————————————————
%
jacob3=zeros(3,3);
%
for i=1:nnel
jacob3(1,1)=jacob3(1,1)+dhdr(i)*xcoord(i);
jacob3(1,2)=jacob3(1,2)+dhdr(i)*ycoord(i);
jacob3(1,3)=jacob3(1,3)+dhdr(i)*zcoord(i);
jacob3(2,1)=jacob3(2,1)+dhds(i)*xcoord(i);
jacob3(2,2)=jacob3(2,2)+dhds(i)*ycoord(i);
jacob3(2,3)=jacob3(2,3)+dhds(i)*zcoord(i);
jacob3(3,1)=jacob3(3,1)+dhdt(i)*xcoord(i);
jacob3(3,2)=jacob3(3,2)+dhdt(i)*ycoord(i);
jacob3(3,3)=jacob3(3,3)+dhdt(i)*zcoord(i);
end
%
%————————————————————————

function [kinmtx3]=fekine3d(nnel,dhdx,dhdy,dhdz)
%————————————————————————
% Purpose:
% determine the kinematic equation between strains and displacements

% for three-dimensional solids
%
% Synopsis:
% [kinmtx3]=fekine3d(nnel,dhdx,dhdy,dhdz)
%
% Variable Description:
% nnel - number of nodes per element
% dhdx - derivatives of shape functions with respect to x
% dhdy - derivatives of shape functions with respect to y
% dhdz - derivatives of shape functions with respect to z
%————————————————————————————————
%
for i=1:nnel
i1=(i-1)*3+1;
i2=i1+1;
i3=i2+1;
kinmtx3(1,i1)=dhdx(i);
kinmtx3(2,i2)=dhdy(i);
kinmtx3(3,i3)=dhdz(i);
kinmtx3(4,i1)=dhdy(i);
kinmtx3(4,i2)=dhdx(i);
kinmtx3(5,i2)=dhdz(i);
kinmtx3(5,i3)=dhdy(i);
kinmtx3(6,i1)=dhdz(i);
kinmtx3(6,i3)=dhdx(i);
end
%
%————————————————————————————————

Nodal displacements are given below and the state of stresses is $\sigma_z = 1000$ and the rest of stresses are zero at every integration point.

```
displace =
d.o.f.   displ.
 1.0000   0.0000      % x-displacement of node 1
 2.0000   0.0000      % y-displacement of node 1
 3.0000   0.0000      % z-displacement of node 1
 4.0000  -0.0030      % x-displacement of node 2
 5.0000   0.0000      % y-displacement of node 2
 6.0000   0.0000      % z-displacement of node 2
 7.0000  -0.0030      % x-displacement of node 3
 8.0000  -0.0030      % y-displacement of node 3
 9.0000   0.0000      % z-displacement of node 3
10.000    0.0000      % x-displacement of node 4
11.000   -0.0030      % y-displacement of node 4
12.000    0.0000      % z-displacement of node 4
13.000    0.0000      % x-displacement of node 5
14.000    0.0000      % y-displacement of node 5
```

15.000	0.0100	% z-displacement of node 5
16.000	-0.0030	% x-displacement of node 6
17.000	0.0000	% y-displacement of node 6
18.000	0.0100	% z-displacement of node 6
19.000	-0.0030	% x-displacement of node 7
20.000	-0.0030	% y-displacement of node 7
21.000	0.0100	% z-displacement of node 7
22.000	0.0000	% x-displacement of node 8
23.000	-0.0030	% y-displacement of node 8
24.000	0.0100	% z-displacement of node 8

‡

Problems

9.1 An orthotropic material under the plane stress condition has the constitutive equation

$$\{\bar{\sigma}\} = [\bar{D}]\{\bar{\epsilon}\}$$

where $\{\bar{\sigma}\}$ and $\{\bar{\epsilon}\}$ are the stress and strain vectors in terms of the material orthotropic directions. Further, $[\bar{D}]$ is given as

$$[\bar{D}] = \begin{bmatrix} Q_{11} & Q_{12} & 0 \\ Q_{12} & Q_{22} & 0 \\ 0 & 0 & Q_{33} \end{bmatrix}$$

where

$$Q_{11} = \frac{E_1}{1 - \nu_{12}\nu_{21}}$$

$$Q_{12} = \frac{\nu_{12} E_2}{1 - \nu_{12}\nu_{21}}$$

$$Q_{22} = \frac{E_2}{1 - \nu_{12}\nu_{21}}$$

$$Q_{33} = G_{12}$$

Subscripts 1 and 2 denote the orthotropic axes, E_1 and E_2 are elastic modulii in the 1- and 2-axis, and G_{12} is the inplane shear modulus. ν_{ij} is Poisson's ratio for the normal strain ϵ_j resulting from the normal strain ϵ_i. If the material orthotropic axes are rotated from the global coordinate axes by angle θ (see Fig. P9.1), show that the strain vectors between the two coordinate systems are related as shown below:

$$\begin{Bmatrix} \epsilon_1 \\ \epsilon_2 \\ \gamma_{12} \end{Bmatrix} = [T] \begin{Bmatrix} \epsilon_x \\ \epsilon_y \\ \gamma_{xy} \end{Bmatrix}$$

where

$$[T] = \begin{bmatrix} \cos^2\theta & \sin^2\theta & \cos\theta\sin\theta \\ \sin^2\theta & \cos^2\theta & -\cos\theta\sin\theta \\ -2\cos\theta\sin\theta & 2\cos\theta\sin\theta & \cos^2\theta - \sin^2\theta \end{bmatrix}$$

9.2 From Prob. 9.1, also prove that $[D] = [T]^T [\bar{D}][T]$ where $[D]$ is the material property matrix in terms of the $x - y$ coordinate system and $[\bar{D}]$ is the material property matrix in terms of the $1 - 2$ coordinate system.

9.3 A two-dimensional elastic body is discretized using six-node triangular elements as shown in Fig. P9.3. Find the equivalent nodal forces for the three nodes located on an element boundary.

9.4 Find the equivalent nodal forces for Fig. P9.4.

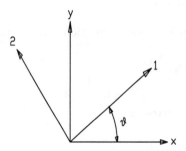

Figure P9.1 Problem 9.1

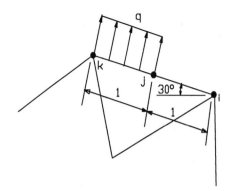

Figure P9.3 Problem 9.3

9.5 Find the equivalent nodal forces for Fig. P9.5.

9.6 Find the deflection of a tapered cantilever beam as shown in Fig. P9.6 using linear triangular elements and the computer programs. Assume unit width of the beam.

9.7 Solve Fig. P9.7 using computer programs. Use different mesh discretization. Some elements are distorted as shown in the figure.

9.8 Modify the computer programs so that linear triangular elements can be used with rectangular elements. Then solve Fig. P9.8.

9.9 Using the computer programs, find the solution for a perforated plate under tension as seen in Fig. P9.9.

9.10 Solve Prob. 9.6 for the transient analysis assuming the beam is initially at rest using the computer programs.

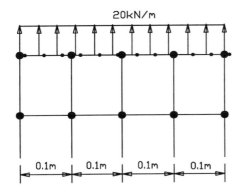

Figure P9.4 Problem 9.4

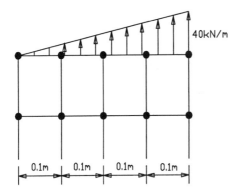

Figure P9.5 Problem 9.5

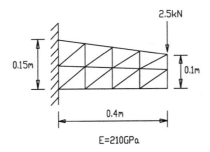

Figure P9.6 Problem 9.6

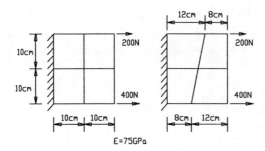

Figure P9.7 Problem 9.7

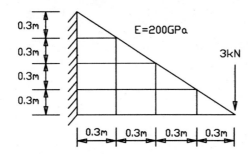

Figure P9.8 Problem 9.8

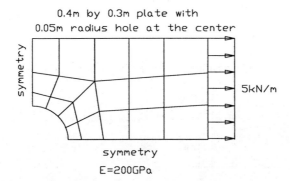

Figure P9.9 Problem 9.9

CHAPTER	TEN

PLATE AND SHELL STRUCTURES

10.0 Chapter Overview

Plates and shells are common structures in engineering and they support transverse loads through bending action. Finite element formulations for plates and shells are more complex than other types of elements. The complexity arises from the nature of high order partial differential equations. In order to overcome the complexity, different formulations have been proposed. As a result, this chapter presents several different formulations for plate and shell elements. First of all, the classical plate bending theory is discussed along with its finite element formulation. Then, other formulations for plate bending elements like shear deformable elements, elements with displacement degrees of freedom only, mixed elements, and hybrid elements are also presented.

Subsequently, two types of shell elements are derived. The first shell element is based on the combination of plate bending and plane stress elements, respectively, with a proper coordinate transformation from a local axis to the global axis. The second shell element is developed by degenerating from 3-D solid elements. Finally, MATLAB programs are provided for different plate and shell elements.

10.1 Classical Plate Theory

The basic assumptions for the classical Kirchhoff plate bending theory are very similar to those for the Euler-Bernoulli beam theory. One of the most important assumptions for both theories is that a straight line normal to the midplane of the plate before deformation remains normal even after deformation. In other words, the transverse deformation is neglected. Therefore, as shown in Fig. 8.1.2, inplane displacements u and v can be expressed as

$$u = -z\frac{\partial w}{\partial x} \quad (10.1.1)$$

$$v = -z\frac{\partial w}{\partial y} \tag{10.1.2}$$

where x and y are the inplane axes located at the midplane of the plate, and z is along the direction of plate thickness as seen in Fig. 10.1.1. In addition, u and v are the displacements in the x- and y-axes, respectively, while w is the transverse displacement (or deflection) along the z-axis.

Because we neglect the transverse shear deformation, inplane strains can be written in terms of the displacements

$$\{\,\epsilon_x \quad \epsilon_y \quad \gamma_{xy}\,\} = -z\{\,\kappa_x \quad \kappa_y \quad \kappa_{xy}\,\} \tag{10.1.3}$$

where

$$\{\kappa\}^T = \{\,\kappa_x \quad \kappa_y \quad \kappa_{xy}\,\} = \{\,\tfrac{\partial^2 w}{\partial x^2} \quad \tfrac{\partial^2 w}{\partial y^2} \quad 2\tfrac{\partial^2 w}{\partial x \partial y}\,\} \tag{10.1.4}$$

is called curvature.

Assuming the plane stress condition for plate bending and substituting Eqs. (10.1.3) and (10.1.4) into Eq. (9.1.5) yield the constitutive equation as given below:

$$\{\sigma\} = -z[D]\{\kappa\} \tag{10.1.5}$$

in which $\{\sigma\} = \{\,\sigma_x \quad \sigma_y \quad \tau_{xy}\,\}^T$ and $[D]$ is defined in Eq. (9.1.6). Moments are defined as

$$\{M\} = \int_{-h/2}^{h/2} \{\sigma\} z \, dz \tag{10.1.6}$$

where $\{M\} = \{\,M_x \quad M_y \quad M_{xy}\,\}^T$ and h is the plate thickness. Substitution of Eq. (10.1.5) into Eq. (10.1.6) gives the relationship between the moment and curvature.

$$\{M\} = -[\bar{D}]\{\kappa\} \tag{10.1.7}$$

where

$$[\bar{D}] = \frac{h^3}{12}[D] \tag{10.1.8}$$

Equilibrium equations are obtained from the free body diagram as shown in Fig. 10.1.1. Moment equilibriums about the y- and x-axes and force equilibrium about the z-axis yield, after neglecting higher order terms,

$$\frac{\partial M_x}{\partial x} + \frac{\partial M_{xy}}{\partial y} - Q_x = 0 \tag{10.1.9}$$

$$\frac{\partial M_{xy}}{\partial x} + \frac{\partial M_y}{\partial y} - Q_y = 0 \tag{10.1.10}$$

$$\frac{\partial Q_x}{\partial x} + \frac{\partial Q_y}{\partial y} + p = 0 \tag{10.1.11}$$

Section 10.1 Classical Plate Theory

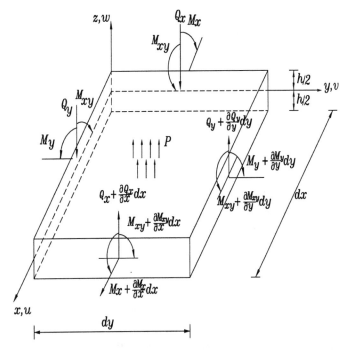

Figure 10.1.1 Free Body Diagram of the Plate Element

where Q_x and Q_y are the shear forces and p is the distributed pressure loading as seen in Fig. 10.1.1. Elimination of the shear forces from Eqs. (10.1.9) through (10.1.11) gives

$$\frac{\partial^2 M_x}{\partial x^2} + 2\frac{\partial^2 M_{xy}}{\partial x \partial y} + \frac{\partial^2 M_y}{\partial y^2} + p = 0 \qquad (10.1.12)$$

Combining Eqs. (10.1.4), (10.1.7), and (10.1.12) finally produces the biharmonic governing equation for plate bending in terms of the transverse displacement w.

$$\frac{\partial^4 w}{\partial x^4} + 2\frac{\partial^4 w}{\partial x^2 \partial y^2} + \frac{\partial^4 w}{\partial y^4} = \frac{p}{D_r} \qquad (10.1.13)$$

where $D_r = \frac{Eh^3}{12(1-\nu^2)}$ is the plate rigidity.

♣ **Example 10.1.1** We want to derive the equilibrium equations, Eqs. (10.1.9) through (10.1.11), from Eqs. (9.4.1) through (9.4.3). Integration of Eqs. (9.4.1) and (9.4.2) over the plate thickness after multiplying them by z yields

$$\int_{-h/2}^{h/2} \left(\frac{\partial \sigma_x}{\partial x} + \frac{\partial \tau_{xy}}{\partial y} + \frac{\partial \tau_{xz}}{\partial z} \right) z\, dz =$$

$$\frac{\partial M_x}{\partial x} + \frac{\partial M_{xy}}{\partial y} - Q_x + [\tau_{xz} z]_{-h/2}^{h/2} = 0 \qquad (10.1.14)$$

$$\int_{-h/2}^{h/2} \left(\frac{\partial \tau_{xy}}{\partial x} + \frac{\partial \sigma_y}{\partial y} + \frac{\partial \tau_{yz}}{\partial z} \right) z\, dz =$$

$$\frac{\partial M_{xy}}{\partial x} + \frac{\partial M_y}{\partial y} - Q_y + [\tau_{yz} z]_{-h/2}^{h/2} = 0 \qquad (10.1.15)$$

in which

$$Q_x = \int_{-h/2}^{h/2} \tau_{xy}\, dz \qquad (10.1.16)$$

$$Q_y = \int_{-h/2}^{h/2} \tau_{yz}\, dz \qquad (10.1.17)$$

If there is no shear stress ($\tau_{xy} = \tau_{yz} = 0$) on the top and bottom surfaces of the plate, Eqs. (10.1.14) and (10.1.15) are equal to Eqs. (10.1.9) and (10.1.10). Integrating Eq. (9.4.3) over the plate thickness gives

$$\int_{-h/2}^{h/2} \left(\frac{\partial \tau_{xz}}{\partial x} + \frac{\partial \tau_{yz}}{\partial y} + \frac{\partial \sigma_z}{\partial z} dz \right) =$$

$$\frac{\partial Q_x}{\partial x} + \frac{\partial Q_y}{\partial y} + \sigma_z(h/2) - \sigma_z(-h/2) = 0 \qquad (10.1.18)$$

If $\sigma_z(h/2) = p$ and $\sigma_z(-h/2) = 0$, Eq. (10.1.18) equals Eq. (10.1.11). ‡

10.2 Classical Plate Bending Element

We derive a three-node plate bending element based on the classical plate theory [34]. The element is shown in Fig. 10.2.1. Each node of the element possesses three degrees of freedom: displacement w in the z direction; a rotation about the x-axis, w_y, (derivative of w with respect to y); and a rotation about the y-axis, w_x (derivative of w with respect to x), as shown in the figure. The displacement function w is assumed to be

$$w(x,y) = a_1 + a_2 x + a_3 y + a_4 x^2 + a_5 xy + a_6 y^2 + a_7 x^3 + a_8(x^2 y + xy^2) + a_9 y^3 = [X]\{a\} \qquad (10.2.1)$$

where

$$[X] = [\,1 \quad x \quad y \quad x^2 \quad xy \quad y^2 \quad x^3 \quad (x^2 y + xy^2) \quad y^3\,] \qquad (10.2.2)$$

$$\{a\} = \{\,a_1 \quad a_2 \quad a_3 \quad a_4 \quad a_5 \quad a_6 \quad a_7 \quad a_8\,\}^T \qquad (10.2.3)$$

Here constants a_i are to be replaced by nodal variables w, w_x, and w_y.

Taking derivatives of the displacement function with respect to x and y yields

$$\frac{\partial w}{\partial x} = a_2 + 2a_4 x + a_5 y + 3a_7 x^2 + a_8(2xy + y^2) \qquad (10.2.4)$$

$$\frac{\partial w}{\partial y} = a_3 + a_5 x + 2a_6 y + a_8(x^2 + 2xy) + 3a_9 y^2 \qquad (10.2.5)$$

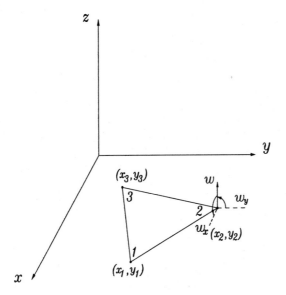

Figure 10.2.1 Three-Node Plate Bending Element

Evaluation of Eqs. (10.2.1) through (10.2.5) at the three nodal points gives the following matrix expression:

$$\{d\} = [\bar{X}]\{a\} \tag{10.2.6}$$

where

$$\{d\} = \{\, w_1 \quad (w_x)_1 \quad (w_y)_1 \quad w_2 \quad (w_x)_2 \quad (w_y)_2 \quad w_3 \quad (w_x)_3 \quad (w_y)_3 \,\}^T \tag{10.2.7}$$

and

$$[\bar{X}] = \begin{bmatrix} 1 & x_1 & y_1 & x_1^2 & x_1 y_1 & y_1^2 & x_1^3 & x_1^2 y_1 + x_1 y_1^2 & y_1^3 \\ 0 & 1 & 0 & 2x_1 & y_1 & 0 & 3x_1^2 & 2x_1 y_1 + y_1^2 & 0 \\ 0 & 0 & 1 & 0 & x_1 & 2y_1 & 0 & x_1^2 + 2x_1 y_1 & 3y_1^2 \\ 1 & x_2 & y_2 & x_2^2 & x_2 y_2 & y_2^2 & x_2^3 & x_2^2 y_2 + x_2 y_2^2 & y_2^3 \\ 0 & 1 & 0 & 2x_2 & y_2 & 0 & 3x_2^2 & 2x_2 y_2 + y_2^2 & 0 \\ 0 & 0 & 1 & 0 & x_2 & 2y_2 & 0 & x_2^2 + 2x_2 y_2 & 3y_2^2 \\ 1 & x_3 & y_3 & x_3^2 & x_3 y_3 & y_3^2 & x_3^3 & x_3^2 y_3 + x_3 y_3^2 & y_3^3 \\ 0 & 1 & 0 & 2x_3 & y_3 & 0 & 3x_3^2 & 2x_3 y_3 + y_3^2 & 0 \\ 0 & 0 & 1 & 0 & x_3 & 2y_3 & 0 & x_3^2 + 2x_3 y_3 & 3y_3^2 \end{bmatrix} \tag{10.2.8}$$

Inverting Eq. (10.2.6) and substituting the result into Eq. (10.2.1) yields

$$\{w\} = [H]\{d\} \tag{10.2.9}$$

where the row vector of shape functions of size 9×1 is computed from

$$[H] = [X][\bar{X}]^{-1} \tag{10.2.10}$$

Inplane strains are computed from Eq. (10.2.9) as

$$\{\epsilon\} = [B]\{d\} \tag{10.2.11}$$

in which

$$\{\epsilon\} = \{\epsilon_x \quad \epsilon_y \quad \gamma_{xy}\}^T \tag{10.2.12}$$

$$[B] = -z[L][\bar{X}]^{-1} \tag{10.2.13}$$

$$[L] = \begin{bmatrix} 0 & 0 & 0 & 2 & 0 & 0 & 6x & 2y & 0 \\ 0 & 0 & 0 & 0 & 0 & 2 & 0 & 2x & 6y \\ 0 & 0 & 0 & 0 & 2 & 0 & 0 & 4(x+y) & 0 \end{bmatrix} \tag{10.2.14}$$

Substitution of the strain-nodal displacement relation, Eq. (10.2.11), into the expression for the element stiffness matrix, Eq. (9.1.24) yields

$$[K^e] = \int_{\Omega^e} \int_z [B]^T [D][B] dz \, d\Omega$$

$$= [\bar{X}]^{-T} \int_{\Omega^e} \int_z z^2 [L]^T [D][L] dz \, d\Omega \, [\bar{X}]^{-1}$$

$$= [\bar{X}]^{-T} \int_{\Omega^e} [L]^T [\bar{D}][L] d\Omega [\bar{X}]^{-1} \tag{10.2.15}$$

where

$$[\bar{D}] = \frac{h^3}{12}[D] \tag{10.2.16}$$

Here $[D]$ is the constitutive matrix of the plane stress condition, Ω^e is the two-dimensional element domain in the xy-plane, and h is the thickness of the plate. The element domain is the triangular shape as seen in Fig. 10.2.1. The element stiffness matrix is of size 9×9 and the corresponding element nodal vector is given in Eq. (10.2.7).

10.3 Shear Deformable Plate Element

The Mindlin/Reissner plate theory includes the effect of transverse shear deformation like the Timoshenko beam theory. Hence, a plane normal to the midplane of the plate before deformation does not remain normal to the midplane any longer after deformation. The internal energy expression for the shear deformable plate should include transverse shear energy as well as bending energy. The internal energy is expressed as

$$U = \frac{1}{2} \int_V \{\sigma_b\}^T \{\epsilon_b\} dV + \frac{\kappa}{2} \int_V \{\sigma_s\}^T \{\epsilon_s\} dV \tag{10.3.1}$$

where

$$\{\sigma_b\} = \{\sigma_x \quad \sigma_y \quad \tau_{xy}\}^T \tag{10.3.2}$$

$$\{\epsilon_b\} = \{\epsilon_x \quad \epsilon_y \quad \gamma_{xy}\}^T \tag{10.3.3}$$

are the bending stress and strain components while

$$\{\sigma_s\} = \{\tau_{xz} \quad \tau_{yz}\}^T \tag{10.3.4}$$

$$\{\epsilon_s\} = \{\gamma_{xz} \quad \gamma_{yz}\}^T \tag{10.3.5}$$

are the transverse shear components. In addition, κ is the shear energy correction factor and equal to $\frac{5}{6}$.

Substitution of the constitutive equations for both bending and shear components yields

$$U = \frac{1}{2}\int_V \{\epsilon_b\}^T [D_b]\{\epsilon_b\} dV + \frac{\kappa}{2}\int_V \{\epsilon_s\}^T [D_s]\{\epsilon_s\} dV \tag{10.3.6}$$

in which

$$[D_b] = \frac{E}{1-\nu^2} \begin{bmatrix} 1 & \nu & 0 \\ \nu & 1 & 0 \\ 0 & 0 & \frac{1-\nu}{2} \end{bmatrix} \tag{10.3.7}$$

is the constitutive equation for the plane stress condition and

$$[D_s] = \begin{bmatrix} G & 0 \\ 0 & G \end{bmatrix} \tag{10.3.8}$$

Further, V is the three-dimensional domain which is equal to $d\Omega \times dz$. The xyz coordinate axis is the same as shown in Fig. 10.1.1.

In order to derive the element stiffness matrix for the shear deformable plate, we need to express the strains in terms of nodal variables. The inplane displacements are given as

$$u = -z\theta_x(x, y) \tag{10.3.9}$$

$$v = -z\theta_y(x, y) \tag{10.3.10}$$

and the transverse displacement is

$$w = w(x, y) \tag{10.3.11}$$

where θ_x and θ_y are rotations of the midplane about the y and x axes, respectively. The midplane is assumed to have no inplane deformation. For the shear deformable plate,

$$\theta_x = \frac{\partial w}{\partial x} - \gamma_{xz} \tag{10.3.12}$$

$$\theta_y = \frac{\partial w}{\partial y} - \gamma_{yz} \tag{10.3.13}$$

where γ is the angle caused by the transverse shear deformation as seen in Fig. 8.2.1.

Because the transverse displacement w and slope θ are independent, we need shape functions to interpolate them independently. As a result, the shear deformable plate element requires C^0 compatibility. Isoparametric shape functions are used for the plate element formulation. The transverse displacement and slopes are interpolated as

$$w = \sum_{i=1}^{n} H_i(\xi, \eta) w_i \tag{10.3.14}$$

$$\theta_x = \sum_{i=1}^{n} H_i(\xi, \eta)(\theta_x)_i \tag{10.3.15}$$

$$\theta_y = \sum_{i=1}^{n} H_i(\xi, \eta)(\theta_y)_i \tag{10.3.16}$$

Here n is the number of nodes per element and the same shape functions are used for the displacement and slope interpolations. For the following presentation, bilinear isoparametric shape functions are used for simplicity. However, higher-order shape functions can be used in the same manner. Both bending and shear strains are computed from displacements.

$$\{\epsilon_b\} = -z[B_b]\{d^e\} \tag{10.3.17}$$

$$\{\epsilon_s\} = [B_s]\{d^e\} \tag{10.3.18}$$

where

$$[B_b] = \begin{bmatrix} \frac{\partial H_1}{\partial x} & 0 & 0 & \frac{\partial H_2}{\partial x} & 0 & 0 & \frac{\partial H_3}{\partial x} & 0 & 0 & \frac{\partial H_4}{\partial x} & 0 & 0 \\ 0 & \frac{\partial H_1}{\partial y} & 0 & 0 & \frac{\partial H_2}{\partial y} & 0 & 0 & \frac{\partial H_3}{\partial y} & 0 & 0 & \frac{\partial H_4}{\partial y} & 0 \\ \frac{\partial H_1}{\partial y} & \frac{\partial H_1}{\partial x} & 0 & \frac{\partial H_2}{\partial y} & \frac{\partial H_2}{\partial x} & 0 & \frac{\partial H_3}{\partial y} & \frac{\partial H_3}{\partial x} & 0 & \frac{\partial H_4}{\partial y} & \frac{\partial H_4}{\partial x} & 0 \end{bmatrix} \tag{10.3.19}$$

$$[B_s] =$$

$$\begin{bmatrix} -H_1 & 0 & \frac{\partial H_1}{\partial x} & -H_2 & 0 & \frac{\partial H_2}{\partial x} & -H_3 & 0 & \frac{\partial H_3}{\partial x} & -H_4 & 0 & \frac{\partial H_4}{\partial x} \\ 0 & -H_1 & \frac{\partial H_1}{\partial y} & 0 & -H_2 & \frac{\partial H_2}{\partial y} & 0 & -H_3 & \frac{\partial H_3}{\partial y} & 0 & -H_4 & \frac{\partial H_4}{\partial y} \end{bmatrix}$$
(10.3.20)

and

$$\{d^e\} =$$

$$\left\{ (\theta_x)_1 \quad (\theta_y)_1 \quad w_1 \quad (\theta_x)_2 \quad (\theta_y)_2 \quad w_2 \quad (\theta_x)_3 \quad (\theta_y)_3 \quad w_3 \quad (\theta_x)_4 \quad (\theta_y)_4 \quad w_4 \right\}^T$$
(10.3.21)

Substitution of Eqs. (10.3.17) and (10.3.18) into the energy expression Eq. (10.3.6) yields for each plate element

$$U = \frac{1}{2}\{d^e\}^T \int_{\Omega^e} \int_z [B_b]^T [D_b][B_b] dz d\Omega \{d^e\} + \frac{\kappa}{2}\{d^e\}^T \int_{\Omega^e} \int_z [B_s]^T [D_s][B_s] dz d\Omega \{d^e\} \tag{10.3.22}$$

As a result, the element stiffness matrix for plate bending can be expressed as

$$[K^e] = \frac{h^3}{12} \int_{\Omega^e} [B_b]^T [D_b][B_b] d\Omega + \kappa h \int_{\Omega^e} [B_s]^T [D_s][B_s] d\Omega \tag{10.3.23}$$

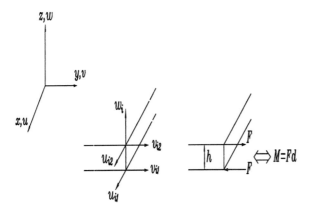

Figure 10.4.1 Plate Element With Displacement Degrees of Freedom

in which h is the plate thickness. One thing to be noted here is that the shear energy becomes dominant compared to the bending energy as the plate thickness becomes very small compared to its side length. This is called *shear locking*. A heuristic explanation for this can be given as below. The bending energy is proportional to h^3 while the shear energy is propotional to h. Therefore, as h gets smaller, the shear energy becomes dominant over the bending energy. To resolve this problem, the *selective* or *reduced* integration technique was proposed. The key of the technique is to underintegrate the shear energy term. In general, the bending term is integrated using the exact integration rule. For example, when four-node bilinear isoparametric elements are used, the 2×2 Gauss-Legendre quadrature is used for the bending term while 1-point integration is used for the shear term. Similarly, the bending term utilizes 3×3 integrations and the shear term uses 2×2 integrations for the nine-node biquadratic isoparametric shape function.

10.4 Plate Element With Displacement Degrees of Freedom

The plate bending element developed in this section is shown in Fig. 10.4.1 where x, y, and z describe the global coordinates of the plate and u, v, and w are the displacements. h is the plate thickness. The xy plane is parallel to the midplane prior to deflection.

The displacement of any point in the plate can be expressed as

$$u = u(x, y, z) \qquad (10.4.1)$$

$$v = v(x, y, z) \qquad (10.4.2)$$

$$w = w(x, y) \qquad (10.4.3)$$

That is, the inplane displacements u and v vary through the plate thickness as well as within the xy-plane while the transverse displacement w remains constant through the plate thickness [35,36]. In order to interpolate the displacements using shape functions and nodal displacements, two different interpolations are needed: one interpolation within the xy-plane and the other in the z-axis. For the xy-plane interpolation, shape functions $N_i(x,y)$ are used where subscript i varies depending on the number of nodes on the xy-plane. On the other hand, shape functions $H_j(z)$ are used for interpolation along the z-axis, where subscript j varies depending on the number of nodes along the plate thickness. Because two inplane displacements are functions of x, y, and z, both shape functions are used while the transverse displacement uses shape functions $N_i(x,y)$. Using isoparametric elements with mapping of the $\xi\eta$-plane onto the xy-plane and the ζ-axis to the z-axis, the three displacements can be expressed as

$$u = \sum_{i=1}^{N_1} \sum_{j=1}^{N_2} N_i(\xi,\eta) H_j(\zeta) u_{ij} \qquad (10.4.4)$$

$$v = \sum_{i=1}^{N_1} \sum_{j=1}^{N_2} N_i(\xi,\eta) H_j(\zeta) v_{ij} \qquad (10.4.5)$$

$$w = \sum_{i=1}^{N_1} N_i(\xi,\eta) w_i \qquad (10.4.6)$$

in which N_1 and N_2 are the numbers of nodes in the xy-plane ($\xi\eta$-plane) and z-axis (ζ-axis), respectively. In addition, the first subscript for u and v denotes the node numbering in terms of the xy-plane ($\xi\eta$-plane) and the second subscript indicates the node numbering in terms of the z-axis (ζ-axis). In the present study, $N_1=4$ and $N_2=2$. That is, four-node quadrilateral shape functions are employed for the xy-plane ($\xi\eta$-plane) interpolation and linear shape functions are employed for the z-axis (ζ-axis) interpolation. Nodal displacements u_{i1} and v_{i1} are displacements on the bottom surface of the plate element and u_{i2} and v_{i2} are displacements on the top surface. As seen in Eqs. (10.4.4) through (10.4.6), there is no rotational degree of freedom for the present plate bending element.

In the present formulation, both bending strain energy and transverse shear strain energy are included. The bending strains and transverse shear strains are expressed in terms of displacements:

$$\{\epsilon_b\} = \begin{Bmatrix} \epsilon_x \\ \epsilon_y \\ \gamma_{xy} \end{Bmatrix} = \begin{bmatrix} \frac{\partial}{\partial x} & 0 & 0 \\ 0 & \frac{\partial}{\partial y} & 0 \\ \frac{\partial}{\partial y} & \frac{\partial}{\partial x} & 0 \end{bmatrix} \begin{Bmatrix} u \\ v \\ w \end{Bmatrix} \qquad (10.4.7)$$

$$\{\epsilon_s\} = \begin{Bmatrix} \gamma_{yz} \\ \gamma_{xz} \end{Bmatrix} = \begin{bmatrix} \frac{\partial}{\partial z} & 0 & \frac{\partial}{\partial x} \\ 0 & \frac{\partial}{\partial z} & \frac{\partial}{\partial y} \end{bmatrix} \begin{Bmatrix} u \\ v \\ w \end{Bmatrix} \qquad (10.4.8)$$

where $\{\epsilon_b\}$ is the bending strains and $\{\epsilon_s\}$ is the transverse shear strains. The normal strain along the plate thickness ϵ_z is omitted here.

Substitution of displacements, Eqs. (10.4.4) through (10.4.6), into the kinematic equations, Eqs. (10.4.7) and (10.4.8), with $N_1=4$ and $N_2=2$ expresses both bending and shear strains in the following way:

$$\{\epsilon_b\} = [B_b]\{d^e\} \qquad (10.4.9)$$

where

$$[B_b] = [[B_{b1}] \quad [B_{b2}] \quad [B_{b3}] \quad [B_{b4}]] \tag{10.4.10}$$

$$[B_{bi}] = \begin{bmatrix} H_1 \frac{\partial N_i}{\partial x} & 0 & H_2 \frac{\partial N_i}{\partial x} & 0 & 0 \\ 0 & H_1 \frac{\partial N_i}{\partial y} & 0 & H_2 \frac{\partial N_i}{\partial y} & 0 \\ H_1 \frac{\partial N_i}{\partial y} & H_1 \frac{\partial N_i}{\partial x} & H_2 \frac{\partial N_i}{\partial y} & H_2 \frac{\partial N_i}{\partial x} & 0 \end{bmatrix} \tag{10.4.11}$$

$$\{d^e\} = \{\{d_1^e\} \quad \{d_2^e\} \quad \{d_3^e\} \quad \{d_4^e\}\}^T \tag{10.4.12}$$

$$\{d_i^e\} = \{u_{i1} \quad v_{i1} \quad u_{i2} \quad v_{i2} \quad w_i\} \tag{10.4.13}$$

$$\{\epsilon_s\} = [B_s]\{d^e\} \tag{10.4.14}$$

where

$$[B_s] = [[B_{s1}] \quad [B_{s2}] \quad [B_{s3}] \quad [B_{s4}]] \tag{10.4.15}$$

$$[B_{si}] = \begin{bmatrix} N_i \frac{\partial H_1}{\partial z} & 0 & N_i \frac{\partial H_2}{\partial z} & 0 & \frac{\partial N_i}{\partial x} \\ 0 & N_i \frac{\partial H_1}{\partial z} & 0 & N_i \frac{\partial H_2}{\partial z} & \frac{\partial H_2}{\partial y} \end{bmatrix} \tag{10.4.16}$$

The constitutive equation for the isotropic material is

$$\{\sigma_b\} = [D_b]\{\epsilon_b\} \tag{10.4.17}$$

where

$$\{\sigma_b\} = \{\sigma_x \quad \sigma_y \quad \tau_{xy}\}^T \tag{10.4.18}$$

$$[D_b] = \frac{E}{1-\nu^2} \begin{bmatrix} 1 & \nu & 0 \\ \nu & 1 & 0 \\ 0 & 0 & \frac{1-\nu}{2} \end{bmatrix} \tag{10.4.19}$$

for the bending components and

$$\{\sigma_s\} = [D_s]\{\epsilon_s\} \tag{10.4.20}$$

where

$$\{\sigma_s\} = \{\tau_{yz} \quad \tau_{xz}\}^T \tag{10.4.21}$$

$$[D_s] = \frac{E}{2(1+\nu)} \begin{bmatrix} 1 & 0 \\ 0 & 1 \end{bmatrix} \tag{10.4.22}$$

where Eq. (10.4.19) is the material property matrix for the plane stress condition as usually assumed for the plate bending theory.

For a unidirectional fibrous composite, the material property matrices become

$$[D_b] = \begin{bmatrix} D_{11} & D_{12} & 0 \\ D_{12} & D_{22} & 0 \\ 0 & 0 & D_{33} \end{bmatrix} \tag{10.4.23}$$

in which

$$D_{11} = \frac{E_1}{1 - \nu_{12}\nu_{21}} \tag{10.4.24}$$

$$D_{12} = \frac{E_1 \nu_{21}}{1 - \nu_{12}\nu_{21}} \qquad (10.4.25)$$

$$D_{22} = \frac{E_2}{1 - \nu_{12}\nu_{21}} \qquad (10.4.26)$$

$$D_{33} = G_{12} \qquad (10.4.27)$$

and

$$[D_s] = \begin{bmatrix} G_{13} & 0 \\ 0 & G_{12} \end{bmatrix} \qquad (10.4.28)$$

Here, 1 and 2 denote the longitudinal and transverse directions of the unidirectional composite, respectively. Further E is the elastic modulus, G_{ij} is the shear modulus of the $i-j$ plane, and ν_{ij} is Poisson's ratio for strain in the j-direction when stressed in the i-direction. There are five independent material properties for Eqs. (10.4.23) through (10.4.28) because of the reciprocal relation $\frac{\nu_{12}}{E_1} = \frac{\nu_{21}}{E_2}$.

The total potential energy can be expressed as

$$\Pi = U - W \qquad (10.4.29)$$

where the internal strain energy U consists of two parts

$$U = U_b + U_s \qquad (10.4.30)$$

The bending strain energy U_b is

$$U_b = \frac{1}{2} \int_\Omega \{\sigma_b\}^T \{\epsilon_b\} d\Omega \qquad (10.4.31)$$

and the transverse shear strain energy is

$$U_s = \frac{1}{2} \int_\Omega \{\sigma_s\}^T \{\epsilon_s\} d\Omega \qquad (10.4.32)$$

where Ω is the plate domain. After finite element discretization, substitution of the previous equations into Eqs. (10.4.31) and (10.4.32) gives

$$U_b = \sum_e \frac{1}{2} \{d^e\}^T \int_{\Omega^e} [B_b]^T [D_b][B_b] d\Omega \{d^e\} \qquad (10.4.33)$$

and

$$U_s = \sum_e \frac{1}{2} \{d^e\}^T \int_{\Omega^e} [B_s]^T [D_s][B_s] d\Omega \{d^e\} \qquad (10.4.34)$$

where summation is performed over the total number of finite elements and superscript e indicates each element. Kinematic matrices $[B_b]$ and $[B_s]$ are provided in Eqs. (10.4.10) and (10.4.15) while the constitutive matrices $[D_b]$ and $[D_s]$ are given in Eqs. (10.4.19) and (10.4.22) for the isotropic material, and in Eqs. (10.4.23) and (10.4.28) for the unidirectional composite. For a laminated composite plate, the material property matrix of each layer must be transformed based on the fiber axis of each layer and the global reference coordinate system.

The external work is written as

$$W = \{d\}^T \{F\} \tag{10.4.35}$$

in which $\{d\}$ is the system nodal displacement vector and $\{F\}$ is the system force vector. Because there is no rotational degree of freedom for the present element, the external moment is applied to the force vector as a couple applied on the top and bottom nodes of the plate element as shown in Fig. 10.4.1. Finally, invoking the stationary value of the total potential energy yields the finite element matrix equation. The element stiffness matrix can be expressed as

$$[K^e] = \int_{\Omega^e} [B_b]^T [D_b][B_b] d\Omega + \int_{\Omega^e} [B_s]^T [D_s][B_s] d\Omega \tag{10.4.36}$$

One thing to be noted here is that the transverse shear strain energy term should be under-integrated numerically to avoid shear locking, especially for a thin plate.

10.5 Mixed Plate Element

The basic equations for the classical plate theory are

$$M_x = -D_r \left(\frac{\partial^2 w}{\partial x^2} + \nu \frac{\partial^2 w}{\partial y^2} \right) \tag{10.5.1}$$

$$M_y = -D_r \left(\frac{\partial^2 w}{\partial y^2} + \nu \frac{\partial^2 w}{\partial x^2} \right) \tag{10.5.2}$$

$$M_{xy} = -D_r(1-\nu) \frac{\partial^2 w}{\partial x \partial y} \tag{10.5.3}$$

$$\frac{\partial^2 M_x}{\partial x^2} + \frac{\partial^2 M_y}{\partial y^2} + 2 \frac{\partial^2 M_{xy}}{\partial x \partial y} = -p \tag{10.5.4}$$

where M is the moment and $D_r = \frac{Eh^3}{12(1-\nu^2)}$ is the flexural rigidity of plate. E is the elastic modulus, h is the plate thickness, ν is Poisson's ratio, and p is the pressure loading. Equations (10.5.1) through (10.5.3) are the constitutive equations and Eq. (10.5.4) is the equilibrium equation of moments.

Applying Galerkin's method to Eqs. (10.5.1) through (10.5.4) does not produce the symmetric matrix. To this end, Eqs. (10.5.1) through (10.5.3) are inverted so that we have

$$S(M_x - \nu M_y) + \frac{\partial^2 w}{\partial x^2} = 0 \tag{10.5.5}$$

$$S(M_y - \nu M_x) + \frac{\partial^2 w}{\partial y^2} = 0 \tag{10.5.6}$$

$$2S(1+\nu)M_{xy} + 2\frac{\partial^2 w}{\partial x \partial y} = 0 \tag{10.5.7}$$

where $S = \frac{12}{Eh^3}$. Now Galerkin's method is applied to Eqs. (10.5.4) through (10.5.7) and integration by parts is performed to develop the weak formulation. The resultant matrix equation for each element is given below [21]:

$$\begin{bmatrix} K_1 & K_2 & 0 & K_3 \\ K_2 & K_1 & 0 & K_4 \\ 0 & 0 & K_5 & K_6 \\ k_3 & K_4 & K_6 & 0 \end{bmatrix} \begin{Bmatrix} M_x \\ M_y \\ M_{xy} \\ w \end{Bmatrix} = \begin{Bmatrix} F_1 \\ F_2 \\ F_3 \\ F_4 \end{Bmatrix} \tag{10.5.8}$$

where

$$K_1 = S \int_{\Omega^e} [N]^T [N] d\Omega \tag{10.5.9}$$

$$K_2 = -\nu K_1 \tag{10.5.10}$$

$$K_3 = -\int_{\Omega^e} \left[\frac{\partial N}{\partial x}\right]^T \left[\frac{\partial N}{\partial x}\right]^T d\Omega \tag{10.5.11}$$

$$K_4 = -\int_{\Omega^e} \left[\frac{\partial N}{\partial y}\right]^T \left[\frac{\partial N}{\partial y}\right]^T d\Omega \tag{10.5.12}$$

$$K_5 = 2(1+\nu)K_1 \tag{10.5.13}$$

$$K_6 = -\int_{\Omega^e} \left(\left[\frac{\partial N}{\partial x}\right]^T \left[\frac{\partial N}{\partial y}\right] + \left[\frac{\partial N}{\partial y}\right]^T \left[\frac{\partial N}{\partial x}\right]\right) d\Omega \tag{10.5.14}$$

$$F_1 = -\int_{\Gamma^e} [N]^T \frac{\partial w}{\partial x} l_x d\Gamma \tag{10.5.15}$$

$$F_2 = -\int_{\Gamma^e} [N]^T \frac{\partial w}{\partial y} l_y d\Gamma \tag{10.5.16}$$

$$F_3 = -\int_{\Gamma^e} [N]^T \left(\frac{\partial w}{\partial y} l_x + \frac{\partial w}{\partial x} l_y\right) d\Gamma \tag{10.5.17}$$

$$F_4 = -\int_{\Gamma^e} [N]^T Q_n d\Gamma + \int_{\Omega^e} [N]^T p d\Omega \tag{10.5.18}$$

$$Q_n = Q_x l_x + Q_y l_y \tag{10.5.19}$$

Here, l_x and l_y are direction cosines of the unit normal vector, and Q is the shear force. $[N]$ is the shape function vector. Any isoparametric element, of either quadrilateral or triangular shape, may be used for these equations.

However, the previous formulation does not include the effect of transverse shear deformation. The mixed plate bending formulation for thick plates is derived below. Equilibrium equations for plate can be written as below including transverse shear forces.

$$\frac{\partial M_x}{\partial x} + \frac{\partial M_{xy}}{\partial y} - Q_x = 0 \tag{10.5.20}$$

$$\frac{\partial M_{xy}}{\partial x} + \frac{\partial M_y}{\partial y} - Q_y = 0 \tag{10.5.21}$$

$$\frac{\partial Q_x}{\partial x} + \frac{\partial Q_y}{\partial y} + p = 0 \tag{10.5.22}$$

The major discrepancy between the thin and thick plate theories is the relations between the rotations and the transverse deflection. In the thin plate theory the rotations are not independent of the transverse deflection but they are independent of the deflection for the thick plate theory. Thus, the displacements in the x, y, and z directions are expressed as

$$u = -z\theta_x(x,y) \tag{10.5.23}$$

$$v = -z\theta_y(x,y) \tag{10.5.24}$$

$$w = w(x,y) \tag{10.5.25}$$

where θ_x and θ_y are rotations about y and x axes, respectively. Substitution of Eqs. (10.5.23) through (10.5.25) into the kinematic equations and use of the constitutive equations give

$$S(M_x - \nu M_y) + \frac{\partial \theta_x}{\partial x} = 0 \tag{10.5.26}$$

$$S(M_y - \nu M_x) + \frac{\partial \theta_y}{\partial y} = 0 \tag{10.5.27}$$

$$2S(1+\nu)M_{xy} + \frac{\partial \theta_x}{\partial y} + \frac{\partial \theta_y}{\partial x} = 0 \tag{10.5.28}$$

If θ_x and θ_y are replaced by $\frac{\partial w}{\partial x}$ and $\frac{\partial w}{\partial y}$, Eqs. (10.5.26) through (10.5.28) are the same as Eqs. (10.5.5) through (10.5.7). Such relations, however, do not hold in the thick plate theory.

Using the constituent and kinematic equations for transverse shear components, the shear forces can be expressed in terms of rotations and the deflection.

$$Q_x = \kappa G h(-\theta_x + \frac{\partial w}{\partial x}) \tag{10.5.29}$$

$$Q_y = \kappa G h(-\theta_y + \frac{\partial w}{\partial y}) \tag{10.5.30}$$

where κ is the shear correction factor equal to 5/6, G is the shear modulus, and h is the plate thickness. Rewritting Eqs. (10.5.29) and (10.5.30) for the rotations yields

$$\theta_x = -\frac{Q_x}{\kappa G h} + \frac{\partial w}{\partial x} \tag{10.5.31}$$

$$\theta_y = -\frac{Q_y}{\kappa G h} + \frac{\partial w}{\partial y} \tag{10.5.32}$$

Putting Eqs. (10.5.31) and (10.5.32) into Eqs. (10.5.26) through (10.5.27) to eliminate the rotations gives

$$S(M_x - \nu M_y) - \frac{1}{\kappa Gh}\frac{\partial Q_x}{\partial x} + \frac{\partial^2 w}{\partial x^2} = 0 \qquad (10.5.33)$$

$$S(M_y - \nu M_x) - \frac{1}{\kappa Gh}\frac{\partial Q_y}{\partial y} + \frac{\partial^2 w}{\partial y^2} = 0 \qquad (10.5.34)$$

$$2S(1+\nu)M_{xy} - \frac{1}{\kappa Gh}\left(\frac{\partial V_x}{\partial y} + \frac{\partial V_y}{\partial x}\right) + 2\frac{\partial^2 w}{\partial x \partial y} = 0 \qquad (10.5.35)$$

Examining Eqs. (10.5.33) through (10.5.35) reveals that the coefficients of shear forces and moments are of order $1/h$ and $1/h^3$, respectively. Thus, as the plate thickness approaches zero, the shear force terms can be neglected compared to the moment terms. This is reasonable because the shear deformation is negligible when the plate thickness is very small compared to its length.

In order to eliminate the shear forces, Eqs. (10.5.20) and (10.5.21) are substituted into Eqs. (10.5.33) through (10.5.35) as well as Eq. (10.5.22). Then, the resultant equations are

$$\left(S - \frac{1}{\kappa Gh}\frac{\partial^2}{\partial x^2}\right)M_x - \nu S M_y - \frac{1}{\kappa Gh}\frac{\partial^2 M_{xy}}{\partial x \partial y} + \frac{\partial^2 w}{\partial x^2} = 0 \qquad (10.5.36)$$

$$-\nu S M_x + \left(S - \frac{1}{\kappa Gh}\frac{\partial^2}{\partial y^2}\right)M_y - \frac{1}{\kappa Gh}\frac{\partial^2 M_{xy}}{\partial x \partial y} + \frac{\partial^2 w}{\partial y^2} = 0 \qquad (10.5.37)$$

$$-\frac{1}{\kappa Gh}\left(\frac{\partial^2 M_x}{\partial x \partial y} + \frac{\partial^2 M_y}{\partial x \partial y}\right) + \left(2S(1+\nu) - \frac{1}{\kappa Gh}\frac{\partial^2}{\partial x^2} - \frac{1}{\kappa Gh}\frac{\partial^2}{\partial y^2}\right)M_{xy} + 2\frac{\partial^2 w}{\partial x \partial y} = 0 \qquad (10.5.38)$$

as well as Eq. (10.5.4). Equations (10.5.36) through (10.5.38) and (10.5.4) have the same four variables, M_x, M_y, M_{xy} and w, as those for the thin plate formulation. If the terms associated with $\frac{1}{\kappa Gh}$ are neglected, these equations are reduced to the thin plate equations. In fact, as the plate thickness approaches zero, these terms are neglected. The shear related terms are proportional to $\frac{1}{h}$ while bending related terms are proportional to $\frac{1}{h^3}$.

Applying Galerkin's method to the four equations yields the following matrix expression:

$$\begin{bmatrix} K_{11} & K_{12} & K_{13} & K_{14} \\ K_{12} & K_{22} & K_{23} & K_{24} \\ K_{13} & K_{23} & K_{33} & K_{34} \\ k_{14} & K_{24} & K_{34} & K_{44} \end{bmatrix} \begin{Bmatrix} M_x \\ M_y \\ M_{xy} \\ w \end{Bmatrix} = \begin{Bmatrix} F_1 \\ F_2 \\ F_3 \\ F_4 \end{Bmatrix} \qquad (10.5.39)$$

where

$$K_{11} = S\int_{\Omega^e}[N]^T[N]d\Omega + \frac{1}{\kappa Gh}\int_{\Omega^e}\left[\frac{\partial N}{\partial x}\right]^T\left[\frac{\partial N}{\partial x}\right]d\Omega \qquad (10.5.40)$$

$$K_{12} = -\nu S\int_{\Omega^e}[N]^T[N]d\Omega \qquad (10.5.41)$$

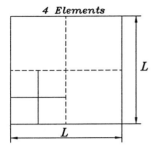
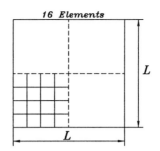

Figure 10.5.1 Square Plate Mesh Using 4 and 16 Elements

$$K_{13} = \frac{1}{\kappa Gh} \int_{\Omega^e} \left[\frac{\partial N}{\partial x}\right]^T \left[\frac{\partial N}{\partial y}\right] d\Omega \qquad (10.5.42)$$

$$K_{14} = -\int_{\Omega^e} \left[\frac{\partial N}{\partial x}\right]^T \left[\frac{\partial N}{\partial x}\right]^T d\Omega \qquad (10.5.43)$$

$$K_{22} = S \int_{\Omega^e} [N]^T [N] d\Omega + \frac{1}{\kappa Gh} \int_{\Omega^e} \left[\frac{\partial N}{\partial y}\right]^T \left[\frac{\partial N}{\partial y}\right] d\Omega \qquad (10.5.44)$$

$$K_{23} = \frac{1}{\kappa Gh} \int_{\Omega^e} \left[\frac{\partial N}{\partial y}\right]^T \left[\frac{\partial N}{\partial x}\right] d\Omega \qquad (10.5.45)$$

$$K_{24} = -\int_{\Omega^e} \left[\frac{\partial N}{\partial y}\right]^T \left[\frac{\partial N}{\partial y}\right]^T d\Omega \qquad (10.5.46)$$

$$K_{33} = 2(1+\nu)S \int_{\Omega^e} [N]^T [N] d\Omega + \frac{1}{\kappa Gh} \int_{\Omega^e} \left[\frac{\partial N}{\partial x}\right]^T \left[\frac{\partial N}{\partial x}\right] d\Omega +$$

$$\frac{1}{\kappa Gh} \int_{\Omega^e} \left[\frac{\partial N}{\partial y}\right]^T \left[\frac{\partial N}{\partial y}\right] d\Omega \qquad (10.5.47)$$

$$K_{34} = -\int_{\Omega^e} \left(\left[\frac{\partial N}{\partial x}\right]^T \left[\frac{\partial N}{\partial y}\right] + \left[\frac{\partial N}{\partial y}\right]^T \left[\frac{\partial N}{\partial x}\right]\right) d\Omega \qquad (10.5.48)$$

$$K_{44} = 0 \qquad (10.5.49)$$

$$F_1 = -\int_{\Gamma^e} [N]^T \frac{\partial w}{\partial x} l_x d\Gamma + \frac{1}{\kappa Gh} \int_{\Gamma^e} [N]^T V_x l_x d\Gamma \qquad (10.5.50)$$

$$F_2 = -\int_{\Gamma^e} [N]^T \frac{\partial w}{\partial y} l_y d\Gamma + \frac{1}{\kappa Gh} \int_{\Gamma^e} [N]^T V_y l_y d\Gamma \qquad (10.5.51)$$

$$F_3 = -\int_{\Gamma^e} [N]^T \left(\frac{\partial w}{\partial y} l_x + \frac{\partial w}{\partial x} l_y\right) d\Gamma +$$

$$\frac{1}{\kappa Gh} \int_{\Gamma^e} [N]^T \left(\frac{\partial}{\partial y}(M_x l_x + M_{xy} l_y) + \frac{\partial}{\partial x}(M_{xy} l_x + M_y l_y)\right) d\Gamma \qquad (10.5.52)$$

$$F_4 = -\int_{\Gamma^e} [N]^T Q_n d\Gamma + \int_{\Omega^e} [N]^T p\, d\Omega \qquad (10.5.53)$$

Table 10.5.1 Comparison of Central Deflections and Bending Moments for a Uniformly Loaded Simply Supported Square Plate.

	Analytic Solution	4 Elem.*	16 Elem.*	4 Elem.**	16 Elem.**
WD/PL^4	0.00406	0.00424	0.00411	0.00409	0.00407
M_x/PL^2	0.0479	0.0525	0.0489	0.0505	0.0485
M_y/PL^2	0.0479	0.0525	0.0489	0.0505	0.0485

4-node quadrilateral element
(*) - Present F.E. Solution, (**) Solution from Ref. [21]

Table 10.5.2 Comparison of Central Deflections and Bending Moments for a Uniformly Loaded Clamped Square Plate.

	Analytic Solution	4 Elem.*	16 Elem.*	4 Elem.**	16 Elem.**
WD/PL^4	0.00126	0.00141	0.00128	0.00148	0.00132
M_x/PL^2	-0.0513	-0.0476	-0.0499	-0.0487	-0.0508
M_y/PL^2	-0.0513	-0.0476	-0.0499	-0.0487	-0.0508

4-node quadrilateral element
(*) - Present F.E. Solution, (**) Solution from Ref. [21]

Some finite element solutions obtained using the present mixed plate bending elements are shown in Tables 10.5.1 through 10.5.5. Isoparametric shape functions were used for both interpolation of moments and displacements. Tables 10.5.1 and 10.5.2 show the results for simply supported and clamped square plates subjected to uniform pressure loads. Because of symmetry, 4 or 16 four-node isoparametric elements were used. The finite element mesh is seen in Fig. 10.5.1. The solutions from the present mixed formulation are also compared with those from another mixed formulation [21]. Table 10.5.3 gives the finite element solutions obtained using 4 eight-node isoparametric elements while Table 10.5.4 compares different isoparametric plate bending elements. The accuracy of each isoparametric element is different even if the total numbers of nodes are almost the same. The elements with more nodes per element give more accurate results.

Both the thin plate theory and the thick plate theory are compared in Table 10.5.5 for an orthotropic plate. The plate is shown in Fig. 10.5.2 with the mesh and material properties. As expected, as the ratio of the plate thickness to the side length increases, there is an increasing difference between the two solutions. The thick plate solutions are very close to the three-dimensional elasticity solutions for thick plates.

Table 10.5.3 Deflections at a Center of Square Plate

Boundary Condition	Analytic Soln	Present Soln
All Edges Supported	0.00406	0.00406
All Edges Clamped	0.00126	0.00125
Two Opposite Edges Simply Supported Two Other Edges Clamped	0.00191	0.00192

8-node quadrilateral element

Table 10.5.4 Comparison of Central Deflections Obtained Using Different Isoparametric Elements for Uniformly Loaded Simply Supported Plates

	W_c	Error (%)	Remark
Analytic Solution	0.2363		Timoshenko
3-Node Triangular	0.1814	-22.81	32 Elements (25 Nodes)
6-Node Triangular	0.2344	-0.80	8 Elements (25 Nodes)
4-Node Quadrilateral	0.2392	1.23	16 Elements (25 Nodes)
8-Node Quadrilateral	0.2365	0.08	4 Elements (21 Nodes)

10.6 Hybrid Plate Element

The hybrid element is based on the assumed strains within the plate element [23]. This element requires C^0 continuity. The formulation is based on a modified potential energy expression as given below for a plate.

$$\Pi = \int_\Omega \left(-\frac{1}{2}\{\epsilon_b\}^T[D_b]\{\epsilon_b\} - \frac{1}{2}\{\epsilon_s\}^T[D_s]\{\epsilon_s\} + \{\epsilon_b\}^T[D_b][L_b]\{d\} + \{\epsilon_s\}^T[D_s][L_s]\{d\}\right)d\Omega - \int_\Gamma \{d\}^T\{p\}d\Gamma \tag{10.6.1}$$

where

$$\{\epsilon_b\} = \left\{ \frac{\partial \theta_x}{\partial x} \quad \frac{\partial \theta_y}{\partial y} \quad \left(\frac{\partial \theta_x}{\partial y} + \frac{\partial \theta_y}{\partial y}\right) \right\}^T \tag{10.6.2}$$

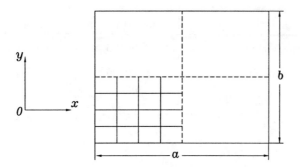

Orthotropic Properties

$$E_{xy} = 0.23319 * E_x$$
$$E_y = 0.543103 * E_x$$
$$G_{xy} = 0.262931 * E_x$$
$$G_{zx} = 0.159914 * E_x$$
$$G_{yz} = 0.26681 * E_x$$

Figure 10.5.2 Orthotropic Plate

$$\{\epsilon_s\} = \left\{ \left(-\theta_x + \frac{\partial w}{\partial x}\right) \quad \left(-\theta_y + \frac{\partial w}{\partial y}\right) \right\}^T \tag{10.6.3}$$

$$\{d\} = \{\theta_x \quad \theta_y \quad w\}^T \tag{10.6.4}$$

Further, $[D_b]$ is the material property matrix for bending strains and $[D_s]$ is the matrix for transverse shear strains. $[L_b]$ is the matrix for the bending strain-displacement operator and $[L_s]$ is the matrix for the shear strain-displacement operator. $\{p\}$ is the pressure loading on the plate.

Invoking a stationary value of the equation results in the equilibrium equation and the generalized strain-displacement relation. In order to obtain the finite element model, generalized strains and displacements are discretized as the following:

$$\{\epsilon_b\} = [B_b]\{\alpha_b\} \tag{10.6.5}$$

$$\{\epsilon_s\} = [B_s]\{\alpha_s\} \tag{10.6.6}$$

$$\{d\} = [N]\{\hat{d}\} \tag{10.6.7}$$

where generalized strains are assumed independently within each element and generalized displacements are interpolated using generalized nodal displacement $\{\hat{d}\}$. Thus, $[B_b]$ and $[B_s]$ are matrices consisting of the polynomial terms of the generalized strain parameter vectors $\{\alpha_b\}$ and $\{\alpha_s\}$, respectively. $[N]$ is the matrix consisting of shape functions. Substituting Eqs. (10.6.5) through (10.6.7) into Eq. (10.6.1) yields

$$\Pi = -\frac{1}{2}\{\alpha_b\}^T[G_b]\{\alpha_b\} - \frac{1}{2}\{\alpha_s\}^T[G_s]\{\alpha_s\} + \{\alpha_b\}^T[H_b]\{\hat{d}\}$$
$$+ \{\alpha_s\}^T[H_s]\{\hat{d}\} - \{\hat{d}\}^T\{F\} \tag{10.6.8}$$

Table 10.5.5 Generalized Central Deflections (E_xW/Pt) for Thin or Thick Uniformly Loaded Simply Supported Square Plates

b/a	t/a	3-D Theory*	Reissner's Theory*	Classical Theory*	Thin Plate Soln**	Thick Plate Soln**
0.5	0.05	21542	21542	21201	21268	21606
	0.1	1408.5	1408.4	1325.1	1329.3	1413.8
	0.14	387.23	387.27	344.93	346.03	389.11
1.0	0.05	10443	10442	10246	10285	10483
	0.1	688.57	688.37	640.39	642.81	692.30
	0.14	191.07	191.02	166.70	167.33	192.49
2.0	0.05	2048.2	2047.9	1988.1	1964.6	2026.8
	0.1	139.08	138.93	124.26	122.79	138.26
	0.14	39.790	39.753	32.345	31.962	39.806

4-node quadrilateral element
(*) Analytical Solution, (**) Present F.E. Solution

where

$$[G_b] = \int_{\Omega^e} [B_b]^T [D_b][B_b] d\Omega \tag{10.6.9}$$

$$[G_s] = \int_{\Omega^e} [B_s]^T [D_s][B_s] d\Omega \tag{10.6.10}$$

$$[H_b] = \int_{\Omega^e} [B_b]^T [D_b][L_b][N] d\Omega \tag{10.6.11}$$

$$[H_s] = \int_{\Omega^e} [B_s]^T [D_s][L_s][N] d\Omega \tag{10.6.12}$$

and

$$\{F\} = \int_{\Gamma^e} [N]^T \{p\} d\Gamma \tag{10.6.13}$$

Invoking stationary values of Eq. (10.6.8) with respect to $\{\alpha_b\}$ and $\{\alpha_s\}$ respectively results in

$$-[G_b]\{\alpha_b\} + [H_b]\{\hat{d}\} = 0 \tag{10.6.14}$$

$$-[G_s]\{\alpha_s\} + [H_s]\{\hat{d}\} = 0 \tag{10.6.15}$$

Eliminating $\{\alpha_b\}$ and $\{\alpha_s\}$ from Eq. (10.6.8) using Eqs. (10.6.14) and (10.6.15) gives

$$\Pi = \frac{1}{2}\{\hat{d}\}^T ([H_b]^T[G_b]^{-T}[H_b] + [H_s]^T[G_s]^{-T}[H_s])\{\hat{d}\} - \{\hat{d}\}^T\{F\} \tag{10.6.16}$$

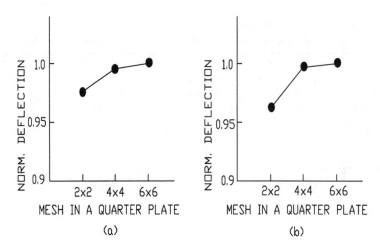

Figure 10.6.1 Uniformly Loaded Square Plate With: (a) Simply Supported Boundary and (b) Clamped Boundary

Equation (10.6.16) finally gives the following finite element system of equations

$$[K]\{\hat{d}\} = \{F\} \qquad (10.6.17)$$

in which

$$[K] = [H_b]^T [G_b]^{-T} [H_b] + [H_s]^T [G_s]^{-T} [H_s] \qquad (10.6.18)$$

For a bilinear plate element, the generalized strain vectors are assumed as

$$[B_b] = \begin{bmatrix} 1 & 0 & 0 & x & 0 & 0 & y & 0 & 0 \\ 0 & 1 & 0 & 0 & x & 0 & 0 & y & 0 \\ 0 & 0 & 1 & 0 & 0 & x & 0 & 0 & y \end{bmatrix} \qquad (10.6.19)$$

and

$$[B_s] = \begin{bmatrix} 1 & 0 \\ 0 & 1 \end{bmatrix} \qquad (10.6.20)$$

These expressions show that the bending strain varies linearly and the shear strain is constant within the bilinear plate element.

Finite element results from the hybrid plate bending elements are provided in Figs. 10.6.1 through 10.6.3. Convergence study for simply supported and clamped square plates subjected to uniform pressure loading is shown in Fig. 10.6.1 while that for a uniformly loaded circular plate with the clamped edge is seen in Fig. 10.6.2. The mesh for the circular plate is shown in Fig. 10.6.3.

10.7 Shell Made of Inplane and Bending Elements

A shell has curvatures along its surface. If the radius of curvature becomes infinity, the shell geometry becomes a flat plate. When a shell is divided into a

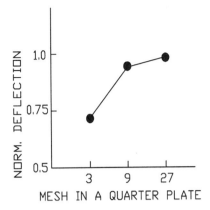

Figure 10.6.2 Uniformly Loaded Circular Plate With Clamped Edge

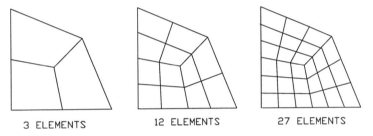

Figure 10.6.3 Meshes for a Quarter of Circular Plate

number of small finite elements, each element may be considered as a flat plate with a different orientation in space. Thus, each element may be modeled as a plate bending element. However, each element has a different orientation so that bending in one element can cause inplane deformation in the next element, as shown for a frame in Fig. 8.7.1.

As a result, a shell element can be obtained by combining a plate bending element and a plane stress element in the same way that a 2-D frame element is obtained from a beam bending element and a bar element. As seen in Fig. 10.7.1, a plate bending element has a transverse deflection and two bending rotations as degrees of freedom per node. On the other hand, a plane stress element has two inplane displacements per node. All together, a shell has five degrees of freedoms per node, three displacements and two rotations. The stiffness matrix of a shell element is given below:

$$\begin{bmatrix} [K_b] & [0] \\ [0] & [K_m] \end{bmatrix} \begin{Bmatrix} \{d_b\} \\ \{d_m\} \end{Bmatrix} = \begin{Bmatrix} \{F_b\} \\ \{F_m\} \end{Bmatrix} \qquad (10.7.1)$$

where K, d, and F indicate the stiffness matrix, nodal displacement/rotation vector, and nodal force/moment vector, respectively. The matrices and vectors consist of two parts, one from plate bending and the other from plate stretching. Subscripts b and m denote bending and membrane (stretching) deformations of the shell element.

When shell elements are oriented differently, for example considering a corner of a box where three elements meet together as seen in Fig. 10.7.2, a bending rotation in one element becomes a so-called drilling rotation in another element. Therefore, in order to assemble the element matrices and vectors into the system matrix and vector, the drilling rotational degrees of freedom should be included. As a result, the degrees of freedom of the element matrix and vector shall increase by one including the drilling degree of freedom per node. Then, Eq. (10.7.1) is rewritten as

$$\begin{bmatrix} [K_b] & [0] & 0 \\ [0] & [K_m] & 0 \\ 0 & 0 & 0 \end{bmatrix} \begin{Bmatrix} \{d_b\} \\ \{d_m\} \\ \theta_z \end{Bmatrix} = \begin{Bmatrix} \{F_b\} \\ \{F_m\} \\ 0 \end{Bmatrix} \quad (10.7.2)$$

The matrix and vectors in Eq. (10.7.2) are expressed in terms of a local coordinate system that has x- and y-axes along the midplane of each flat shell element and the z-axis normal to the plane. So as to assemble those matrices and vectors into the system matrix and vectors, the nodal degrees of freedom in terms of each local axis must be transformed into the corresponding nodal degrees of freedom in terms of the global coordinate axes common to all elements. If the transformation matrix is called $[T]$, then it can be written as

$$\{d^{local}\} = [T]\{d^{global}\} \quad (10.7.3)$$

Thus, matrix $[T]$ transforms the global degrees of freedom into the local degrees of freedom. It consists of direction cosines between the global and local coordinate systems.

At each node, the relation between the local and global degrees of freedom is expressed as

$$\begin{Bmatrix} u^{local} \\ v^{local} \\ w^{local} \\ \theta_x^{local} \\ \theta_y^{local} \\ \theta_z^{local} \end{Bmatrix} = \begin{bmatrix} l_{11} & l_{12} & l_{13} & 0 & 0 & 0 \\ l_{21} & l_{22} & l_{23} & 0 & 0 & 0 \\ l_{31} & l_{32} & l_{33} & 0 & 0 & 0 \\ 0 & 0 & 0 & l_{11} & l_{12} & l_{13} \\ 0 & 0 & 0 & l_{11} & l_{12} & l_{13} \\ 0 & 0 & 0 & l_{11} & l_{12} & l_{13} \end{bmatrix} \begin{Bmatrix} u^{global} \\ v^{global} \\ w^{global} \\ \theta_x^{global} \\ \theta_y^{global} \\ \theta_z^{global} \end{Bmatrix} \quad (10.7.4)$$

where l_{ij} is the direction cosine between the local axis x_i and the global axis X_j. This relationship can be used for each node. Thus, the transformation matrix $[T]$ for a four-node element becomes

$$[T] = \begin{bmatrix} [T_d] & 0 & 0 & 0 \\ 0 & [T_d] & 0 & 0 \\ 0 & 0 & [T_d] & 0 \\ 0 & 0 & 0 & [T_d] \end{bmatrix} \quad (10.7.5)$$

where the matrix $[T_d]$ is that used in Eq. (10.7.4) of size 6×6.

Using the transformation matrix, the transformed stiffness matrix and load vector are given as below:

$$[K^{global}] = [T]^T [K^{local}] [T] \quad (10.7.6)$$

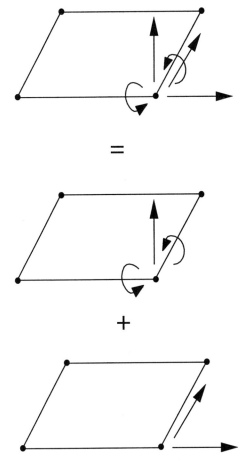

Figure 10.7.1 Shell Composed of Bending and Stretching

$$\{F^{global}\} = [T]^T \{F^{local}\} \qquad (10.7.7)$$

In a case that a shell is actually flat, the shell stiffness matrix becomes singular because of the singular terms associated with the drilling degrees of freedom. In order to avoid such a problem, a small number can be added to the diagonal term of the matrix in Eq. (10.7.2) associated with the drilling degree of freedom. This number should not be so small so that the modified matrix may not be numerically singular or near singular. On the other hand, the number should not be so large to affect the accuracy of the solution because it is an arbitrary addition.

10.8 Shell Degenerated from 3-D Solid

Geometry. A point in a shell structure can be expressed by a vector sum of two vectors. The first vector is a position vector from the origin of the coordinate system

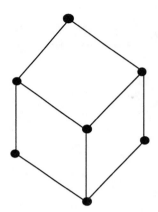

Figure 10.7.2 Three Shell Elements Perpendicular to Each Other

to a point on the reference surface of the shell element. The second vector is a position vector from this reference surface to the point under consideration. The midsurface of the shell is usually taken as the reference surface, but it is not required in this formulation. The second vector described above is usually normal to the reference surface.

Two shape functions are used to describe a position in the element: N^k is the two-dimensional shape function in the ξ-η plane, and H^k is the one-dimensional shape function along the ζ axis, where (ξ,η,ζ) describes a point in the natural coordinate system. A generic point of a shell may now be described in terms of the position vectors of the nodes and the shape functions:

$$x_i(\xi,\eta,\zeta) = \sum_{k=1}^{n} N^k(\xi,\eta) x_i^k + \sum_{k=1}^{n} N^k(\xi,\eta) H^k(\zeta) V_{3i}^k \quad (i=1,2,3) \qquad (10.8.1)$$

where x_i^k is the position vector of node k in the reference surface; V_{3i} is the unit vector at the node k; and n is the number of nodes per element. In the present formulation, a four-noded shell element is considered.

The unit vector V_{3i}^k is defined as

$$V_{3i}^k = \frac{(x_i^k)^{top} - (x_i^k)^{bottom}}{\| (x_i^k)^{top} - (x_i^k)^{bottom} \|} \qquad (10.8.2)$$

where *top* and *bottom* indicate the top and bottom surfaces of the shell, and $\| \ \|$ denotes the Euclidean norm. The one-dimensional shape function H^k is expressed as

$$H^k(\zeta) = \left[\frac{1}{4}(1+\zeta)(1-\bar{\zeta}) - \frac{1}{4}(1-\zeta)(1+\bar{\zeta})\right] \| (x_i^k)^{top} - (x_i^k)^{bottom} \| \qquad (10.8.3)$$

in which $\bar{\zeta}$ indicates the location of the reference surface and varies from -1 to 1. $\bar{\zeta}=0$ denotes the midsurface. The 2-D shape function N^k is provided in Chapter 6.

Section 10.8 Shell Degenerated from 3-D Solid

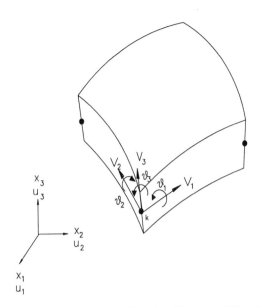

Figure 10.8.1 Shell Element Showing Degrees of Freedom

Displacement. The displacement field in a shell can be written as

$$u_i(\xi,\eta,\zeta) = \sum_{k=1}^{n} N^k(\xi,\eta) u_i^k + \sum_{k=1}^{n} N^k(\xi,\eta) H^k(\zeta)\bigl(-V_{2i}^k \theta_1^k + V_{1i}^k \theta_2^k\bigr) \quad (i=1,2,3)$$
(10.8.4)

in which u_i is the displacement along the x_i axis, u_i^k is the nodal displacement at the node k, and unit vectors V_{1i}^k and V_{2i}^k lie along the reference surface. V_{1i}^k, V_{2i}^k, and V_{3i}^k are perpendicular to one another. θ_1^k and θ_2^k are rotational degrees of freedom along the unit vectors V_{1i}^k and V_{2i}^k, respectively. The right-hand rule is assumed for the positive direction of each rotation. That is, as the thumb of the right-hand is in the direction of each unit vector, the rotational direction of the hand is the positive rotation. (See Fig. 10.8.1.)

Strain-Displacement Relation. Six strain components are computed from Eq. (10.8.4) by taking the derivative with respect to the x_i-axis. The result can be written in matrix form as

$$\{\epsilon\} = [B]\{d\} \tag{10.8.5}$$

where

$$\{\epsilon\} = \{\epsilon_{11} \quad \epsilon_{22} \quad \epsilon_{33} \quad \gamma_{12} \quad \gamma_{23} \quad \gamma_{13}\}^T \tag{10.8.6}$$

$$[B] = [B^1 \quad B^2 \quad \cdots \quad B^n] \tag{10.8.7}$$

and

$$\{d\} = \{d^1 \quad d^2 \quad \cdots \quad d^n\}^T \tag{10.8.8}$$

The detailed expression for $[B^k]$ is

$$[B^k] = \begin{bmatrix} \frac{\partial N^k}{\partial x_1} & 0 & 0 & -g_1^k V_{21}^k & g_1^k V_{11}^k \\ 0 & \frac{\partial N^k}{\partial x_2} & 0 & -g_2^k V_{22}^k & g_2^k V_{12}^k \\ 0 & 0 & \frac{\partial N^k}{\partial x_3} & -g_3^k V_{23}^k & g_3^k V_{13}^k \\ \frac{\partial N^k}{\partial x_2} & \frac{\partial N^k}{\partial x_1} & 0 & -g_2^k V_{21}^k - g_1^k V_{22}^k & g_2^k V_{11}^k + g_1^k V_{12}^k \\ 0 & \frac{\partial N^k}{\partial x_3} & \frac{\partial N^k}{\partial x_2} & -g_3^k V_{22}^k - g_2^k V_{23}^k & g_3^k V_{12}^k + g_2^k V_{13}^k \\ \frac{\partial N^k}{\partial x_3} & 0 & \frac{\partial N^k}{\partial x_1} & -g_3^k V_{21}^k - g_1^k V_{23}^k & g_3^k V_{11}^k + g_1^k V_{13}^k \end{bmatrix} \quad (10.8.9)$$

in which

$$g_i^k = \frac{\partial N^k}{\partial x_i} H^k + N^k \frac{\partial H^k}{\partial x_i} \quad (10.8.10)$$

In addition, the vector $\{d^k\}$ is

$$\{d^k\} = \{\, u_1 \quad u_2 \quad u_3 \quad \theta_1 \quad \theta_2 \,\} \quad (10.8.11)$$

Jacobian Matrix. In order to compute the derivatives such as $\frac{\partial N^k}{\partial x_i}$ and $\frac{\partial H^k}{\partial x_i}$, the Jacobian matrix is required and is defined as

$$[J] = \begin{bmatrix} x_{1,\xi} & x_{2,\xi} & x_{3,\xi} \\ x_{1,\eta} & x_{2,\eta} & x_{3,\eta} \\ x_{1,\zeta} & x_{2,\zeta} & x_{3,\zeta} \end{bmatrix} \quad (10.8.12)$$

where

$$\frac{\partial x_i}{\partial \xi} = \sum_{k=1}^{n} \frac{\partial N^k}{\partial \xi} x_i^k + \sum_{k=1}^{n} \frac{\partial N^k}{\partial \xi} H^k V_{3i}^k \quad (i = 1, 2, 3) \quad (10.8.13)$$

$$\frac{\partial x_i}{\partial \eta} = \sum_{k=1}^{n} \frac{\partial N^k}{\partial \eta} x_i^k + \sum_{k=1}^{n} \frac{\partial N^k}{\partial \eta} H^k V_{3i}^k \quad (i = 1, 2, 3) \quad (10.8.14)$$

$$\frac{\partial x_i}{\partial \zeta} = \sum_{k=1}^{n} N^k \frac{\partial H^k}{\partial \zeta} V_{3i}^k \quad (i = 1, 2, 3) \quad (10.8.15)$$

The inverse of the Jacobian matrix is called matrix $[R]$. Then,

$$\frac{\partial N^k}{\partial x_i} = R_{i1} \frac{\partial N^k}{\partial \xi} + R_{i2} \frac{\partial N^k}{\partial \eta} \quad (i = 1, 2, 3) \quad (10.8.16)$$

$$\frac{\partial H^k}{\partial x_i} = R_{i3} \frac{\partial H^k}{\partial \zeta} \quad (i = 1, 2, 3) \quad (10.8.17)$$

Here R_{ij} is the component of the matrix $[R]$.

Constitutive Equations. Stresses and strains can be expressed as

$$\{\sigma'\} = [D']\{\epsilon'\} \quad (10.8.18)$$

Shell Degenerated from 3-D Solid

where $\{\sigma'\}$ and $\{\epsilon'\}$ are the stress and strain components in the local axes which are set along the reference plane made of vectors V_1 and V_2 as shown in Fig. 10.8.1. The constitutive matrix $[D']$ is given below for an isotropic material:

$$[D'] = \begin{bmatrix} \frac{E}{(1-\nu^2)} & \frac{E\nu}{(1-\nu^2)} & 0 & 0 & 0 & 0 \\ \frac{E\nu}{(1-\nu^2)} & \frac{E}{(1-\nu^2)} & 0 & 0 & 0 & 0 \\ 0 & 0 & 0 & 0 & 0 & 0 \\ 0 & 0 & 0 & \frac{E}{2(1+\nu)} & 0 & 0 \\ 0 & 0 & 0 & 0 & \frac{\kappa E}{2(1+\nu)} & 0 \\ 0 & 0 & 0 & 0 & 0 & \frac{\kappa E}{2(1+\nu)} \end{bmatrix} \quad (10.8.19)$$

in which E and ν are the elastic and Poisson's ratio, respectively. This matrix includes the transverse shear deformation. The fifth and sixth rows are for the transverse shear deformation. κ is the shear correction factor which is chosen as 5/6 for an isotropic material.

The material property matrix $[D']$ is transformed into a matrix in terms of the stresses and strains of the global axes. The transformed material property matrix is

$$[D] = [T_\epsilon]^T [D'][T_\epsilon] \quad (10.8.20)$$

where the transformation matrix $[T_\epsilon]$ is

$$[T_\epsilon] = \begin{bmatrix} l_{11}^2 & l_{12}^2 & l_{13}^2 & l_{11}l_{12} & l_{12}l_{13} & l_{13}l_{11} \\ l_{21}^2 & l_{22}^2 & l_{23}^2 & l_{21}l_{22} & l_{22}l_{23} & l_{23}l_{21} \\ l_{31}^2 & l_{32}^2 & l_{33}^2 & l_{31}l_{32} & l_{32}l_{33} & l_{33}l_{31} \\ 2l_{11}l_{21} & 2l_{12}l_{22} & 2l_{13}l_{23} & (l_{11}l_{22}+l_{21}l_{12}) & (l_{12}l_{23}+l_{22}l_{13}) & (l_{13}l_{21}+l_{23}l_{11}) \\ 2l_{21}l_{31} & 2l_{22}l_{32} & 2l_{23}l_{33} & (l_{21}l_{32}+l_{31}l_{22}) & (l_{22}l_{33}+l_{32}l_{23}) & (l_{23}l_{31}+l_{33}l_{21}) \\ 2l_{31}l_{11} & 2l_{32}l_{12} & 2l_{33}l_{13} & (l_{31}l_{12}+l_{11}l_{32}) & (l_{32}l_{13}+l_{12}l_{33}) & (l_{33}l_{11}+l_{13}l_{31}) \end{bmatrix}$$

$$(10.8.21)$$

Here, l_{ij} is the direction cosines of the unit vector V_i with respect to the x_j-axis.

Element Stiffness Matrix. The element stiffness matrix is computed from

$$[K] = \int_{\Omega^e} [B]^T [D][B] d\Omega \quad (10.8.22)$$

However, the rotational degrees of freedom are expressed in terms of the local vectors and they should be expressed in terms of the global axes so that they can be assembled properly. Such a transformation is obtained from

$$\{d^k\}_{global} = [T_{rot}]\{d^k\} \quad (10.8.23)$$

where

$$[T_{rot}] = \begin{bmatrix} 1 & 0 & 0 & 0 & 0 & 0 \\ 0 & 1 & 0 & 0 & 0 & 0 \\ 0 & 0 & 1 & 0 & 0 & 0 \\ 0 & 0 & 0 & l_{11} & l_{21} & l_{31} \\ 0 & 0 & 0 & l_{12} & l_{22} & l_{32} \\ 0 & 0 & 0 & l_{13} & l_{23} & l_{33} \end{bmatrix} \quad (10.8.24)$$

For a four-node shell element, the transformation matrix for rotational degrees of freedom becomes

$$[\tilde{T}_{rot}] = \begin{bmatrix} [T_{rot}] & 0 & 0 & 0 \\ 0 & [T_{rot}] & 0 & 0 \\ 0 & 0 & [T_{rot}] & 0 \\ 0 & 0 & 0 & [T_{rot}] \end{bmatrix} \qquad (10.8.25)$$

The local nodal degrees of freedom vector in Eq. (10.8.23) includes θ_3 for the proper coordinate transformation as shown below:

$$\{d^k\} = \{\, u_1 \;\; u_2 \;\; u_3 \;\; \theta_1 \;\; \theta_2 \;\; \theta_3 \,\} \qquad (10.8.26)$$

As a result, the element stiffness matrix $[K]$ should be expanded to incorporate the degree of freedom at each node, as shown in Eq. (10.7.2). Then, the transformed element stiffness matrix is

$$[\tilde{K}] = [\tilde{T}_{rot}]^{-T}[K][\tilde{T}_{rot}]^{-1} \qquad (10.8.27)$$

10.9 MATLAB Application to Plates

A static finite element analysis of plate bending is performed using the shear deformable plate bending formulation discussed in Sec. 10.3. Some example problems are solved using MATLAB programs below.

♣ **Example 10.9.1** A simply supported square plate is subjected to a concentrated load at the center. Find the deflection of the plate using the shear deformable displacement formulation. The size of the plate is 10 in. by 10 in. and its thickness is 0.1 in. It is made of a steel whose elastic modulus is 30×10^6 psi and Poisson's ratio 0.3. The applied force is 40 lb at the center. A quarter of the plate is modeled due to symmetry and it is divided into 4 four-node elements (see Fig. 10.9.1).

The MATLAB program is written for the finite element analysis. Two point integration is used for the bending term while one point integration is used for the shear term for the selective integration technique. As far as the boundary conditions are concerned, two edges are simply supported and two edges are symmetric. As a result, nodes 1, 2, and 3 are constrained for θ_x and w. Nodes 1, 4, and 7 are constrained for θ_y and w. Nodes 3, 6, and 9 are constrained for θ_x while nodes 7, 8, and 9 are constrained for θ_y. The resultant constrained degrees of freedom are 1, 2, 3, 4, 6, 7, 9, 11, 12, 16, 20, 21, 23, 25, and 26. The external force is applied at node 9 with the third degree of freedom. Hence, the concentrated force is applied at the 27th degree of freedom of the load vector. Because of the quarter symmetry, a quarter of the force is applied to the load vector. The finite element solution gives the center deflection of 0.0168 in. while the analytical solution is 0.0169in.

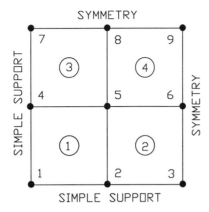

Figure 10.9.1 A Quarter of Square Plate With 4 Elements

```
%------------------------------------------------------------
% Example 10.9.1
% A simply supported square plate is subjected to a concentrated load
% at the center. Find the deflection of the plate using 4 four-node
% isoparametric elements of the shear deformable displacement
% formulation. The size of the plate is 10 in. by 10 in. and its
% thickness is 0.1 in. It is made of steel and the applied
% force is 40 lb.
% (see Fig. 10.9.1 for the finite element mesh)
%
% Variable descriptions
% k = element matrix
% kb = element matrix for bending stiffness
% ks = element matrix for shear stiffness
% f = element vector
% kk = system matrix
% ff = system vector
% disp = system nodal displacement vector
% gcoord = coordinate values of each node
% nodes = nodal connectivity of each element
% index = a vector containing system dofs associated with each element
% pointb = matrix containing sampling points for bending term
% weightb = matrix containing weighting coefficients for bending term
% points = matrix containing sampling points for shear term
% weights = matrix containing weighting coefficients for shear term
% bcdof = a vector containing dofs associated with boundary conditions
% bcval = a vector containing boundary condition values associated with
%         the dofs in bcdof
% kinmtpb = matrix for kinematic equation for bending
% matmtpb = matrix for material property for bending
% kinmtps = matrix for kinematic equation for shear
```

```
% matmtps = matrix for material property for shear
%---------------------------------------------------------------
%
%-------------------------------------------
% input data for control parameters
%-------------------------------------------
nel=4;                          % number of elements
nnel=4;                         % number of nodes per element
ndof=3;                         % number of dofs per node
nnode=9;                        % total number of nodes in system
sdof=nnode*ndof;                % total system dofs
edof=nnel*ndof;                 % degrees of freedom per element
emodule=30e6;                   % elastic modulus
poisson=0.3;                    % Poisson's ratio
t=0.1;                          % plate thickness
nglxb=2; nglyb=2;    % 2x2 Gauss-Legendre quadrature for bending
nglxs=1; nglys=1;    % 1x1 Gauss-Legendre quadrature for shear
%
%---------------------------------------------------------
% input data for nodal coordinate values
% gcoord(i,j) where i-> node no. and j-> x or y
%---------------------------------------------------------
gcoord=[0.0 0.0; 2.5 0.0; 5.0 0.0;
0.0 2.5; 2.5 2.5; 5.0 2.5;
0.0 5.0; 2.5 5.0; 5.0 5.0];
%
%-----------------------------------------------------------
% input data for nodal connectivity for each element
% nodes(i,j) where i-> element no. and j-> connected nodes
%-----------------------------------------------------------
nodes=[1 2 5 4; 2 3 6 5; 4 5 8 7; 5 6 9 8];
%
%---------------------------------------
% input data for boundary conditions
%---------------------------------------
bcdof=[1 2 3 4 6 7 9 11 12 16 20 21 23 25 26];    % constrained dofs
bcval=zeros(1,15);          % whose described values are zeros
%
%-----------------------------------------
% initialization of matrices and vectors
%-----------------------------------------
ff=zeros(sdof,1);               % system force vector
kk=zeros(sdof,sdof);            % system matrix
disp=zeros(sdof,1);             % system displacement vector
index=zeros(edof,1);            % index vector
kinmtpb=zeros(3,edof);          % kinematic matrix for bending
matmtpb=zeros(3,3);             % constitutive matrix for bending
kinmtps=zeros(2,edof);          % kinematic matrix for shear
```

Section 10.9 MATLAB Application to Plates

```
matmtps=zeros(2,2);                      % constitutive matrix for shear
%
%---------------------
% force vector
%---------------------
ff(27)=10;                               % applied concentrated force
%
%-----------------------------------------------------
% computation of element matrices and vectors and their assembly
%-----------------------------------------------------
%
% for bending stiffness
%
[pointb,weightb]=feglqd2(nglxb,nglyb);
%                                        % sampling points & weights
matmtpb=fematiso(1,emodule,poisson)*t^3/12;
%                                        % material property matrix
%
% for shear stiffness
%
[points,weights]=feglqd2(nglxs,nglys);
%                                        % sampling points & weights
shearm=0.5*emodule/(1.0+poisson);        % shear modulus
shcof=5/6;                               % shear correction factor
matmtps=shearm*shcof*t*[1 0; 0 1];       % material property matrix
%
for iel=1:nel                            % loop for the total number of elements
%
for i=1:nnel
nd(i)=nodes(iel,i);                      % extract nodes for (iel)-th element
xcoord(i)=gcoord(nd(i),1);               % extract x value of the nodes
ycoord(i)=gcoord(nd(i),2);               % extract y value of the nodes
end
%
k=zeros(edof,edof);                      % initialization of element matrix
kb=zeros(edof,edof);                     % initialization of bending matrix
ks=zeros(edof,edof);                     % initialization of shear matrix
%
%-----------------------------------------
% numerical integration for bending term
%-----------------------------------------
for intx=1:nglxb
x=pointb(intx,1);                        % sampling point in x-axis
wtx=weightb(intx,1);                     % weight in x-axis
for inty=1:nglyb
y=pointb(inty,2);                        % sampling point in y-axis
wty=weightb(inty,2) ;                    % weight in y-axis
%
```

```
[shape,dhdr,dhds]=feisoq4(x,y);          % compute shape functions and
%                                        % derivatives at sampling point
%
jacob2=fejacob2(nnel,dhdr,dhds,xcoord,ycoord);     % Jacobian matrix
%
detjacob=det(jacob2);                    % determinant of Jacobian
invjacob=inv(jacob2);                    % inverse of Jacobian matrix
%
[dhdx,dhdy]=federiv2(nnel,dhdr,dhds,invjacob);     % derivatives w.r.t.
%                                        % physical coordinate
%
kinmtpb=fekinepb(nnel,dhdx,dhdy);        % bending kinematic matrix
%
%————————————————————
% compute bending element matrix
%————————————————————
kb=kb+kinmtpb'*matmtpb*kinmtpb*wtx*wty*detjacob;
%
end
end                                      % end of integration loop for bending term
%
%————————————————————
% numerical integration for bending term
%————————————————————
for intx=1:nglxs
x=points(intx,1);                        % sampling point in x-axis
wtx=weights(intx,1);                     % weight in x-axis
for inty=1:nglys
y=points(inty,2);                        % sampling point in y-axis
wty=weights(inty,2) ;                    % weight in y-axis
%
[shape,dhdr,dhds]=feisoq4(x,y);          % compute shape functions and
%                                        % derivatives at sampling point
%
jacob2=fejacob2(nnel,dhdr,dhds,xcoord,ycoord);     % Jacobian matrix
%
detjacob=det(jacob2);                    % determinant of Jacobian
invjacob=inv(jacob2);                    % inverse of Jacobian matrix
%
[dhdx,dhdy]=federiv2(nnel,dhdr,dhds,invjacob);     % derivatives w.r.t.
%                                        % physical coordinate
%
kinmtps=fekineps(nnel,dhdx,dhdy,shape);  % shear kinematic matrix
%
%————————————————————
% compute shear element matrix
%————————————————————
ks=ks+kinmtps'*matmtps*kinmtps*wtx*wty*detjacob;
```

```
%
end
end                                    % end of integration loop for shear term
%
%------------------------
% compute element matrix
%------------------------
k=kb+ks;
%
index=feeldof(nd,nnel,ndof);           % extract associated system dofs
%
kk=feasmbl1(kk,k,index);               % assemble element matrices
%
end
%
%------------------------
% apply boundary conditions
%------------------------
[kk,ff]=feaplyc2(kk,ff,bcdof,bcval);
%
%------------------------
% solve the matrix equation
%------------------------
disp=kk\ff;
%
num=1:1:sdof;
displace=[num' disp]                   % print nodal displacements
%
%------------------------
```

```
function [kinmtpb]=fekinepb(nnel,dhdx,dhdy)
%------------------------
% Purpose:
% determine the kinematic matrix expression relating bending curvatures
% to rotations and displacements for shear deformable plate bending
%
% Synopsis:
% [kinmtpb]=fekinepb(nnel,dhdx,dhdy)
%
% Variable Description:
% nnel - number of nodes per element
% dhdx - derivatives of shape functions with respect to x
% dhdy - derivatives of shape functions with respect to y
%------------------------
%
for i=1:nnel
```

```
i1=(i-1)*3+1;
i2=i1+1;
i3=i2+1;
kinmtpb(1,i1)=dhdx(i);
kinmtpb(2,i2)=dhdy(i);
kinmtpb(3,i1)=dhdy(i);
kinmtpb(3,i2)=dhdx(i);
kinmtpb(3,i3)=0;
end
%
%----------------------------------------------------


function [kinmtps]=fekineps(nnel,dhdx,dhdy,shape)
%----------------------------------------------------
% Purpose:
% determine the kinematic matrix expression relating shear strains
% to rotations and displacements for shear deformable plate bending
%
% Synopsis:
% [kinmtps]=fekineps(nnel,dhdx,dhdy,shape)
%
% Variable Description:
% nnel - number of nodes per element
% dhdx - derivatives of shape functions with respect to x
% dhdy - derivatives of shape functions with respect to y
% shape - shape function
%----------------------------------------------------
%
for i=1:nnel
i1=(i-1)*3+1;
i2=i1+1;
i3=i2+1;
kinmtps(1,i1)=-shape(i);
kinmtps(1,i3)=dhdx(i);
kinmtps(2,i2)=-shape(i);
kinmtps(2,i3)=dhdy(i);
end
%
%----------------------------------------------------
```

♣ **Example 10.9.2** The same square plate as that used in Example 10.9.1 is analyzed here. However, the boundary of the plate is clamped and the plate is subjected to a uniform pressure of 2 psi. Using the same number of elements as before, determine the center deflection of the plate.

Because of different boundary conditions and loads compared to the previous example, the following vectors substitute those in Example 10.9.1. Otherwise, the rest of the program is the same. The finite element result shows the center deflection of 0.0088 in. while the analytical solution is 0.0092 in.

```
bcdof=[1 2 3 4 5 6 7 8 9 10 11 12 16 19 20 21 23 25 26];
bcval=zeros(1,19);
%
ff(3)=3.125; ff(6)=6.25; ff(9)=3.125;
ff(12)=6.25; ff(15)=12.5; ff(18)=6.25;
ff(21)=3.125; ff(24)=6.25; ff(27)=3.125;
```

‡

10.10 MATLAB Application to Shells

The following two examples show static shell analysis using two different shell elements as discussed in Sec. 10.7 and Sec. 10.8.

♣ **Example 10.10.1**   A barrel vault has radius of 25 ft, subtended angle of 80 degrees, length of 50 ft, and thickness of 3 in. The structure has elastic modulus of 3 msi, Poisson's ratio of 0, and weight of 90 lb/ft^2. Two curved edges are assumed to be supported by rigid diaphragms, and the other two edges are free. The figure of the structure is shown in Fig 10.10.1. Because of symmetry, a quarter of the structure is modeled. When 4x4 elements are used, the vertical deflection at the center of the free edge is 3.5088 in. The analytical solution is 3.7033 in. as reported in Ref. [37]. Thus, the finite element solution agrees well with the analytical solution.

```
%----------------------------------------------------
% Example 10.10.1
% A barrel vault has radius r=25 ft, subtended angle of 80 degrees,
% length 50 ft, and thickness 3 in.
% Elastic modulus is 3 msi, Poisson's ratio is 0, and
% weight of the shell is 90 lb per square ft. Two curved edges are
% supported by rigid diaphragm and the other two edges are free.
% Solve using 4 by 4 elements of a quarter of the shell.
% (See Fig. 10.10.1 for the structure.)
%
% Variable descriptions
```

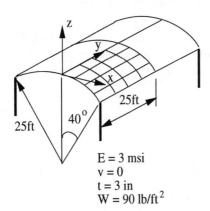

Figure 10.10.1 A Quarter of Barrel Vault

```
% k = element matrix in the local axes
% ke = element matrix in the global axes
% kb = element matrix for bending stiffness
% ks = element matrix for shear stiffness
% km = element matrix for membrane stiffness
% f = element vector
% kk = system matrix
% ff = system vector
% disp = system nodal displacement vector
% gcoord = coordinate values of each node
% nodes = nodal connectivity of each element
% index = a vector containing system dofs associated with each element
% pointb = matrix containing sampling points for bending term
% weightb = matrix containing weighting coefficients for bending term
% points = matrix containing sampling points for shear term
% weights = matrix containing weighting coefficients for shear term
% bcdof = a vector containing dofs associated with boundary conditions
% bcval = a vector containing boundary condition values associated with
%              the dofs in 'bcdof'
% kinmtsb = matrix for kinematic equation for bending
% matmtsb = matrix for material property for bending
% kinmtss = matrix for kinematic equation for shear
% matmtss = matrix for material property for shear
% kinmtsm = matrix for kinematic equation for membrane
% matmtsm = matrix for material property for membrane
% tr3d = transformation matrix from local to global axes
%
%----------------------------------------------------------------
%
%----------------------------------
% input data for control parameters
```

```
%----------------------------------
%
clear
nel=16;                              % number of elements
nnel=4;                              % number of nodes per element
ndof=6;                              % number of dofs per node
nnode=25;                            % total number of nodes in system
sdof=nnode*ndof;                     % total system dofs
edof=nnel*ndof;                      % degrees of freedom per element
emodule=3e6;                         % elastic modulus
poisson=0.0;                         % Poisson's ratio
t=3;                                 % plate thickness
nglxb=2; nglyb=2;                    % 2x2 Gauss-Legendre quadrature for bending
nglb=nglxb*nglyb;                    % number of sampling points for bending
nglxs=1; nglys=1;                    % 1x1 Gauss-Legendre quadrature for shear
ngls=nglxs*nglys;                    % number of sampling points for shear
%
%----------------------------------
% input data for nodal coordinate values
% gcoord(i,j) where i->node no. and j->x or y
%----------------------------------
%
gcoord=[0.0 0.0 0.0; 0.0 6.25 0.0; 0.0 12.5 0.0; 0.0 18.75 0.0;0.0 25.0 0.0;
4.34 0.0 -0.38;4.34 6.25 -0.38;4.34 12.5 -0.38;4.34 18.75 -0.38;4.34 25.0 -0.38;
8.55 0.0 -1.51;8.55 6.25 -1.51;8.55 12.5 -1.51;8.55 18.75 -1.51;8.55 25.0 -1.51;
12.5 0.0 -3.35;12.5 6.25 -3.35;12.5 12.5 -3.35;12.5 18.75 -3.35;12.5 25.0 -3.35;
16.1 0.0 -5.85;16.1 6.25 -5.85;16.1 12.5 -5.85;16.1 18.75 -5.85;16.1 25.0 -5.85];
gcoord=12.0*gcoord;                  % unit change from foot to inch
%
%----------------------------------
% input data for nodal connectivity for each element
% nodes(i,j) where i-> element no. and j-> connected nodes
%----------------------------------
%
nodes=[6 7 2 1; 7 8 3 2; 8 9 4 3; 9 10 5 4;
11 12 7 6; 12 13 8 7; 13 14 9 8; 14 15 10 9;
16 17 12 11; 17 18 13 12; 18 19 14 13; 19 20 15 14;
21 22 17 16; 22 23 18 17; 23 24 19 18; 24 25 20 19];
%
%----------------------------------
% input data for boundary conditions
%----------------------------------
%
bcdof=[1 3 5 6 7 11 12 13 17 18 19 23 24 25 26 28 29 30 ...
31 33 56 58 60 61 63 86 88 90 91 93 116 118 120 ...
121 122 123 146 148 150];            % constrained dofs
bcval=zeros(size(bcdof));            % whose values are 0
%
```

```
%------------------------------------------
% initialization of matrices and vectors
%------------------------------------------
%
ff=zeros(sdof,1);                  % system force vector
kk=zeros(sdof,sdof);               % system matrix
disp=zeros(sdof,1);                % system displacement vector
index=zeros(edof,1);               % index vector
kinmtsb=zeros(3,edof);             % kinematic matrix for bending
matmtsb=zeros(3,3);                % constitutive matrix for bending
kinmtsm=zeros(3,edof);             % kinematic matrix for membrane
matmtsm=zeros(3,3);                % constitutive matrix for membrane
kinmtss=zeros(2,edof);             % kinematic matrix for shear
matmtss=zeros(2,2);                % constitutive matrix for shear
tr3d=zeros(edof,edof);             % transformation matrix
%
%-----------------------
% force vector
%-----------------------
%
ff(3)=-613.6;                      % transverse force at node 1
ff(9)=-1227.2;                     % transverse force at node 2
ff(15)=-1227.2;                    % transverse force at node 3
ff(21)=-1227.2;                    % transverse force at node 4
ff(27)=-613.6;                     % transverse force at node 5
ff(33)=-1227.2;                    % transverse force at node 6
ff(39)=-2454.4;                    % transverse force at node 7
ff(45)=-2454.4;                    % transverse force at node 8
ff(51)=-2454.4;                    % transverse force at node 9
ff(57)=-1227.2;                    % transverse force at node 10
ff(63)=-1227.2;                    % transverse force at node 11
ff(69)=-2454.4;                    % transverse force at node 12
ff(75)=-2454.4;                    % transverse force at node 13
ff(81)=-2454.4;                    % transverse force at node 14
ff(87)=-1227.2;                    % transverse force at node 15
ff(93)=-1227.2;                    % transverse force at node 16
ff(99)=-2454.4;                    % transverse force at node 17
ff(105)=-2454.4;                   % transverse force at node 18
ff(111)=-2454.4;                   % transverse force at node 19
ff(117)=-1227.2;                   % transverse force at node 20
ff(123)=-613.6;                    % transverse force at node 21
ff(129)=-1227.2;                   % transverse force at node 22
ff(135)=-1227.2;                   % transverse force at node 23
ff(141)=-1227.2;                   % transverse force at node 24
ff(147)=-613.6;                    % transverse force at node 25
%
%--------------------------------------------------------
% computation of element matrices and vectors and their assembly
```

```
%----------------------------------------------
%
% for bending and membrane stiffness
%
[pointb,weightb]=feglqd2(nglxb,nglyb);      % sampling points and weights
matmtsm=fematiso(1,emodule,poisson)*t;       % membrane property
matmtsb=fematiso(1,emodule,poisson)*t^3/12;  % bending property
%
% for shear stiffness
%
[points,weights]=feglqd2(nglxs,nglys);      % sampling points and weights
shearm=0.5*emodule/(1.0+poisson);            % shear modulus
shcof=5/6;                                   % shear correction factor
matmtss=shearm*shcof*t*[1 0; 0 1];           % shear material property
%
for iel=1:nel                 % loop for the total number of elements
%
for i=1:nnel
nd(i)=nodes(iel,i);           % extract connected node for (iel)-th element
xcoord(i)=gcoord(nd(i),1);    % extract x value of the node
ycoord(i)=gcoord(nd(i),2);    % extract y value of the node
zcoord(i)=gcoord(nd(i),3);    % extract z value of the node
end
%
%
% compute the local direction cosines and local axes
%
[tr3d,xprime,yprime]=fetransh(xcoord,ycoord,zcoord,nnel);
%
k=zeros(edof,edof);            % element matrix in local axes
ke=zeros(edof,edof);           % element matrix in global axes
km=zeros(edof,edof);           % element membrane matrix
kb=zeros(edof,edof);           % element bending matrix
ks=zeros(edof,edof);           % element shear matrix
%
%----------------------------------------------
% numerical integration for bending term
%----------------------------------------------
%
for intx=1:nglxb
x=pointb(intx,1);                            % sampling point in x-axis
wtx=weightb(intx,1);                         % weight in x-axis
for inty=1:nglyb
y=pointb(inty,2);                            % sampling point in y-axis
wty=weightb(inty,2) ;                        % weight in y-axis
%
[shape,dhdr,dhds]=feisoq4(x,y);              % compute shape functions and
                                             % derivatives at sampling point
```

```
%
jacob2=fejacob2(nnel,dhdr,dhds,xprime,yprime);    % compute Jacobian
%
detjacob=det(jacob2);                             % determinant of Jacobian
invjacob=inv(jacob2);                             % inverse of Jacobian matrix
%
[dhdx,dhdy]=federiv2(nnel,dhdr,dhds,invjacob);    % derivatives w.r.t.
                                                  % physical coordinate
%
kinmtsb=fekinesb(nnel,dhdx,dhdy);                 % bending kinematic matrix
kinmtsm=fekinesm(nnel,dhdx,dhdy);                 % membrane kinematic matrix
%
%----------------------------------
% compute bending element matrix
%----------------------------------
%
kb=kb+kinmtsb'*matmtsb*kinmtsb*wtx*wty*detjacob;
km=km+kinmtsm'*matmtsm*kinmtsm*wtx*wty*detjacob;
end
end                    % end of numerical integration loop for bending term
%
%----------------------------------
% numerical integration for bending term
%----------------------------------
%
for intx=1:nglxs
x=points(intx,1);                                 % sampling point in x-axis
wtx=weights(intx,1);                              % weight in x-axis
for inty=1:nglys
y=points(inty,2);                                 % sampling point in y-axis
wty=weights(inty,2) ;                             % weight in y-axis
%
[shape,dhdr,dhds]=feisoq4(x,y);                   % compute shape functions and
                                                  % derivatives at sampling point
%
jacob2=fejacob2(nnel,dhdr,dhds,xprime,yprime);    % compute Jacobian
%
detjacob=det(jacob2);                             % determinant of Jacobian
invjacob=inv(jacob2);                             % inverse of Jacobian matrix
%
[dhdx,dhdy]=federiv2(nnel,dhdr,dhds,invjacob);    % derivatives w.r.t.
                                                  % physical coordinate
%
kinmtss=fekiness(nnel,dhdx,dhdy,shape);           % shear kinematic matrix
%
%----------------------------------
% compute shear element matrix
%----------------------------------
```

```
%
ks=ks+kinmtss'*matmtss*kinmtss*wtx*wty*detjacob;
%
end
end                         % end of numerical integration loop for shear term
%
%----------------------------
% compute element matrix
%----------------------------
%
k=km+kb+ks;
%
%----------------------------
% transform from local to global systems
%----------------------------
%
ke=tr3d'*k*tr3d;
%
index=feeldof(nd,nnel,ndof);        % extract dofs associated with element
%
kk=feasmbl1(kk,ke,index);           % assemble element matrices
%
end
%
%----------------------------
% check the singular drilling dof
%----------------------------
%
for i=1:sdof
if(abs(kk(i,i)) < 1e-5)
%
sum=0.0;
for j=1:sdof
sum=sum+abs(kk(i,j));
end
%
if (sum < 1e-5)
kk(i,i)=1;
end
%
end
%
end
%
%----------------------------
% apply boundary conditions
%----------------------------
%
```

```
[kk,ff]=feaplyc2(kk,ff,bcdof,bcval);
%
%----------------------
% solve the matrix equation
%----------------------
%
disp=kk\ff;
%
num=1:1:sdof;
displace=[num' disp]                    % print nodal displacements
%
%------------------------------------------------------------
```


```
function [kinmtsb]=kwkinesb(nnel,dhdx,dhdy)
%------------------------------------------------------------
% Purpose:
% Determine the kinematic matrix expression relating bending curvatures
% to rotations and displacements for shear deformable plate bending
%
% Synopsis:
% [kinmtsb]=kwkinesb(nnel,dhdx,dhdy)
%
% Variable Description:
% nnel - number of nodes per element
% dhdx - derivatives of shape functions with respect to x
% dhdy - derivatives of shape functions with respect to y
%------------------------------------------------------------
%
for i=1:nnel
i1=(i-1)*6+1;
i2=i1+1;
i3=i2+1;
i4=i3+1;
i5=i4+1;
i6=i5+1;
kinmtsb(1,i5)=dhdx(i);
kinmtsb(2,i4)=-dhdy(i);
kinmtsb(3,i5)=dhdy(i);
kinmtsb(3,i4)=-dhdx(i);
kinmtsb(3,i6)=0;
end
%
%------------------------------------------------------------
```

```
function [kinmtsm]=kwkinesm(nnel,dhdx,dhdy)
%------------------------------------------------------------
% Purpose:
% Determine the kinematic equation between strains and displacements
% for two-dimensional solids
%
% Synopsis:
% [kinmtsm]=kwkinesm(nnel,dhdx,dhdy)
%
% Variable Description:
% nnel - number of nodes per element
% dhdx - derivatives of shape functions with respect to x
% dhdy - derivatives of shape functions with respect to y
%------------------------------------------------------------
%
for i=1:nnel
i1=(i-1)*6+1;
i2=i1+1;
i3=i2+1;
i4=i3+1;
i5=i4+1;
i6=i5+1;
kinmtsm(1,i1)=dhdx(i);
kinmtsm(2,i2)=dhdy(i);
kinmtsm(3,i1)=dhdy(i);
kinmtsm(3,i2)=dhdx(i);
kinmtsm(3,i6)=0.0;
end
%
%------------------------------------------------------------


function [kinmtss]=kwkiness(nnel,dhdx,dhdy,shape)
%------------------------------------------------------------
% Purpose:
% Determine the kinematic matrix expression relating shear strains
% to rotations and displacements for shear deformable plate bending
%
% Synopsis:
% [kinmtss]=kwkiness(nnel,dhdx,dhdy,shape)
%
% Variable Description:
% nnel - number of nodes per element
% dhdx - derivatives of shape functions with respect to x
% dhdy - derivatives of shape functions with respect to y
% shape - shape function
%------------------------------------------------------------
```

```
%
for i=1:nnel
i1=(i-1)*6+1;
i2=i1+1;
i3=i2+1;
i4=i3+1;
i5=i4+1;
i6=i5+1;
kinmtss(1,i3)=dhdx(i);
kinmtss(1,i5)=shape(i);
kinmtss(2,i3)=dhdy(i);
kinmtss(2,i4)=-shape(i);
kinmtss(2,i6)=0.0;
end
%
%————————————————————————


function [tr3d,xprime,yprime]=kwtransh(xcoord,ycoord,zcoord,n)
%————————————————————————
% Purpose:
% Compute direction cosines between three-dimensional
% local and global coordinate axes
%
% Synopsis:
% [tr3d,xprime,yprime]=kwtransh(xcoord,ycoord,zcoord,n)
%
% Variable Description:
% xcoord - nodal x coordinates (4x1)
% ycoord - nodal y coordinates (4x1)
% zcoord - nodal z coordinates (4x1)
% n - number of nodes per element
% tr3d - 3d transformation matrix from local to global axes
% xprime - coordinate in terms of the local axes (4x1)
% yprime - coordinate in terms of the global axes (4x1)
%
% Note:
% The local x-axis is defined in the direction from the first node
% to the second node. Nodes 1, 2, and 4 define the local xy-plane.
% The local z-axis is defined normal to the local xy-plane.
% The local y-axis is defined normal to the x and z axes.
%————————————————————————
%
% compute direction cosines
%
v12x=xcoord(2)-xcoord(1);
v12y=ycoord(2)-ycoord(1);
```

```
v12z=zcoord(2)-zcoord(1);
l12=sqrt(v12x$^2$+v12y$^2$+v12z$^2$);
v23x=xcoord(3)-xcoord(2);
v23y=ycoord(3)-ycoord(2);
v23z=zcoord(3)-zcoord(2);
l23=sqrt(v23x$^2$+v23y$^2$+v23z$^2$);
v34x=xcoord(4)-xcoord(3);
v34y=ycoord(4)-ycoord(3);
v34z=zcoord(4)-zcoord(3);
l34=sqrt(v34x$^2$+v34y$^2$+v34z$^2$);
v14x=xcoord(4)-xcoord(1);
v14y=ycoord(4)-ycoord(1);
v14z=zcoord(4)-zcoord(1);
l14=sqrt(v14x$^2$+v14y$^2$+v14z$^2$);
v13x=xcoord(3)-xcoord(1);
v13y=ycoord(3)-ycoord(1);
v13z=zcoord(3)-zcoord(1);
l13=sqrt(v13x$^2$+v13y$^2$+v13z$^2$);
v1tx=v12y*v14z-v12z*v14y;
v1ty=v12z*v14x-v12x*v14z;
v1tz=v12x*v14y-v12y*v14x;
v1yx=v1ty*v12z-v1tz*v12y;
v1yy=v1tz*v12x-v1tx*v12z;
v1yz=v1tx*v12y-v1ty*v12x;
vxx=v12x/l12;
vxy=v12y/l12;
vxz=v12z/l12;
vyx=v1yx/sqrt(v1yx$^2$+v1yy$^2$+v1yz$^2$);
vyy=v1yy/sqrt(v1yx$^2$+v1yy$^2$+v1yz$^2$);
vyz=v1yz/sqrt(v1yx$^2$+v1yy$^2$+v1yz$^2$);
vzx=v1tx/sqrt(v1tx$^2$+v1ty$^2$+v1tz$^2$);
vzy=v1ty/sqrt(v1tx$^2$+v1ty$^2$+v1tz$^2$);
vzz=v1tz/sqrt(v1tx$^2$+v1ty$^2$+v1tz$^2$);
%
% transformation matrix
%
for i=1:2*n
i1=(i-1)*3+1;
i2=i1+1;
i3=i2+1;
tr3d(i1,i1)=vxx;
tr3d(i1,i2)=vxy;
tr3d(i1,i3)=vxz;
tr3d(i2,i1)=vyx;
tr3d(i2,i2)=vyy;
tr3d(i2,i3)=vyz;
tr3d(i3,i1)=vzx;
```

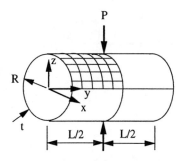

R=5.0 in. t=0.094 in. L=10.35 in.

E=10.5e7 psi v=0.3125

Figure 10.10.2 Pinched Cylinder

```
        tr3d(i3,i2)=vzy;
        tr3d(i3,i3)=vzz;
        end
        %
        % compute nodal values in terms of local axes
        %
        alpa213=acos((l12^2+l13^2-l23^2)/(2*l12*l13));
        alpa314=acos((l13^2+l14^2-l34^2)/(2*l13*l14));
        alpa41y=2*atan(1)-alpa213-alpa314;
        xprime(1)=0; yprime(1)=0;
        xprime(2)=l12; yprime(2)=0;
        xprime(3)=l13*cos(alpa213); yprime(3)=l13*sin(alpa213);
        xprime(4)=l14*sin(alpa41y); yprime(4)=l14*cos(alpa41y);
        %----------------------------------------------------------
```

‡

♣ **Example 10.10.2**   A cylinder is pinched across its diagonal direction with a force of 100 lb at the center lengthwise. It has radius of 5 in., length of 10.35 in., and thickness of 0.094 in. The material has elastic modulus of 10.5 msi and Poisson's ratio 0.3125. One-eighth of the cylinder is modeled due to symmetry using 25 four-node shell elements. Figure 10.10.2 shows the cylinder.

```
        %----------------------------------------------------------
        % Example 10.10.2
        % A cylinder is pinched across its diagonal direction with a force of
        % 100 lb at center lengthwise. It has radius of 5 in., length of
        % 10.35 in., and thickness of 0.094 in.
        % The material has elastic modulus of 10.5 msi and Poisson's ratio 0.3125.
```

% One-eighth of the cylinder is modeled due to symmetry using
% 25 four-node shell elements.
% (see Fig. 10.10.2 for the finite element mesh)
%
% Variable descriptions
% k = element matrix in terms of local axes
% kt = element matrix in terms of global axes
% f = element vector
% kk = system matrix
% ff = system vector
% disp = system nodal displacement vector
% gcoord = coordinate values of each node
% nodes = nodal connectivity of each element
% index = a vector containing system dofs associated with each element
% point1 = matrix containing sampling points for inplane axes
% weight1 = matrix containing weighting coefficients for inplane axes
% pointz = matrix containing sampling points for transverse axis
% weightz = matrix containing weighting coefficients for transverse axis
% bcdof = a vector containing dofs associated with boundary conditions
% bcval = a vector containing boundary condition values associated with
%   the dofs in 'bcdof'
% bmtx = matrix for kinematic equation
% dmtx = matrix for material property
% trsh = nodal variable transformation matrix
% rot6 = strain transformation matrix
% aj = 3-D Jacobian matrix
%
%————————————————————————————
%
%——————————————————
% input data for control parameters
%——————————————————
%
clear
nel=25;   % number of elements
nnel=4;   % number of nodes per element
ndof=6;   % number of dofs per node
nnode=36;   % total number of nodes in system
sdof=nnode*ndof;   % total system dofs
edof=nnel*ndof;   % degrees of freedom per element
emodule=10.5e6;   % elastic modulus
poisson=0.3125;   % Poisson's ratio
thick=0.094;   % shell thickness
nglx=1; ngly=1; nglz=2;   % 1x1x2 Gauss-Legendre quadrature
ngl=nglx*ngly*nglz;   % number of sampling points per element
%
%——————————————————————
% input data for nodal coordinate values

% gcoord(i,j) where i->node no. and j->x,y,z
%————————————————————
%
gcoord=[
0.0000 0.0000 5.0000
0.0000 1.0350 5.0000
0.0000 2.0700 5.0000
0.0000 3.1050 5.0000
0.0000 4.1400 5.0000
0.0000 5.1750 5.0000
1.5451 0.0000 4.7553
1.5451 1.0350 4.7553
1.5451 2.0700 4.7553
1.5451 3.1050 4.7553
1.5451 4.1400 4.7553
1.5451 5.1750 4.7553
2.9389 0.0000 4.0451
2.9389 1.0350 4.0451
2.9389 2.0700 4.0451
2.9389 3.1050 4.0451
2.9389 4.1400 4.0451
2.9389 5.1750 4.0451
4.0451 0.0000 2.9389
4.0451 1.0350 2.9389
4.0451 2.0700 2.9389
4.0451 3.1050 2.9389
4.0451 4.1400 2.9389
4.0451 5.1750 2.9389
4.7553 0.0000 1.5451
4.7553 1.0350 1.5451
4.7553 2.0700 1.5451
4.7553 3.1050 1.5451
4.7553 4.1400 1.5451
4.7553 5.1750 1.5451
5.0000 0.0000 0.0000
5.0000 1.0350 0.0000
5.0000 2.0700 0.0000
5.0000 3.1050 0.0000
5.0000 4.1400 0.0000
5.0000 5.1750 0.0000];
%
%————————————————————————
% input data for nodal connectivity for each element
% nodes(i,j) where i-> element no. and j-> connected nodes
%————————————————————————
%
nodes=[
7 8 2 1 ;

```
8 9 3 2 ;
9 10 4 3 ;
10 11 5 4 ;
11 12 6 5 ;
13 14 8 7 ;
14 15 9 8 ;
15 16 10 9 ;
16 17 11 10 ;
17 18 12 11 ;
19 20 14 13 ;
20 21 15 14 ;
21 22 16 15 ;
22 23 17 16 ;
23 24 18 17 ;
25 26 20 19 ;
26 27 21 20 ;
27 28 22 21 ;
28 29 23 22 ;
29 30 24 23 ;
31 32 26 25 ;
32 33 27 26 ;
33 34 28 27 ;
34 35 29 28 ;
35 36 30 29 ];
%
%————————————————
% input data for boundary conditions
%————————————————
%
bcdof=[1 5 6 ...
7 11 12 ...
13 17 18 ...
19 23 24 ...
25 29 30 ...
31 32 34 35 36 ...
38 ...
68 70 72 ...
104 106 108 ...
140 142 144 ...
176 178 180 ...
183 184 185 ...
189 190 191 ...
195 196 197 ...
201 202 203 ...
207 208 209 ...
212 213 214 215 216 ];          % constrained dofs
bcval=zeros(size(bcdof));       % whose values are 0
%
```

```
%--------------------------------------
% initialization of matrices and vectors
%--------------------------------------
%
ff=zeros(sdof,1);                       % system force vector
kk=zeros(sdof,sdof);                    % system matrix
disp=zeros(sdof,1);                     % system displacement vector
index=zeros(edof,1);                    % index vector
dmtx=zeros(6,6);                        % constitutive matrix for shear
%
%--------------------
% force vector
%--------------------
%
ff(33)=-25;                             % transverse force at node 3
%
%--------------------------------------
% compute material property matrix
%--------------------------------------
%
dmtx(1,1)=emodule/(1.0-poisson$^2$);
dmtx(1,2)=poisson*dmtx(1,1);
dmtx(2,1)=dmtx(1,2);
dmtx(2,2)=dmtx(1,1);
dmtx(4,4)=emodule/(2.0*(1.0+poisson));
dmtx(5,5)=(5/6)*dmtx(4,4);
dmtx(6,6)=dmtx(5,5);
%
%--------------------------------------------------
% computation of element matrices and vectors and their assembly
%--------------------------------------------------
%
% for inplane integration
%
[point1,weight1]=feglqd2(nglx,ngly);    % sampling points and weights
%
% for transverse integration
%
[pointz,weightz]=feglqd1(nglz);         % sampling points and weights
%
for iel=1:nel                           % loop for the total number of elements
%
for i=1:nnel
nd(i)=nodes(iel,i);                     % extract connected node for (iel)-th element
xcoord(i)=gcoord(nd(i),1);              % extract x value of the node
ycoord(i)=gcoord(nd(i),2);              % extract y value of the node
zcoord(i)=gcoord(nd(i),3);              % extract z value of the node
end
```

```
%
k=zeros(edof,edof); kt=zeros(edof,edof); trsh=zeros(edof,edof); %
%————————————————————————
% compute direction cosine vectors
% v1 and v2: tangent to the element
% v3: normal to the element
%————————————————————————
%
d1(1)=xcoord(2)-xcoord(1);    % define the vector from node 1 to node 2
d1(2)=ycoord(2)-ycoord(1);
d1(3)=zcoord(2)-zcoord(1);
%
d2(1)=xcoord(4)-xcoord(1);    % define the vector from node 4 to node 1
d2(2)=ycoord(4)-ycoord(1);
d2(3)=zcoord(4)-zcoord(1);
%
v3(1)=d1(2)*d2(3)-d1(3)*d2(2); % vector v3 normal to vectors d1 and d2
v3(2)=d1(3)*d2(1)-d1(1)*d2(3);
v3(3)=d1(1)*d2(2)-d1(2)*d2(1);
sum=sqrt(v3(1)^2+v3(2)^2+v3(3)^2);
v3(1)=v3(1)/sum;              % make vector v3 a unit vector
v3(2)=v3(2)/sum;
v3(3)=v3(3)/sum;
%
if v3(2) > 0.999999    % if v3 is along the y-axis, v1 is set along x-axis
v1(1)=1.0;
v1(2)=0.0;
v1(3)=0.0;
else                   % otherwise, v1 is the cross product of y-axis and v3
v1(1)=v3(3);
v1(2)=0.0;
v1(3)=-v3(1);
sum=sqrt(v1(1)^2+v1(2)^2+v1(3)^2);
v1(1)=v1(1)/sum;              % make vector v1 a unit vector
v1(2)=v1(2)/sum;
v1(3)=v1(3)/sum;
end
%
v2(1)=v3(2)*v1(3)-v3(3)*v1(2);   % v2 is cross product of v3 and v1
v2(2)=v3(3)*v1(1)-v3(1)*v1(3);
v2(3)=v3(1)*v1(2)-v3(2)*v1(1);
sum=sqrt(v2(1)^2+v2(2)^2+v2(3)^2);
v2(1)=v2(1)/sum;              % make vector v2 a unit vector
v2(2)=v2(2)/sum;
v2(3)=v2(3)/sum;
%
%————————————————————————
```

```
% construct nodal variable transformation matrix
%------------------------------------------------
%
a(1,1)=v1(1);
a(2,1)=v1(2);
a(3,1)=v1(3);
a(1,2)=v2(1);
a(2,2)=v2(2);
a(3,2)=v2(3);
a(1,3)=v3(1);
a(2,3)=v3(2);
a(3,3)=v3(3);
%
ainv=inv(a);
%
for i=1:nnel
i1=(i-1)*ndof+1;
i2=i1+1;
i3=i2+1;
i4=i3+1;
i5=i4+1;
i6=i5+1;
%
trsh(i1,i1)=1.0;
trsh(i2,i2)=1.0;
trsh(i3,i3)=1.0;
trsh(i4,i4)=ainv(1,1);
trsh(i4,i5)=ainv(1,2);
trsh(i4,i6)=ainv(1,3);
trsh(i5,i4)=ainv(2,1);
trsh(i5,i5)=ainv(2,2);
trsh(i5,i6)=ainv(2,3);
trsh(i6,i4)=ainv(3,1);
trsh(i6,i5)=ainv(3,2);
trsh(i6,i6)=ainv(3,3);
end
%
%------------------------------------------------
% strain transformation matrix
%------------------------------------------------
%
rot6(1,1)=v1(1)$^2$;
rot6(1,2)=v1(2)$^2$;
rot6(1,3)=v1(3)$^2$;
rot6(1,4)=v1(1)*v1(2);
rot6(1,5)=v1(2)*v1(3);
rot6(1,6)=v1(1)*v1(3);
```

rot6(2,1)=v2(1)^2;
rot6(2,2)=v2(2)^2;
rot6(2,3)=v2(3)^2;
rot6(2,4)=v2(1)*v2(2);
rot6(2,5)=v2(2)*v2(3);
rot6(2,6)=v2(1)*v2(3);
rot6(3,1)=v3(1)^2;
rot6(3,2)=v3(2)^2;
rot6(3,3)=v3(3)^2;
rot6(3,4)=v3(1)*v3(2);
rot6(3,5)=v3(2)*v3(3);
rot6(3,6)=v3(1)*v3(3);
rot6(4,1)=2.0*v1(1)*v2(1);
rot6(4,2)=2.0*v1(2)*v2(2);
rot6(4,3)=2.0*v1(3)*v2(3);
rot6(4,4)=v1(1)*v2(2)+v2(1)*v1(2);
rot6(4,5)=v1(2)*v2(3)+v2(2)*v1(3);
rot6(4,6)=v1(3)*v2(1)+v2(3)*v1(1);
rot6(5,1)=2.0*v2(1)*v3(1);
rot6(5,2)=2.0*v2(2)*v3(2);
rot6(5,3)=2.0*v2(3)*v3(3);
rot6(5,4)=v2(1)*v3(2)+v3(1)*v2(2);
rot6(5,5)=v2(2)*v3(3)+v3(2)*v2(3);
rot6(5,6)=v2(3)*v3(1)+v3(3)*v2(1);
rot6(6,1)=2.0*v3(1)*v1(1);
rot6(6,2)=2.0*v3(2)*v1(2);
rot6(6,3)=2.0*v3(3)*v1(3);
rot6(6,4)=v3(1)*v1(2)+v1(1)*v3(2);
rot6(6,5)=v3(2)*v1(3)+v1(2)*v3(3);
rot6(6,6)=v3(3)*v1(1)+v1(3)*v3(1);
%
%————————————————
% material property matrix transformation
%————————————————
%
dmtxt=rot6'*dmtx*rot6;
%
%————————————————
% numerical integration loop
%————————————————
%
for intx=1:nglx
r=point1(intx);   % sampling point in x-axis
wtx=weight1(intx,1);   % weight in x-axis
for inty=1:ngly
s=point1(inty,2);   % sampling point in y-axis
wty=weight1(inty,2) ;   % weight in y-axis

```
for intz=1:nglz
t=pointz(intz);                        % sampling point in z-axis
wtz=weightz(intz) ;                    % weight in z-axis
%
[shape2,dhdr,dhds]=feisoq4(r,s);       % compute 2-D shape functions and
                                       % derivatives at sampling point
%
                                       % derivatives at sampling point
shape1(1)=0.5*(1-t);
shape1(2)=0.5*(1+t);
dhdt(1)=-0.5;
dhdt(2)=0.5;
%
%―――――――――――――――――――――――――
% compute jacobian matrix and its inverse
%―――――――――――――――――――――――――
%
hz=thick*0.5*t;
hzdt=thick*0.5;
%
aj(1,1)=dhdr(1)*xcoord(1)+dhdr(2)*xcoord(2)+dhdr(3)*xcoord(3)+ ...
dhdr(4)*xcoord(4)+dhdr(1)*hz*v3(1)+dhdr(2)*hz*v3(1)+ ...
dhdr(3)*hz*v3(1)+dhdr(4)*hz*v3(1);
aj(2,1)=dhds(1)*xcoord(1)+dhds(2)*xcoord(2)+dhds(3)*xcoord(3)+ ...
dhds(4)*xcoord(4)+dhds(1)*hz*v3(1)+dhds(2)*hz*v3(1)+ ...
dhds(3)*hz*v3(1)+dhds(4)*hz*v3(1);
aj(3,1)=hzdt*v3(1);
aj(1,2)=dhdr(1)*ycoord(1)+dhdr(2)*ycoord(2)+dhdr(3)*ycoord(3)+ ...
dhdr(4)*ycoord(4)+dhdr(1)*hz*v3(2)+dhdr(2)*hz*v3(2)+ ...
dhdr(3)*hz*v3(2)+dhdr(4)*hz*v3(2);
aj(2,2)=dhds(1)*ycoord(1)+dhds(2)*ycoord(2)+dhds(3)*ycoord(3)+ ...
dhds(4)*ycoord(4)+dhds(1)*hz*v3(2)+dhds(2)*hz*v3(2)+ ...
dhds(3)*hz*v3(2)+dhds(4)*hz*v3(2);
aj(3,2)=hzdt*v3(2);
aj(1,3)=dhdr(1)*zcoord(1)+dhdr(2)*zcoord(2)+dhdr(3)*zcoord(3)+ ...
dhdr(4)*zcoord(4)+dhdr(1)*hz*v3(3)+dhdr(2)*hz*v3(3)+ ...
dhdr(3)*hz*v3(3)+dhdr(4)*hz*v3(3);
aj(2,3)=dhds(1)*zcoord(1)+dhds(2)*zcoord(2)+dhds(3)*zcoord(3)+ ...
dhds(4)*zcoord(4)+dhds(1)*hz*v3(3)+dhds(2)*hz*v3(3)+ ...
dhds(3)*hz*v3(3)+dhds(4)*hz*v3(3);
aj(3,3)=hzdt*v3(3);
%
ajinv=inv(aj);
det3=det(aj);
%
%―――――――――――――――――――――
% compute global derivatives
%―――――――――――――――――――――
```

```
%
for i=1:nnel
derivg(1,i)=ajinv(1,1)*dhdr(i)+ajinv(1,2)*dhds(i);
derivg(2,i)=ajinv(2,1)*dhdr(i)+ajinv(2,2)*dhds(i);
derivg(3,i)=ajinv(3,1)*dhdr(i)+ajinv(3,2)*dhds(i);
dhdz(1,i)=ajinv(1,3)*hzdt;
dhdz(2,i)=ajinv(2,3)*hzdt;
dhdz(3,i)=ajinv(3,3)*hzdt;
end
%
%------------------------------------------------
% compute strain-displacement matrix called bmtx
%------------------------------------------------
%
bmtx=zeros(ndof,edof);
%
for i=1:nnel
i1=(i-1)*ndof+1;
i2=i1+1;
i3=i2+1;
i4=i3+1;
i5=i4+1;
i6=i5+1;
%
gk1=derivg(1,i)*hz+shape2(i)*dhdz(1,i);
gk2=derivg(2,i)*hz+shape2(i)*dhdz(2,i);
gk3=derivg(3,i)*hz+shape2(i)*dhdz(3,i);
%
bmtx(1,i1)=derivg(1,i);                % elements associated with epsilon_x
bmtx(1,i4)=gk1*(-v2(1));
bmtx(1,i5)=gk1*v1(1);
bmtx(2,i2)=derivg(2,i);                % elements related to epsilon_y
bmtx(2,i4)=gk2*(-v2(2));
bmtx(2,i5)=gk2*v1(2);
bmtx(3,i3)=derivg(3,i);                % elements related to epsilon_z
bmtx(3,i4)=gk3*(-v2(3));
bmtx(3,i5)=gk3*v1(3);
bmtx(4,i1)=derivg(2,i);                % elements related to gamma_{xy}
bmtx(4,i2)=derivg(1,i);
bmtx(4,i4)=gk2*(-v2(1))+gk1*(-v2(2));
bmtx(4,i5)=gk2*v1(1)+gk1*v1(2);
bmtx(5,i2)=derivg(3,i);                % elements related to gamma_{yz}
bmtx(5,i3)=derivg(2,i);
bmtx(5,i4)=gk3*(-v2(2))+gk2*(-v2(3));
bmtx(5,i5)=gk3*v1(2)+gk2*v1(3);
bmtx(6,i1)=derivg(3,i);                % elements related to gamma_{xz}
bmtx(6,i3)=derivg(1,i);
bmtx(6,i4)=gk3*(-v2(1))+gk1*(-v2(3));
```

```
bmtx(6,i5)=gk3*v1(1)+gk1*v1(3);
end
%
detwt=det3*wtx*wty*wtz;
%
k=k+bmtx'*dmtxt*bmtx*detwt;           % element matrix in local axes
%
end
end
end                                    % end of numerical integration loop
%
kt=trsh'*k*trsh;                       % element matrix in global axis
%
index=feeldof(nd,nnel,ndof);           % extract dofs associated with element
%
kk=feasmbl1(kk,kt,index);              % assemble element matrices
%
end
%
%----------------------
% apply boundary conditions
%----------------------
%
[kk,ff]=feaplyc2(kk,ff,bcdof,bcval);
%
%----------------------
% solve the matrix equation
%----------------------
%
disp=kk\ff;
%
num=1:1:sdof;
displace=[num' disp]                   % print nodal displacements
%
%------------------------------------------------------------


function [shape,dhdr]=feisol2(rvalue)
%------------------------------------------------------------
%Purpose
% Compute isoparametric 2-node shape functions
% and their derivatives at the selected
% point in terms of the natural coordinate.
%
%Synopsis:
% [shape,dhdr]=kwisol2(rvalue)
%
```

```
%Variable Description:
% shape - shape functions for the linear element
% dhdr - derivatives of shape functions
% rvalue - r coordinate value of the selected point
%
%Notes:
% 1st node at rvalue=-1
% 2nd node at rvalue=1
%---------------------------------------------------
%
shape(1)=0.5*(1.0-rvalue);          % first shape function
shape(2)=0.5*(1.0+rvalue);          % second shape function
%
dhdr(1)=-0.5;                       % derivative of the first shape function
dhdr(2)=0.5;                        % derivative of the second shape function
%
%---------------------------------------------------
```

The analysis shows that the deflection under the load is 0.1087 in the load direction. The analytical solution from Ref. [37] is 0.1139. Thus, the two solutions agree well.

‡

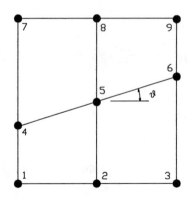

Figure P10.4 Problem 10.4

Problems

10.1 Redo Example 10.7.1 for uniform pressure of 0.4 psi instead of the concentrated load using the computer programs. Compare the present solution to that in Example 10.7.1.

10.2 Redo Example 10.7.2 for a center load of 200 lb instead of the pressure load using the computer programs. Compare the present solution to that in Example 10.7.2.

10.3 Redo Example 10.7.1 for a plate with two opposite edges simply supported and the other two opposite edges clamped.

10.4 Redo Example 10.7.1 for a mesh shown in Fig. P10.4. Change the angle θ in the figure from 5 degrees to 30 degrees by increments of 5 degrees. Compare the present solutions to that in Example 10.7.1.

10.5 Consider a clamped circular plate which has elastic modulus of 200GPa, radius of 0.2m, and thickness of 10mm. The plate is subjected to a center load of 2.0kN. Find the deflection using the computer programs for meshes shown in Fig. 10.6.3.

10.6 Redo Prob. 10.5 for a simply supported plate.

10.7 Find the deflection of a triangular shape of plate with simple support. The plate dimension is given in Fig. P10.7 and its thickness is 2mm. Its elastic modulus is 70GPA and it is subjected to a center force of 100N. Find the deflection of the plate using the computer programs.

10.8 Redo Prob. 10.7 for the clamped plate.

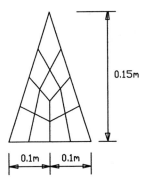

Figure P10.7 Problem 10.7

CHAPTER **ELEVEN**

CONTROL OF FLEXIBLE STRUCTURES

11.0 Chapter Overview

Control of flexible structures is discussed in this chapter. Especially, the finite element application to structural control problems is presented along with fundamental control theories. Many MATLAB programs are also presented for demonstration of the control system design.

11.1 Introduction

The subject of flexible structures control consists of both dynamic analysis and control theory. Usually, these two disciplines are mingled together in such a way that we have to understand both disciplines to an equal extent in order to achieve our goal. The dynamic analysis of flexible structures is dominated by the finite element analysis as discussed in other chapters of this book. The control theory, on the other hand, is introduced in this chapter. The essence of each exemplary control theory is discussed in this chapter. In-depth discussion on control theories is available in a number of literature sources. The control theories are introduced here to help the readers of this book to understand the key features in conjunction with the finite element analysis of flexible structures. Presented in Fig. 11.1.1 is a flow diagram representing the relationship between mathematical modeling and control system design for a given structural system. The mathematical modeling represented by finite element method has been discussed so far. The control system design in this chapter will mainly make use of the finite element modeling results. The control system design is demonstrated also using MATLAB. MATLAB *m-files* are generated in order to solve example problems. The example problems in this chapter do not need a specific MATLAB *Toolbox*.

There are two distinct approaches for control system design. One is called the *frequency domain approach* or classical control technique and the other one is the *time domain approach* mainly adopted in the modern control technique. The *frequency domain approach* relies upon analytical tools, and is still popular in the majority of

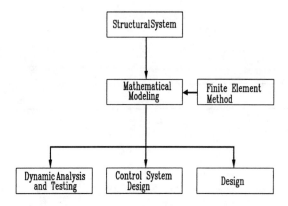

Figure 11.1.1 Flow Diagram for Structural Analysis and Control

existing control systems. For a given system, the frequency domain approach focuses on the relationship between input and output. The input to the system is modified to improve the output performance of a system. On the other hand, the modern control technique is motivated by the rapid advance of computational power. The system behavior is described by a set of variables over time domain. The control input then tries to control each variable in order to satisfy desired system responses.

Each method has its own unique features. For the *frequency domain approach*, plenty of analytical results and tools are available. On the other hand, the modern control theory is easy to implement with the help of the abundant computational software available. Modern control techniques impose another important issue on estimation of all degrees of freedom with a limited number of sensors. The number of sensors and actuators are usually less than the degrees of freedom of truncated (approximate) systems. Estimating all the flexible modes is critical to designing an active control law in the time domain.

Active control of flexible structures is mainly represented by vibration control using mechanical, electrical, and/or electromechanical devices. Inherent flexibility of the structure raises a number of issues in the area of active vibration control. The majority of flexible structures are distributed parameter systems. Therefore, they are essentially infinite dimensional dynamic systems. Obviously, the infinite dimensional systems are not practical for a control law design. Mathematical approaches like finte element analysis can be used to derive finite dimensional systems which closely duplicate the original infinite dimensional systems.

Before we work on the dynamic analysis and control system design for flexible structures, we introduce a basic stability theory. The stability theory is a key concept establishing the goal of a control law design. By stability, we mean the dynamic characteristic of a given dynamic system representing the behavior of dynamic motion of the system, for example, whether the motion is decaying or growing with respect to time.

The Lyapunov stability theory has been considered as a background for understanding the stability of a dynamic system.

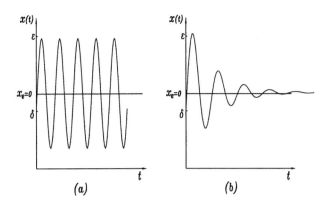

Figure 11.2.1 Time History of Stable(a) and Asymptotically Stable(b) System

11.2 Stability Theory

The Lyapunov stability theory is one of the most frequently referred tools for stability analysis and control system design of a dynamic system [38].

Definition of Stability

Consider a general form of the nonlinear system

$$\dot{x} = f(x,t), \qquad x(t_0) = x_0 \tag{11.2.1}$$

where x is a vector and $f(x,t)$ represents a general nonlinear function. The above system is stable in the Lyapunov sense with respect to the equilibrium state if for any given value $\epsilon > 0$, there exists a number $\delta(\epsilon, t_0) > 0$ for which the $||x(t)|| < \epsilon$ for all $t > t_0$ and $||x(t_0)|| < \delta$.

The above condition implies that the magnitude of $x(t)$ remains within a finite small value in the presence of small initial perturbation. This definition includes also undamped pure oscillatory motions.

Asymptotic Stability

The system is *asymptotically stable* if it satisfies the stability condition and $\lim_{t \to \infty} ||x(t)|| = 0$.

The asymptotic stability implies that the state vector converges to the equilibrium point which is assumed zero at steady state in this case. The difference in the time response of both stable and asymptotically stable cases is displayed in Fig. 11.2.1.

Lyapunov Second Stability Theory

The second Lyapunov stability theory uses a nonnegative energy function for a given system. If the energy function decreases, then the system is stable toward an equilibrium point. In other words, the system energy is taken to be minimum at the equilibrium point.

Theorem :   Let $V(x)$ be an energy function or a *Lyapunov function*. The system is stable if $V(x) > 0$ and $\dot{V}(x) \leq 0$ for all values of x. If the time rate of change of $V(x)$ is less than zero, that is $\dot{V}(x) < 0$, then the system is *asymptotically stable*.

It is important to note that we cannot draw any conclusion about stability when a desired Lyapunov function is not found. In this case, we should try to find a Lyapunov function or apply Lyapunov's instability theorem, which is not discussed here. Another drawback of Lyapunov's approach is that there is no systematic way of finding an appropriate Lyapunov function candidate for stability proof. As it will be explained later in this chapter, the Lyapunov second stability theory is also used for a control law design; the control law tries to decrease a Lyapunov function suppressing the motion of a system.

♣ **Example 11.2.1**   Consider a set of coupled first order systems given by

$$\dot{x}_1 = x_2 - x_1(x_1^2 + x_2^2)$$
$$\dot{x}_2 = -x_1 - x_2(x_1^2 + x_2^2)$$

We want to check stability of the system. The equilibrium points of the system can be obtained from

$$\dot{x}_1 = 0, \qquad \dot{x}_2 = 0$$

Based upon the equilibrium point, we select a trial Lyapunov function as $V = x_1^2 + x_2^2$ which is always positive. The time derivative of the Lyapunov function in conjunction with the above set of equations turns out to be

$$\dot{V} = -2(x_1^2 + x_2^2)$$

Obviously, the $\dot{V}$ is guaranteed to be negative, and the system is asymptotically stable.

‡

♣ **Example 11.2.2**   Consider a simple second order system whose governing equation is given as

$$m\ddot{q} + c\dot{q} + kq = 0$$

where $m > 0$, $c > 0$, $k > 0$. There are many ways to check the stability of the above system. In order to apply the Lyapunov theory, we transform the above equation into a set of first order equations as

$$\dot{x}_1 = x_2$$
$$m\dot{x}_2 = -cx_2 - kx_1$$

where $x_1 = q$, $x_2 = \dot{q}$. Now we take a Lyapunov function as

$$V(x) = \frac{1}{2}(mx_2^2 + kx_1^2)$$

and the time derivative of the Lyapunov function becomes

$$\dot{V}(x) = mx_2\dot{x}_2 + kx_1\dot{x}_1$$
$$= -cx_2^2$$

Therefore, $\dot{V}(x) < 0$ as long as $x_2 \neq 0$, and the system is asymptotically stable. Note that x_2 can be instantaneously zero but converges to zero only at the steady equilibrium point.

‡

Stability of a Linear First Order System

Consider a linear first order system given as

$$\{\dot{x}\} = [A]\{x\} \tag{11.2.2}$$

where $\{x\}$ is an n by 1 vector. In order to set up the stability condition of the system, we choose the following Lyapunov function

$$U = \{x\}^T[P]\{x\} \tag{11.2.3}$$

The time derivative of the Lyapunov function becomes

$$\dot{U} = \{\dot{x}\}^T[P]\{x\} + \{x\}^T[P]\{\dot{x}\}$$
$$= \{x\}^T(A^T[P] + [P]A)\{x\} \tag{11.2.4}$$

For stability we require
$$-[Q] = [A]^T[P] + [P][A] \tag{11.2.5}$$

where $[Q]$ is a positive definite matrix satisfying the property $\{x\}^T[Q]\{x\} > 0$ for $\{x\} \neq 0$ and $[Q] = [Q]^T$. Thus,

$$\dot{U} = -\{x\}^T[Q]\{x\} < 0 \tag{11.2.6}$$

Therefore, we have the following theorem for stability of the linear system.

Theorem : The linear system $\{\dot{x}\} = [A]\{x\}$ is stable if and only if there exists a positive definite $[P]$ matrix which satisfies Eq. (11.2.5) for a given positive definite matrix $[Q]$.

♣ **Example 11.2.3** Let us assume a two degree of freedom system

$$\{\dot{x}\} = [A]\{x\} = \begin{bmatrix} 0 & 1 \\ -2 & -1 \end{bmatrix} \{x\}$$

In order to check the stability of the system, first we assume a $[Q]$ matrix in Eq. (11.2.5) as

$$[Q] = \begin{bmatrix} 1 & 0 \\ 0 & 1 \end{bmatrix}$$

Once we have $[Q]$, then we solve Eq. (11.2.5)

$$-\begin{bmatrix} 1 & 0 \\ 0 & 1 \end{bmatrix} = \begin{bmatrix} p_1 & p_2 \\ p_2 & p_3 \end{bmatrix} \begin{bmatrix} 0 & 1 \\ -2 & -1 \end{bmatrix} + \begin{bmatrix} 0 & -2 \\ 1 & -1 \end{bmatrix} \begin{bmatrix} p_1 & p_2 \\ p_2 & p_3 \end{bmatrix}$$

The resultant $[P]$ matrix, therefore, becomes

$$[P] = \frac{1}{4} \begin{bmatrix} 7 & 1 \\ 1 & 3 \end{bmatrix}$$

It is not difficult to check that $[P]$ is positive definite; thus, the system is stable.
‡

Bounded Input Bounded Output(BIBO) Stability

In the above definition of stability, we were concerned only about a system itself without including an external input. When there is a certain external input to the system, the stability of the system obviously should take the magnitude of the input into account. This is defined as, in general, the BIBO stability of the system as described below:

> When a system is under excitation by an external input with bounded magnitude, it is called BIBO stable if the output of the system is also bounded.

Detailed mathematical description on BIBO stability will be provided later when we discuss the transfer function analysis of a system.

11.3 Stability of Multiple Degrees of Freedom Systems

In the previous section, we discussed the stability theory, especially stability definition and Lyapunov function approach. Now, we want to discuss the stability of linearized multiple degrees of freedom system which is the main outcome of finite element analysis. Understanding the stability property of the multiple degrees of

Section 11.3   Stability of Multiple Degrees of Freedom Systems

freedom system is important before we make any attempt to design a feedback control law for a system.

System without Damping

Using the Lyapunov stability theory, we want to analyze the stability of a linearized multiple degrees of freedom system. Let us consider an n dimensional finite dimensional dynamic system which is usually produced by finite element analysis. The governing equations of motion without damping are described by

$$[M]\{\ddot{\mathbf{q}}\} + [K]\{\mathbf{q}\} = [F]\{\mathbf{u}\} \tag{11.3.1}$$

where $[M]$ is the system mass matrix, $[K]$ is the stiffness matrix, $\{\mathbf{q}\}$ is the generalized coordinate vector, $[F]$ is the input influence matrix, and $\{\mathbf{u}\}$ is the control input vector. In case there is no forcing function, the free vibrational motion satisfies

$$[M]\{\ddot{\mathbf{q}}\} + [K]\{\mathbf{q}\} = 0 \tag{11.3.2}$$

and the solution to Eq. (11.3.2) is a pure sinusoidal motion as explained in Chapter 8

$$\{\mathbf{q}(t)\} = \sum_{k=1}^{n} c_k \{\phi_k\} e^{i\omega_k t} \tag{11.3.3}$$

where $\{\phi_k\}$ and ω_k are system parameters. The stability of Eq. (11.3.2) can be proved in various ways. One of them is the Lyapunov approach. Considering the fact that the total system energy (kinetic plus potential energy) is a direct indicator of system stability, a Lyapunov function candidate is suggested as

$$U = \frac{1}{2}\{\dot{\mathbf{q}}\}^T [M]\{\dot{\mathbf{q}}\} + \frac{1}{2}\{\mathbf{q}\}^T [K]\{\mathbf{q}\} \tag{11.3.4}$$

Note that the Lyapunov function is always positive ($U > 0$) since the mass and stiffness matrices satisfy

$$\{\mathbf{x}\}^T [M]\{\mathbf{x}\} > 0, \qquad \{\mathbf{x}\}^T [K]\{\mathbf{x}\} \geq 0 \tag{11.3.5}$$

for $\{\mathbf{x}\} \neq 0$, and they are symmetric ($[M] = [M]^T$ and $[K] = [K]^T$). Next, the time derivative of the Lyapunov function in Eq. (11.3.4) in conjunction with Eq. (11.3.1) yields

$$\begin{aligned} \dot{U} = \frac{dU}{dt} &= \{\dot{\mathbf{q}}\}^T ([M]\{\ddot{\mathbf{q}}\} + [K]\{\mathbf{q}\}) \\ &= \{\dot{\mathbf{q}}\}^T [F]\{\mathbf{u}\} \end{aligned} \tag{11.3.6}$$

Without the external forcing input, that is $\{\mathbf{u}\} = 0$, Eq. (11.3.6) becomes

$$\dot{U} = 0, \quad \text{or} \quad U = const \tag{11.3.7}$$

In other words, the energy is conserved, therefore the motion should be a pure sinusoidal type. It is important to note that the stability argument does not depend upon the system property itself. That is, the mass and stiffness matrices are dropped from the final expression of $\dot{U}$. This will be discussed again in the later part of this chapter when we deal with deriving a stabilizing control law for infinite dimensional systems.

If we want to design the input $\{\mathbf{u}\}$ so that the system is stabilized, then one possible solution will be to select $\{\mathbf{u}\}$ in such a way that

$$\dot{U} < 0 \qquad (11.3.8)$$

In other words, the energy decreases toward an equilibrium point with the judicious selection of the control input $\{\mathbf{u}\}$.

System with Damping

A linearized multiple degrees of freedom dynamic system with damping is described in the form

$$[M]\{\ddot{\mathbf{q}}\} + [D]\{\dot{\mathbf{q}}\} + [K]\{\mathbf{q}\} = [F]\{\mathbf{u}\} \qquad (11.3.9)$$

where $[D]$ is a nonnegative definite damping matrix. The above system is intuitively stable by the damping term introduced. In order to prove stability, we take a candidate Lyapunov function

$$U = \frac{1}{2}(\{\dot{\mathbf{q}}\}^T[M]\{\dot{\mathbf{q}}\} + \{\mathbf{q}\}^T[K]\{\mathbf{q}\}) \qquad (11.3.10)$$

As it is shown, the Lyapunov function form is the same as that of Eq. (11.3.4). It is not surprising that both expressions are identical considering the Lyapunov function represents total energy (kinetic and potential energies) in both cases. The time derivative of the Lyapunov function becomes

$$\dot{U} = \{\dot{\mathbf{q}}\}^T([M]\{\ddot{\mathbf{q}}\} + [K]\{\mathbf{q}\}) \qquad (11.3.11)$$

Using the governing equations of motion Eq. (11.3.9), we obtain

$$\dot{U} = \{\dot{\mathbf{q}}\}^T(-[D]\{\dot{\mathbf{q}}\} + [F]\{\mathbf{u}\}) \qquad (11.3.12)$$

With the external control input ignored($\{\mathbf{u}\} = 0$)

$$\dot{U} = -\{\dot{\mathbf{q}}\}^T[D]\{\dot{\mathbf{q}}\} \qquad (11.3.13)$$

Therefore, the time derivative of the Lyapunov function is a quadratic form in $[D]$ and $\{\dot{\mathbf{q}}\}$. If the damping matrix $[D]$ is positive definite with $\{\dot{\mathbf{q}}\}^T[D]\{\dot{\mathbf{q}}\} > 0$, then the system is asymptotically stable. In case the damping matrix is only semidefinite,

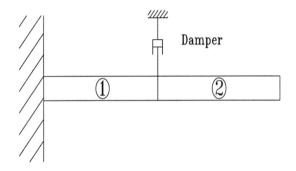

Figure 11.3.1 A Finite Element Beam Model with a Damper

i.e. $\{\dot{\mathbf{q}}\}^T[D]\{\dot{\mathbf{q}}\} \geq 0$, then the asymptotic stability of the system is not guaranteed. In this case, we can use another technique in order to prove the stability.

♣ **Example 11.3.1** Consider a finite element beam model in Fig. 11.3.1. The material properties are given as $EI = 1.112 \times 10^4$, $\rho = 0.003$, $l = 20$ with consistent units. A set of dampers is assumed at a nodal point in order to add damping to the system. The mathematical modeling of this system is given by

$$[M]\{\ddot{\mathbf{q}}\} + [D]\{\dot{\mathbf{q}}\} + [K]\{\mathbf{q}\} = 0$$

where

$$[M] = \begin{bmatrix} 0.0223 & 0 & 0.0039 & -0.0093 \\ 0 & 0.0571 & 0.0093 & -0.0214 \\ 0.0039 & 0.0093 & 0.0111 & -0.0157 \\ -0.0093 & -0.0214 & -0.0157 & 0.0286 \end{bmatrix}$$

$$[K] = 10^3 \times \begin{bmatrix} 0.2668 & 0 & -0.1334 & 0.6671 \\ 0 & 8.8946 & -0.6671 & 2.2236 \\ -0.1334 & -0.6671 & 0.1334 & -0.6671 \\ 0.6671 & 2.2236 & -0.6671 & 4.4473 \end{bmatrix}, \quad [D] = \begin{bmatrix} 0 & 0 & 0 & 0 \\ 0 & 0 & 0 & 0 \\ 0 & 0 & c & 0 \\ 0 & 0 & 0 & 0 \end{bmatrix}$$

where c is the damping coefficient of the damper. When the Lyapunov function is taken as the total energy of the system, the time derivative of the Lyapunov function becomes

$$\dot{U} = -\{\dot{\mathbf{q}}\}^T[D]\{\dot{\mathbf{q}}\}$$
$$= -c\dot{q}_3^2$$

As long as $\dot{q}_3 \neq 0$, $\dot{U} < 0$ and the system is asymptotically stable. Even if the damping matrix has zero diagonal values, the system is still asymptotically stable since $\dot{q}_3 \neq 0$ except for the equilibrium point.

‡

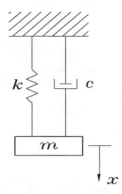

Figure 11.4.1 An Example of a Second Order System

11.4 Analysis of a Second Order System

A scalar second order system is frequently adopted as a reference explaining fundamental concepts of system responses. In fact, the majority of dynamical systems can be explained by using a scalar second order system. In this section, we want to introduce the natural frequency and damping ratio of a scalar system and key parameters associated with time responses. One of the most typical examples of the scalar second order system is a spring mass and damper system in Fig. 11.4.1. The governing equation of motion is given by

$$m\ddot{x} + c\dot{x} + kx = f(t) \tag{11.4.1}$$

where $f(t)$ is the external force applied to the mass. Dividing both sides by the mass yields

$$\ddot{x} + (c/m)\dot{x} + (k/m)x = f(t)/m \tag{11.4.2}$$

At this point, we define two parameters

$$\omega_n = \sqrt{k/m}, \quad \zeta = c/c_{cr} \tag{11.4.3}$$

where ω_n is the natural frequency and ζ is the damping ratio. In addition, $c_{cr} = 2\sqrt{mk}$ is defined as the critical damping ratio. As a consequence, Eq. (11.4.2) can be rewritten as

$$\ddot{x} + 2\zeta\omega_n\dot{x} + \omega_n^2 x = \omega_n^2 F(t) \tag{11.4.4}$$

where we used $f(t)/k \equiv F(t)$. In order to derive the solution, we use the Laplace transform technique ignoring initial conditions temporarily. This is also motivated by the fact that, for linear systems, the stability condition is independent of the initial conditions.

$$(s^2 + 2\zeta\omega_n s + \omega_n^2)X(s) = \omega_n^2 F(s) \tag{11.4.5}$$

Therefore,

$$X(s)/F(s) = \omega_n^2/(s^2 + 2\zeta\omega_n s + \omega_n^2) \tag{11.4.6}$$

Section 11.4 Analysis of a Second Order System

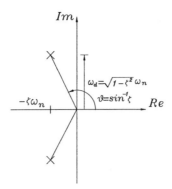

Figure 11.4.2 Characteristic Roots of the Second Order System

where $X(s) = L[x(t)]$ and $F(s) = L[F(t)]$ are Laplace transforms of $x(t)$ and $F(t)$, respectively. The equation obtained by setting the denominator of the transfer function equal to zero is called the *characteristic equation*. The characteristic equation of the above transfer function, therefore, is given by

$$s^2 + 2\zeta\omega_n s + \omega_n^2 = 0 \tag{11.4.7}$$

and the solution is given by

$$s_{1,2} = -\zeta\omega_n \pm \omega_n\sqrt{\zeta^2 - 1} \tag{11.4.8}$$

The above characteristic root is plotted on a complex plane in Fig. 11.4.2. The dynamic behavior is dependent upon the magnitude of the damping ratio ζ and the natural frequency ω_n.

As a special case, we analyze a step response of the system by selecting

$$F(t) = 1, \quad t > 0 \quad \text{or} \quad F(s) = 1/s \tag{11.4.9}$$

that is, *a unit step input*. By substituting $F(s)$ into the above equation, the output can be found by the inverse Laplace transform technique $x(t) = L^{-1}[X(s)]$. The result turns out to be dependent upon the magnitude of the damping ratio ζ.

i). *Underdamped case*, $0 < \zeta < 1$. In this case, $s_{1,2}$ are complex conjugate, and the motion turns out to be a damped oscillatory one.

$$x(t) = 1 - \frac{e^{-\zeta\omega_n t}}{\sqrt{1-\zeta^2}} sin(\omega_d t + \phi) \tag{11.4.10}$$

where $\phi = tan^{-1}\sqrt{(1-\zeta^2)}/\zeta$ and $\omega_d = \sqrt{(1-\zeta^2)}\omega_n$ is the damped natural frequency.

ii). *Critically damped case*, $\zeta = 1$. In this case, $s_{1,2}$ are repeated real numbers. The motion is monotonically increasing toward the steady state value

$$x(t) = 1 - e^{-\omega_n t}(1 + \omega_n t) \tag{11.4.11}$$

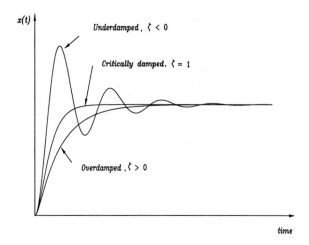

Figure 11.4.3 Step Responses with Different Damping Ratios

iii). Overdamped case, $\zeta > 1$. In this case $s_{1,2}$ are all negative real. The motion is also monotonically increasing toward the steady state

$$x(t) = 1 + \frac{\omega_n e^{-\zeta\omega_n t}}{2\sqrt{\zeta^2-1}} \left(\frac{e^{-(\sqrt{\zeta^2-1})\omega_n t}}{(\zeta + \sqrt{\zeta^2-1})\omega_n} - \frac{e^{(\sqrt{\zeta^2-1})\omega_n t}}{(\zeta - \sqrt{\zeta^2-1})\omega_n} \right) \qquad (11.4.12)$$

Three different motions are presented in Fig. 11.4.3. As seen in the figure, the usual response trends are exponentially decaying motions dictated by $e^{-\zeta\omega_n t}$ superimposed by damped sinusoidal motion with damped frequency, ω_d. In particular, the constant $\tau = 1/\zeta\omega_n$ is called the *time constant* which represents how long it takes for the response to reach a certain level from an initial condition. The shorter the *time constant* the quicker the response tends to be.

For an underdamped motion, some parameters are introduced characterizing the transient response of the motion. Those parameters are sometimes used to prescribe design specifications. In Fig. 11.4.4, different labels are used to denote those specifications.

i) Rise time, t_r:

The rise time is the required time for the response to start from zero value and cross the unit steady state value. The rise time is found from

$$x\Big|_{t=t_r} = 1 \qquad (11.4.13)$$

It turns out that

$$t_r = \frac{1}{\omega_d} tan^{-1}\left(-\frac{\sqrt{1-\zeta^2}}{\zeta}\right) \qquad (11.4.14)$$

Section 11.4   Analysis of a Second Order System

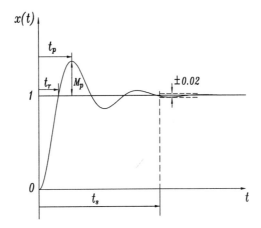

Figure 11.4.4 Key Parameters in the Step Response

ii) Peak time, t_p:

Peak time is the instance when the response reaches a maximum value. The peak time can be obtained as

$$\left.\frac{dx}{dt}\right|_{t=t_p} = 0 \tag{11.4.15}$$

which produces

$$t_p = \frac{\pi}{\omega_d} = \frac{\pi}{\omega_n\sqrt{1-\zeta^2}} \tag{11.4.16}$$

iii) Maximum overshoot, M_p:

The maximum overshoot represents the amount of maximum deviation of the response from the steady value. From Fig. 11.4.4, the maximum overshoot satisfies

$$M_p = x(t_p) - 1$$
$$= e^{-(\zeta/\sqrt{1-\zeta^2})\pi} \tag{11.4.17}$$

iv) Settling time, t_s:

The step response of the underdamped system experiences a transient response finally reaching a steady state value. The settling time represents the amount of time it takes for the response to stay within a certain band prescribed around the steady state value. The size of the band can be selected as, for example, 2% or 5%. For a 2% band, the settling time turns out to be approximated by the multiple of the time constant, τ.

$$t_s = 4\tau = \frac{4}{\zeta\omega_n} \tag{11.4.18}$$

♣ **Example 11.4.1** A second order system is given by

$$X(s)/F(s) = \frac{\omega_n^2}{s^2 + 2\zeta\omega_n s + \omega_n^2}$$

where $\omega=2.0$ (rad/sec) and $\zeta=0.2$. The unit step response results are calculated by calling the $fesecnd(\zeta,\omega_n)$ command.

$$[t_p, t_r, t_s, M_p] = [1.603,\quad 1.772,\quad 10,\quad 0.527]$$

```
function [t_p, t_r, t_s, M_p]=fesecnd(zeta, w_n)
%----------------------------------------------------------------
% Purpose:
%    The function subroutine fesecnd.m calculates dynamic characteristics
%    of a typical standard second order system.
%
%                        w_n^2
%          H(s)= ---------------------
%                s^2+2*zeta*w_n*s+w_n^2
%
% Synopsis:
%    [t_p, t_r, M_p, t_s]=fsecond(zeta, w_n)
%
% Variable Description:
%    Input parameters : zeta - damping ratio
%                       w_n - natural frequency
%    Output parameters : t_p - peak time,    t_r - rise time
%                        t_s - settling time,  M_p - maximum overshoot,
%----------------------------------------------------------------
w_d=sqrt(1-zeta^2)*w_n;        % Calculate undamped natural frequency
t_p=pi/w_d;                    % Calculate peak time
t_r=atan2(sqrt(1-zeta^2), -zeta);  % Calculate rise time
t_s=4/zeta/w_n;                % Calculate settling time
M_p=exp(-zeta*pi/sqrt(1-zeta^2));  % Calculate maximum overshoot
%----------------------------------------------------------------
```

‡

The same analysis and definition can be applied to multiple degree of freedom systems which may be generated by finite element analysis. From the original governing equation

$$[M]\{\ddot{\mathbf{q}}\} + [D]\{\dot{\mathbf{q}}\} + [K]\{\mathbf{q}\} = [F]\{\mathbf{u}\} \qquad (11.4.19)$$

and assuming Rayleigh damping with $[C] = \alpha[M] + \beta[K]$, we obtain the modal coordinate form governing equation

$$\ddot{\eta}_i + 2\zeta_i\omega_i\dot{\eta}_i + \omega_i^2\eta_i = f_i, \qquad i=1,2,\ldots,n \qquad (11.4.20)$$

Now for each modal coordinate parameters(ω_i, ζ_i) we can check the dynamic characteristics.

11.5 State Space Form Description

In general, the equations of motion of dynamic systems are described by second order differential equations. Finite element modeling of dynamic systems also results in second order differential equations of motion. The second order differential equations cover a generic class of dynamic systems. The analytical solution of second order equations of motion is essentially equivalent to solving ordinary differential equations.

On the other hand, the first order state space form description of dynamic systems has certain advantages over the second order form description. The second order equations can be transformed into first order equations and the first order forms also can be transformed into second order equations. The majority of existing computer software tools are written for the first order systems. This is due to the inherent nature of first order equations which are more convenient for numerical computations. Another significant advantage of the first order form descriptions is that we can analyze the equations in an explicit form.

Consider a linearized second order dynamic system

$$[M]\{\ddot{\mathbf{q}}\} + [D]\{\dot{\mathbf{q}}\} + [K]\{\mathbf{q}\} = [F]\{\mathbf{u}\} \qquad (11.5.1)$$

In order to write the above equation in the first order form we introduce a vector which is usually called the *state vector*

$$\{x\} = \left\{\begin{array}{c} \mathbf{q} \\ \dot{\mathbf{q}} \end{array}\right\} \qquad (11.5.2)$$

Now we have the following relationship

$$\begin{aligned}
\{\dot{x}\} = \frac{d}{dt}\left\{\begin{array}{c} \mathbf{q} \\ \dot{\mathbf{q}} \end{array}\right\} &= \left\{\begin{array}{c} \dot{\mathbf{q}} \\ -[M]^{-1}[D]\dot{\mathbf{q}} - [M]^{-1}[K]\mathbf{q} + [M]^{-1}[F]u \end{array}\right\} \\
&= \left[\begin{array}{cc} 0 & I \\ -[M]^{-1}[K] & -[M]^{-1}[D] \end{array}\right]\left\{\begin{array}{c} \mathbf{q} \\ \dot{\mathbf{q}} \end{array}\right\} + \left[\begin{array}{c} 0 \\ [M]^{-1}[F] \end{array}\right]u \quad (11.5.3) \\
&\equiv [A]\{x\} + [B]\{u\}
\end{aligned}$$

In other words, we rewrite the original second order differential equation in the first order form by introducing the *state vector* $\{x\}$. The *state vector* and associated properties constitute the so-called *state space*. The *state space* and related subjects are well described in Refs. [39-42].

The size of the first order system, however, increases by twofold compared to the original second order system. This may seem to be a drawback; however, modern computational capability resolves this concern to a considerable extent. As mentioned earlier, the first order form has certain advantages being adopted in the majority

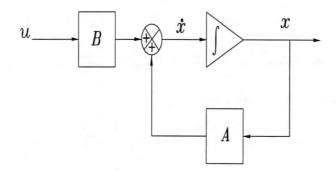

Figure 11.5.1 Graphical Representation of the State Space Equation

of engineering applications. Graphical representation of the first order state space equation is presented in Fig. 11.5.1.

Eigenvalue Problem and Free Response

Let us consider a case without an external input, that is $\{u\} = 0$, sometimes called the *autonomous system*. Thus, Eq. (11.5.3) reduces to

$$\{\dot{x}\} = [A]\{x\} \tag{11.5.4}$$

In order to find the solution $\{x(t)\}$ to the above equation, we assume

$$\{x(t)\} = ce^{\lambda t}\{\phi\} \tag{11.5.5}$$

where c is an arbitrary constant and $\{\phi\}$ is a vector of consistent size with $\{x(t)\}$. Then we substitute Eq. (11.5.5) into Eq. (11.5.4) arriving at

$$c(\lambda I - [A])e^{\lambda t}\{\phi\} = 0 \tag{11.5.6}$$

In other words,

$$(\lambda I - [A])\{\phi\} = 0 \tag{11.5.7}$$

Since $\{\phi\}$ should be a nonzero vector for the existence of a nontrivial solution, it should follow

$$|\lambda I - [A]| = 0 \tag{11.5.8}$$

The above equation can be solved for λ for the given system matrix $[A]$. There are n λs as the size of $[A]$. For an arbitrary i^{th} λ_i, we can rewrite Eq. (11.5.7) as

$$\lambda_i\{\phi_i\} = [A]\{\phi_i\} \tag{11.5.9}$$

Equation (11.5.9) is called the *eigenvalue problem*, which is a crucial concept in linear dynamic system analysis. Even if we eliminate the control input temporarily for the eigenvalue analysis, the eigenvalues and eigenvectors are used for stability analysis and computing mode shapes of the system.

For a given system, there may exist repeated eigenvalues, for example, a system with rigid body motion such as pure translational and rotational motions. This situation is treated by somewhat different approaches. Momentarily, the analysis is restricted to the case where the eigenvalues are all distinct. Let us write the eigenvalue problem for each index

$$\lambda_1\{\phi_1\} = [A]\{\phi_1\}$$
$$\lambda_2\{\phi_2\} = [A]\{\phi_2\}$$
$$\vdots \qquad (11.5.10)$$
$$\lambda_n\{\phi_n\} = [A]\{\phi_n\}$$

The above set of equations can be combined into a single matrix equation form

$$[\Gamma][U] = [U][A] \qquad (11.5.11)$$

where $[U]$ is a matrix whose columns consist of eigenvectors and $[\Gamma]$ is a diagonal matrix for which each diagonal term consists of eigenvalues.

$$[U] = [\{\phi_1\}, \{\phi_2\}, \ldots, \{\phi_n\}], \qquad [\Lambda] = diag[\lambda_i], \qquad i = 1, 2, \ldots \lambda_n \qquad (11.5.12)$$

Providing that the eigenvalues are all distinct, Eq. (11.5.11) turns into a relationship [39-42]

$$[A] = [U]^{-1}[\Lambda][U] \qquad (11.5.13)$$

That is, the matrix $[A]$ can be rewritten as a combination of a matrix which consists of eigenvectors and a diagonal matrix of eigenvalues. As discussed above, when there are repeated eigenvalues, we should use a modified form of equation. Equation (11.5.13) is named as a *similarity* transformation of A.

Note that the eigenvalue problem is invariant under the *similarity* transformation

$$|\lambda I - [U]^{-1}[\Lambda][U]| = |[U]^{-1}\lambda[U] - [U]^{-1}[\Lambda][U]|$$
$$= |[U]^{-1}||\lambda I - [\Lambda]||[U]|$$
$$= |\lambda I - [\Lambda]| \qquad (11.5.14)$$

where we used $|[U]| = 1/|[U]^{-1}|$.

As we might remember, the eigenvalue problem for second order systems in Chapter 8 can be similarly applied to a first order system in this case. The solution of the eigenvalue problem leads us to the analytical expression for $\{x(t)\}$

$$\{x(t)\} = c_1 e^{\lambda_1 t}\{\phi_1\} + c_2 e^{\lambda_2 t}\{\phi_2\} + \cdots + c_n e^{\lambda_n t}\{\phi_n\} \qquad (11.5.15)$$

The constants $(c_1, c_2, \cdots, c_n)$ are obtained from the initial condition. Equation (11.5.15) tells us that once we compute eigenvalues and eigenvectors, then we obtain the expression for the response with respect to the initial condition.

♣ **Example 11.5.1** Let us consider a finite element model for a beam with only one element as shown in Fig. 11.5.2. The mass and stiffness matrices for this system are

$$[M] = \frac{\rho h}{420}\begin{bmatrix} 156 & -22h \\ -22h & 4h^2 \end{bmatrix}, [K] = \frac{EI}{h^3}\begin{bmatrix} 12 & -6h \\ -6h & 4h^2 \end{bmatrix}$$

where $\rho = 0.002 kg/m$ is the linear mass density, $EI = 10\ Nm^2$ is the beam rigidity, and $h = 1\ m$ is the element length. The second order system is converted into the first order system in accordance with Eq. (11.5.3). The result is

$$[A] = \begin{bmatrix} 0 & 0 & 1.0000 & 0 \\ 0 & 0 & 0 & 1.0000 \\ 6.30 \times 10^4 & -4.8 \times 10^4 & 0 & 0 \\ 5.04 \times 10^5 & -3.69 \times 10^5 & 0 & 0 \end{bmatrix}$$

Now we have a MATLAB built-in function $[V,D]=eig([A])$ command, as explained in Chapter 1, in order to compute the eigenvalue and eigenvector of $[A]$. Consequently,

$$\lambda_{1,2} = \pm 550.35i, \qquad \lambda_{3,4} = \pm 55.86i$$

$$\{\phi_1\} = \begin{Bmatrix} 2.3635 \times 10^{-4} \\ 1.8016 \times 10^{-3} \\ 0.1301i \\ 0.9915i \end{Bmatrix}, \quad \{\phi_2\} = \begin{Bmatrix} 2.3635 \times 10^{-4} \\ 1.8016 \times 10^{-3} \\ -0.1301i \\ -0.9915i \end{Bmatrix}$$

$$\{\phi_3\} = \begin{Bmatrix} -1.0516 \times 10^{-2} \\ -1.4485 \times 10^{-2} \\ -0.5874i \\ -0.8091i \end{Bmatrix}, \quad \{\phi_4\} = \begin{Bmatrix} -1.0516 \times 10^{-2} \\ -1.4485 \times 10^{-2} \\ 0.5874i \\ 0.8091i \end{Bmatrix}$$

The time response of the system due to the initial condition can be written as

$$\{x(t)\} = c_1\{\phi_1\}e^{\lambda_1 t} + c_2\{\phi_2\}e^{\lambda_2 t} + c_3\{\phi_3\}e^{\lambda_3 t} + c_4\{\phi_4\}e^{\lambda_4 t}$$

In order to find the constants (c_1, c_2, c_3, c_4), we assume an initial condition as

$$\{x(0)\} = \begin{Bmatrix} 0.05 \\ -0.01 \\ 0.1 \\ 0.0 \end{Bmatrix}$$

Therefore, the constants are calculated as

$$c_1, c_2 = -26.719 \pm 8.4788 \times 10^{-2}i, \quad c_3, c_4 = -2.9779 \pm 0.1039i$$

Substituting the constants into $\{x(t)\}$ and using the famous Euler's formula as [39]

$$e^{i\theta} = cos(\theta) + isin(\theta)$$

we obtain the final form for $\{x(t)\}$ analytically

$$\{x(t)\} = \begin{Bmatrix} -1.2630 \times 10^{-2} \\ -9.6274 \times 10^{-2} \\ -2.2062 \times 10^{-2} \\ -1.6813 \times 10^{-1} \end{Bmatrix} cos(550.35t) + \begin{Bmatrix} -4.0079 \times 10^{-5} \\ -3.0551 \times 10^{-4} \\ 6.9523 \\ 5.2984 \times 10^{1} \end{Bmatrix} sin(550.35t) +$$

$$\begin{Bmatrix} 6.2673 \times 10^{-2} \\ 8.6328 \times 10^{-2} \\ 1.2206 \times 10^{-1} \\ 1.6813 \times 10^{-1} \end{Bmatrix} cos(55.86t) + \begin{Bmatrix} 2.1852 \times 10^{-3} \\ 3.0100 \times 10^{-3} \\ -3.5008 \\ -4.8221 \end{Bmatrix} sin(55.86t)$$

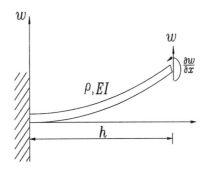

Figure 11.5.2 A Finite Element Model with a Single Element

‡

Forced Response and State Transition Matrix

The solution to the first order equation Eq. (11.5.3) can be also found by using the Laplace transform technique

$$sX(s) - \{x(0)\} = [A]X(s) + [B]U(s) \tag{11.5.16}$$

where $X(s) \equiv L[\{x(t)\}]$ is the Laplace transform of $\{x(t)\}$, and $U(s) \equiv L[\{u(t)\}]$ is the Laplace transform of $\{u(t)\}$. Alternatively,

$$X(s) = (sI - [A])^{-1}x(0) + (sI - [A])^{-1}[B]U(s) \tag{11.5.17}$$

In order to find the time response, we take the inverse Laplace transform of $X(s)$. First, the inverse Laplace transform of $(sI - [A])^{-1}$ should be evaluated. If a is a scalar, we know

$$L^{-1}[(s-a)^{-1}] = e^{at} \tag{11.5.18}$$

Generalizing Eq. (11.5.18) to a matrix $[A]$, we can derive a similar relationship as shown below.

$$L^{-1}[(sI - [A])^{-1}] = e^{[A]t} \tag{11.5.19}$$

Therefore,

$$\{x(t)\} = e^{[A]t}\{x(0)\} + \int_0^t e^{[A](t-\tau)}[B]\{u(\tau)\}d\tau \tag{11.5.20}$$

where the second term on the right-hand side represents a convolution integral. The response consists of two parts; one is due to the initial condition ($\{x(0)\}$) and the other one is due to the control input ($\{u\}$).

It is not as straightforward to understand $e^{[A]t}$ as the scalar case e^{at}. In order to analyze $e^{[A]t}$, first we introduce the Taylor series expansion.

$$e^{[A]t} = [I] + [A]t + \frac{[A]^2 t^2}{2!} + \frac{[A]^3 t^3}{3!} + \cdots \tag{11.5.21}$$

The first fundamental question is whether the infinite series converge or not. The answer is yes from a physical intuition; the factorial term in the denominator dominates the exponential term in the numerator. Efficient numerical algorithms have been developed and a MATLAB built-in routine *expm* is also available.

Another interesting property can be derived from Eq. (11.5.13) which calculates the exponential of a matrix

$$e^{[A]t} = [U]^{-1} e^{[\Lambda]t} [U] \qquad (11.5.22)$$

where

$$e^{[\Lambda t]} = diag[e^{\lambda_i t}], \qquad i = 1, 2, \ldots, n$$

The proof follows as

$$\begin{aligned}
e^{[A]t} &= [I] + [A]t + [A]^2 t^2/2! + [A]^3 t^3/3! + \cdots \\
&= [U]^{-1}[U] + ([U]^{-1}[\Lambda][U])t + ([U]^{-1}[\Lambda][U])([U]^{-1}[\Lambda][U])t^2/2! \\
&\quad + ([U]^{-1}[\Lambda][U])([U]^{-1}[\Lambda][U])([U]^{-1}[\Lambda][U])t^3/3! + \cdots \\
&= [U]^{-1}([I] + [\Lambda]t + [\Lambda]^2 t^2/2! + [\Lambda^3]t^3/3! + \cdots)[U] \\
&= [U]^{-1} e^{[\Lambda]t} [U] \qquad (11.5.23)
\end{aligned}$$

Therefore, the eigenvalue solution can be used again to calculate $e^{[A]t}$.

Let's go back to the first order equation without the control input term. Thus,

$$\{\dot{x}\} = [A]\{x\} \qquad (11.5.4)$$

In order to derive the solution, we assume

$$\{x(t)\} = [\Phi(t, \tau)]\{x(\tau)\} \qquad (11.5.24)$$

where $[\Phi(t, \tau)]$ is the so-called *state transition matrix* which relates the state variable at different instants, that is, $\{x(t)\}$ with $\{x(\tau)\}$ for $t > \tau$. Obviously, $[\Phi(t, \tau)]$ is a time varying matrix. Now substitute Eq. (11.5.24) into Eq. (11.5.4).

$$[\dot{\Phi}(t, \tau)]\{x(\tau)\} = [A][\Phi(t, \tau)]\{x(\tau)\} \qquad (11.5.25)$$

Therefore,

$$[\dot{\Phi}(t, \tau)] = [A][\Phi(t, \tau)] \qquad (11.5.26)$$

By solving Eq. (11.5.26) we can derive the analytical expression of $\{x(t)\}$. The above differential equation is combined with the initial condition of $[\Phi(t, \tau)]$ by noting that

$$\{x(\tau)\} = \Phi(\tau, \tau)\{x(\tau)\} \qquad (11.5.27)$$

In other words,

$$[\Phi(\tau, \tau)] = I \qquad (11.5.28)$$

and the combination of Eqs. (11.5.26) and (11.5.28) constitutes a matrix differential equation. Obviously, a numerical integration technique can be applied to the differential equation. There are other useful properties of $[\Phi(t, \tau)]$ [39-42].

$$\begin{aligned}
&(i) \quad [\Phi(t, \tau)] = [\Phi(t, t_2)][\Phi(t_2, \tau)], \qquad t < t_2 < \tau \\
&(ii) \quad [\Phi(t, \tau)] = [\Phi(\tau, t)]^{-1} \qquad\qquad\qquad\qquad\qquad (11.5.29)
\end{aligned}$$

Section 11.5 — State Space Form Description

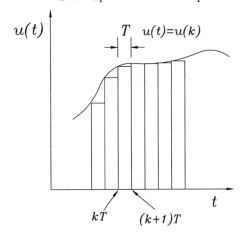

Figure 11.5.3 Graphical Representation of a Zero Order Hold(ZOH)

As is the case, when $[A]$ is a constant

$$[\Phi(t,\tau)] = e^{[A](t-\tau)} \qquad (11.5.30)$$

Equation (11.5.20) is generalized using the *state transition matrix*

$$\{x(t)\} = [\Phi(t,\tau)]\{x(\tau)\} + \int_\tau^t [\Phi(t,\xi)][B]\{u(\xi)\}d\xi \qquad (11.5.31)$$

The *state transition matrix* is a useful tool understanding a linear first order system. It also represents propagation of the initial condition which could be a perturbation due to external disturbance. In celestial mechanics, orbit perturbation phenomenon is analyzed quite often by the *state transition matrix* concept.

Time Response by Numerical Technique

Frequently, the time response of a system due to the initial condition or external control input is needed to analyze the behavior of the system. For nonlinear systems, numerical techniques are used, and for linear systems we can use other approaches. One of the useful techniques for linear system analysis and digital computation is to set the control input constant, called *zero order hold*, during a certain interval of time. The result becomes more accurate as the time interval, during which the control is set to be constant, decreases.

The control input, for instance, is set to be [39]

$$u(t) = u(k) = constant, \quad kT < t < (k+1)T \qquad (11.5.32)$$

where T is the sampling period, and $u(k)$ is the magnitude of control input between k^{th} and $(k+1)^{th}$ step as shown in Fig. 11.5.3. Let us assume the state vector is evaluated at k^{th} step. We are interested in evaluating the time response of the state vector at $(k+1)^{th}$ step by utilizing information at k^{th} step. Using Eq. (11.5.30)

$$\{x((k+1)T)\} = [\Phi((k+1)T, kT)]\{x(kT)\} + \int_0^T [\Phi(\xi,0)][B]\{u(k)\}d\xi \qquad (11.5.33)$$

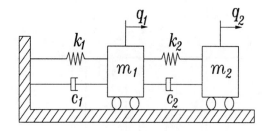

Figure 11.5.4 A Spring Mass and Damper System for Time Response Example

The state vector $[\Phi((k+1)T, kT)]$ is written as

$$[\Phi((k+1)T, kT)] = e^{[A]T} \qquad (11.5.34)$$

In addition

$$\int_0^T [\Phi(\xi, 0)]u(k)d\xi = [A]^{-1}[e^{[A]T} - I]u(k) \qquad (11.5.35)$$

Thus, Eq. (11.5.33) becomes

$$\{x(k+1)\} = e^{[A]T}\{x(k)\} + [A]^{-1}[e^{[A]T} - I][B]u_k = [\Phi]\{x(k)\} + [\Gamma]u(k) \qquad (11.5.36)$$

where we used simple notation $k+1 \equiv (k+1)T$ and $k \equiv kT$. Equation (11.5.36) is a discretized state space equation by holding the control input constant during each sampling period(T). The time response is easily computed by sequential substitution for the discrete form of the equation [39].

$$\begin{aligned}
\{x(1)\} &= [\Phi]\{x(0)\} + [\Gamma]\{u(0)\} \\
\{x(2)\} &= [\Phi]\{x(1)\} + [\Gamma]\{u(1)\} = [\Phi]^2\{x(0)\} + [\Phi][\Gamma]\{u(0)\} + [\Gamma]\{u(1)\} \\
\{x(3)\} &= [\Phi]^3\{x(0)\} + [\Phi]^2[\Gamma]\{u(0)\} + [\Phi][\Gamma]\{u(1)\} + [\Gamma]\{u(2)\} \\
&\vdots \\
\{x(n)\} &= [\Phi]^n\{x(0)\} + \sum_{j=0}^{n-1}[\Phi]^j[\Gamma]\{u(n-1-j)\}
\end{aligned} \qquad (11.5.37)$$

♣ **Example 11.5.2** In this example, we apply the time response formula to an example structural system. A spring, mass, and damper system is given in Fig. 11.5.4. The governing equation of motion for the system is given as

$$[M]\{\ddot{q}\} + [C]\{\dot{q}\} + [K]\{q\} = [F]\{u\}$$

where $[M]$, $[K]$ are 2 by 2 matrices, and $[F]$ is the input influence matrix given as

$$[M] = \begin{bmatrix} m_1 & 0 \\ 0 & m_2 \end{bmatrix} = \begin{bmatrix} 0.5 & 0 \\ 0 & 0.5 \end{bmatrix}$$

$$[C] = \begin{bmatrix} c_1+c_2 & -c_2 \\ -c_2 & c_2 \end{bmatrix} = \begin{bmatrix} 4 & -2 \\ -2 & 2 \end{bmatrix}$$

$$[K] = \begin{bmatrix} k_1+k_2 & -k_2 \\ -k_2 & k_2 \end{bmatrix} = \begin{bmatrix} 16 & -8 \\ -8 & 8 \end{bmatrix}$$

$$[F] = \begin{bmatrix} 1 & 0 \\ 0 & 1 \end{bmatrix}$$

The second order governing equation is transformed into a first order state space form. In other words,

$$\{\dot{x}\} = [A]\{x\} + [B]\{u\}$$
$$\{y\} = [C]\{x\} + [D]\{u\}$$

where $[A]$ and $[B]$ matrices are based upon Eq. (11.5.3). A MATLAB *m-file* *felresp.m* is written for this example. The *felresp.m* computes time responses by converting the original state space equation into a discrete equation by a *zero order hold* approximation for the control input. The initial condition vector is assumed as $\{x(0)\} = \begin{bmatrix} 1 & 0.1 & -2 & 2 \end{bmatrix}^T$. The external control inputs are prescribed as

$$u_1(t) = sin(10t), \quad u_2(t) = 3cos(10t)$$

Figure 11.5.5 presents the time response result by *felresp.m*. As we can see, the motion is oscillatory affected by the harmonic external input.

```
function [x,y]=felresp(A,B,C,D,x0,u,t)
%---------------------------------------------------------------
% Purpose:
%    find the time response of a linear system driven by initial condition
%    and external input. The numerical algorithm used in this program is
%    zero order hold approximation for control input for discretized system.
%
% Synopsis:
%    [x,y]=felresp(A,B,C,D,x0,u,t)
%
% Variable Description:
%    A, B, C, D; system matrices in
%
%    xdot = Ax + Bu, y = Cx + Du
%
%    x0; initial condition vector for the state variables
%    t ; integration time at equal distance as t=0:dt:tf
%    dt- time step, tf - final time
%    u ; control input vector with as many rows as the size of t
```

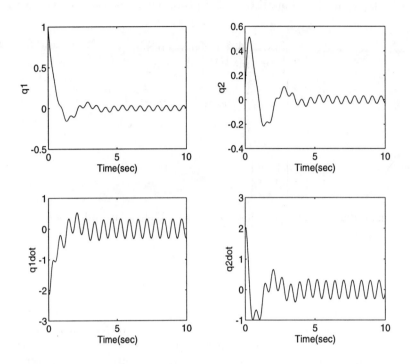

Figure 11.5.5 Time Response Result for Example 11.5.2

```
%     x(y) ; state(output) vector
%
%  Notes:
%     The control input vector must have as many columns as
%     the number of input
%------------------------------------------------------------
[n,m]=size(B);
%------------------------------------------------------------
% Transform into discrete equation by zero-holder approximation
%------------------------------------------------------------
Ts=t(2)-t(1);
Phi=expm(A*Ts);
Gamma=inv(A)*(Phi-eye(n))*B;
nc=max(size(t));
x=zeros(nc,n);
tx=zeros(n,1);
xi=x0;
tx=xi;
% Calculate time responses x first
for i=1:nc
x(i,:)=tx';
```

```
tx=Phi*tx+Gamma*U(i,:)';
end
% Calculate the output response by using y=Cx+Du
y=(C*x'+D*U')';
%------------------------------------------------
```

‡

11.6 Transfer Function Analysis

Transfer function analysis of linear dynamic systems is another useful tool analyzing the characteristics of the linear systems. This approach has been widely applied to various areas such as control system design, modal testing of structures, and so forth. The basic idea of transfer function analysis is to convert a system described in the time domain into the Laplace transform domain (s $domain$) so that the relationship between input and output is written as an algebraic expression of the Laplace transform variable.

Let us consider a state space representation of a dynamic system

$$\{\dot{x}\} = [A]\{x\} + [B]\{u\} \tag{11.6.1}$$
$$\{y\} = [C]\{x\} + [D]\{u\} \tag{11.6.2}$$

Performing Laplace transform on both sides of Eqs. (11.6.1) and (11.6.2) yields

$$sX(s) - \{x(0)\} = [A]X(s) + [B]U(s) \tag{11.6.3}$$
$$Y(s) = [C]X(s) + [D]U(s) \tag{11.6.4}$$

where $X(s) = L[\{x(t)\}]$ and $U(s) = L[\{u(t)\}]$, and L is the Laplace transform operator. Here, we drop the initial condition vector $\{x(0)\}$ since the dynamic characteristics of linear systems are independent of the initial condition. By collecting common terms, Eq. (11.6.3) becomes

$$X(s) = (sI - [A])^{-1}[B]U(s) \tag{11.6.5}$$

Substitution of Eq. (11.6.5) into Eq. (11.6.4) results in [41]

$$Y(s) = [[C](sI - [A])^{-1}[B] + [D]]U(s) = [H(s)]U(s) \tag{11.6.6}$$

It should be noted that $[H(s)]$ may be a matrix if the number of inputs and/or outputs is greater than one. As a special case of a single input and single output system, $[H(S)]$ becomes a scalar quantity representing the ratio between input and output as

$$H(s) = \frac{Y(s)}{U(s)} \tag{11.6.7}$$

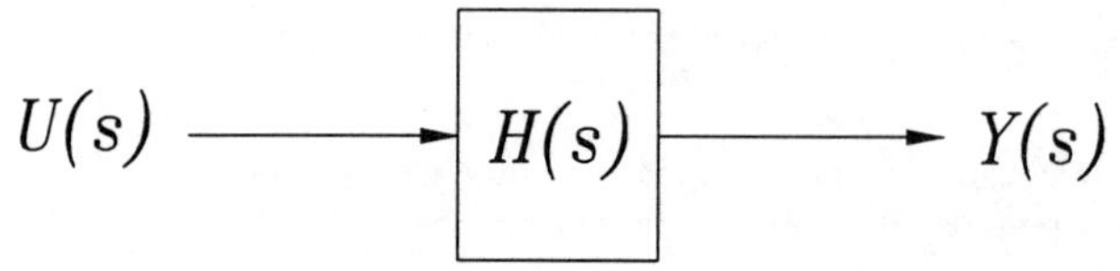

$$U(s) \longrightarrow \boxed{H(s)} \longrightarrow Y(s)$$

Figure 11.6.1 Representation of a Transfer Function using a Block Diagram

In other words, the transfer function relates the output($Y(s)$) to the given input ($U(s)$) in an algebraic expression. Figure 11.6.1 shows graphically the transfer function based upon the input-output relationship. This is a significant advantage over the original equation in the time domain, where the analytical expression for the input and output is not readily available.

For the open-loop transfer function described as

$$H(s) = \frac{N(s)}{D(s)} \qquad (11.6.8)$$

the solutions of $D(s) = 0$ are called *poles* and those of $N(s) = 0$ are called *zeros* of the system [39-42]. The *poles* determine the stability of the system and the *zeros* usually determine a time domain response shape. The *zeros* and *poles* of a given system in state space form can be obtained by a MATLAB built-in function *poly*. The detailed explanation for the *poly* command is provided in Chapter 1.

♣ **Example 11.6.1** Let us consider a finite element model of a Euler-Bernoulli beam as shown in Fig. 11.6.2. There is an actuator located at the tip of the beam. Applying the standard beam element using Hermite polynomials yields

$$[M]\{\ddot{\mathbf{q}}\} + [K]\{\mathbf{q}\} = [F]\{u\}$$

where $[M]$, $[K]$ are 4 by 4 matrices, and $[F]$ is the input influence matrix given as

$$[M] = \begin{bmatrix} 4.46 & 0.00 & 0.77 & -2.23 \\ 0.00 & 16.5 & 2.23 & -6.17 \\ 0.77 & 2.23 & 2.23 & -3.77 \\ -2.23 & -6.17 & -3.77 & 8.23 \end{bmatrix} \times 10^{-2}$$

and

$$[K] = \begin{bmatrix} 4.722 & 0.000 & -2.361 & 14.17 \\ 0.000 & 226.7 & -14.17 & 56.67 \\ -2.361 & -14.17 & 2.361 & -14.17 \\ 14.17 & 56.67 & -14.17 & 113.3 \end{bmatrix}, \quad [F] = \begin{bmatrix} 0 \\ 0 \\ 0 \\ 1 \end{bmatrix}$$

The second order governing equation is transformed into a first order state space form. In other words,

$$\{\dot{x}\} = [A]\{x\} + [B]\{u\}$$
$$\{y\} = [C]\{x\} + [D]\{u\}$$

Section 11.6 Transfer Function Analysis

Table 11.6.1 Coefficients of Numerator Matrices

Output variables	Coefficients of the numerator
v_1	$[0 \quad 21.10 \quad 0 \quad -6.287 \times 10^4 \quad 0 \quad -1.292 \times 10^8 \quad 0 \quad 1.228 \times 10^{10}]$
θ_1	$[0 \quad 14.55 \quad 0 \quad -7.509 \times 10^4 \quad 0 \quad 1.656 \times 10^7 \quad 0 \quad 1.841 \times 10^9]$
v_2	$[0 \quad 0 \quad 228.0 \quad 0 \quad 1.303 \times 10^6 \quad 0 \quad 8.556 \times 10^8 \quad 3.928 \times 10^{10}]$
θ_2	$[0 \quad 0 \quad 121.1 \quad 0 \quad 3.825 \times 10^5 \quad 0 \quad 1.295 \times 10^8 \quad 0 \quad 2.455 \times 10^9]$

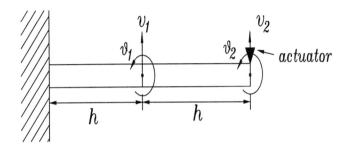

Figure 11.6.2 A Finite Beam Model with an Actuator Input

where $[A]$ and $[B]$ matrices are generated from $[M], [K]$, and $[F]$. A MATLAB m-file *festotf.m* is written which converts a state space form into a transfer function. The transfer functions are found between the actuator input and different outputs. The denominator polynomial $D(s)$ is given by

$$D(s) = s^8 + 1.101 \times 10^4 s^6 + 1.242 \times 10^7 s^4 + 1.174 \times 10^9 s^2 + 2.898 \times 10^9$$

The coefficients of the numerator matrix are provided in Table 11.6.1.

```
function [Num, Den] = festotf(A,B,C,D,iu)
%-----------------------------------------------------------
% Purpose:
%    The function subroutine fsstotf.m converts a state space form of
%    system into a transfer function form.
%
%    For a given system
%
%           xdot = Ax+Bu      y = Cx+Du
%
%    The transfer function becomes
%
%              N(s)          -1
%    H(s) = ---------  = C(sI-A) B + D
%              D(s)
```

```
%
% Synopsis:
%    [Num,Den]=festotf(A,B,C,D,iu)
%
% Variable Description:
%    Input parameters -
%       System matrices [A,B,C,D]
%       iu - Index for control input(iu-th input)
%    Output parameters -
%       D(s) - Vector of coefficients of the denominator polynomial
%       N(s) - Vector of coefficients of the numerator polynomials
%    Note -
%       There are same number of rows in N(s) as the number of output
%---------------------------------------------------------------
Den = poly(A);              % Determine denominator polynomial
B = B(:,iu);                % Select the corresponding column
D = D(:,iu);
[m,n] = size(C);
Num = ones(m, n+1);
for i=1:m
Num(i,:) = poly(A-B*C(i,:)) + (D(i) - 1) * Den;
end
%---------------------------------------------------------------
```

‡

From Eq. (11.6.5), $(sI-[A])^{-1}$ plays a key role determining stability and other dynamic characteristics of the system. According to a linear algebra theory, it is rewritten as

$$(sI-[A])^{-1} = \frac{Adj(sI-[A])}{|sI-[A]|} \tag{11.6.9}$$

where $|sI-[A]|$ represents the determinant of $(sI-[A])$, which should be an nth order polynomial in s

$$|sI-[A]| = s^n + a_1 s^{n-1} + a_2 s^{n-2} + \cdots + a_{n-1}s + a_n \tag{11.6.10}$$

and $Adj(\ )$ denotes the *adjoint* of a matrix. The equation $|sI-[A]|=0$ is called the *characteristic equation*. Equation (11.6.10) also can be expressed as

$$|sI-[A]| = (s-\lambda_1)(s-\lambda_2)\cdots(s-\lambda_n) \tag{11.6.11}$$

and the solutions of the *characteristic equation* are

$$s = \lambda_1,\ \lambda_2,\ \ldots, \lambda_n \tag{11.6.12}$$

where λ_i is called the ith eigenvalue or characteristic root of the system.

The transfer function can also be written as a combination of each first order term in s called *partial fraction expansion* [39-42]. That is

$$H(s) = \frac{c_1}{(s - \lambda_1)} + \frac{c_2}{(s - \lambda_2)} + \cdots + \frac{c_n}{(s - \lambda_n)} \qquad (11.6.13)$$

where the coefficient c_i can be determined from

$$c_i = H(s)(s - \lambda_i)\Big|_{s=\lambda_i} \qquad (11.6.14)$$

Here we consider a unit impulsive input. The impulsive input is a popular choice in modal testing and structural system analysis. Mathematically, it is represented in terms of the *Dirac Delta* function. For example, if the input is an impulse input with unit magnitude, then

$$u(t) = \delta(t - t_0) \qquad (11.6.15)$$

for which

$$\begin{cases} \int_{-\infty}^{\infty} \delta(t - t_0) dt = 1, & t = t_0 \\ \delta(t - t_0) = 0, & t \neq t_0 \end{cases} \qquad (11.6.16)$$

The Laplace transform of the unit impulse function at $t_0 = 0$ is unity, that is $U(s) = 1$. Therefore, the output is equal to the transfer function itself. Thus,

$$\begin{aligned} Y(s) &= H(s)U(s) \\ &= H(s) \end{aligned} \qquad (11.6.17)$$

Now the time domain response for $Y(s)$ or $H(s)$ can be obtained by taking the inverse Laplace transform of $Y(s)$. Based upon the expression in Eq. (11.6.13), the response is expressed as

$$y(t) = L^{-1}[Y(s)] = c_1 e^{\lambda_1 t} + c_2 e^{\lambda_2 t} + \cdots + c_n e^{\lambda_n t} \qquad (11.6.18)$$

Equation (11.6.18) is sometimes called the *impulse response* by the nature of the applied impulsive input. It is obvious from the expression in Eq. (11.6.18) that the response is stable if

$$Re[\lambda_i] < 0, \quad \text{for,} \quad i = 1, 2 \ldots n \qquad (11.6.19)$$

where $Re[\lambda_i]$ denotes a real part of λ_i. In other words, the solution of the characteristic equation should have negative real parts for stability.

When the input applied is a step function as

$$u(t) = U_0 \qquad (11.6.20)$$

the system output becomes

$$Y(s) = H(s)U_0 \frac{1}{s} \qquad (11.6.21)$$

The output $y(t)$ created by the step input is called the *step response*

$$y(t) = L^{-1}[Y(s)]$$
$$= L^{-1}[H(s)U_0 \frac{1}{s}] \qquad (11.6.22)$$

Bounded Input Bounded Output (BIBO) Stability

At this point, we want to go back to the stability discussion of BIBO stability which was introduced in Sec. 11.2. For simplicity, we take a single-input and single-output system. From Eq. (11.6.6), the time domain solution for the input and output is given by the convolution integral

$$y(t) = \int_0^t h(t-\tau)u(\tau)d\tau \qquad (11.6.23)$$

where $h(t) = L^{-1}[H(s)]$ is the *impulse response* of the system. In case the input to the system is bounded

$$|u(\tau)| < M \qquad (11.6.24)$$

Then the output equation satisfies

$$|y(t)| = |\int_0^t h(t-\tau)u(\tau)d\tau|$$
$$\leq \int_0^t |h(t-\tau)u(\tau)|d\tau$$
$$\leq \int_0^t |h(t-\tau)||u(\tau)|d\tau \qquad (11.6.25)$$
$$\leq M \int_0^t |h(t-\tau)|d\tau$$

Thus, the output of the system is bounded if $\int_0^t |h(t-\tau)|d\tau$ is bounded [38]. In particular, the stability is based upon the steady-state condition, and the output is bounded at steady-state when

$$\int_0^\infty |h(\infty-\tau)|d\tau < N \qquad (11.6.26)$$

Since the impulse response ($h(t)$) of the system depends upon poles or characteristic roots of the system transfer function, the system BIBO stability depends upon the system poles and the magnitude of the input which must be bounded.

Basic Concept of Feedback Control

The basic feedback control concept in frequency domain is represented in terms of a block diagram in Fig. 11.6.3. It is represented by the transfer function description using Laplace transform. The system is under three different external

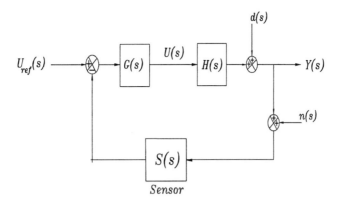

Figure 11.6.3 Feedback Control Block Diagram

inputs: reference command input $U_{ref}(s)$, disturbance $d(s)$, and sensor noise $n(s)$. The three inputs affect the system behavior in combination. The output ($Y(s)$) of the system is measured by a sensor, and the output from the sensor is compared to the reference input. The compared signal called *error signal* is fed to the actuator, and the actuator applies corrective signal to the system. By the actuator signal, we hope the error signal goes to zero asymptotically. Here we can derive a relationship between the reference input $U_{ref}(s)$ and the output $Y(s)$. Note

$$E(s) = U_{ref}(s) - Y(s) \tag{11.6.27}$$

Since

$$\begin{aligned} Y(s) &= H(s)G(s)E(s) \\ &= H(s)G(s)(U_{ref}(s) - Y(s)) \end{aligned} \tag{11.6.28}$$

then,

$$Y(s)/U_{ref}(s) = \frac{H(s)G(s)}{1 + G(s)H(s)} \tag{11.6.29}$$

Similarly, we can derive the transfer functions such as $Y(s)/d(s)$ and $Y(s)/n(s)$.

$$Y(s)/d(s) = \frac{1}{1 + G(s)H(s)} \tag{11.6.30}$$

$$Y(s)/n(s) = \frac{-G(s)H(s)}{1 + G(s)H(s)} \tag{11.6.31}$$

Note that as we increase the feedback gain $H(s)$, the output due to disturbance decreases while the output due to noise increases. This is a typical aspect of a feedback control law with a varying feedback gain. The feedback control law produces a new transfer function and corresponding characteristic equation [40]

$$1 + G(s)H(s) = 0 \tag{11.6.32}$$

The new transfer function is called a *closed-loop transfer function* and the corresponding system is called a *closed-loop system*. As discussed earlier, the solution of the characteristic equation determines the stability of a system and the controller block($G(s)$) is designed so that the closed-loop system becomes stable. Obviously, the closed-loop system stability depends on not only $G(s)$ but also $H(s)$. Depending on the nature of $H(s)$, it is sometimes easy or difficult to design a stabilizing controller $G(s)$.

Also, an important performance criterion in the transfer function description of a system is the *error* defined as the difference between the reference input and the actual output

$$E(s) = U_{ref}(s) - Y(s) \tag{11.6.33}$$

which follows as

$$E(s) = U_{ref}(s) - \frac{G(s)H(s)}{1+G(s)H(s)} U_{ref}(s)$$

$$= \frac{1}{1+G(s)H(s)} U_{ref}(s) \tag{11.6.34}$$

Similarly, the error due to disturbance and measurement noise is represented as

$$E(s)/d(s) = \frac{-1}{1+G(s)H(s)} \tag{11.6.35}$$

$$E(s)/n(s) = \frac{G(s)H(s)}{1+G(s)H(s)} \tag{11.6.36}$$

The steady-state error can be obtained by the *final value theorem* of the Laplace transform technique.

$$e(\infty) = \lim_{s \to 0} sE(s) \tag{11.6.37}$$

One thing interesting in Eq. (11.6.34) is that the size of $G(s)H(s)$ increases as the size of the error signal decreases. This is a typical aspect of a control system using a high feedback gain for performance improvement.

Proportional plus Derivative Control Law

One of the popular classical control law techniques is the Proportional plus Derivative (PD) control law. The controller $G(s)$ uses error signal and provides command input to the system, and the input signal is a combination of proportional plus derivative of the error signal. That is,

$$G(s) = K_p(\tau s + 1) \tag{11.6.38}$$

where the operator s represents a derivative operator in Laplace transform as $L[\dot{e}(t)] = sE(s)$, and K_p is a constant. The constant K_p term contributes to eliminating a steady state error and τ contributes to a better transient response. For example, a spring mass system is given by

$$m\ddot{x} + kx = f(t) \tag{11.6.39}$$

or in transfer function form

$$X(s)/F(s) = H(s) = \frac{1}{s^2 + \omega_n^2} \tag{11.6.40}$$

Section 11.6 Transfer Function Analysis 459

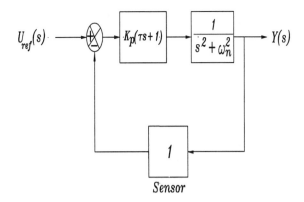

Figure 11.6.4 Proportional plus Derivative(PD) Control

where $F(s) = f(s)/m$ and $\omega_n = (k/m)^{1/2}$. The open-loop characteristic equation follows as

$$s^2 + \omega_n^2 = 0 \tag{11.6.41}$$

The solutions are $s = \pm\omega_n i$ and the system is neutrally stable or oscillatory. The response never reaches a steady-state and it does not track input command. Now we add a PD controller as in Fig.11.6.4, and the closed-loop characteristic equation becomes

$$1 + K_p(\tau s + 1)\frac{1}{s^2 + \omega_n^2} = 0 \tag{11.6.42}$$

Thus,

$$s^2 + K_p \tau s + (K_p + \omega_n^2) = 0 \tag{11.6.43}$$

The closed-loop system poles are governed by the parameters (τ, K_p) of the controller. The parameters directly control the closed-loop transient response through damping ratio and natural frequency.

$$\bar{\omega}_n^2 = K_p + \omega_n^2, \qquad 2\zeta\bar{\omega}_n = K_p \tau \tag{11.6.44}$$

where $\bar{\omega}_n$ and ζ are the desired natural frequency and damping ratio, respectively. Also, we consider a *step-input* response of the system with $U_{ref}(s) = 1/s$. Thus,

$$Y(s) = \frac{G(s)H(s)}{1 + G(s)H(s)}\frac{1}{s} \tag{11.6.45}$$

According to the Laplace *final value theorem*,

$$\lim_{t \to \infty} y(t) = \lim_{s \to 0} sY(s) \tag{11.6.46}$$

In this case, Eq.(11.6.29) yields

$$y(\infty) = \lim_{s \to 0} \frac{G(s)H(s)}{1 + G(s)H(s)}$$

$$= \lim_{s \to 0} \frac{K_p(\tau s + 1)}{s^2 + K_p \tau s + (\omega_n^2 + K_p)} = 1 \tag{11.6.47}$$

Now the steady state error, using Eq. (11.6.34), becomes

$$e(\infty) = \lim_{s \to 0} sE(s)$$

$$= \lim_{s \to 0} \frac{s(s^2 + \omega_n^2)}{s^2 + K_p \tau s + (K_p \tau + \omega_n^2)} = 0 \qquad (11.6.48)$$

In other words, by introducing a PD controller, the steady state output tracks the input command with zero steady state error.

There are other control actions such as integral and proportional plus integral control actions. The integral control action is generally known to improve steady-state performance by eliminating steady error due to external disturbances.

Proportional plus Integral Control Law

The Proportional plus Integral (PI) control law is useful in eliminating the source of external disturbances. The PI control law is represented by the following form

$$G(s) = K_p(1 + \tau_I/s) \qquad (11.6.49)$$

where K_p and τ_I are constants. That is, the control signal is developed as a combination of the proportional and integral of error signal. In order to examine the PI control law we use the same spring mass system used for the PD control law. Therefore,

$$G(s)H(s) = K_p(1 + \tau_I/s)\frac{1}{s^2 + \omega_n^2} \qquad (11.6.50)$$

We assume a situation where the system is under a constant disturbance of magnitude W so that

$$d(s) = W/s \qquad (11.6.51)$$

Now the error signal due to the disturbance is given as

$$E(s) = \frac{-1}{1 + G(s)H(s)} d(s)$$

$$= \frac{-(s^2 + \omega_n^2)s}{s^3 + (\omega_n^2 + K_p)s + K_p \tau_I} \frac{W}{s} \qquad (11.6.52)$$

Application of the *final value theorem* to $E(s)$ yields

$$e(\infty) = \lim_{s \to 0} sE(s)$$

$$= \lim_{s \to 0} \frac{-Ws(s^2 + \omega_n^2)}{s^3 + (\omega_n^2 + K_p)s + K_p \tau_I} = 0 \qquad (11.6.53)$$

Therefore, the PI control law achieves zero steady-state error in spite of the external disturbance. Note that if the control law is a PD control law, then the steady-state error is a nonzero value. For disturbances such as linearly changing and parabolic types, the integration order of the PI control law should change consequently.

11.7 Control Law Design for State Space Systems

In the previous section, we discussed the state space form formulation of dynamic systems, for example, finite element modeling of structural systems. There are certain advantages in the state space form approach compared to its counterpart which is the frequency domain technique. In this section, control law design issues based upon the state space form of dynamic systems are discussed. Different control laws are introduced accompanied by examples programmed in MATLAB. Since the majority of modern computational tools in MATLAB *Toolboxes* are written for the state space form of equations, we elect to put more emphasis on the state space representation of systems.

For a typical linear system, we start with

$$\{\dot{x}\} = [A]\{x\} + [B]\{u\}$$

The open-loop stability of the above system without external input is determined by

$$|\lambda I - [A]| = 0$$

The solution to the above equation does not always ensure stability of the system, and it is our goal to design the control input $\{u\}$ so that the desired behavior of the system can be achieved. Before we discuss the control law design, the important *Controllability* definition of dynamic systems is discussed.

Controllability of System

Basically, the controllability represents the ability of a control input to control or change all the state variables of a system. An example case is an actuator located at the nodal point of a specific mode of a flexible structure. In this case, the particular mode is not controllable by the actuator. For flexible structures, the actuator location, therefore, is a significant factor for controllability.

It is important to design a control system under a condition where a pole is cancelled by a zero in a transfer function. In this case we lose the cancelled pole, under the same principle as losing controllability in a state space equation. The order of a transfer function is directly related to the number of state variables in a state space form representation and the reduced order of the transfer function also reduces the order of the state space form equation. The precise definition of controllability is as follows.

Definition of Controllability

The system is controllable if there exists a control input $\{u(t)\}$ and time t_f by which an arbitrary $\{x(t_f)\}$ can be reached from $\{x(t_0)\}$ with $t_0 < t \leq t_f$.

In a mathematical theorem, it is stated as:

Theorem : The controllability condition for a given linear system

$$\{\dot{x}\} = [A]\{x\} + [B]\{u\} \tag{11.7.1}$$

is prescribed by the condition:

$$[G_c(t_f, t_0)] = \int_{t_0}^{t_f} e^{[A](t_f-\tau)}[B][B]^T e^{[A]^T(t_f-\tau)} d\tau \qquad (11.7.2)$$

should be positive definite [40].

The above matrix, $[G_c(t_f, t_0)]$, is called the *Controllability Grammian*. In order to verify the above theorem, we start with the solution

$$\{x(t_f)\} = [\Phi(t_f, t_0)]\{x(t_0)\} + \int_{t_0}^{t_f} [\Phi(t, \tau)][B]\{u(\tau)\} d\tau \qquad (11.7.3)$$

The control input $\{u(\tau)\}$ can be rewritten as

$$\{u(\tau)\} = [B]^T [\Phi(t_f, \tau)]^T [G_c(t_f, t_0)]^{-1} [\{x(t_f)\} - \Phi(t_f, t_0)\{x(t_0)\}] \qquad (11.7.4)$$

This is verified by plugging $\{u(\tau)\}$ into Eq.(11.7.3). Hence,

$$\{x(t_f)\} = [\Phi(t_f, t_0)]\{x(t_0)\} +$$
$$\int_{t_0}^{t_f} [\Phi(t_f, \tau)][B][B]^T [\Phi(t_f, \tau)]^T d\tau \left[[G_c]^{-1}\{x(t_f)\} - [G_c]^{-1}[\Phi(t, t_0)]\{x(t_0)\} \right]$$
$$= \{x(t_f)\} \qquad (11.7.5)$$

It is easily shown that the *Controllability Grammian* should be positive definite in order to guarantee the existence of $\{u(\tau)\}$. In other words, $[G_c]^{-1}$ must exist, and $[G_c]$ is naturally symmetric.

Without loss of generality, we take $t_f = \infty$. In this case the *Controllability* is tested throughout the steady-state. The *Controllability Grammian*, therefore, turns into

$$[G_c(\xi)] = \int_0^\infty e^{[A]\xi}[B][B]^T e^{[A]^T \xi} d\xi \qquad (11.7.6)$$

where $\xi = t_f - \tau$. We can show that $[G_c(\xi)]$ satisfies

$$[A][G_c] + [G_c][A]^T + [B][B]^T = 0 \qquad (11.7.7)$$

The above equation is called the *Lyapunov equation*. In order to prove the above equation we take

$$\frac{d}{d\xi} e^{[A]\xi}[B][B]^T e^{[A]^T \xi} = [A] e^{[A]\xi}[B][B]^T e^{[A]^T \xi} + e^{[A]\xi}[B][B]^T e^{[A]^T \xi} [A]^T \qquad (11.7.8)$$

Integrating both sides over $[0, \infty]$ yields

$$e^{[A]\xi}[B][B]^T e^{[A]^T \xi} \Big|_0^\infty = [A] \int_0^\infty e^{[A]\xi}[B][B]^T e^{[A]^T \xi} d\xi + \int_0^\infty e^{[A]\xi}[B][B]^T e^{[A]^T \xi} d\xi A^T$$
$$(11.7.9)$$

Section 11.7 Control Law Design for State Space Systems

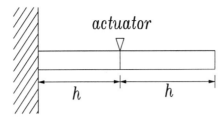

Figure 11.7.1 Example Model for Controllability Test

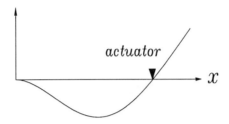

Figure 11.7.2 An Actuator Located at the Nodal Point

Assuming that $[G_c]$ has a bounded value, we arrive at the final expression

$$-[B][B]^T = [A][G_c] + [G_c][A]^T \qquad (11.7.10)$$

In order to solve the above equation, we can use a MATLAB command *lyap.m*. It is not easy to compute $[G_c]$ even if we use the *Lyapunov equation*. There is an alternative form of controllability condition. That is, the rank of a *controllability matrix*

$$[P] = [[B], [A][B], [A]^2[B], \ldots, [A]^{n-1}[B]] \qquad (11.7.11)$$

should be the same as the order of the system [39-42]. The rank test is more attractive than computing the *Controllability Grammian* in the sense that we avoid the numerical computational work.

♣ **Example 11.7.1** In this example, we test the controllability of a given dynamic system. A MATLAB *m-file fectobt.m* is written for this purpose. A finite element model and an assumed set of sensors and actuators are shown in Fig. 11.7.1. Using a standard beam element, both mass and stiffness matrices are generated.

$$[M] = \begin{bmatrix} 0.093 & 0.000 & 0.016 & -0.097 \\ 0.000 & 1.488 & 0.097 & -0.558 \\ 0.016 & 0.097 & 0.046 & -0.164 \\ -0.097 & -0.558 & -0.164 & 0.744 \end{bmatrix}$$

$$[K] = \begin{bmatrix} 0.522 & 0.000 & -0.261 & 3.264 \\ 0.000 & 108.8 & -3.264 & 27.20 \\ -0.261 & -3.264 & 0.261 & -3.264 \\ 3.264 & 27.20 & -3.264 & 54.40 \end{bmatrix}, \qquad [F] = \begin{bmatrix} 1 \\ 0 \\ 0 \\ 0 \end{bmatrix}$$

For simplicity, a modal truncation technique is used including only the first two flexible modes.

$$\ddot{\eta}_1 + 0.135\eta_1 = -1.359u$$
$$\ddot{\eta}_2 + 5.373\eta_2 = -2.916u$$

The above set of equations is transformed into the first order form with the result

$$[A] = \begin{bmatrix} 0 & 0 & 1 & 0 \\ 0 & 0 & 0 & 1 \\ -0.134 & 0 & 0 & 0 \\ 0 & -5.373 & 0 & 0 \end{bmatrix}, \quad [B] = \begin{bmatrix} 0 \\ 0 \\ -1.359 \\ -2.916 \end{bmatrix}$$

The controllability matrix is calculated as

$$[P] = \begin{bmatrix} 0 & -1.359 & 0 & 0.184 \\ 0 & -2.916 & 0 & 15.668 \\ -1.359 & 0 & 0.184 & 0 \\ -2.916 & 0 & 15.668 & 0 \end{bmatrix}$$

and the rank of the $[P]$ matrix turns out to be equal to the order of the system. Therefore, the given system is controllable. The MATLAB *m-file* source program for this example is provided below. It produces yes/no type answers and the condition number of the controllability matrix. The condition number is an index which is equal to unity for an identity matrix and very large for a singular or a near singular matrix.

```
function [Ctobty,rrank,ccond]=fectobt(A,B)
%----------------------------------------------------------------
%  Purpose:
%     The function subroutine fectobt.m calculates controllability matrix
%     and/or observability of a system described in state space form
%
%              xdot = Ax + Bu
%
%  Synopsis:
%     [Ctobty,rrank,ccond]=fectobt(A,B)
%
%     i) For controllability test, the input argument should follow as
%
%              fctobty(A,B)
%
%     ii) For observability test, we should provide the input argument as
%
%              fctobty(A^T, C^T) : ( )^T is transpose of ( )
%
%  Variable Description:
%     Output parameters - Ctobty : Controllability or observability matrix
```

```
%                       rrank : rank of Ctobty which determines yes or no
%                       ccond : Condition number of Ctobty
%----------------------------------------------------------------
n=max(size(A));              % Find out the size of the system matrix
%----------------------------------------------------------------
% Build the controllability/observability matrix (Ctobty)
%----------------------------------------------------------------
Ctobty=B;
Ao=A;
for i=1:n-1
Ctobty=[Ctobty Ao*B];
Ao=Ao*A;
end
rrank=rank(Ctobty);
ccond=cond(Ctobty);
%----------------------------------------------------------------
```

‡

♣ **Example 11.7.2**   The system was controllable in the previous example due to the location of the actuator. Therefore the actuator controls at least the first two flexible modes. In this example, we examine an uncontrollable case by selecting a specific actuator location. This is possible by precalculating the nodal point of a flexible mode. Figure 11.7.2 presents the actuator location which is at the nodal point of the second flexible mode. The governing equations consequently are given as

$$\ddot{\eta}_1 + 0.135\eta_1 = 2.816u$$
$$\ddot{\eta}_2 + 6.789\eta_2 = -0.411u$$

The size of the coefficient in front of the control input for the second mode is relatively small (0.411) compared to that of the first mode (2.816). This is because of the actuator located at the second mode nodal point. Theoretically, this coefficient should be equal to zero. The numerical inaccuracy is due to the number of elements which is equal to two and the finite element modeling algorithm using Hermite polynomials.

The above set of equations is transformed into the first order form with the result

$$[A] = \begin{bmatrix} 0 & 0 & 1 & 0 \\ 0 & 0 & 0 & 1 \\ -0.134 & 0 & 0 & 0 \\ 0 & -6.789 & 0 & 0 \end{bmatrix}, \quad [B] = \begin{bmatrix} 0 \\ 0 \\ 2.816 \\ -0.411 \end{bmatrix}$$

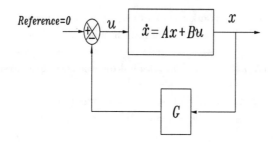

Figure 11.7.3 Feedback Control Concept for State Space Equation

The controllability matrix is calculated as

$$[P] = \begin{bmatrix} 0.000 & 2.816 & 0.000 & -0.377 \\ 0.000 & -0.411 & 0.000 & 2.790 \\ 2.816 & 0.000 & -0.377 & 0.000 \\ -0.411 & 0.000 & 2.790 & 0.000 \end{bmatrix}$$

Application of the rank test to the above matrix yields a full rank, and the system is controllable. This is not surprising considering the numerical error caused by the small number of elements. In case the size of coefficient -0.411 is small enough by refining the modeling technique, the system will be uncontrollable. This is also observable from the modal coordinate equation, and the second mode is completely uncontrollable when the coefficient in front of $\{u\}$ is equal to zero.

‡

Feedback Control Law Design in State Space

Feedback control laws have various applications in many dynamic systems. The key idea of a feedback control law is to utilize the measurement of the current state of a system, and use the measured signal to construct an actuator signal. The feedback control law design concept in state space form equation is presented in Fig. 11.7.3. The state vector of a system is directly used for the actuator signal. Depending upon the sensor available, sometimes the combination of state variables are used for the actuator command.

The feedback control law has some inherent advantages such as overcoming unknown external disturbance and initial condition off-set error. A number of linear control system design tools for state space form of systems are developed in MATLAB *Control Toolbox* [43,44], so that we can use them efficiently to design desired control laws.

Suppose we have a dynamic system described as

$$\{\dot{x}\} = [A]\{x\} + [B]\{u\} \qquad (11.7.12)$$
$$\{y\} = [C]\{x\} \qquad (11.7.13)$$

where we assumed a disturbance-free condition and dropped the $[D]\{u\}$ term which is normally unused for control system design. By a feedback control, we mean a control input $\{u\}$ prescribed in the form

$$\{u\} = -[G]\{y\}, \qquad \{y\} = [C]\{x\} \qquad (11.7.14)$$

or

$$\{u\} = -[G]\{x\} \qquad (11.7.15)$$

The first control law in Eq. (11.7.14) uses the direct sensor output $\{y\}$, called *output feedback*, while the second control law in Eq.(11.7.15) is called *full state feedback* since it uses all state variables $\{x\}$. In the case of the output feedback, the closed-loop system becomes

$$\{\dot{x}\} = ([A] - [B][G][C])\{x\} \qquad (11.7.16)$$

Thus, the system is stable if

$$\lambda_i([A] - [B][G][C]) < 0, \qquad i = 1, 2, \ldots, n \qquad (11.7.17)$$

Since usually the number of sensors or outputs is limited compared to that of state variables, it is not straightforward to satisfy the stability condition in the output feedback control law. Usually, a stabilizing output feedback control law design is technically more involved than a full state feedback control law design. There is no unified tool for an output feedback design compared with full state feedback, and numerical iterative techniques are frequently used for an output feedback law design.

Full State Feedback Law

On the other hand, the state feedback uses all state variables in order to stabilize a system. In spite of the difficulty estimating the state variables, the state feedback has elegant properties and has received significant attention. Therefore, we elect to put more emphasis on the full state feedback rather than the output feedback. The essence of the full state feedback law design is to find a feedback gain which makes the closed-loop system stable. In other words,

$$\lambda_i(A - BG) \leq 0, \qquad i = 1, 2, \ldots, n \qquad (11.7.18)$$

In a full state feedback control law design, there are two distinct approaches: one is the *pole placement* technique and the other is the *Linear Quadratic Regulator(LQR)* method based upon the optimal control theory.

Single Input System

For simplicity and better understanding of the fundamental of feedback control laws design, we start to form a single input system. In this case, we replace the input distribution matrix, $[B]$ by a column vector $[\,b\,]$. As a consequence, we start with

$$\{\dot{x}\} = [A]\{x\} + [\,b\,]\{u\} \qquad (11.7.19)$$

and the input is prescribed as

$$u = -[\,g\,]^T\{x\} \qquad (11.7.20)$$

where [g] is a gain vector
$$[\,g\,] = [g_1, g_2, \ldots, g_n] \tag{11.7.21}$$
The closed-loop system is
$$\{\dot{x}\} = ([A] - [\,b\,][\,g\,])\{x\} \tag{11.7.22}$$
and the characteristic equation
$$|\lambda I - [A] + [\,b\,][\,g\,]| = (\lambda - \lambda_1)(\lambda - \lambda_2)\cdots(\lambda - \lambda_n) \tag{11.7.23}$$
where λ_i is a closed-loop eigenvalue or pole. Suppose we want to have the closed-loop eigenvalues placed at certain desired locations as
$$\lambda_1^d, \lambda_2^d, \ldots, \lambda_n^d \tag{11.7.24}$$
Therefore, it should follow as
$$(\lambda - \lambda_1)(\lambda - \lambda_2)\cdots(\lambda - \lambda_n) = (\lambda - \lambda_1^d)(\lambda - \lambda_2^d)\cdots(\lambda - \lambda_n^d) \tag{11.7.25}$$
The above equation can be rewritten as
$$\lambda^n + a_1\lambda^{n-1} + \cdots + a_{n-1}\lambda + a_n = \lambda^n + a_1^d\lambda^{n-1} + \cdots + a_{n-1}^d\lambda + a_n^d \tag{11.7.26}$$
Since the coefficients of the left-hand side of the polynomial equations are functions of the feedback gains, there are n set nonlinear algebraic equations
$$\begin{aligned} a_1(g_1, g_2, \ldots, g_n) &= a_1^d \\ a_2(g_1, g_2, \ldots, g_n) &= a_2^d \\ &\vdots \\ a_n(g_1, g_2, \ldots, g_n) &= a_n^d \end{aligned} \tag{11.7.27}$$
The right-hand side of the equations are given, and we can find a unique set of feedback gains. Many algorithms are suggested in connection with the pole placement technique.

Here, an exemplary algorithm is introduced.

Bass-Gura Formula

The Bass-Gura formula makes use of the determinant properties [41]. It starts with the closed-loop characteristic equation
$$\begin{aligned} a_c(s) &= |sI - [A] + [\,b\,][\,g\,]| \\ &= |[sI - [A]][I + (sI - [A])^{-1}[\,b\,][\,g\,]]| \\ &= |sI - A||I + (sI - [A])^{-1}[\,b\,][\,g\,]| \\ &= a(s)(1 + [\,g\,](sI - [A])^{-1}[\,b\,]) \end{aligned} \tag{11.7.28}$$

where we used

$$|I + (sI - [A])^{-1}[\,b\,][\,g\,]| = 1 + [\,g\,](sI - [A])^{-1}[\,b\,] \qquad (11.7.29)$$

Therefore, Eq. (11.7.25) can be rewritten as

$$a_c(s) - a(s) = a(s)[\,g\,](sI - [A])^{-1}[\,b\,] \qquad (11.7.30)$$

At this point, a special relationship for $(sI - [A])^{-1}$ is introduced as [41]

$$(sI-[A])^{-1} = \frac{1}{a(s)}[s^{n-1}I + s^{n-2}([A]+a_1 I) + s^{n-3}([A]^2 + a_1[A] + a_2 I) + \cdots] \qquad (11.7.31)$$

Equation (11.7.30), therefore, becomes

$$a_c(s) - a(s) = [\,g\,][s^{n-1}I + s^{n-2}([A]+a_1 I) + s^{n-3}([A]^2 + a_1[A] + a_2 I) + \cdots][\,b\,] \qquad (11.7.32)$$

Comparing both sides of the polynomials, we obtain

$$\alpha_1 - a_1 = [\,g\,][\,b\,], \qquad \alpha_2 - a_2 = [\,g\,][A][\,b\,] + a_1[\,g\,][\,b\,]$$
$$\alpha_3 - a_3 = [\,g\,][A]^2[\,b\,] + a_1[\,g\,][A][\,b\,] + a_2[\,g\,][\,b\,] \qquad (11.7.33)$$

or in matrix form [41]

$$[\,\hat{\alpha}\,] - [\,\hat{a}\,] = [\,g\,][P][\Psi_-]^T \qquad (11.7.34)$$

where

$$[\,\hat{\alpha}\,] = [\alpha_1 \; \alpha_2 \; \cdots \; \alpha_n], \qquad [\,\hat{a}\,] = [a_1 \; a_2 \; \cdots \; a_n] \qquad (11.7.35)$$

Furthermore, $[P]$ is the controllability matrix and $[\Psi_-]$ is a *lower triangular Toeplitz* matrix given as

$$\Psi_- = \begin{bmatrix} 1 & & & & \\ a_1 & 1 & & & \\ a_2 & a_1 & 1 & & \\ \vdots & \vdots & \vdots & 1 & \\ a_n & a_{n-1} & \cdots & a_1 & 1 \end{bmatrix} \qquad (11.7.36)$$

Assuming that the system is controllable so that the controllability matrix $[P]$ is full rank and invertible, the feedback gain is given by

$$[\,g\,] = \Psi_-^{-T} P^{-1}([\,\hat{\alpha}\,] - [\,\hat{a}\,]) \qquad (11.7.37)$$

♣ **Example 11.7.3**   The Bass-Gura algorithm is applied to an example system. The system matrices are generated from a finite element analysis for a beam model.

$$[M] = \begin{bmatrix} 0.223 & 0.000 & 0.039 & -0.139 \\ 0.000 & 1.286 & 0.139 & -0.482 \\ 0.039 & 0.139 & 0.111 & -0.236 \\ -0.139 & -0.482 & -0.236 & 0.643 \end{bmatrix}$$

and

$$[K] = \begin{bmatrix} 2.418 & 0.000 & -1.209 & 9.067 \\ 0.000 & 181.3 & -9.067 & 45.33 \\ -1.209 & -9.067 & 1.209 & -9.067 \\ 9.067 & 45.33 & -9.067 & 90.67 \end{bmatrix}, \quad [F] = \begin{bmatrix} 1 \\ 0 \\ 0 \\ 0 \end{bmatrix}$$

The second order differential equation is transformed into a first order state form as

$$\{\dot{x}\} = [A]\{x\} + [B]\{u\}$$

where

$$[A] = \begin{bmatrix} 0 & I \\ -[M]^{-1}[K] & 0 \end{bmatrix}, \quad [B] = \begin{bmatrix} 0 \\ [M]^{-1}[F] \end{bmatrix}$$

Now the desired closed-loop poles are specified as

$$-0.5 \pm 0.6i, \quad -1.0 \pm 3.0i, \quad -2.0 \pm 2.0i, \quad -0.7 \pm 0.4i$$

A MATLAB m-file $febasgr.m$ is written to implement the Bass-Gura formula. The resultant feedback gain vector turns out to be

$$[\,g\,] = [\,-2.0881 \times 10^1, \quad -2.6291 \times 10^2, \quad 2.8713 \times 10^1, \quad -2.4912 \times 10^2,$$
$$2.1851 \times 10^{-2}, \quad 5.5844, \quad -2.8757 \times 10^{-1}, \quad 1.3877]$$

```
function [g]=febasgr(A,B,dc)
%---------------------------------------------------------------
% Purpose:
%    The function subroutine febasgr.m calculates a feedback gain for a
%    single input system by Bass-Gura formula.
%
% Synopsis:
%    [g]=febasgr(A,B,dc)
%
%    System equation : xdot = Ax + bu
%
% Variable Description:
%    Input variable : dc - A vector consisting of desired closed-loop poles
%    Output : g - A feedback gain vector.
%---------------------------------------------------------------
ao= poly(A);                  % Calculate coefficient of the given system
alpha = poly(dc);             % Calculate coefficient of the desired polynomial
[P,rank,cond]=fctobty(A,B);            % Compute controllability matrix
```

```
n=max(size(A));
%----------------------------------------
% Build a Toeplitz matrix
%----------------------------------------
Toep=zeros(n,n);
for i=1:n
Toep(i:n,i)=[ao(1:n-i+1)]';
end
g=[alpha(2:n+1)-ao(2:n+1)]*(inv(Toep))'*inv(P);   % Calculate the gain
g=real(g);                                        % Take the real part of the gain
%----------------------------------------
```

‡

Multi-Input System

The pole placement technique for multi-input systems is rather different from the single input system case. The feedback gains are not uniquely determined due to the number of feedback gain elements which is greater than that of state variables of the system.

Consider a multiple input system with the governing equation

$$\{\dot{x}\} = [A]\{x\} + [B]\{u\} \tag{11.7.38}$$

where $[B]$ is an $n \times m (\geq 2)$ input influence matrix. The control law is assumed to be a full state feedback law

$$\{u\} = -[G]\{x\} \tag{11.7.39}$$

where $[G]$ is an $m \times n$ gain matrix. The closed-loop system stability is determined by

$$|\lambda I - [A] + [B][G]| = 0 \tag{11.7.40}$$
$$= \lambda^n + a_1 \lambda^{n-1} + \cdots + a_{n-1}\lambda + a_n$$

Therefore, the characteristic equation is a polynomial of order n, the same as the order of the system. The size of the gain matrix $[G]$, however, exceeds n: there are exactly $n \times m$ elements in $[G]$. For example, let us consider a system with the following system matrices

$$[A] = \begin{bmatrix} a_{11} & a_{12} & a_{13} \\ a_{21} & a_{22} & a_{23} \\ a_{31} & a_{33} & a_{33} \end{bmatrix}, \quad [B] = \begin{bmatrix} b_{11} & b_{12} \\ b_{21} & b_{22} \\ b_{31} & b_{32} \end{bmatrix} \tag{11.7.41}$$

Then, the full state feedback law is suggested as $\{u\} = -[G]\{x\}$, where the gain matrix, in order to satisfy dimensionality, should have the form

$$[G] = \begin{bmatrix} g_{11} & g_{12} & g_{13} \\ g_{21} & g_{22} & g_{23} \end{bmatrix} \tag{11.7.42}$$

The feedback control design goal is to produce the closed-loop poles of a system. There are three closed-loop poles while the number of gain elements is six. Since we have more parameters than the number of equations to be satisfied, the extra degree of freedom can be used for other purposes such as improving system robustness. By system robustness, we mean the property of a system; a system is called robust when the performance of a controlled system is invariant with respect to system uncertainty.

11.8 Linear Quadratic Regulator

Linear Quadratic Regular (LQR) theory is originated from the optimal control theory. The key idea of this method is to take a performance function and design a control law which minimizes the performance function.

In order to understand the LQR technique, first we should discuss the basic principle of optimal control theory. The generic optimal control theory starts from finding the control input $\{u(t)\}$ which minimizes a performance index [45,46]

$$J = h(\{x(t_f)\}, t_f) + \int_{t_0}^{t_f} \phi[\{x(t)\}, \{u(t)\}, t] dt \qquad (11.8.1)$$

where t_0 and t_f are starting and final times, respectively. On the other hand, the state vector $\{x(t)\}$ and control input $\{u(t)\}$ satisfy the nonlinear governing equations of motion

$$\{\dot{x}\} = f(\{x(t)\}, \{u(t)\}, t) \qquad (11.8.2)$$

For notational simplicity, we temporarily drop the { } sign for vector notation. Since the control input must satisfy the governing equation while trying to minimize the performance index, Eqs. (11.8.1) and (11.8.2) are combined together by the *Lagrange multiplier* (λ) as

$$J = h(x(t_f), t_f) + \int_{t_0}^{t_f} \left[\phi(x(t), u(t), t) + \lambda^T [f(x(t), u(t), t) - \dot{x}]\right] dt \qquad (11.8.3)$$

For some reasons which are not explained here, we define the *Hamiltonian* of the system as

$$H = \phi(x(t), u(t), t) + \lambda^T f(x(t), u(t), t) \qquad (11.8.4)$$

so that Eq. (11.8.3) is rewritten as

$$J = h(x(t_f), t_f) + \int_{t_0}^{t_f} \left[H(x(t), u(t), \lambda(t), t) - \lambda^T \dot{x}\right] dt \qquad (11.8.5)$$

One of the optimal control theories is the *variational principle*, for which we assume the variation of the state vector and control input from the optimal one. The graphical representation for the variational principle is presented in Fig. 11.8.1. We

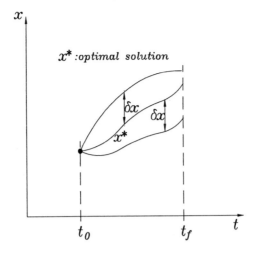

Figure 11.8.1 Variation of Trajectory about the Optimal One

assume an optimal trajectory and allow variations about the optimal trajectory. The performance index is also varied about the optimal value.

The variation of the state vector and control input vector results in the variation of the performance index. In order to satisfy the optimality condition, the variation of the performance index should be equal to zero. That is, the performance index is an optimal one with no variation [45,46].

$$\delta J(\delta x, \delta x(t_f), \delta u, \delta \lambda) = 0 \qquad (11.8.6)$$

where $\delta(\ )$ represents variation of $(\ )$ from the optimal value.

Based upon the variational principle we take the variation of the expression in Eq. (11.8.5)

$$\delta J = \delta h(x(t_f), t) + \int_{t_0}^{t_f} [\delta H(x(t), u(t), \lambda(t), t) + \delta \dot{\lambda}^T x(t)] dt \qquad (11.8.7)$$

When the final time (t_f) is fixed

$$\delta h(x(t_f), t_f) = \left[\frac{\partial h}{\partial x(t_f)}\right]^T \delta x(t_f) \qquad (11.8.8)$$

Further properties of variation are given as

$$\delta H(x(t), u(t), \lambda(t), t) = \left[\frac{\partial H}{\partial x}\right]^T \delta x + \left[\frac{\partial H}{\partial u}\right]^T \delta u + \left[\frac{\partial H}{\partial \lambda}\right]^T \delta \lambda \qquad (11.8.9)$$

Also,

$$\int_{t_0}^{t_f} \delta \lambda^T \dot{x} dt = \int_{t_0}^{t_f} [\delta \lambda^T \dot{x} + \lambda^T \delta \dot{x}] dt$$

$$= \lambda^T \delta x \big|_{t_0}^{t_f} + \int_{t_0}^{t_f} [\delta \lambda^T \dot{x} - \dot{\lambda}^T \delta x] dt \qquad (11.8.10)$$

where we used integration by parts on $\lambda^T \delta\dot{x}$. Therefore, the combination of Eqs. (11.8.8) through (11.8.10) yields

$$\delta J = \left[\frac{\partial h}{\partial x(t_f)} - \lambda(t_f)\right]^T \delta x(t_f)$$
$$+ \int_{t_0}^{t_f} \left[(\frac{\partial H}{\partial x} + \dot{\lambda})^T \delta x + (\frac{\partial H}{\partial u})^T \delta u + (\frac{\partial H}{\partial \lambda} - \dot{x})^T \delta\lambda\right] dt \quad (11.8.11)$$

where we assumed that $\delta x(t_0) = 0$ since the initial condition is usually fixed allowing no variation. Once again, for optimality, $\delta J = 0$ must be satisfied. Therefore, we obtain the following conditions from the optimality condition

$$\lambda(t_f) = \frac{\partial h}{\partial x(t_f)} \quad (11.8.12)$$

$$\dot{\lambda} + \frac{\partial H}{\partial x} = 0 \quad (11.8.13)$$

$$\frac{\partial H}{\partial \lambda} - \dot{x} = f(x, u, t) - \dot{x}(t) = 0 \quad (11.8.14)$$

$$\frac{\partial H}{\partial u} = 0 \quad (11.8.15)$$

The above set of equations is solved numerically because of the nature of the problem. The boundary condition on λ at the final time (t_f) and initial condition on the state vector $\{x(t_0)\}$ turn into the so-called Two Point Boundary Value Problem (TPBVP). Numerous solution techniques and applications have been developed to solve the optimal control problem. For instance, in robotics areas, the rotational motion of a robot arm is analyzed in terms of optimal performance such as minimum-time, minimum-fuel, and minimum vibration during maneuver.

Due to the limited space we elect to directly jump into linear optimal control theory. The linear optimal control theory also has wide applications. First, a frequently used performance index is prescribed as

$$J = \frac{1}{2} \int_0^\infty (\{x\}^T [Q]\{x\} + \{u\}^T [R]\{u\}) dt \quad (11.8.16)$$

where $[Q]$ is a positive definite or semidefinite weighting matrix with $\{x\}^T [Q]\{x\} \geq 0$ for $\{x\} \neq 0$, and $[R]$ is a positive definite weighting matrix such that $\{u\}^T [R]\{u\} > 0$, for $\{u\} \neq 0$. The upper limit of the performance index is ∞ which implies that we are interested in the steady-state behavior of the system. In other words, the system should be stabilized at the steady-state so that the peformance index is bounded within a value.

Our goal is to find a control function for which the performance index is minimum while the original system equation

$$\{\dot{x}\} = [A]\{x\} + [B]\{u\}, \quad \text{for given} \ \{x(0)\} \quad (11.8.17)$$

is satisfied. In order to apply the optimal control theory in the previous part, we define the *Hamiltonian* of the system as

$$H = \frac{1}{2}(\{x\}^T[Q]\{x\} + \{u\}^T[R]\{u\}) + \{\lambda\}^T([A]\{x\} + [B]\{u\}) \qquad (11.8.18)$$

Next we apply the optimality condition in Eqs. (11.8.12) through (11.8.15) in such a way that

$$\{\dot{\lambda}\} = -\frac{\partial H}{\partial \{x\}} = -[Q]\{x\} - [A]^T\{\lambda\}, \qquad \{\lambda(\infty)\} = 0 \qquad (11.8.19)$$

$$\frac{\partial H}{\partial \{u\}} = [R]\{u\} + [B]^T\{\lambda\} = 0 \qquad (11.8.20)$$

Thus, the optimal control input is a function of λ as

$$\{u\} = -[R]^{-1}[B]^T\{\lambda\} \qquad (11.8.21)$$

In other words, once we solve for $\{\lambda\}$ the control input is obtained. However, it is not easy to compute $\{\lambda\}$ since the boundary condition of $\{\lambda\}$ is given at the steady-state as $\{\lambda(\infty)\} = 0$ while the initial condition of $\{x\}$ is given at the initial time $t = 0$. There are different approaches solving the above set of equations. A popular method is to start with

$$\{\lambda\} = [S]\{x\} \qquad (11.8.22)$$

where $[S]$ is a positive definite matrix called the Ricatti matrix. Therefore, the control input can be written as

$$\{u\} = -[G]\{x\} \qquad (11.8.23)$$

where $[G]$ is a feedback gain matrix

$$[G] = [R]^{-1}[B]^T[S] \qquad (11.8.24)$$

Substituting Eq.(11.8.22) into Eq.(11.8.19) and dropping [ ] notation temporarily yields

$$\dot{S}\{x\} + S\{\dot{x}\} = -Q\{x\} - A^T S\{x\}$$
$$\dot{S}\{x\} + S(A\{x\} + B\{u\}) = -Q\{x\} - A^T S\{x\}$$
$$\dot{S}\{x\} + S(A\{x\} - BR^{-1}B^T S\{x\}) = -Q\{x\} - A^T S\{x\} \qquad (11.8.25)$$

Therefore,

$$-[\dot{S}] = [S][A] + [A]^T[S] - [S][B][R]^{-1}[B]^T[S] + [Q] \qquad (11.8.26)$$

The above equation is a matrix differential equation, and we can integrate numerically from $[S(t_f)] = 0$ where t_f is far enough. An alternative strategy of solving Eq. (11.8.26) is to use the steady-state solution. When the system reaches a steady-state the Ricatti matrix satisfies $[\dot{S}] = 0$. Therefore, we obtain the so-called *Algebraic Ricatti Equation (ARE)*

$$[0] = [S][A] + [A]^T[S] - [S][B][R]^{-1}[B]^T[S] + [Q] \tag{11.8.27}$$

There are many algorithms studied to solve the ARE [47]. They are dominated by numerical techniques due to the nature of the problem, a nonlinear algebraic matrix equation. It turns out that the feedback gain matrix($[G]$) is also found from the Hamiltonian matrix by Potter [48].

$$[H] = \begin{bmatrix} [A] & -[B][R]^{-1}[B]^T \\ -[Q] & -[A]^T \end{bmatrix} \tag{11.8.28}$$

The size of the Hamiltonian matrix is now $2n \times 2n$ where n is the size of the original system. Using the Hamiltonian matrix, we can solve the eigenvalue problem

$$[H]\{\phi_i\} = \lambda_i \{\phi_i\} \tag{11.8.29}$$

We can prove that there are two sets of eigenvalues for $[H]$: one set with negative real parts, and the other set with positive real parts. For each eigenvalue, we arrange the corresponding eigenvector as

$$[\Phi] = \begin{bmatrix} [\Phi_{11}], & [\Phi_{12}] \\ [\Phi_{21}], & [\Phi_{22}] \end{bmatrix} \tag{11.8.30}$$

Therefore, $[\Phi_{11}], [\Phi_{21}]$ correspond to eigenvalues with positive real parts and $[\Phi_{12}], [\Phi_{22}]$ correspond to eigenvalues with negative real parts. The solution of ARE turns out be a function of the eigenvectors as

$$[S] = [\Phi_{22}][\Phi_{12}]^{-1} \tag{11.8.31}$$

Potter's method is quite popular, and a MATLAB *felqr.m* is written based upon this algorithm. The feedback gain is computed from the result of the eigenvalue solution of the Hamiltonian matrix. The optimal cost function satisfies

$$J_{opt} = \{x_0\}^T [S] \{x_0\} \tag{11.8.32}$$

where $\{x_0\}$ is the initial condition of $\{x\}$. The proof is provided as

$$J_{opt} = \frac{1}{2} \int_0^\infty \left[\{x\}^T [Q]\{x\} + \{u\}^T [R]\{u\} \right] dt$$

$$= \frac{1}{2} \int_0^\infty \left[\{x\}^T (-[S][A] - [A]^T[S] + [S][B][R]^{-1}[B]^T[S])\{x\} + \{u\}^T [R]\{u\} \right] dt \tag{11.8.33}$$

where we used the ARE for $[Q]$. Next, Eqs. (11.8.23) and (11.8.24) are utilized so that

$$\begin{aligned}
J_{opt} &= -\frac{1}{2}\int_0^\infty \left[([A]\{x\}+[B]\{u\})^T[S]\{x\}+\{x\}^T[S]([A]\{x\}+[B]\{u\})\right]dt \\
&= -\frac{1}{2}\int_0^\infty \left[\{\dot{x}\}^T[S]\{x\}+\{x\}^T[S]\{\dot{x}\}\right]dt \quad (11.8.34)\\
&= -\frac{1}{2}\{x\}^T[S]\{x\}\Big|_0^\infty \\
&= \frac{1}{2}\{x_0\}^T[S]\{x_0\}
\end{aligned}$$

where the steady-state value $\{x(\infty)\}$ is assumed to be zero for a stable closed-loop system. The optimum cost function is a function of the initial condition and Riccati matrix $[S]$.

In the LQR approach, once we solve the ARE, the feedback gain $[G]$ is automatically obtained. This is a significant advantage over the pole placement technique where we have to specify the desired poles. In particular, there is no essential difference for multiple input and output systems for the LQR approach. The weighting matrices $[Q]$ and $[R]$ are the only design parameters. The closed-loop system poles are determined by $[Q]$ and $[R]$. It is not easy to select those matrices in general, however. The trial and error procedure is usually taken. Significant research effort has been made in the LQR related subject. The solution of ARE is now readily available in computational tools in MATLAB Control *Toolbox*.

♣ **Example 11.8.1** The LQR technique is applied to an example system which is to be stabilized by a full state feedback. The finite element beam model is given by

$$[M] = \begin{bmatrix} 6.240 & 0.000 & 1.080 & -3.120 \\ 0.000 & 23.04 & 3.120 & -8.640 \\ 1.080 & 3.120 & 3.120 & -5.280 \\ -3.120 & -8.640 & -5.280 & 11.520 \end{bmatrix} \times 10^{-2}$$

$$[K] = \begin{bmatrix} 4.722 & 0.000 & -2.361 & 14.17 \\ 0.000 & 226.7 & -14.17 & 56.67 \\ -2.361 & -14.17 & 2.361 & -14.17 \\ 14.17 & 56.67 & -14.17 & 113.3 \end{bmatrix}, \quad [F] = \begin{bmatrix} 0 & 1 \\ 0 & 0 \\ 0 & 0 \\ 1 & 0 \end{bmatrix}$$

The state space form equation is developed using the same convention as in Eq.(11.5.3). The state and control input weighting matrices are chosen as

$$[Q] = I_{8\times 8}, \qquad [R] = 0.1 \times I_{2\times 2}$$

A MATLAB *m-file felqr.m* is written using Potter's algorithm. The final feedback gain matrix is computed as

$$[G] = \begin{bmatrix} 8.636 & 33.44 & -5.040 & 63.10 & 0.404 & -0.067 & 3.124 & 0.884 \\ 7.266 & 30.70 & -5.961 & 47.55 & 3.354 & 0.337 & -0.337 & 0.807 \end{bmatrix}$$

```
function [G,S]=felqr(A,B,Q,R);
%------------------------------------------------------------
% Purpose:
%    The function subroutine felqr.m calculates the feedback gain matrix
%    by Linear Quadratic Regulator(LQR) technique.
%    The given system is
%
%              xdot = Ax + Bu,    u = -Gx
%
% and the performance index to be minimized is defined as
%
%              J=(1/2)integral(x'Qx+u'Ru)dt
% Synopsis:
%    [G,S]=felqr(A,B,Q,R)
%
% Variable Description:
%    Input arguments - A, B, Q, R
%    Output parameters - G = R^{-1}G'S : feedback gain matrix
%              S : Solution of the Algebraic Ricatti Equation (ARE)
%                   AS+A'S-SBR^{-1}S+Q=0
%
% Notes:
%    i). (A,B) should be controllable.
%    ii). Q is at least positive semidefinite.
%       R is at least positive definite.
%------------------------------------------------------------
H=[A -B*inv(R)*B';              % Build the Hamiltonian matrix
-Q -A'];
[V,D]=eig(H);                   % Solve eigenvalue problem
n=size(A); twon=max(size(H));
% Normalized each eigenvector to unity magnitude
av=abs(V);
magav=av'*av;
dmagav=diag(magav);
V=V*sqrt(inv(diag(dmagav)));    % Normalize the eigenvector
%------------------------------------------------------------
% Sort the eigenvalues with stable real parts
%------------------------------------------------------------
rel=real(diag(D));
nindex=[];pindex=[];
for i=1:twon
if(rel(i)<=0)
nindex=[nindex i];
else
pindex=[pindex i];
end
end
V=V(:, [pindex,nindex]);        % Rearrange the eigenvector order
```

Section 11.8 Linear Quadratic Regulator

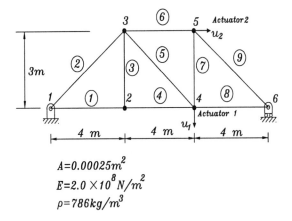

$A = 0.00025 m^2$
$E = 2.0 \times 10^8 N/m^2$
$\rho = 786 kg/m^3$

Figure 11.8.2 A Truss Structure Example for LQR Controller Design

```
S=real(V(n+1:twon,n+1:twon)*inv(V(1:n,n+1:twon)));
G=real(inv(R)*B'*S);
```

‡

♣ **Example 11.8.2** Another example of LQR design is applied to a two-dimensional truss structure. The geometric configuration including two actuator locations is presented in Fig. 11.8.2. As is shown, the actuators are acting both in vertical and horizontal directions. The structural parameters are selected as $E = 2.0 \times 10^8 N/m^2, \rho = 786 kg/m^3, A = 0.00025 m^2$. Using the consistent mass matrix and the finite element modeling technique in Chapter 7 (See Example 7.5.2), we obtain

$$[M] = \begin{bmatrix} [M]_{11} & [M]_{12} \\ [M]_{12}^T & [M]_{22} \end{bmatrix}, \qquad [K] = \begin{bmatrix} [K]_{11} & [K]_{12} \\ [K]_{12}^T & [K]_{22} \end{bmatrix}$$

where

$$[M]_{11} = \begin{bmatrix} 0.7205 & 0 & 0.0983 & 0 & 0.1310 \\ 0 & 0.7205 & 0 & 0.0983 & 0 \\ 0.0983 & 0 & 1.1135 & 0 & 0.1638 \\ 0 & 0.0983 & 0 & 1.1135 & 0 \\ 0.1310 & 0 & 0.1638 & 0 & 1.0480 \end{bmatrix}$$

$$[M]_{12} = \begin{bmatrix} 0 & 0 & 0 & 0 \\ 0.1310 & 0 & 0 & 0 \\ 0 & 0.1310 & 0 & 0 \\ 0.1638 & 0 & 0.1310 & 0 \\ 0 & 0.0983 & 0 & 0.1310 \end{bmatrix}$$

$$[M]_{22} = \begin{bmatrix} 1.0480 & 0 & 0.0983 & 0 \\ 0 & 0.7860 & 0 & 0.1638 \\ 0.0983 & 0 & 0.7860 & 0 \\ 0 & 0.1638 & 0 & 0.5895 \end{bmatrix}$$

Furthermore,

$$[K]_{11} = 10^4 \times \begin{bmatrix} 2.5000 & 0.0000 & 0.0000 & 0.0000 & -1.2500 \\ 0.0000 & 1.6667 & 0.0000 & -1.6667 & 0 \\ 0.0000 & 0.0000 & 2.5300 & 0.0000 & -0.6400 \\ 0.0000 & -1.6667 & 0.0000 & 2.3867 & 0.4800 \\ -1.2500 & 0 & -0.6400 & 0.4800 & 3.1400 \end{bmatrix}$$

$$[K]_{12} = 10^4 \times \begin{bmatrix} 0 & 0 & 0 & 0 \\ 0 & 0 & 0 & 0 \\ 0.4800 & -1.2500 & 0 & 0 \\ -0.3600 & 0 & 0 & 0 \\ -0.4800 & 0.0000 & 0.0000 & -1.2500 \end{bmatrix}$$

$$[K]_{22} = 10^4 \times \begin{bmatrix} 2.0267 & 0.0000 & -1.6667 & 0 \\ 0.0000 & 1.8900 & -0.4800 & -0.6400 \\ -1.6667 & -0.4800 & 2.0267 & 0.4800 \\ 0 & -0.6400 & 0.4800 & 1.8900 \end{bmatrix}$$

The open-loop eigenvalues of the first order state space system based upon Eq. (11.5.3) turn out to be

$$100 \times [\pm 0.2409i, \pm 0.4679i, \pm 0.7398i, \pm 1.2434i,$$
$$\pm 1.6334i, \pm 2.1022i, \pm 2.1801i, \pm 2.3101i, \pm 2.8022i]$$

In order to design the LQR control law, first we select the weighting matrices, $[Q]$ and $[R]$, which appear in the performance index.

$$[Q] = I_{18 \times 18}, \quad [R] = 0.01 \times I_{2 \times 2}$$

where I is an identity matrix. The resultant feedback gain matrix $([G])$ is computed using *felqr.m* as

$$[G] = [G_1 \quad G_2]$$

where

$[G_1] =$
$$\begin{bmatrix} 1.427 & -58.65 & 10.52 & 63.55 & 0.501 & 12.99 & -11.25 & -12.42 & 7.278 \\ -10.78 & 28.01 & -20.19 & -35.49 & 24.68 & 9.153 & 17.54 & -15.53 & -2.489 \end{bmatrix}$$

$[G_2] =$
$$\begin{bmatrix} -0.57 & -0.47 & -0.47 & 0.14 & -0.39 & 6.49 & -0.49 & 0.06 & -0.04 \\ -0.83 & 0.38 & -1.43 & -0.45 & 0.25 & -0.61 & 6.74 & -0.29 & 0.04 \end{bmatrix}$$

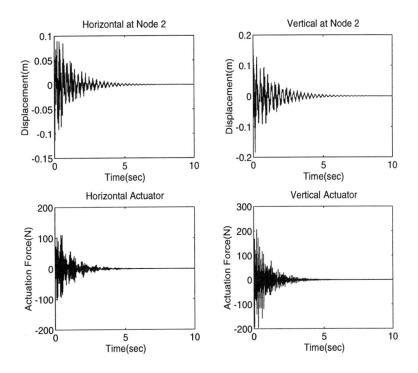

Figure 11.8.3 Simulation Results for the Truss Structure by LQR Controller

In addition, the closed-loop system eigenvalues by the feedback control action result in

$$[-1.1 \pm 280.2i, \ -1.3 \pm 231.0i, \ -0.8 \pm 218.0i, \ -1.0 \pm 210.2i,$$
$$-0.6 \pm 163.3i, \ -1.0 \pm 124.3i, \ -0.8 \pm 74.0i, \ -0.9 \pm 46.8i,$$
$$-0.59 \pm 24.08i]$$

Thus, the system is stabilized by the full state feedback action. The closed-loop system stability is also verified by the numerical simulation. For this, we assume the initial condition as

$$x(0) = [0.1, \ 0, \ 0, \ 0.2, \ -0.2, \ 0, \ 0, \ 0, \ 0, \ 0, \ 0, \ 0, \ 0, \ 0, \ 0, \ 0, \ 0, \ 0]^T$$

The simulation result is presented in Fig. 11.8.3. The horizontal and vertical displacements at the nodal point #2 are displayed. Also, the resultant actuator responses in both directions are presented. The motions are stabilized, decaying as time passes.

‡

11.9 Modal Control for Second Order Systems

The modal control approach is generally applied to linearized second order structural systems even if it is applicable to first order systems [49,50]. The linearized second order systems are transformed into a decoupled set of modal coordinate equations. The decoupled equations are controlled independently. A drawback of the modal control is the number of inputs which is usually less than number of modal coordinates. In order to have independent control over all modal coordinates, from a physical sense, we should have the same number of actuators as that of modal coordinates.

Consider an undamped second order system

$$[M]\{\ddot{q}\} + K\{q\} = [F]\{u\} \quad (11.9.1)$$

Using the earlier results, the above equation is transformed into modal coordinate equations by introducing a coordinate transformation

$$\{\ddot{\eta}\} + [\Omega]\{\eta\} = [\Phi]^T[F]\{u\} \quad (11.9.2)$$

Let us assume that the size of the control input vector $\{u\}$ is $m \times 1$, being prescribed as a linear combination of the position and velocity vector.

$$\{u\} = [K_p]\{q\} + [K_v]\{\dot{q}\} \quad (11.9.3)$$

where $[K_p]$ and $[K_v]$ are gain matrices with appropriate dimensions. The above control input is in physical coordinate systems which can be rewritten in modal coordinates by using $\{q\} = [\Phi]^T\{\eta\}$.

$$\begin{aligned}\{u\} &= [K_p]\{q\} + [K_v]\{\dot{q}\} \\ &= [K_p][\Phi]\{\eta\} + [K_v][\Phi]\{\dot{q}\} \\ &= [\bar{K}_p]\{\eta\} + [\bar{K}_v]\{\dot{\eta}\}\end{aligned} \quad (11.9.4)$$

Equation (11.9.4), therefore, becomes

$$\{\ddot{\eta}\} + [\Omega]\{\eta\} = [\Phi]^T[F](\bar{K}_p\{\eta\} + \bar{K}_v\{\dot{\eta}\}) \quad (11.9.5)$$

Unfortunately, the left hand side of the equation is in decoupled form for each coordinate while the right hand side expressions are coupled. Further, if the size of control input is less than that of the given system, it is not feasible to control each coordinate independently. When the number of control input vectors is equal to that of the system, it is possible to have independent control over each modal coordinate.

For further development, we go back to Eq. (11.9.2)

$$\{\ddot{\eta}\} + [\Omega]\{\eta\} = \{\bar{f}\} \quad (11.9.2)$$

where $\{\bar{f}\} \equiv [f_1, f_2, \cdots, f_n]^T = [\Phi]^T[F]\{u\}$ represents an input force vector whose elements match with each modal coordinate. Now, for an ith modal coordinate

$$\ddot{\eta}_i + \omega_i^2 \eta_i = f_i \quad (11.9.6)$$

Section 11.9   Modal Control for Second Order Systems   483

The modal input force vector is selected as

$$f_i = -g_v^i \dot{\eta}_i - g_p^i \eta_i \tag{11.9.7}$$

so that the closed-loop system becomes

$$\ddot{\eta}_i + g_v^i \dot{\eta}_i + (\omega_i^2 + g_p^i)\eta_i = 0 \tag{11.9.8}$$

As we can see, the feedback gains (g_v^i, g_p^i) directly affect the closed-loop system of a modal coordinate. Therefore, we can control the dynamics of each modal coordinate independently.

The modal input force vector does not have physical meaning unless it is transformed into the physical input. The original relationship between the modal input force vector and the physical control input vector is

$$\{\bar{f}\} = [\Phi]^T [F]\{\mathbf{u}\} \tag{11.9.9}$$

The modal matrix is obtained from

$$[\Phi]^T [M][\Phi] = [I] \tag{11.9.10}$$

Therefore,

$$[\Phi^T]^{-1} = [M][\Phi] \tag{11.9.11}$$

and the control input vector satisfies

$$[F]\{\mathbf{u}\} = [M][\Phi]\{\bar{f}\} \tag{11.9.12}$$

Assuming that $[F]$ is invertible, it follows as [48,49]

$$\{\mathbf{u}\} = [F]^{-1}[M][\Phi]\{\bar{f}\} \tag{11.9.13}$$

The invertability of the $[F]$ matrix depends upon the rank, and there should be at least the same number of actuators in order to ensure the independent modal control.

♣ **Example 11.9.1** A finite element Euler-Bernoulli beam model is used to demonstrate the modal control technique. The model is presented in Fig. 11.9.1. There are four actuators located at each nodal point. Application of the finite element method yields mass, stiffness, and input influence matrices as

$$[M] = \begin{bmatrix} 0.743 & 0.000 & 0.129 & -0.619 \\ 0.000 & 7.619 & 0.619 & -2.857 \\ 0.129 & 0.619 & 0.371 & -1.048 \\ -0.619 & -2.857 & -1.048 & 3.810 \end{bmatrix}$$

and

$$[K] = \begin{bmatrix} 1.020 & 0.000 & -0.510 & 5.100 \\ 0.000 & 136.0 & -5.100 & 34.00 \\ -0.510 & -5.100 & 0.510 & -5.100 \\ 5.100 & 34.000 & -5.100 & 68.000 \end{bmatrix}, \quad [F] = \begin{bmatrix} 1 & 0 & 0 & 0 \\ 0 & 1 & 0 & 0 \\ 0 & 0 & 1 & 0 \\ 0 & 0 & 0 & 1 \end{bmatrix}$$

The modal matrix which consists of normalized eigenvectors is computed as

$$[\Phi] = \begin{bmatrix} -0.481 & -1.031 & -0.162 & 0.675 \\ -0.041 & 0.016 & 0.304 & 0.347 \\ -1.416 & 1.428 & -1.588 & 2.667 \\ -0.049 & 0.172 & -0.383 & 1.289 \end{bmatrix}$$

The physical system is transformed into modal coordinate equations as in Eq. (11.9.6).

$$\ddot{\eta}_1 + 0.033\eta_1 = f_1$$
$$\ddot{\eta}_2 + 1.312\eta_2 = f_2$$
$$\ddot{\eta}_3 + 15.00\eta_3 = f_3$$
$$\ddot{\eta}_4 + 126.4\eta_4 = f_4$$

where f_i is the ith component of the modal input force vector as $\{\bar{f}\} = [\Phi][F]\{u\}$. For each modal coordinate, we can design a proportional plus derivative type feedback control law. That is,

$$\left\{ \begin{array}{l} f_1 = -0.018\eta_1 - 0.080\dot{\eta}_1 \\ f_2 = -0.115\eta_2 - 0.504\dot{\eta}_2 \\ f_3 = -0.387\eta_3 - 1.704\dot{\eta}_3 \\ f_4 = -1.124\eta_4 - 4.947\dot{\eta}_4 \end{array} \right\}$$

where each control input added a 10% increase in natural frequency and 0.2 damping ratio. Once the modal control input is specified, the physical control input is obtained from Eq. (11.9.12) as

$$\{u\} = [F]^{-1}[M]\Phi\{\bar{f}\}$$

where

$$[F]^{-1}[M][\Phi] = \begin{bmatrix} -0.509 & -0.689 & -0.087 & 0.047 \\ -1.051 & 0.511 & 2.425 & 0.613 \\ -0.562 & 0.227 & -0.021 & -0.058 \\ 1.713 & -0.248 & -0.563 & 0.706 \end{bmatrix}$$

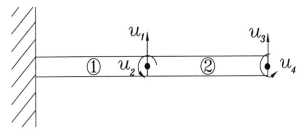

Figure 11.9.1 A Finite Element Model for Modal Control

11.10 Dynamic Observer

Observability

Another important issue in modern control design is observability. The observability of a dynamic system represents the ability of reconstructing all state variables using a finite number of sensor outputs. In the previous discussion of full state feedback control laws design, we assumed that the state variables are available which is not guaranteed in general situations. In the majority of modern control system designs and analyses, the number of sensors is less than that of state variables due to actual constraints. Also, it will be a significant advantage if we can estimate all state variables using only a limited number of sensors or measurement devices.

The *Observability* is a primary requirement estimating the state variables out of direct sensor output. Mathematical description of the *Observability* condition is similar to that of *Controllability*.

Definition of Observability

A system is observable if and only if any state $\{x(t)\}$ can be determined by using a finite output $\{y(\tau)\}$, for $t \leq \tau \leq T$.

The mathematical theorem is stated below.

Theorem :   A system is observable if and only if the matrix

$$[G_o(T,t)] = \int_t^T [\Phi(\tau,t)]^T [C]^T [C][\Phi(\tau,t)] d\tau \qquad (11.10.1)$$

is positive definite [41]. The above matrix is called *Observability Grammian*.

Proving the above theorem is similar to the controllability case and omitted here for brevity. In the same context as the controllability case, if we take $T = \infty$, then it follows

$$[G_o] = \int_0^\infty [\Phi(\xi)]^T [C]^T [C][\Phi(\xi)] d\xi \qquad (11.10.2)$$

Also, $[G_o(\xi)]$ turns out to satisfy the *Lyapunov equation*.

$$[A]^T [G_o] + [G_o][A] + [C]^T [C] = 0 \qquad (11.10.3)$$

Also, a MATLAB command *lyap* is available to find the solution to the Lyapuov equation. An alternative condition for observability is provided as follows.

The system is observable if and only if the *observability matrix*

$$[Q] = \left[[C]^T, [A]^T[C]^T, \ldots, [A^{n-1}]^T[C]^T\right]^T \qquad (11.10.4)$$

has rank n, the order of the system [39-42].

The observability test by rank is similar to the *Controllability* test as we examine both of them. That is, the observability test of $([A], [C])$ can be replaced by the controllability test of $([A]^T, [C]^T)$. In other words, there exists *duality* between controllability and $([A], [C])$ and the observability test of $([A], [C])$.

♣ **Example 11.10.1** In this example, we apply the observability test to a spring mass system in Fig. 11.10.1. A sensor measuring displacement is assumed. The equations of motion are established as

$$[M]\{\ddot{\mathbf{q}}\} + [K]\{\mathbf{q}\} = \{0\}$$

where the system mass and stiffness matrices are

$$[M] = \begin{bmatrix} m_1 & 0 & 0 \\ 0 & m_2 & 0 \\ 0 & 0 & m_3 \end{bmatrix}, \quad [K] = \begin{bmatrix} k_1 + k_2 & -k_2 & 0 \\ -k_2 & k_2 + k_3 & -k_3 \\ 0 & -k_3 & k_3 \end{bmatrix}$$

and $\mathbf{q} = [q_1, q_2, q_3]^T$, $[m_1, m_2, m_3] = [0.5, 1.0, 0.5](kg)$, $[k_1, k_2, k_3] = [7, 3, 9](N/m)$. The above second order equation can be rewritten in the first order form using Eq. (11.5.3). The output equation becomes

$$y = q_1 = [1, 0, 0, 0, 0, 0]\{x\} = [C]\{x\}$$

where $\{x\} = [q_1, q_2, q_3, \dot{q}_1, \dot{q}_2, \dot{q}_3]^T$. The system matrix is given by

$$[A] = \begin{bmatrix} 0 & 0 & 0 & 1 & 0 & 0 \\ 0 & 0 & 0 & 0 & 1 & 0 \\ 0 & 0 & 0 & 0 & 0 & 1 \\ -20 & 6 & 0 & 0 & 0 & 0 \\ 3 & -12 & 9 & 0 & 0 & 0 \\ 0 & 18 & -18 & 0 & 0 & 0 \end{bmatrix}$$

Therefore, the observability matrix is

$$[Q] = \begin{bmatrix} 1 & 0 & -20 & 0 & 418 & 0 \\ 0 & 0 & 6 & 0 & -192 & 0 \\ 0 & 0 & 0 & 0 & 54 & 0 \\ 0 & 1 & 0 & -20 & 0 & 418 \\ 0 & 0 & 0 & 6 & 0 & -192 \\ 0 & 0 & 0 & 0 & 0 & 54 \end{bmatrix}$$

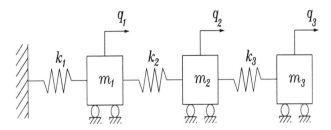

Figure 11.10.1 Spring Mass Model for Observability Test

A MATLAB *m-file fectobt.m* which has been used for the controllability test can be used again for the observability test. Thus, the rank of the $[Q]$ matrix was estimated as six, and the condition number of $[Q]$ was 2.5104×10^3. These results prove that the system is observable. In other words, using one physical sensing device, we can estimate all six variables.

‡

Dynamic Observer Design

As discussed in the previous sections, one key aspect of modern control technique is the so-called full state feedback. In other words, the system is described by a set of state variables and all state variables are combined into a feedback control law named *full state feedback*.

In general, it is not easy, if not impossible, to measure all the state variables. This is mainly due to the number of sensors available compared to the number of state variables. The ideal situation is probably implementing as many sensors as state variables. This approach, however, is neither practical nor cost-effective. Therefore, the fundamental question is how to estimate the state variables and one feasible solution is a dynamic observer.

The dynamic observer is a popular mathematical algorithm in modern control theory and analysis. Basically, the dynamic observer is a mathematical model based upon the given physical system. The mathematical model is used to construct the physical system based upon the sensor input.

Let us assume a dynamic system described in the form

$$\{\dot{x}\} = [A]\{x\} + [B]\{u\}$$
$$\{y\} = [C]\{x\}$$
(11.10.5)

Then a dynamic observer can be prescribed by the following set of equations

$$\{\dot{\hat{x}}\} = [A]\{\hat{x}\} + [B]\{u\} + [L](\{y\} - \{\hat{y}\})$$
$$\{\hat{y}\} = [C]\{\hat{x}\}$$
(11.10.6)

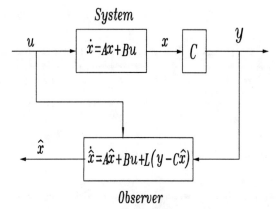

Figure 11.10.2 Graphical Representation of a Dynamic Observer

where $[L]$ is the observer gain to be determined and $\{y\}$ is the sensor output which is to be provided into the observer dynamics. Figure 11.10.2 shows us the graphical representation of a dynamic observer.

In order to help understand the observer, we combine Eqs. (11.10.4) and (11.10.5), obtaining

$$\{\dot{x}\} - \{\dot{\hat{x}}\} = [A]\{x\} - [A]\{\hat{x}\} - [L](\{y\} - \{\hat{y}\}) \qquad (11.10.7)$$

Let us introduce a vector defined as $\{e\} = \{x\} - \{\hat{x}\}$ which represents error between the physical system and dynamic observer. Also, by using the output equation as $\{\hat{y}\} = [C]\{\hat{x}\}$, it follows that

$$\{\dot{e}\} = ([A] - [L][C])\{e\} \qquad (11.10.8)$$

Therefore, the error vector satisfies the equation and

$$\{e(t)\} = exp^{([A]-[L][C])t}\{e(0)\} \qquad (11.10.9)$$

As is shown, the error vector response is explicitly represented as a function of time. The desired behavior of the dynamic observer will be the one with zero error, which means a perfect matching between the system and the observer.

For stability of the error vector response, or $\lim_{t \to \infty}\{e(t)\} = \{0\}$, it should follow that

$$\lambda_i([A] - [L][C]) < 0 \qquad (11.10.10)$$

where λ_i is an ith eigenvalue of $[A] - [L][C]$. Since $[A]$ and $[C]$ are already defined, the main strategy of a dynamic observer design is to design the observer gain matrix $[L]$ in such a way that the closed-loop system $([A] - [L][C])$ is stable. It is interesting to see that the error vector remains trivially zero all the time when the system and observer have the perfectly same initial condition $\{e(0)\} = \{0\}, \to \{e(t)\} = 0, \ t > 0$.

From the closed-loop system matrix $[A] - [L][C]$, we find a useful property designing the observer gain. Since the eigenvalues of the transpose of a matrix are the same as those of the original matrix

$$\lambda_i([A] - [L][C]) \equiv \lambda_i([A]^T - [C]^T[L]^T) \qquad (11.10.11)$$

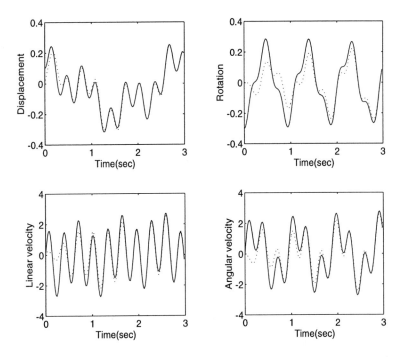

Figure 11.10.3 Simulation Result of a Dynamic Observer(Solid Line - System, Dotted Line - Observer)

the observer gain can be obtained by treating $([A]^T, [C]^T)$ as if they were $([A], [B])$ in the feedback control law design. This *duality* [40,41] between a control law design and an observer design thus eliminates the need for writing a separate observer gain design algorithm.

♣ **Example 11.10.2** A dynamic observer design is demonstrated for the same finite element model used in the previous example. A single sensor measuring displacement is assumed and we want to estimate four state variables. The observer gain is designed by a pole placement technique. The observer gain matrix becomes, as expected, a 1×4 matrix.

$$[L] = [18.31 \ \ 0.620 \ \ 61.64 \ \ 13.43 \ \ 117.7 \ \ 0.984 \ \ 466.3 \ \ 239.3]$$

Also, simulation results using the designed gain matrix and impulsive input at the tip of the structure are presented in Fig. 11.10.3. A MATLAB source code for simulation is provided below. The initial discrepancies between the actual system and observer are shown to diminish asymptotically.

```
%————————————————————————————
%
```

```
% This program obssim.m demonstrates a dynamic observer
% design and displays the simulation result. A finite beam element is
% adopted as a system. For observer gain design, the LQR technique
% is used, and the dynamic simulation is performed using felresp.m
%
%-----------------------------------------------------------------
% Provide the system mass and stiffness matrix
M =[0.5571 0 0.0964 -0.3482
0 3.2143 0.3482 -1.2054
0.0964 0.3482 0.2786 -0.5893
-0.3482 -1.2054 -0.5893 1.6071];
K =[2.4178 0 -1.2089 9.0667
0 181.3333 -9.0667 45.3333
-1.2089 -9.0667 1.2089 -9.0667
9.0667 45.3333 -9.0667 90.6667];
F=[1;0;0;0];
% Transform into the first order state space form equation
A=[0*eye(4),eye(4);-inv(M)*K,0*eye(4)];
B=[0*ones(4,1);inv(M)*F];
C=[1 0 0 0 0 0 0 0];
% Use the felqr.m function to design the observer gain
[L,S]=felqr(A',C',eye(8),0.01);
% Now build the total closed-loop system for both system and observer
Atot=[A, 0*eye(8);L'*C, A-L'*C];
Btot=[B;B];
Ctot=eye(16);
Dtot=eye(16,1);
% Define simulation time, control input, and initial conditions
t=0:0.01:3.0-0.01;
u=zeros(300,1);
x0=zeros(16,1);
x0(1,1)=0.1; x0(2,1)=-0.3; x0(3,1)=0.2;
% Use felresp.m to simulate the total system
[x,y]=felresp(Atot,Btot,Ctot,Dtot,x0,u,t);
%-----------------------------------------------------------------
```

11.11 Compensator Design

In the previous sections, we discussed the control law design and the observer design separately. The observer design is mainly needed in order to provide the feedback control law with estimated state variables. Therefore, the control law and observer are combined into a complete system. The combined system is called a

Section 11.11 Compensator Design

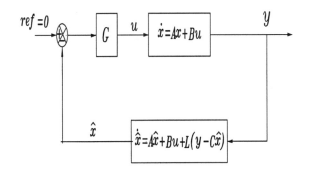

Figure 11.11.1 Combined Dynamic Observer and Feedback Control Law

compensator. The block diagram representation of a compensator is provided in Fig. 11.11.1.

Consider a feedback control law and a dynamic observer as

$$\{\dot{x}\} = [A]\{x\} + [B]\{u\} \qquad (11.11.1)$$

At this point we assume the control input $\{u\}$ as a full state feedback form

$$\{u\} = -[G]\{x\} \qquad (11.11.2)$$

and the observer is given by

$$\{\dot{\hat{x}}\} = [A]\{\hat{x}\} + [B]\{u\} + [L](\{y\} - [C]\{\hat{x}\}) \qquad (11.11.3)$$

As mentioned above, we want to use the estimated state variables from the observer in the control law. In other words,

$$\{u\} = -[G]\{\hat{x}\} \qquad (11.11.4)$$

The suggested control law is substituted into Eqs. (11.11.1) and (11.11.3) producing

$$\{\dot{x}\} = [A]\{x\} - [B][G]\{\hat{x}\} \qquad (11.11.5)$$
$$\{\dot{\hat{x}}\} = ([A] - [B][G] - [L][C])\{\hat{x}\} + [L]\{y\}$$

Substituting the output equation $\{y\} = [C]\{x\}$ and rearranging the above equations yield

$$\begin{Bmatrix} \dot{x} \\ \dot{\hat{x}} \end{Bmatrix} = \begin{bmatrix} [A] & -[B][G] \\ [L][C] & [A] - [B][G] - [L][C] \end{bmatrix} \begin{Bmatrix} x \\ \hat{x} \end{Bmatrix} \qquad (11.11.6)$$

In order to check the stability of the combined system, we solve

$$\begin{vmatrix} \lambda I - [A] & [B]G \\ -[L][C] & \lambda I - [A] + [B][G] + [L][C] \end{vmatrix} = 0 \qquad (11.11.7)$$

Here, for simplification, we want to make use of a property in linear algebra: the determinant of a matrix is invariant by adding a constant multiple of a row (column) to another row (column) [40,41]. Thus, it follows

Table 11.11.1 Closed-Loop Eigenvalues for the Compensator

Number	Eigenvalue
1	-1024.2
2	$-2.075 \pm 20.22i$
3	$-8.059 \pm 12.86i$
4	-9.9833
5	$-0.497 \pm 7.177i$
6	$-0.775 \pm 6.749i$
7	$-0.810 \pm 1.623i$
8	$-1.036 \pm 0.901i$
9	-1.010
10	-1.000

$$\begin{vmatrix} \lambda I - [A] + [B][G] & [B][G] \\ 0 & \lambda I - [A] + [L][C] \end{vmatrix} \quad (11.11.8)$$

Consequently, the characteristic equation of the combined system is simplified into

$$|\lambda I - [A] + [B][G]||\lambda I - [A] + [L][C]| = 0 \quad (11.11.9)$$

The characteristic equation consists of two separate parts: the feedback control law and the dynamic observer, respectively. This implies that one can design the feedback gain $[G]$ first, then the observer gain or vice versa. This property is very elegant, and sometimes called the *separation principle*.

♣ **Example 11.11.1** A compensator design is demonstrated with simulation results. The same finite element model is used as in the previous section. According to the *separation principle*, the compensator is designed in two stages. The first stage is a feedback control law

$$[G] = [2.273 \ 2.769 \ -1.147 \ 11.69 \ 0.594e \ 1.314 \ 0.504 - 1.196] \times 10^3$$

and the second stage is to design a dynamic observer with the result

$$[L] = [18.31 \ 0.621 \ 61.64 \ 13.43 \ 117.7 \ 0.984 \ 466.3 \ 239.3 \ ]$$

from the result of Example 11.10.2. The closed-loop system eigenvalues are provided in Table 11.11.1. Also, simulation is performed with the results provided in Fig. 11.11.2. As is shown, the system is stabilized by the feedback control law which makes use of estimated state variables from the dynamic observer.

```
%----------------------------------------
%
% This program compen.m demonstrates a dynamic observer
% design and displays the simulation result. A finite typical element is
% adopted as a system. For observer gain design, the LQR technique
% is used, and the dynamic simulation is performed using felresp.m.
%
%----------------------------------------
% Provide the system mass and stiffness matrix
M =[0.5571 0 0.0964 -0.3482
0 3.2143 0.3482 -1.2054
0.0964 0.3482 0.2786 -0.5893
-0.3482 -1.2054 -0.5893 1.6071];
K =[2.4178 0 -1.2089 9.0667
0 181.3333 -9.0667 45.3333
-1.2089 -9.0667 1.2089 -9.0667
9.0667 45.3333 -9.0667 90.6667];
F=[1;0;0;0];
% Transform into the first order state space form equation
A=[0*eye(4),eye(4);-inv(M)*K,0*eye(4)];
B=[0*ones(4,1);inv(M)*F];
C=[1 0 0 0 0 0 0 0];
% Use the felqr.m function to design the observer gain and the
% full state feedback gain
[G,Sc]=felqr(A, B, 1000*eye(8), 0.01);
[L,So]=felqr(A',C', eye(8),0.01);
% Now build the total closed-loop system for both closed-loop system
and observer
Atot=[A, -B*G;L'*C, A-L'*C-B*G];
Btot=[B;B];
Ctot=eye(16);
Dtot=eye(16,1);
% Define simulation time, control input, and initial conditions
t=0:0.01:6.0-0.01;
u=zeros(600,1);
x0=zeros(16,1);
x0(1,1)=0.1; x0(2,1)=-0.3; x0(3,1)=0.2;
% Use felresp.m to simulate the total system
[x,y]=felresp(Atot,Btot,Ctot,Dtot,u,t,x0);
%----------------------------------------
```

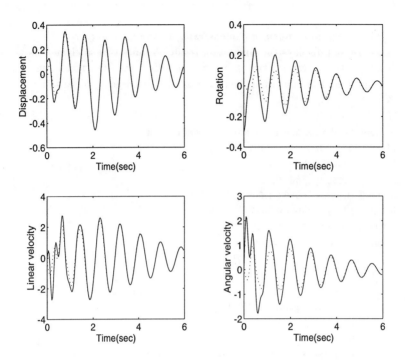

Figure 11.11.2 Simulation Results of a Compensator

11.12 Output Feedback Design by Using Collocated Sensor/Actuator

Discrete Multiple Degrees of Freedom System

A control law design and implementation for flexible structures involves a number of technical issues. One of them is the effect of sensor and actuator locations. Since the vibrational motion of the structure induces phase error at different locations of the structure, the sensor and actuator placement should take the phase difference into account. The best strategy is to place the sensor and the actuator at the same location called a collocated sensor/actuator system. The collocated sensor and actuator pair has some inherent advantages such as stability guarantee despite potential technical problems which may arise in the collocation process.

Consider a linearized undamped dynamic system

$$[M]\{\ddot{q}\} + [K]\{q\} = [F]\{u\} \qquad (11.12.1)$$

and assume a sensor measurement

$$\{y\} = [C]\{\dot{q}\} \qquad (11.12.2)$$

The input influence matrix $[F]$ and the output distribution matrix $[C]$ represent location of actuators and sensors, respectively. For a collocated sensor/actuator pair, they are identical [51]. In other words,

$$[C] = [F] \qquad (11.12.3)$$

Section 11.12   Output Feedback Design by Using Collocated Sensor/Actuator   495

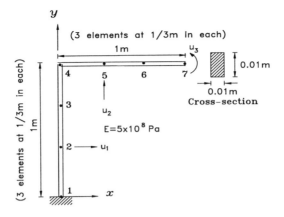

Figure 11.12.1 An L-Shape Frame Structure with Three Actuators

Consider an output feedback control law defined as

$$\{u\} = -[G]\{y\} \tag{11.12.4}$$
$$= -[G][F]\{\dot{\mathbf{q}}\}$$

where $[G]$ is a gain matrix. In order to check the stability of the system, we take a Lyapunov function

$$U = \frac{1}{2}(\{\dot{\mathbf{q}}\}^T[M]\{\dot{\mathbf{q}}\} + \{\mathbf{q}\}^T[K]\{\mathbf{q}\}) \tag{11.12.5}$$

and the time derivative of the Lyapunov function follows as

$$\dot{U} = \{\dot{\mathbf{q}}\}^T[F]\{\mathbf{u}\} \tag{11.12.6}$$

Substituting the suggested control law [51]

$$\dot{U} = -\{\dot{\mathbf{q}}\}^T[F]^T[G][F]\{\dot{\mathbf{q}}\} \tag{11.12.7}$$

and the condition for stability depends upon the feedback gain matrix $[G]$. The gain matrix should be positive definite for stability guarantee $\dot{U} < 0$ for $\{\dot{\mathbf{q}}\} \neq 0$. In fact, $\{\dot{\mathbf{q}}\}$ becomes zero only instantaneously; hence, $\dot{U}$ remains negative except for the perfect equilibrium point where $(\{\mathbf{q}\},\{\dot{\mathbf{q}}\}) = (0,0)$. Therefore, the control law design is rather simple for a collocated sensor/actuator pair, guaranteeing the stability of the system.

♣ **Example 11.12.1** In this example, the output feedback law design example with collocated sensors and actuators is demonstrated for an L-shaped frame structure. The L-shaped frame structure in Fig. 11.12.1 is modeled by six finite elements. Detailed finite element analysis is provided in Example 8.10.3. Each node has three degrees of freedom: vertical and horizontal displacements, and rotation. The material property is selected as $E = 5 \times 10^8 N/m^2$,

$\rho = 1000 kg/m^3$ and $A = 0.0001 m^2$. There are three actuators located at node numbers 2, 5, and 7. The direction of each actuator is also described in Fig. 11.12.1 - horizontal at node $2(u_1)$, vertical at node $5(u_2)$, and rotational at node $7(u_3)$. The finite element analysis result gives us

$$[M]\{\ddot{q}\} + [K]\{q\} = [F]\{u\}$$

where $[M]$ and $[K]$ are 18 by 18 mass and stiffness matrices, respectively. The control input vector $\{u\} = [u_1, u_2, u_3]^T$ includes the three actuators. The input influence matrix $([F])$ is easily computed as

$$[F] = \begin{bmatrix} 1 & 0 & 0 & 0 & 0 & 0 & 0 & 0 & 0 & 0 & 0 & 0 & 0 & 0 & 0 & 0 & 0 & 0 \\ 0 & 0 & 0 & 0 & 0 & 0 & 0 & 0 & 0 & 0 & 1 & 0 & 0 & 0 & 0 & 0 & 0 & 0 \\ 0 & 0 & 0 & 0 & 0 & 0 & 0 & 0 & 0 & 0 & 0 & 0 & 0 & 0 & 0 & 0 & 0 & 1 \end{bmatrix}$$

By assuming sensor measurement at the same location where the actuators are located, the control law is designed as simply

$$u_1 = -g_1 \dot{q}_h^1, \quad u_2 = -g_2 \dot{q}_v^5, \quad u_3 = -g_3 \dot{\theta}^7$$

where $g_1 > 0$, $g_2 > 0$, $g_3 > 0$ are feedback gains. In addition, q_h^1, q_v^5 represent horizontal and vertical displacements at node numbers 1 and 5, respectively, and θ^7 represents rotation at the 7th node. The control law is easily rewritten as

$$\{u\} = -[G][F]\{\dot{q}\}$$

where $[G]$ and $[F]$ are matrices with appropriate dimension. The control law should stabilize the system. In order to verify this, we apply the control law to the orginal second order system, and the resultant closed-loop system has the following form

$$[M]\{\ddot{q}\} + [F][G][F]^T\{\dot{q}\}[K]\{q\} = 0$$

The above system is easily transformed into a first order system by Eq. (11.5.3) as

$$\{\dot{x}\} = [A]\{x\} + [B]\{u\}$$

Note $[B] = 0$ in this case, since the equation is developed from the closed-loop system. The system now is simplified as $\{\dot{x}\} = [A]\{x\}$. The initial conditions are selected in an arbitrary manner as

> Horizontal displacement at node 1 = 0.1m
>
> Vertical displacement at node 5 = 0.1m
>
> Angular rotation at node 18 = 0.05 rad

The MATLAB m-file *felresp.m* is used for the simulation for 6 seconds. The time responses of the horizontal and vertical displacement at node numbers 2 and 5, and of the angular displacement at node number 7 are displayed in Fig. 11.12.2 including the actuator response at node number 2. The simulation results indicate that the system is stabilized by the feedback control law. The

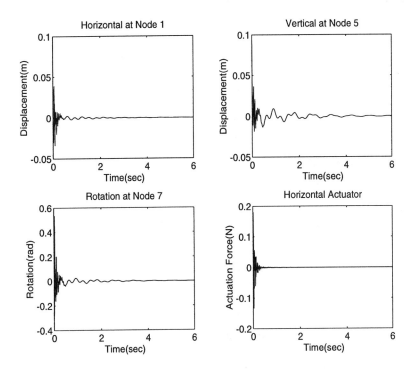

Figure 11.12.2 Simulation Results for the L-Shape Frame Structure by DVF Controller

big advantage of the control law used in this example is the robustness of the control law. We used only three feedback gains, which is in significant contrast with the LQR design where we need as many feedback gains as the size of the system.

‡

Infinite Dimensional Continuous System

As mentioned earlier, control law design for flexible structures is complicated. In particular, the controller performance relies upon the accuracy of mathematical modeling. There are many issues in mathematical modeling such as finite dimensional truncation, model uncertainty, and estimation of state variables.

One promising method for control law design is to retain the original governing equations of motion. Dynamics of the flexible structure are usually described by partial differential equations of motion or a combination of partial and ordinary differential equations of motion. A significant advantage of using the original governing equations is the robustness of the control law without involving finite dimensional approximation, and there is no modeling error issue in this approach.

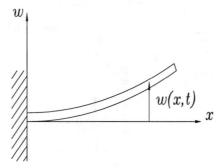

Figure 11.12.3 A Typical Continuous System

Let us consider a flexible beam with one end fixed to the base and the other end free as in Fig. 11.12.3. The governing equation of motion is given as

$$\rho \frac{\partial^2 w}{\partial t^2} + EI \frac{\partial^4 w}{\partial x^4} = F\delta(x - x_c) \tag{11.12.8}$$

and the boundary conditions are given by

$$w = \frac{\partial w}{\partial x} = 0 \quad \text{at} \quad x = 0 \tag{11.12.9}$$

$$EI \frac{\partial w^2}{\partial x^2} = EI \frac{\partial w^3}{\partial x^3} = 0 \quad \text{at} \quad x = l$$

where ρ is the linear mass density and EI is the beam rigidity. The actuator located at $x = x_c$ is represented by the *Dirac Delta* function ($\delta(x - x_c)$).

In order to design a control law, the Lyapunov approach is adopted with the candidate Lyapunov function

$$U = \frac{1}{2} \int_0^l \left[\rho \left(\frac{\partial w}{\partial t} \right)^2 + EI \left(\frac{\partial^2 w}{\partial x^2} \right)^2 \right] dx \tag{11.12.10}$$

The Lyapunov function represents the total kinetic plus potential energy of the structure. The time derivative of the Lyapunov function is taken in combination with Eqs. (11.12.8) and (11.12.9), and the final result becomes

$$\dot{U} = \int_0^l \dot{w} F \delta(x - x_c) dx \tag{11.12.11}$$

Integration over the *Dirac Delta* function becomes

$$\dot{U} = \dot{w}(x_c, t) F \tag{11.12.12}$$

The natural choice of a stabilizing control input results in

$$F = -g\dot{w}(x_c, t) \tag{11.12.13}$$

Section 11.12 Output Feedback Design by Using Collocated Sensor/Actuator   499

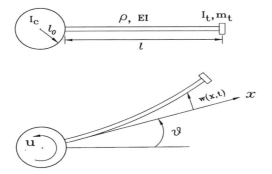

Figure 11.12.4 A Rotating Flexible Beam Attached at the Base

where $g > 0$ is the feedback gain which guarantees stability with $\dot{U} < 0$. Therefore, the derived control law globally stabilizes the structure, and there is no need for modeling the structure. This control law is usually called Direct Velocity Feedback(DVF) control law.

Output Feedback Control Law Design for a Rotating Beam

Rotating flexible beams have been used in many engineering applications such as robotics and space engineering. The mathematical modeling of a flexible beam can be done using the finite element method or other methods. Once we have the mathematical modeling, then we can develop a control law based upon the desired control objective.

The governing equations of motion of a rotating flexible beam in Fig. 11.12.4 are derived as

$$I_c\ddot{\theta} + \int_{l_0}^{l} \rho x(\ddot{w} + x\ddot{\theta})dx + m_t l(l\ddot{\theta} + \ddot{w}(l,t)) + I_t(\ddot{\theta} + \ddot{w}'(l,t)) = u$$

$$\rho(\ddot{w} + x\ddot{\theta}) + EI\frac{\partial^4 w}{\partial x^4} = 0 \qquad (11.12.14)$$

with the boundary conditions

$$w = \frac{\partial w}{\partial x} = 0, \qquad at \qquad x = l_0 \qquad (11.12.15)$$

$$EI\frac{\partial^2 w}{\partial x^2} = -I_t(\ddot{\theta} + \ddot{w}'), \qquad EI\frac{\partial^3 w}{\partial x^3} = m_t(l\ddot{\theta} + \ddot{w}), \qquad at \qquad x = l$$

where $(\ddot{\ }) \equiv \frac{\partial^2(\ )}{\partial t^2}$ and $(\ )' \equiv \frac{\partial(\ )}{\partial x}$.

Finite Element Modeling

For finite element modeling, the *extended Hamilton's principle* is used for each element as [51-53]

$$\int_{t_1}^{t_2} (\delta L + \delta W)dt = 0 \qquad (11.12.16)$$

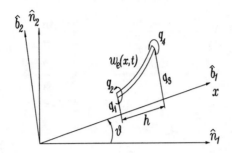

Figure 11.12.5 A Finite Element of A Rotating Flexible Beam

where $L = T_e - V_e$ is the system Lagrangian which is the difference between the kinetic (T_e) and the potential (V_e) energies. The element kinetic and potential energies are given by

$$T_e = \int_{x_e}^{x_e+h} \rho(\dot{w}_e + x\dot{\theta})^2 dx, \quad V_e = \int_{x_e}^{x_e+h} EI\left(\frac{\partial^2 w}{\partial x^2}\right)^2 dx \qquad (11.12.17)$$

Now we introduce a finite dimensional discretization for the w_e over the ith element as shown in Fig. 11.12.5.

$$w_e(x,t) = \phi_1(x)q_1(t) + \phi_2(x)q_2(t) + \phi_3(x)q_3(t) + \phi_4(x)q_4(t) \qquad (11.12.18)$$

where $q_1(t)$ is the vertical displacement and $q_2(t)$ is the rotation at the left-end of the element, and similar definitions exist for $q_3(t)$ and $q_4(t)$ for the right-end of the element. The shape functions are Hermite polynomials

$$\begin{aligned}
\phi_1(x) &= 1 - 3\left(\frac{x-x_i}{h}\right)^2 + 2\left(\frac{x-x_i}{h}\right)^2 \\
\phi_2(x) &= (x-x_i) - 2h\left(\frac{x-x_i}{h}\right)^2 + h\left(\frac{x-x_i}{h}\right)^3 \\
\phi_3(x) &= 3\left(\frac{x-x_i}{h}\right)^2 - 2\left(\frac{x-x_i}{h}\right)^3 \\
\phi_4(x) &= -h\left(\frac{x-x_i}{h}\right)^2 + h\left(\frac{x-x_i}{h}\right)^3
\end{aligned} \qquad (11.12.19)$$

Substitution of the above expressions into Eq. (11.12.17) and integration by parts yield [51-53]

$$M_e^i \ddot{\mathbf{q}}_e + K_e^i \mathbf{q}_e = 0 \qquad (11.12.20)$$

where the element mass and stiffness matrices are presented in Table 11.12.1. The same principle can be applied to the elements which include the tip and the rigid center body. Once we set up the equations for every finite element, then we combine all equations for each element into global governing equations as a typical finite element

Section 11.12 Output Feedback Design by Using Collocated Sensor/Actuator

Table 11.12.1 Element Matrix for the Rotating Beam

$$M_e^{(i)} = \begin{bmatrix} M_{11}^i & M_{12}^i & M_{13}^i \\ M_{21}^i & M_{22}^i & M_{23}^i \\ M_{31}^i & M_{32}^i & M_{33}^i \end{bmatrix}, \quad K_e^{(i)} = \begin{bmatrix} 0 & 0 & 0 \\ 0 & K_{22}^i & K_{23}^i \\ 0 & K_{32}^i & K_{33}^i \end{bmatrix}, \quad M_t = \begin{bmatrix} M_{11}^t & M_{12}^t \\ M_{21}^t & M_{22}^t \end{bmatrix}$$

$$M_{11}^i = \frac{\rho}{3}\{(x_i + l_0)^2 + (x_i + l_0 + h)(x_i + l_0) + (x_i + l_0 + h)^2\}$$

$$M_{12}^i = [M_{21}^i]^T = \rho h \left[\frac{3}{20} + \frac{1}{2}(x_i + l_0) \quad \frac{1}{30}h^2 + \frac{1}{12}h(x_i + l_0) \right]$$

$$M_{13}^i = [M_{31}^i]^T = \rho h \left[\frac{7}{20}h + \frac{1}{2}(x_i + l_0) \quad -\frac{1}{20}h^2 - \frac{1}{12}h(x_i + l_0) \right]$$

$$M_{22}^i = \frac{\rho h}{420} \begin{bmatrix} 156 & 22h \\ 22h & 4h^2 \end{bmatrix}$$

$$M_{23}^i = [M_{32}^i]^T = \frac{\rho h}{420} \begin{bmatrix} 54 & -13h \\ 13h & -3h^2 \end{bmatrix}$$

$$M_{33}^i = \frac{\rho h}{420} \begin{bmatrix} 156 & -22h \\ -22h & 4h^2 \end{bmatrix}$$

$$K_{22}^i = \frac{EI}{h^3} \begin{bmatrix} 12 & 6h \\ 6h & 4h^2 \end{bmatrix}$$

$$K_{23}^i = [K_{32}^i]^T = \frac{EI}{h^3} \begin{bmatrix} -12 & 6h \\ -6h & 2h^2 \end{bmatrix}$$

$$K_{33}^i = \frac{EI}{h^3} \begin{bmatrix} 12 & -6h \\ -6h & 4h^2 \end{bmatrix}$$

$$M_{11}^t = J_t + m_t(l_0 + l)^2$$

$$M_{12}^t = [M_{21}^t]^T = [m_t(l_0 + l) \quad I_t]$$

$$M_{22}^t = \begin{bmatrix} m_t & 0 \\ 0 & I_t \end{bmatrix}$$

analysis procedure. Therefore, the resultant governing equations of motion are derived as

$$[M]\{\ddot{q}\} + [K]\{q\} = \{F\}u \qquad (11.12.21)$$

where

$$\{q\} = [\theta, q_1, q_2, \ldots, q_N]^T, \qquad \{F\} = [1, 0, \ldots, 0]^T$$

and the global mass and stiffness matrices are given by

$$[M] = \begin{bmatrix} I_c + M_{\theta\theta} & M_{\theta q} \\ M_{q\theta} & M_{qq} \end{bmatrix}, \quad [K] = \begin{bmatrix} 0 & \underline{0} \\ \underline{0}^T & K_{qq} \end{bmatrix} \qquad (11.12.22)$$

where $\underline{0}$ is a zero vector of size $1 \times N$, and

$$M_{\theta\theta} = \sum_{i=1}^{N} M_{11}^i + M_{11}^t$$

$$M_{\theta \mathbf{q}} = [M_{13}^1 + M_{12}^2 \quad M_{13}^2 + M_{12}^3 \quad M_{13}^3 + M_{12}^4 \ldots M_{13}^{N-1} + M_{12}^N \quad M_{13}^N + M_{12}^t]$$

Furthermore,

$$M_{\mathbf{qq}} = \begin{bmatrix} M_{33}^1 + M_{22}^2 & M_{23}^2 & & & & \\ M_{32}^2 & M_{33}^2 + M_{22}^3 & M_{23}^3 & & & \\ & M_{32}^3 & M_{33}^3 + M_{22}^4 & M_{23}^4 & & \\ & & & \ddots & & \\ & & & & M_{32}^{N-1} & \bar{M} & M_{23}^N \\ & & & & & M_{32}^N & M_{33}^N \end{bmatrix}$$

and

$$K_{\mathbf{qq}} = \begin{bmatrix} K_{33}^1 + K_{22}^2 & K_{23}^2 & & & & \\ K_{32}^2 & K_{33}^2 + K_{22}^3 & K_{23}^3 & & & \\ & K_{32}^3 & K_{33}^3 + K_{22}^4 & K_{23}^4 & & \\ & & & \ddots & & \\ & & & & K_{32}^{N-1} & \bar{K} & K_{23}^N \\ & & & & & K_{32}^N & K_{33}^N \end{bmatrix}$$

where $\bar{M} = M_{33}^{N-1} + M_{22}^N$ and $\bar{K} = K_{33}^{N-1} + K_{22}^N$. The second order system in Eq. (11.12.20) can be transformed into a first order system for analysis purposes.

♣ **Example 11.12.2** An example rotating beam model is used to verify the modeling procedure discussed above. The material properties of the example model are $\rho = 0.003$, $EI = 1.1118 \times 10^4$, $l_0 = 3.5$, $L = 47.57$, $I_t = 0.0018$, $m_t = 0.156$, and $I_c = 9.06$ with consistent units. A MATLAB *m-file* *ferobem.m* is written and the output results are system mass/stiffness matrices and natural frequencies. The first nine natural frequencies are listed in Table 11.12.2.

```
function [w,M,K]=ferobem(N,EI,rho,I_c,I_t,m_t,l_0,L)
%----------------------------------------------------------------
% Purpose:
%     The MATLAB function subroutine ferobem.m produces a finite
%     element modeling of a rotating flexible beam attached
%     to a rigid base.
%
% Synopsis:
%     [w,M,K]=ferobem(N,EI,rho,I_c,I_t,m_t,l_0,L)
%
% Variable Description:
%     Input parameters :
%           N - number of elements
%           EI - beam rigidity
%           rho - linear mass density
```

Section 11.12 Output Feedback Design by Using Collocated Sensor/Actuator   503

```
%              I_c(I_t) - moment of inertia of the center body (tip mass)
%              m_t - tip mass
%              l_0 - radius of the center body
%              L - beam length
%    Output : M, K - system mass, stiffness matrices
%              w - natural frequency
%--------------------------------------------------------------
%--------------------------------------------------------------
% Calculate pure rigid body portion
%--------------------------------------------------------------
h=L/N;                                              % Element length
M11t=I_t+m_t*(L+l_0)^2;M12t=[m_t*(l_0+L) I_t];
M22t=[m_t 0; 0 I_t];M21t=M12t';
xi=0;
for i=1:N
M11r(1,i)=rho*h*((xi+l_0)^2+(xi+l_0+h)*(xi+l_0)+(xi+l_0+h)^2)/3;
xi=xi+h;
end
Mthth=M11t+sum(M11r)+I_c;
%--------------------------------------------------------------
% Calculate element mass and stiffness matrices
%--------------------------------------------------------------
M33=rho*h*[156,-22*h;-22*h,4*h^2]/420;
M22=rho*h*[156, 22*h; 22*h,4*h^2]/420;
M23=rho*h*[54,-13*h;13*h,-3*h^2]/420;M32=M23';
K22=EI*[ 12, 6*h; 6*h,4*h^2]/h^3;
K23=EI*[-12, 6*h;-6*h,2*h^2]/h^3;K32=K23';
K33=EI*[ 12,-6*h;-6*h,4*h^2]/h^3;
%--------------------------------------------------------------
% Calculate global mass and stiffness matrices
%--------------------------------------------------------------
Mqq(1:2,1:2)=M33+M22;Mqq(1:2,3:4)=M23;
for i=1:N-2
NI=2*i+1;
Mqq(NI:NI+1,NI-2:NI-1)=M32;Mqq(NI:NI+1,NI:NI+1)=M33+M22;
Mqq(NI:NI+1,NI+2:NI+3)=M23;
end
Mqq(2*N-1:2*N,2*N-3:2*N-2)=M32;
Mqq(2*N-1:2*N,2*N-1:2*N)=M33+M22t;
Kqq(1:2,1:2)=K33+K22;Kqq(1:2,3:4)=K23;
for i=1:N-2
NI=2*i+1;
Kqq(NI:NI+1,NI-2:NI-1)=K32;Kqq(NI:NI+1,NI:NI+1)=K33+K22;
Kqq(NI:NI+1,NI+2:NI+3)=K23;
end
Kqq(2*N-1:2*N,2*N-3:2*N-2)=K32; Kqq(2*N-1:2*N,2*N-1:2*N)=K33;
%--------------------------------------------------------------
% Compute rigid and flexible coupling terms
```

504   Control of Flexible Structures   Chapter 11

Table 11.12.2 Natural Frequencies of the Rotating Beam

Mode	Natural Frequencies(Hz)
0	0.000
1	1.030
2	3.030
3	7.360
4	14.82
5	27.38
6	44.36
7	70.27
8	100.8

```
%————————————————————————————————
xi=0;
for i=1:N-1,
M13=rho*h*[7*h/20+(xi+l_0)/2 -h^2/20-h*(xi+l_0)/12];
xi=xi+h;
M12=rho*h*[3*h/20+(xi+l_0)/2 h^2/30+h*(xi+l_0)/12];
Mthq(1,2*i-1:2*i)=M13+M12;
end
M13=rho*h*[7*h/20+(xi+l_0)/2 -h^2/20-h*(xi+l_0)/12];
Mthq(1,2*N-1:2*N)=M13+M12t;
%————————————————————————————————
% Combine rigid, flexible, and coupling terms
%————————————————————————————————
M=[Mthth,Mthq;Mthq',Mqq];        % Global mass and stiffness matrices
K=[0,zeros(1,2*N);zeros(2*N,1),Kqq];
wo=sqrt(eig(K,M));               % Compute natural frequencies
w=sort(wo);
%————————————————————————————————
```

‡

Control Law Design by Lyapunov Approach

As a special case, we assume a collocated sensor/actuator pair for the rotating beam. The actuator is located at the center body producing torque about the vertical axis and the sensor is also located at the center body measuring the angular displacement and/or angular velocity of the center body. With the collocated sensor/actuator set, the control law design is relatively simple. In the previous section,

Section 11.12 Output Feedback Design by Using Collocated Sensor/Actuator

we used an original partial differential equation deriving a stabilizing control law, and the same principle can be applied to the rotating beam case.

First, we select a candidate Lyapunov function as [54-56]

$$2U = a_1 I_c \dot{\theta}^2 + a_2 \left[\int_{l_0}^{l} [\rho(x\dot{\theta} + \dot{w})^2 + EI\left(\frac{\partial^2 y}{\partial x^2}\right)^2] dx + m_t(l\dot{\theta} + \dot{w}(l,t))^2 \right. $$
$$\left. + I_t(\dot{\theta} + \dot{w}'(l,t))^2 \right] + a_3(\theta - \theta_f)^2 \qquad (11.12.23)$$

The Lyapunov function is shown as the combination of each substructure's energy: center body, beam and tip mass. Furthermore, a_1, a_2, and a_3 are positive weighting constants determining the relative importance of the substructure's energy, and θ_f is a constant final desired angle. The Lyapunov function is positive definite with respect to the steady equilibrium point

$$(\theta, \dot{\theta}, y, \dot{y})_f = (\theta_f, 0, 0, 0) \qquad (11.12.24)$$

In other words, the Lyapunov function, initially being positive, approaches zero at the steady equilibrium point. The control torque at the center body should be designed in such a way that the Lyapunov function decreases asymptotically toward the equilibrium point. For this purpose, we take the time derivative of the given Lyapunov function, and make use of the governing equation and boundary condition finally arriving at [54-56]

$$\dot{U} = a_1 \left[u + g_1(\theta - \theta_f) + g_3(l_0 S_0 - M_0) \right] \dot{\theta} \qquad (11.12.25)$$

where $g_1 = a_3/a_1 > 0$, $g_3 = (a_2 - a_1)/a_1 > -1$ are design parameters or feedback gains of the control law. Furthermore, M_0 and S_0 are the internal bending moment and shear force, respectively, at the root of the beam.

$$M_0 = EI \frac{\partial^2 w}{\partial x^2}\bigg|_{x=l_0}, \qquad S_0 = EI \frac{\partial^3 w}{\partial x^3}\bigg|_{x=l_0} \qquad (11.12.26)$$

Since our goal is to design a stabilizing control torque input, the most natural choice is to make the time derivative of the Lyapunov function negative in such a way that

$$u + g_1(\theta - \theta_f) + g_3(l_0 S_0 - M_0) = -g_2 \dot{\theta}, \qquad g_2 > 0 \qquad (11.12.27)$$

Therefore,

$$u = -g_1(\theta - \theta_f) - g_2 \dot{\theta} - g_3(l_0 S_0 - M_0) \qquad (11.12.28)$$

so that

$$\dot{U} = -a_1 g_2 \dot{\theta}^2 < 0 \qquad (11.12.29)$$

As we can see, $\dot{U} < 0$ as long as $\dot{\theta} \neq 0$. At $\dot{\theta} = 0$, the Lyapunov function is equal to zero which does not mean that the system is at equilibrium condition due to other nonzero motions like angular position error and flexible motion. In order to reach the steady equilibrium state, the Lyapunov function continues to decrease as dictated by Eq. (11.12.24).

The derived control law in Eq. (11.13.28) globally stabilizes the flexible structure with respect to the desired equilibrium point. Since we do not use any finite approximation, the control law is free of usual issues such as robustness, truncation error, and modeling uncertainty. The boundary force and moment term are modeled using the Hermite polynomials as [57]

$$l_0 S_0 - M_0 = l_0 EI \frac{\partial^3 y}{\partial x^3}\bigg|_{l_0} - EI \frac{\partial^2 y}{\partial x^2}\bigg|_{l_0}$$
$$= l_0 \left(-\frac{12}{h^3} v_2^1 - \frac{6}{h^2} \theta_2^1\right) - \left(\frac{6}{h^2} v_2^1 - \frac{2}{h} \theta_2^1\right) \quad (11.12.30)$$

where v_2^1 and θ_2^1 represent displacement and rotation at the right-hand side nodal point of the first element.

Simulation results with the control law employed is provided in Figs. 11.12.6 and 11.12.7 together with a MATLAB *m-file*. One significant difference in the two simulation results is the effect of boundary force feedback gain. The boundary force is shown to be a sensitive parameter which improves the closed-loop response performance.

♣ **Example 11.12.3** In this example, the feedback control law in Eq. (11.12.18) is demonstrated for the same model used in Example 11.12.2. The flexible beam was modelled with three finite elements. The feedback gains are chosen as $g_1 = 100$, $g_2 = 200, g_3 = 0, -0.5$. The desired final angle is 1 radian. The feedback gain (g_3) on the boundary force feedback is tested to investigate its effect on the closed-loop performance.

As we can see, the center body angle converges to the final angle within 40 seconds of simulation time. Also, the feedback on boundary force with $g_3 = -0.5$ enhances the closed-loop performance.

```
function [y]=ferbsim(M,K,F,g1,g2,g3,EI,h,l0,thf,tf)
%---------------------------------------------------------------
% Purpose:
%    This MATLAB m-file ferbsim.m is a simulation program for
%    a rotating flexible beam attached to a base. The mathematical model
%    is created from frobfem.m as system mass and stiffness matrices
%
% Synopsis:
%    [y] = ferbsim(M,K,F,g1,g2,g3,EI,h,l0,thf,tf)
%
% Variable Description:
%    Input parameters - M, K, F - System matrices
%                       g1, g2, g3 - Feedback gains
%                       EI, h, l0 - Parameters for boundary force calculation
%                       thf, tf - Final angle and final simulation time
```

```
%       Output parameter - y
%
%--------------------------------------------------
[n,n]=size(M);
I=eye(n);
% Build closed-loop system matrices
K(1,1)=g1;
K1I=EI*(10*[-12/h^3,6/h^2]-[6/h^2,-2/h]);
K(1,2:3)=K(1,2:3)+g3*K1I;
Damp=0*I;
Damp(1,1)=g2;
% Generate state space form for simulation purpose
A=[0*I,I;-inv(M)*K,-inv(M)*Damp];
B=[zeros(n,1);inv(M)*F];
C=eye(2*n);
D=zeros(2*n,1);
% Perform simulation using felresp.m MATLAB function
nstep=1000;
dt=tf/nstep;
t=0:dt:tf-dt;
u= g1*thf*ones(1000,1);
x0=zeros(2*n,1);
y=felresp(A,B,C,D,x0,u,t);
%--------------------------------------------------
```

‡

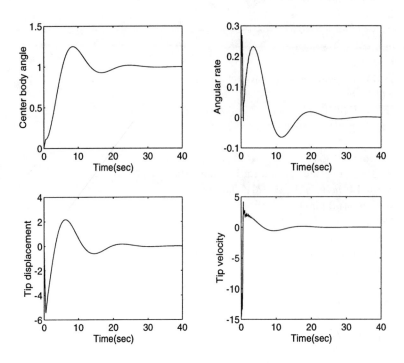

Figure 11.12.6 Simulation Results of a Rotating Beam with $g_1 = 100$, $g_2 = 200$, $g_3 = 0$

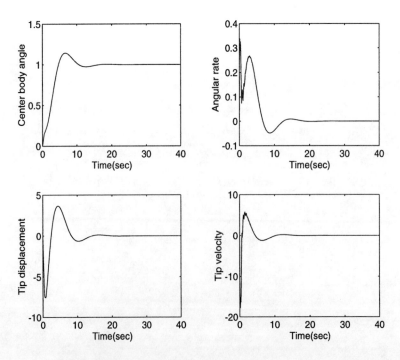

Figure 11.12.7 Simulation Results of a Rotating Beam with $g_1 = 100$, $g_2 = 200$, $g_3 = -0.5$

Problems

11.1 For a given system

$$\dot{x}_1 = x_2$$
$$\dot{x}_2 = -x_2 - x_1 - (2x_2 + x_1)^2 x_2$$

i) Find out the equilibrium point by setting $\dot{x}_1 = 0, \dot{x}_2 = 0$.
ii) Using a Lyapunov function $V(\mathbf{x}) = x_1^2 + x_2^2$, find out the stability of the system.

11.2 Calculate the state transition matrix for a system

$$\begin{Bmatrix} \dot{x}_1 \\ \dot{x}_2 \end{Bmatrix} = \begin{bmatrix} -1 & 0 \\ 1 & 0 \end{bmatrix} \begin{Bmatrix} x_1 \\ x_2 \end{Bmatrix} + \begin{bmatrix} 0 \\ -1 \end{bmatrix} u(t)$$

11.3 A second order system is given as

$$\ddot{x} + 2\dot{x} + 10x = 10f(t)$$

with an unit step input $f(t) = 1, (t \geq 0)$ applied, find out parameters such as
i) Rise time(t_r),
ii) Peak time(t_p),
iii) Maximum overshoot(M_p), and
iv) Settling time(t_s)

11.4 A system is described as

$$\begin{Bmatrix} \dot{x}_1 \\ \dot{x}_2 \end{Bmatrix} = \begin{bmatrix} 0 & 1 \\ -4 & 0 \end{bmatrix} \begin{Bmatrix} x_1 \\ x_2 \end{Bmatrix} + \begin{bmatrix} 0 \\ 1 \end{bmatrix} u$$

i) Derive the analytical expression for the impulse response due to $u(t) = \delta(t)$ and $[x_1, x_2](0) = [0, 0]$.
ii) Use the same data in part i) to derive the analytical expression of the response with respect to the unit step input.

11.5 Assume a transfer function is given by

$$H(s) = \frac{1}{(s+1)(s+3)}$$

i) Find out the unit impulse response.
ii) Do the same thing as in i) for unit step input.

11.6 Using the state transition matrix property, show that

$$\frac{\partial \Phi(t, \tau)}{\partial \tau} = -\Phi(t, \tau) A(\tau)$$

11.7 A typical linearized second order system is given as

$$\begin{bmatrix} m_1 & 0 \\ 0 & m_2 \end{bmatrix} \begin{Bmatrix} \ddot{q}_1 \\ \ddot{q}_2 \end{Bmatrix} + \begin{bmatrix} k_1 + k_2 & -k_2 \\ -k_2 & k_2 \end{bmatrix} \begin{Bmatrix} q_1 \\ q_2 \end{Bmatrix} = \begin{bmatrix} 1 \\ 0 \end{bmatrix} u$$

i) Find out the analytical expression of the transfer function between $q_1(t)$, $q_2(t)$, and $u(t)$. In other words, find $Q_1(s)/U(s)$ and $Q_2(s)/U(s)$.

ii) For the given values $[m_1, m_2] = [0.5, 0.5](kg)$, and $[k_1, k_2] = [2, 4]$, plot the poles and zeros of $Q_1(s)/U(s)$ and $Q_2(s)/U(s)$ on the complex plane. Also, compare the zeros of the two transfer function.

iii) We want to design a feedback control law as

$$u(t) = -g_1 q_1(t) - g_2 \dot{q}_1(t)$$

where g_1 and g_2 are positive position and velocity feedback gain, respectively. Show that the control law stabilizes the closed-loop system.

11.8 A set of spring mass and damper system is given in Fig. P11.8.

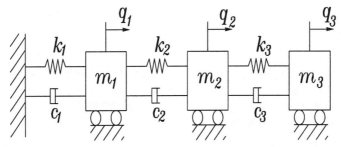

Figure P11.8 A Spring Mass and Damper System

The governing equations of motion are

$$[M]\{\ddot{q}\} + [D]\{\dot{q}\} + [K]\{q\} = [F]u$$

where

$$[M] = \begin{bmatrix} m_1 & 0 & 0 \\ 0 & m_2 & 0 \\ 0 & 0 & m_3 \end{bmatrix}, \quad [D] = \begin{bmatrix} c_1 + c_2 & -c_2 & 0 \\ -c_2 & c_2 + c_3 & -c_3 \\ 0 & -c_3 & c_3 \end{bmatrix}$$

and

$$[K] = \begin{bmatrix} k_1 + k_2 & -k_2 & 0 \\ -k_2 & k_2 + k_3 & -k_3 \\ 0 & -k_3 & k_3 \end{bmatrix}, \quad F = \begin{bmatrix} 0 \\ 0 \\ 1 \end{bmatrix}, \quad y = q_1(t)$$

The parameter values are given by $[m_1, m_2, m_3] = [1, 1, 1](kg)$, $[c_1, c_2, c_3] = [0.1, 0.5, 0.3](N-sec/m)$, and $[k_1, k_2, k_3] = [5, 8, 9](N/m)$.

i) Is this system controllable?
ii) Is this system observable?

11.9 A first order system is given by

$$\{\dot{x}\} = [A]\{x\} + [b]u, \quad u = -[k]\{x\}$$

where $[b]$ is a column vector and u is a scalar control input. If a system $([A],[b])$ is controllable, show that the system with feedback $([A]-[b][k],[b])$ is also controllable. (Hint: Check the rank of the controllability matrix of both systems.)

11.10 A dynamic system is modeled by finite element analysis, and the results are given as

$$[M] = \begin{bmatrix} 0.5 & 0 & 0 \\ 0 & 0.5 & 0 \\ 0 & 0 & 0.5 \end{bmatrix}, \quad [K] = \begin{bmatrix} 3 & -2 & 0 \\ -2 & 5 & -3 \\ 0 & -3 & 3 \end{bmatrix}$$

and

$$[F] = [1, 0, 0]^T$$

where the notations are identical to those in the main text. The system is transformed into the first order form of equation as

$$\{\dot{x}\} = [A]\{x\} + [B]\{u\}$$

i) Find out the open-loop eigenvalues of the system using the matrix $[A]$. The desired eigenvalues are prescribed as

$$-1.5 \pm 2.0i, \quad -2.0 \pm 4.0i, \quad -1.0 \pm 3.0i$$

ii) Use the Bass-Gura formula to compute the feedback gains which achieve the desired closed-loop eigenvalues.

11.11 A finite element model of the longitudinal vibration of a beam is developed as a lumped-mass model. The first order description of the system is given by

$$\{\dot{x}\} = [A]\{x\} + [B]\{u\}, \quad y = [C]\{x\}$$

where

$$[A] = \begin{bmatrix} 0 & 0 & 0 & 1 & 0 & 0 \\ 0 & 0 & 0 & 0 & 1 & 0 \\ 0 & 0 & 0 & 0 & 0 & 1 \\ -36 & 36 & 0 & 0 & 0 & 0 \\ 18 & -36 & 18 & 0 & 0 & 0 \\ 0 & 36 & -36 & 0 & 0 & 0 \end{bmatrix}, \quad [B] = \begin{bmatrix} 0 \\ 0 \\ 0 \\ 2 \\ 0 \\ 0 \end{bmatrix}$$

Use the MATLAB *m-file felqr.m* in order to design an optimal feedback gain matrix. Use

$$[Q] = I_{6 \times 6}, \quad [R] = 0.1$$

11.12 Using the same data in problem 11, and

$$[C] = [1 \ 0 \ 0 \ 0 \ 0 \ 0]$$

design a dynamic observer. Use the same $[Q]$ and $[R]$.

11.13 For a given scalar system

$$\dot{x} = ax + u$$

and a performance index

$$J = \frac{1}{2}\int_0^\infty (qx^2 + ru^2)dt$$

where $q > 0$, $r > 0$ are constant weighting factors: i) Find out the optimal feedback gain by solving the Ricatti equation in Eq. (11.8.5). ii) Prove that the closed-loop system is stable using the result of part i).

11.14 A flexible structure is under longitudinal vibration as in Fig. P11.14.

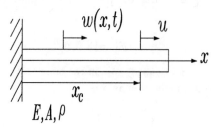

Figure P11.14 A Bar under Longitudinal Vibration

The governing equation of motion is derived as

$$\rho \frac{\partial^2 w(x,t)}{\partial t^2} = EA\frac{\partial^2 w(x,t)}{\partial x^2}$$

If we want to place an actuator at $x = x_c$ of the beam, show that control law $u(t) = -g_2 \frac{\partial w(x_c,t)}{\partial t}$, $g_2 > 0$ stabilizes the system. Hint : Use the Lyapunov function approach.

11.15 A feedback control law for a rotating beam is derived in Eq. (11.12.28) as

$$u = -g_1(\theta - \theta_f) - g_2\dot{\theta} - g_3(l_0 S_0 - M_0)$$

Verify the above expression taking the Lyapunov function in Eq. (11.12.20). You may have to use the governing equations and boundary conditions in order to prove the control law expression.

CHAPTER	TWELVE

SPECIAL TOPICS

12.0 Chapter Overview

Finite element formulations are presented for various special problems. Singular elements are presented for crack analysis in order to model the crack tip singularity accurately. The first two sections consider modeling of stationary cracks while the third section discusses modeling of propagating cracks. Another problem is a semi-infinite domain because finite elements cannot encompass such a domain in a conventional way. As a result, semi-infinite elements are developed. Other topics include thermal stress analysis of layered media, static and dynamic buckling analysis, and nonlinear analysis. MATLAB example programs are also provided.

12.1 Stationary Singular Elements

The stress and strain fields at the crack tip are singular in the linear elastic analysis. The singularity is expressed as $r^{-1/2}$ where r is the radial distance from the crack tip. In order to capture the solution around the crack tip accurately without using so many elements, so-called singular finite elements can be used at the crack tip. The elements can represent the singular stress and strain fields at the crack tip.

Let us consider one-dimensional singular shape functions first of all. Let $\{\xi_i | i = 1, 2, ..n\}$ denote nodal points on the natural coordinate ξ, whose end points are defined by $\xi = \pm 1$. For simplicity the nodal points are assumed to be separated by a uniform distance as shown in Fig. 12.1.1. The standard shape functions are given in Eqs. (6.1.1), (6.1.2), (6.1.13), (6.1.14), and (6.1.15).

In order to generate singular shape functions, the polynomial expression of ξ is assumed to be

$$u(\xi) = a_0 + a_1\xi + ... + a_n\xi^n + a_{n+1}(\xi - \xi_o)^p \qquad (12.1.1)$$

where p is between 0 and 1. The last term in the expression gives the desired order of derivative singularity at $\xi = \xi_o$. If a derivative singularity of order 1/2 is needed at the first node at $\xi = -1$, then p=1/2 and ξ_o=-1.

Two–Node Element

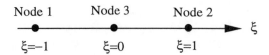

Three–Node Element

Figure 12.1.1 One-Dimensional Singular Elements

For the two-node element, the polynomial is assumed

$$u(\xi) = a_0 + a_1(\xi - \xi_o)^p \tag{12.1.2}$$

Then, the singular shape functions with $\xi=-1$ are

$$N_1(\xi) = 1 - (\xi + 1)^p/2^p \tag{12.1.3}$$

$$N_2(\xi) = (\xi + 1)^p/2^p \tag{12.1.4}$$

For the three-node element, the assumed polynomial is

$$u(\xi) = a_0 + a_1\xi + a_2(\xi - \xi_o)^p \tag{12.1.5}$$

Then, the singular shape functions with $\xi=-1$ are

$$N_1(\xi) = \{1 + (2^p - 1)\xi - (\xi + 1)^p\}/(2 - 2^p) \tag{12.1.6}$$

$$N_2(\xi) = \{1 + \xi - (\xi + 1)^p\}/(2 - 2^p) \tag{12.1.7}$$

$$N_3(\xi) = \{-2^p - 2^p\xi + 2(\xi + 1)^p\}/(2 - 2^p) \tag{12.1.8}$$

Other higher order shape functions can be derived in the same way.

Two-dimensional shape functions can be derived from one-dimensional shape functions. In particular, the two-dimensional shape functions, which have derivative singularity at a corner node, are presented below [58]. In the following derivation, the derivative singularity is assumed to occur at $\xi=-1$ and $\eta=-1$. The derivative singular shape functions for the four-node isoparametric element as shown in Fig. 12.1.2 are developed below.

Along four sides of the element, the following one-dimensional interpolation expressions are used:

$$u_{12}(\xi) = N_1(\xi)u_1 + N_2(\xi)u_2 \quad \text{on edge 12} \tag{12.1.9}$$

$$u_{23}(\eta) = H_1(\eta)u_2 + H_2(\eta)u_3 \quad \text{on edge 23} \tag{12.1.10}$$

$$u_{43}(\xi) = H_1(\xi)u_4 + H_2(\xi)u_3 \quad \text{on edge 43} \tag{12.1.11}$$

$$u_{14}(\eta) = N_1(\eta)u_1 + N_2(\eta)u_4 \quad \text{on edge 12} \tag{12.1.12}$$

where H_i and N_i denote the standard and singular shape functions. The standard shape functions are given in Eqs. (6.1.1) and (6.1.2), and the singular shape functions are given in Eqs. (12.1.3) and (12.1.4). u_i is the nodal variable at node i while u_{ij} is the variable along the edge ij. The variable inside the element is obtained using the following interpolation:

$$u(\xi,\eta) = \frac{1}{2}\big[H_2(\xi)u_{23} + H_1(\xi)u_{14}\big] + \frac{1}{2}\big[H_2(\xi)u_{43} + H_1(\xi)u_{12}\big] \tag{12.1.13}$$

Substitution of Eqs. (12.1.9) through (12.1.12) into Eq. (12.1.13) yields

$$u(\xi,\eta) = \sum_{i=1}^{4} L_i(\xi,\eta)u_i \tag{12.1.14}$$

where

$$L_1(\xi,\eta) = \frac{1}{2}\big[H_1(\xi)N_1(\eta) + N_1(\xi)H_1(\eta)\big] \tag{12.1.15}$$

$$L_2(\xi,\eta) = \frac{1}{2}\big[H_2(\xi)H_1(\eta) + N_2(\xi)H_1(\eta)\big] \tag{12.1.16}$$

$$L_3(\xi,\eta) = H_2(\xi)H_2(\eta) \tag{12.1.17}$$

$$L_4(\xi,\eta) = \frac{1}{2}\big[H_1(\xi)N_2(\eta) + H_1(\xi)H_2(\eta)\big] \tag{12.1.18}$$

These shape functions satisfy the following conditions:

$$L_i(\xi_j,\eta_j) = \delta_{ij} \tag{12.1.19}$$

and

$$\sum_{i=1}^{4} L_{ij}(\xi,\eta) = 1 \tag{12.1.20}$$

Here δ_{ij} is the Kronecker delta.

A similar technique can be used for the nine-node quadrilateral element as shown in Fig. 12.1.2. The following interpolations are used along the lines of the element.

$$u_{125}(\xi) = N_1(\xi)u_1 + N_2(\xi)u_2 + N_3(\xi)u_5 \quad \text{on line 152} \tag{12.1.21}$$

$$u_{236}(\eta) = H_1(\eta)u_2 + H_2(\eta)u_3 + H_3(\eta)u_6 \quad \text{on line 263} \tag{12.1.22}$$

$$u_{473}(\xi) = H_1(\xi)u_4 + H_2(\xi)u_3 + H_3(\xi)u_7 \quad \text{on line 473} \tag{12.1.23}$$

$$u_{148}(\eta) = N_1(\eta)u_1 + N_2(\eta)u_4 + N_3(\eta)u_8 \quad \text{on line 184} \tag{12.1.24}$$

$$u_{579}(\eta) = H_1(\eta)u_5 + H_2(\eta)u_7 + H_3(\eta)u_9 \quad \text{on line 597} \tag{12.1.25}$$

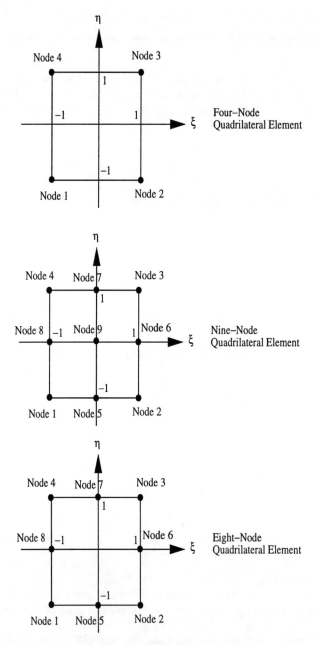

Figure 12.1.2 Two-Dimensional Singular Elements

$$u_{869}(\xi) = H_1(\xi)u_8 + H_2(\xi)u_6 + H_3(\xi)u_9 \quad \text{on line 896} \tag{12.1.26}$$

where H_i and N_i are the standard and singular shape functions of the three-node one-dimensional element. The variable $u(\xi, \eta)$ inside the element is interpolated as follows:

$$u(\xi,\eta) = \frac{1}{2}[H_1(\xi)u_{148} + H_2(\xi)u_{236} + H_3(\xi)u_{597}]$$
$$+ \frac{1}{2}[H_1(\eta)u_{125} + H_2(\eta)u_{437} + H_3(\eta)u_{869}] \qquad (12.1.27)$$

Substitution of Eqs. (12.1.21) through (12.1.26) into Eq. (12.1.27) results in the following shape functions.

$$u(\xi,\eta) = \sum_{i=1}^{9} L_i(\xi,\eta) u_i \qquad (12.1.28)$$

where

$$L_1(\xi,\eta) = \frac{1}{2}[H_1(\xi)N_1(\eta) + N_1(\xi)H_1(\eta)] \qquad (12.1.29)$$

$$L_2(\xi,\eta) = \frac{1}{2}[H_2(\xi)H_1(\eta) + N_2(\xi)H_1(\eta)] \qquad (12.1.30)$$

$$L_3(\xi,\eta) = H_2(\xi)H_2(\eta) \qquad (12.1.31)$$

$$L_4(\xi,\eta) = \frac{1}{2}[H_1(\xi)N_2(\eta) + H_1(\xi)H_2(\eta)] \qquad (12.1.32)$$

$$L_5(\xi,\eta) = \frac{1}{2}[H_3(\xi)H_1(\eta) + N_3(\xi)H_1(\eta)] \qquad (12.1.33)$$

$$L_6(\xi,\eta) = H_2(\xi)H_3(\eta) \qquad (12.1.34)$$

$$L_7(\xi,\eta) = H_3(\xi)H_2(\eta) \qquad (12.1.35)$$

$$L_8(\xi,\eta) = \frac{1}{2}[H_1(\xi)N_3(\eta) + H_1(\xi)H_3(\eta)] \qquad (12.1.36)$$

$$L_9(\xi,\eta) = H_3(\xi)H_3(\eta) \qquad (12.1.37)$$

These shape functions satisfy the conditions given in Eqs. (12.1.19) and (12.1.20).

Singular eight-node shape functions can be developed from the following interpolation:

$$u(\xi,\eta) = \frac{1}{2}[H_1(\xi)u_{148} + H(\xi)u_{236}] + [H_1(\eta)u_{125} + H_2(\eta)u_{437}]$$
$$+ \frac{1}{4}[H_3(\xi)(1-\eta)u_5 + H_3(\eta)(1+\xi)u_6 + H_3(\xi)(1+\eta)u_7$$
$$+ H_3(\eta)(1-\xi)u_8] \qquad (12.1.38)$$

where u_{ijk} is defined in Eqs. (12.1.21) through (12.1.26). The shape functions are shown below:

$$u(\xi,\eta) = \sum_{i=1}^{8} L_i(\xi,\eta) u_i \qquad (12.1.39)$$

in which

$$L_1(\xi,\eta) = \frac{1}{2}\big[H_1(\xi)N_1(\eta) + N_1(\xi)H_1(\eta)\big] \qquad (12.1.40)$$

$$L_2(\xi,\eta) = \frac{1}{2}\big[H_2(\xi)H_1(\eta) + N_2(\xi)H_1(\eta)\big] \qquad (12.1.41)$$

$$L_3(\xi,\eta) = H_2(\xi)H_2(\eta) \qquad (12.1.42)$$

$$L_4(\xi,\eta) = \frac{1}{2}\big[H_1(\xi)N_2(\eta) + H_1(\xi)H_2(\eta)\big] \qquad (12.1.43)$$

$$L_5(\xi,\eta) = \frac{1}{2}\big[N_3(\xi)H_1(\eta) + \frac{1}{2}H_3(\xi)(1-\eta)\big] \qquad (12.1.44)$$

$$L_6(\xi,\eta) = \frac{1}{2}\big[H_2(\xi)H_3(\eta) + \frac{1}{2}(1+\eta)H_3(\eta)\big] \qquad (12.1.45)$$

$$L_7(\xi,\eta) = \frac{1}{2}\big[H_3(\xi)H_2(\eta) + \frac{1}{2}H_3(\xi)(1+\eta)\big] \qquad (12.1.46)$$

$$L_8(\xi,\eta) = \frac{1}{2}\big[H_1(\xi)N_3(\eta) + \frac{1}{2}(1-\eta)H_3(\eta)\big] \qquad (12.1.47)$$

The eight-node singularity element is compatible with standard eight-node elements along the element boundaries 23 and 34. On the other hand, the singular element is compatible with the same kind of singular elements along the boundaries 12 and 14. All singular elements located at the crack tip should have singularity at their nodes coincident with the crack tip.

Previously developed derivative singular shape functions are used to interpolate the variables like the displacements while standard shape functions are used for interpolation of the geometric coordinate variables like x and y for the mapping from the natural coordinate system to the physical coordinate system. Previous derivations can be applied to higher order singular elements in 2-D or 3-D with either point singularity or line singularity.

♣ **Example 12.1.1**   Consider plates with a center crack, a single edge crack, or double edge cracks as seen in Fig. 12.1.3. The ratios of the plate width to crack size as well as the plate length to width are shown in the figure. The center cracked or double edge cracked plate was modeled as a quarter of the plate because of symmetry while the single edge cracked plate was considered half of it. All those modeled domains were divided into 10×10 equal size eight-node elements. Singular elements were used at the crack tip while standard elements were used for the rest of the domains. Normalized stress intensity factors $K_1/\sigma\sqrt{\pi a}$ are compared in Table 12.1.1 using 10×10 eight-node elements. Here, K_1 is the stress intensity factor of the first mode, σ is the applied tensile traction, and a is the crack size as shown in Fig. 12.1.3. In the comparison, so-called quarter-point eight-node elements, which are described in the next section, were also included. The quarter-point element replaced the singular elements in the mesh. The results show that both singular elements developed in either this or the next section resulted in very comparable solutions compared to the analytical solutions. Overall, the singular element developed here was better than the quarter-point singular element. ‡

Crack Analysis in Plate Bending. When the shear deformable plate bending theory is used for crack analysis, the singular solution field has $w = O(r^{3/2})$ and $\theta_x = \theta_y = O(r^{1/2})$, where w and θ indicate the transverse deflection and rotation, respectively, and r is the radial distance from the crack tip. This result suggests that two different finite element interpolation functions need to be utilized for w and θ, respectively, in the application to fracture problems of plate bending. The commonly used quarter-point singular elements cannot satisfy the condition stated above.

In order to meet the requirement, the singular elements developed in this section can be applied. For example, an eight-node singular element of order p is used with $p = 3/2$ for w and $p = 1/2$ for θ.

In order to evaluate the crack tip behavior, the J-integral can be used. The J-integral for plate bending can be written as

$$J = \int_\Gamma \left(W dy - p_i \frac{\partial \theta_i}{\partial x} dx \right) \tag{12.1.48}$$

where

$$w = \int_0^\gamma M_{ij} d\gamma_{ij} \tag{12.1.49}$$

and p_i, M_{ij}, and γ_{ij} are the generalized tractions, stresses, and strains. Substituting the crack tip solutions into Eq. (12.1.48) and limiting r to zero yield the following relation between J-integral and stress intensity factor K_I

$$J = \frac{K_I h^3}{12E} \tag{12.1.50}$$

where h is the plate thickness, E is Young's modulus, and

$$K_I = \frac{12}{h^3} \phi(1) M \sqrt{\pi a} \tag{12.1.51}$$

Here, M is the uniform bending moment applied at the boundary, and a is half of the crack length. The coefficient $\phi(1)$ is a function of h/a. A numerical example is given in Ref. [59].

12.2 Quarter-Point Singular Elements

One way to generate singularity at the crack tip is using so-called quarter-point isoparametric elements. Those elements are very easy to use with isoparametric elements. Let's consider the eight-node isoparametric element. As shown in Fig. 12.2.1, the element shape and the shape functions in the natural coordinate system

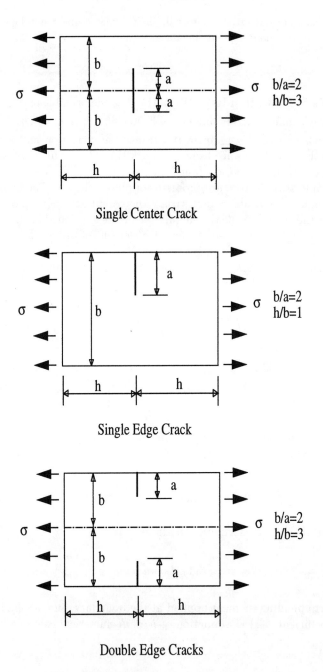

Figure 12.1.3 Plates with Cracks

are the same as those in Ch. 6. That is, standard shape functions are used for the crack tip elements. On the other hand, the crack element in the physical domain is different from the standard shape. One corner node is located at the crack tip and two mid-nodes close to the crack tip are placed at the quarter distance along the element

Section 12.2					Quarter-Point Singular Elements					521

Table 12.1.1 Comparison of Stress Intensity Factors $\frac{K_1}{\sigma\sqrt{\pi a}}$

	Analytic	Present Singular	Quarter-Point Singular
Single Center Crack	1.187	1.170	1.088
Single Edge Crack	2.818	2.949	2.719
Double Edge Crack	1.163	1.154	1.075

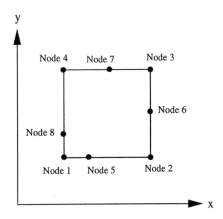

crack tip at node 1
distance between nodes 1 and 5 =
  1/4 of distance between nodes 1 and 2
distance between nodes 1 and 8 =
  1/4 of distance between nodes 1 and 4

Figure 12.2.1 Quarter-Point Eight-Node Isoparametric Element

side length, respectively as shown in Fig. 12.2.1. By doing this, singularity of $r^{-1/2}$ can be achieved at the crack tip.

In order to show how such a singularity can be developed at the crack tip node, let's consider an element boundary which contains the crack tip node. For example, the element boundary 152 in Fig. 12.2.1 is considered here. The corresponding shape functions for the three nodes along the element boundary are the same as the quadratic shape functions like

$$H_1(\xi) = \frac{1}{2}(\xi - \xi^2) \qquad (12.2.1)$$

$$H_2(\xi) = \frac{1}{2}(\xi + \xi^2) \qquad (12.2.2)$$

$$H_5(\xi) = 1 - \xi^2 \qquad (12.2.3)$$

For simplicity, let $x_1 = 0$, $x_2 = a$, and $x_5 = a/4$ with all their y-coordinate values the same. This does not restrict the generality of the following development. Along the element boundary, du/dx is computed as

$$\frac{du}{dx} = \frac{dH_1}{d\xi}\frac{d\xi}{dx}u_1 + \frac{dH_2}{d\xi}\frac{d\xi}{dx}u_2 + \frac{dH_5}{d\xi}\frac{d\xi}{dx}u_5 \tag{12.2.4}$$

where

$$\frac{dH_1(\xi)}{d\xi} = \frac{1}{2}(1 - 2\xi) \tag{12.2.5}$$

$$\frac{dH_2(\xi)}{d\xi} = \frac{1}{2}(1 + 2\xi) \tag{12.2.6}$$

$$\frac{dH_5(\xi)}{d\xi} = 2\xi \tag{12.2.7}$$

All these derivatives do not possess any singularity. On the other hand, $d\xi/dx$ is computed from the inverse of $dx/d\xi$ which is

$$\frac{dx}{d\xi} = \frac{dH_1}{d\xi}\frac{d\xi}{dx}x_1 + \frac{dH_2}{d\xi}\frac{d\xi}{dx}x_2 + \frac{dH_5}{d\xi}\frac{d\xi}{dx}x_5$$
$$= \frac{1}{2}a(1 + \xi) \tag{12.2.8}$$

Thus, $d\xi/dx$ is singular at $\xi = -1$ which corresponds to the crack tip node 1. Then, du/dx becomes singular at the crack node.

This development is also true for the element boundary 184 in Fig. 12.2.1. Thus, the quarter-point eight-node element has singularity $r^{-1/2}$ along two element boundaries. However, such an singularity does not hold inside the element. The same concept developed above can be applied to the six-node isoparametric element as illustrated in Fig. 12.2.2.

When quarter-point singular elements are compared to the singular elements developed in Sec. 12.1, there are clear distinctions between the two families. The former group introduces singularity caused by the coordinate transformation while the latter has singularity directly in the shape functions. Quarter-point singular elements are very easy to use with any standard program with isoparametric elements while the derivative singular shape functions are easy to apply to any order of singular problem. For example, the singularity order is other than $-1/2$ when the crack tip is located at the interface of two different materials and normal to the materials' interface.

12.3 Moving Singular Elements

As a crack grows, its crack tip moves with the crack. In finite element analysis, the moving crack can be modeled discretely with a finite jump from one node to the next node as shown in Fig. 12.3.1. In order to make the crack grow smoothly, the element size or nodal spacing should be very small. Otherwise, there is a

Section 12.3   Moving Singular Elements   523

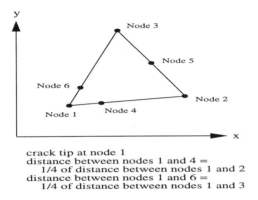

crack tip at node 1
distance between nodes 1 and 4 =
    1/4 of distance between nodes 1 and 2
distance between nodes 1 and 6 =
    1/4 of distance between nodes 1 and 3

Figure 12.2.2 Quarter-Point Six-Node Isoparametric Element

large oscillation in the energy as the crack tip jumps from one node to another. However, using a large number of small elements along the crack tip is computationally expensive.

In order to make the crack propagation more continuous in the finite element model, either nodes or elements should be reconfigured as a crack propagates. Especially, remeshing elements is very cumbersome even if it is sometimes necessary as shown in Fig. 12.3.2. This section describes a singular element with a moving node so that a continuously moving singular crack tip can be modeled in the finite element analysis.

In order to develop a singular eight-node element, which has a movable node with any order of derivative singularity, the singularity is assumed to occur at node number 5 in Fig. 12.3.3. If the direction of a crack growth is along the edge 152 of the element, node 5 locating on the crack tip will move along the edge depending on the amount of crack propagation based on an assumed crack growth criterion. Let's assume that node 5 is located in $(\xi_o, -1)$ in the natural coordinate system (ξ, η). As the crack tip moves, ξ_o changes from -1 to 1. On the edge the *Jacobian* is constant and there is a linear mapping from the crack tip, say x_o, to the local tip ξ_o. If the magnitude of the crack propagation is larger than the element edge size, the next element will be modeled in the same way. In this manner, any size of crack propagation can be modeled. However, when the crack tip is located at any corner node, i.e. $\xi_o = \pm 1$, this singular element may not be used. Instead the singular element developed in Sec. 12.1 can be used.

Other side nodes, i.e. nodes 6 through 8, may be located at the midpoint of each edge. Then this singular element is compatible with other standard elements along those edges. However, derivative singular shape functions based on this configuration did not provide accurate solutions from numerical experiments. Thus, node 7 is allowed to move to $(\xi_o, 1)$, like node 5. In order to make the present singular element compatible with a standard element on edge 473 of Fig. 12.3.3, a transient element is used as shown in Fig. 12.3.4. Using the transition element on the edge 473 of the singular element, the set of singular and transition elements become edge compatible with standard elements.

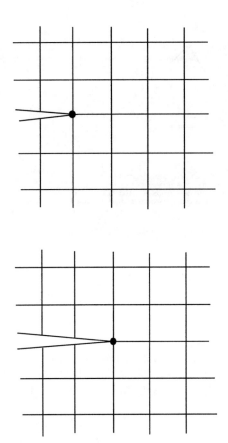

Figure 12.3.1 Finite Jump of Crack Tip from One Node to the Next

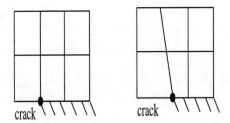

Figure 12.3.2 Remeshing for Change of Nodes or Elements for a Moving Crack

The eight-node singular element at node 5 has shape functions as given below:

$$L_1(\xi, \eta) = [C_1(\xi) + 0.5 E_1(\xi)] D_1(\eta) \qquad (12.3.1)$$

$$L_2(\xi, \eta) = [C_2(\xi) + 0.5 E_2(\xi)] D_1(\eta) \qquad (12.3.2)$$

Section 12.3　　　　　　　　Moving Singular Elements　　　　　　　　525

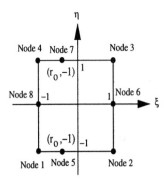

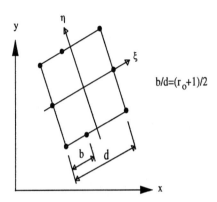

Figure 12.3.3 Moving Singular Element with Eight Nodes

$$L_3(\xi, \eta) = \left[C_2(\xi) + 0.5 F_2(\xi)\right] D_2(\eta) \qquad (12.3.3)$$

$$L_4(\xi, \eta) = \left[C_1(\xi) + 0.5 F_1(\xi)\right] D_2(\eta) \qquad (12.3.4)$$

$$L_5(\xi, \eta) = 0.5 E_3(\xi) D_1(\eta) + 0.5 F_3(\xi) G_1(\eta) \qquad (12.3.5)$$

$$L_6(\xi, \eta) = \left[C_2(\xi) + 0.25(1+\xi)\right] D_1(\eta) \qquad (12.3.6)$$

$$L_7(\xi, \eta) = 0.5 F_3(\xi) \left[D_2(\eta) + G_2(\eta)\right] \qquad (12.3.7)$$

$$L_8(\xi, \eta) = \left[C_1(\xi) + 0.25(1-\xi)\right] D_3(\eta) \qquad (12.3.8)$$

where

$$C_1(\xi) = \frac{(\xi-1)(\xi-\xi_o)}{4(1+\xi_o)} \qquad (12.3.9)$$

$$C_2(\xi) = \frac{(\xi+1)(\xi-\xi_o)}{4(1-\xi_o)} \qquad (12.3.10)$$

$$D_1(\eta) = \frac{(\eta^2-\eta)}{2} \qquad (12.3.11)$$

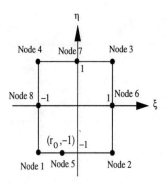

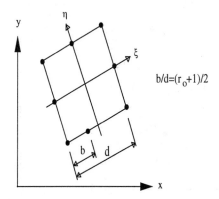

Figure 12.3.4 Transition Element with Eight Nodes

$$D_2(\eta) = \frac{(\eta^2 + \eta)}{2} \tag{12.3.12}$$

$$D_3(\eta) = 1 - \eta^2 \tag{12.3.13}$$

$$E_1(\xi) = \left[\xi_o(1-\xi_o)^p - (1-\xi_o)^p \xi + (1-\xi_o)|\xi - \xi_o|^p\right]/DE \tag{12.3.14}$$

$$E_2(\xi) = \left[-\xi_o(1+\xi_o)^p + (1+\xi_o)^p \xi + (1+\xi_o)|\xi - \xi_o|^p\right]/DE \tag{12.3.15}$$

$$E_3(\xi) = \left[(1+\xi_o)^p + (\xi - \xi_o)^p\{(1-\xi_o)^p - (1+\xi_o)^p\}\xi - 2|\xi - \xi_o|^p\right]/DE \tag{12.3.16}$$

$$DE = (1+\xi_o)^p(1-\xi_o) + (1-\xi_o)^p(1+\xi_o) \tag{12.3.17}$$

$$F_1(\xi) = \frac{\xi_o}{2(1+\xi_o)} - \frac{\xi}{2} + \frac{\xi^2}{2(1+\xi_o)} \tag{12.3.18}$$

$$F_2(\xi) = -\frac{\xi_o}{2(1-\xi_o)} + \frac{\xi}{2} + \frac{\xi^2}{2(1-\xi_o)} \tag{12.3.19}$$

$$F_3(\xi) = \frac{1-\xi^2}{1-\xi_o^2} \tag{12.3.20}$$

$$G_1(\eta) = 1 - \frac{1}{2^p}(1+\eta)^p \qquad (12.3.21)$$

$$G_2(\eta) = \frac{1}{2^p}(1+\eta)^p \qquad (12.3.22)$$

In these expressions, p depends on the desired order of derivative singularity. If one over the square root r singularity is needed, p is set to $1/2$.

The transition element shape functions are given below:

$$H_1(\xi,\eta) = \frac{1}{4}(1-\xi)(1-\eta) - \frac{1-\xi_o}{2}H_5(\xi,\eta) - \frac{1}{2}H_8(\xi,\eta) \qquad (12.3.23)$$

$$H_2(\xi,\eta) = \frac{1}{4}(1+\xi)(1-\eta) - \frac{1+\xi_o}{2}H_5(\xi,\eta) - \frac{1}{2}H_6(\xi,\eta) \qquad (12.3.24)$$

$$H_3(\xi,\eta) = \frac{1}{4}(1+\xi)(1+\eta) - \frac{1}{2}H_6(\xi,\eta) - \frac{1}{2}H_7(\xi,\eta) \qquad (12.3.25)$$

$$H_4(\xi,\eta) = \frac{1}{4}(1-\xi)(1+\eta) - \frac{1}{2}H_7(\xi,\eta) - \frac{1}{2}H_8(\xi,\eta) \qquad (12.3.26)$$

$$H_5(\xi,\eta) = \frac{1}{2(1-\xi_o^2)}(1-\xi^2)(1-\eta) \qquad (12.3.27)$$

$$H_6(\xi,\eta) = \frac{1}{2}(1+\xi)(1-\eta^2) \qquad (12.3.28)$$

$$H_7(\xi,\eta) = \frac{1}{2}(1-\xi^2)(1+\eta) \qquad (12.3.29)$$

$$H_8(\xi,\eta) = \frac{1}{2}(1-\xi)(1-\eta^2) \qquad (12.3.30)$$

Both singular and transition shape functions satisfy the conditions given in Eqs. (12.1.19) and (12.1.20).

♣ **Example 12.3.1** A plate with a single edge crack as shown in Fig. 12.3.5 was solved using the present singular element at the crack tip along with the transition element. The static stress intensity factor is plotted as a function of a/b in Fig. 12.3.6, where a is the crack size and b is the width of the plate. The stress intensity factors were computed using J-integral [60]. In computing them, the side nodes 5 and 7 in Fig. 12.3.3 were moved depending on the crack size. Therefore, the 10×10 mesh remained the same. The finite element solution was quite compatible with the analytical solution. ‡

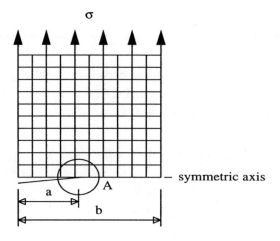

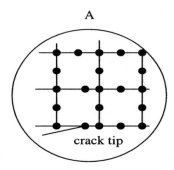

Figure 12.3.5 Plates with a Single Edge Crack

12.4 Semi-Infinite Element

The finite element method is applied to infinite domains to analyze some physical problems. There are two ways to model infinite domains using the finite element method. The first method is to truncate infinite domains into finite sizes of domains. In this case, a truncated finite domain should be large enough to represent the infinite domain. The other method is to use semi-infinite elements.

A systematic approach is given here to derive semi-infinite elements. The semi-infinite elements are parametric elements which map finite domains onto infinite domains. A one-dimensional case is derived in detail below, and its extension to two- and three-dimensional cases is straightforward. Let ξ and x be the natural and physical coordinates as shown in Fig. 12.4.1. The natural domain $-1 \leq \xi \leq 1$ maps onto the physical domain $x_1 \leq x < \infty$ if the semi-infinite element extends to the positive x-direction. To accomplish such a transformation, the following equation is assumed.

$$x = a_0 + a_1\xi + a_2\xi^2 + ... + a_{n-1}f(\xi) \qquad (12.4.1)$$

where n is the number of nodes per element, and $f(\xi)$ approaches infinity as ξ goes to 1. The specific form of $f(\xi)$ depends on the given problem. For simplicity, a

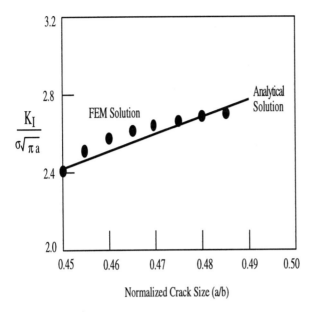

Figure 12.3.6 Normalized Stress Intensity Factor vs. Normalized Crack Size

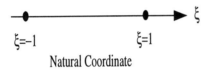

Figure 12.4.1 Natural and Physical Coordinates

three-noded element is considered from now on. This does not reduce generality of the following derivation. For a three-noded element, Eq. (12.4.1) becomes

$$x = a_0 + a_1\xi + a_2 f(\xi) \tag{12.4.2}$$

In order to map the points $\xi=-1$, 0, and 1 in the natural coordinate system onto $x=x_1$, x_2, and ∞ in the physical coordinate system, Eq. (12.4.2) results in

$$x = [x_2 - a_2 f(0)] + [x_2 - x_1 + a_2\{f(-1) - f(0)\}]\xi + a_2 f(\xi) \tag{12.4.3}$$

where a_2 and $f(\xi)$ are selected to make x go to infinity as ξ goes to 1.

♣ **Example 12.4.1** As the first example, the following differential equation is considered.

$$\frac{d^2u}{dx^2} = \frac{2}{x^3} \qquad (12.4.4)$$

The boundary conditions are $u(1) = 1$ and $u \to 0$ as $x \to \infty$. The example is the same as that given by Bettess [61]. The analytical solution is $u = 1/x$. Equation (12.4.4) gives the variational form

$$\pi = \int_1^\infty \left[\left(\frac{du}{du}\right)^2 + \frac{4u}{x^3}\right] dx \qquad (12.4.5)$$

Using the three-noded semi-infinite element,

$$u = H_1(\xi)u_1 + H_2(\xi)u_2 + H_3(\xi)u_3 \qquad (12.4.6)$$

in which

$$H_1(\xi) = (-\xi + \xi^2)/2 \qquad (12.4.7)$$

$$H_2(\xi) = (1 - \xi^2) \qquad (12.4.8)$$

$$H_3(\xi) = (\xi + \xi^2)/2 \qquad (12.4.9)$$

The three nodal points are located at $x_1=1$, $x_2=2$ and $x_3=\infty$, respectively. Applying the boundary conditions

$$u = H_1(\xi) + H_2(\xi)u_2 \qquad (12.4.10)$$

Taking a derivative of u with respective to x,

$$\frac{du}{dx} = \frac{dH_1}{d\xi}\frac{d\xi}{dx} + \frac{dH_2}{d\xi}\frac{d\xi}{dx}u_2 \qquad (12.4.11)$$

To compute du/dx, Eq. (12.4.3) is used. Function $\frac{\alpha}{1-\xi}$ is selected for $f(\xi)$ where α is an arbitrary positive constant. The effect of α on u_2 is shown in Fig. 12.4.2 with $a_2=1$. When $\alpha=2$ is used, $u_2=0.5$ and $u = 1/x$ which is the same as the analytical solution. ‡

♣ **Example 12.4.2** The next example is shown in Fig. 12.4.3. A half of the beam is modeled due to symmetry. Three Hermitian elements and one three-noded semi-infinite element are used. Function $f(\xi) = exp(\alpha/(1-\xi)^p)$ is chosen. Figure 12.4.4 shows the transverse deflection normalized with respect to the analytical solution. Figures 12.4.5 through 12.4.7 show the transverse deflections for different values of α, a_2 and p. Parameters α and a_2 have less effect than p as shown in those figures because p controls the decay more dominantly. ‡

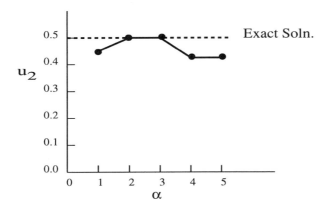

Figure 12.4.2 Effect of α on Solution u_2

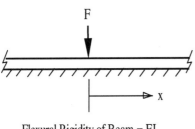

Flexural Rigidity of Beam = EI
Modulus of Foundation = k

Figure 12.4.3 An Infinite Beam on Elastic Foundation

12.5 Thermal Stress in Layered Beams

Let us consider a multilayered beam made of different materials as shown in Fig. 12.5.1. Each layer is assumed to be homogeneous and isotropic. However, elastic moduli and coefficients of thermal expansion are different for each layer. Because displacements should be continuous along the interface of any two neighboring layers, the beam formulation given in Sec. 8.3 is useful for this study. Each layer is modeled with the beam elements with only displacement degrees of freedom. The finite element mesh of beams throughout the whole layer must be compatible like a 2-D mesh. That is, beam elements have the same length through layers even if they may have different lengths along the beam direction.

For each beam element, the element stiffness matrix and column vector are computed and assembled, paying attention to the shared degrees of freedom among beam elements. The element stiffness matrix can be calculated using Eq. (8.3.7). For thermal analysis, the thermal load vector is computed in a similar way as described

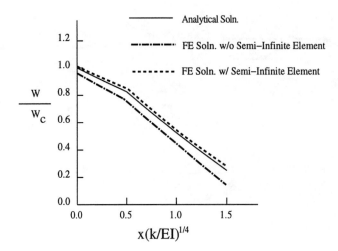

Figure 12.4.4 Deflection of Infinite Beam on Elastic Foundation

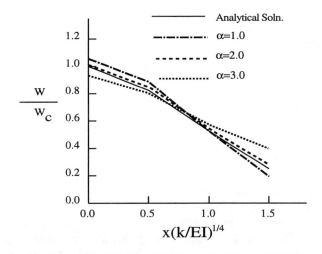

Figure 12.4.5 Effect of α on Deflection ($p=0.1$, $a_2=0.05$)

in Sec. 9.7. In other words,

$$\{F^{th}\} = E\alpha\Delta T \int_0^l \int_0^h \{B_b\}^T dy dx \tag{12.5.1}$$

where $\{F^{th}\}$ is the thermal load vector for the beam element, E is the elastic modulus, α is the coefficient of thermal expansion, and ΔT is the change of temperature of a beam element. The matrix $\{B_b\}$ is defined in Eq. (8.3.12), and the integration limits l and h denote the element length and height, respectively. Carrying out the integration

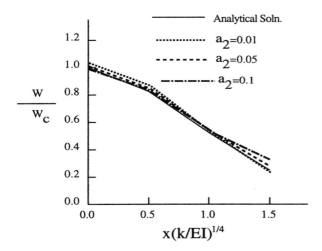

Figure 12.4.6 Effect of a_2 on Deflection ($p=1.0$, $\alpha=2.0$)

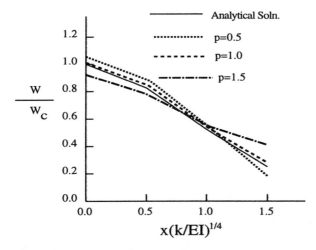

Figure 12.4.7 Effect of p on Deflection ($\alpha=2.0$, $a_2=0.05$)

of Eq. (12.5.1) yields

$$\{F^{th}\} = 0.5Eh\alpha\Delta T\{-1 \quad -1 \quad 0 \quad 1 \quad 1 \quad 0\}^T \tag{12.5.2}$$

After solving the nodal displacements, the bending stress at the bottom and top of an element are obtained from

$$\sigma_x^b = \frac{E}{l}(u_2^b - u_1^b) \tag{12.5.3}$$

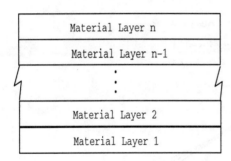

Figure 12.5.1 Leyer Beam with Different Materials

$$\sigma_x^t = \frac{E}{l}\left(u_2^t - u_1^t\right) \tag{12.5.4}$$

Here, subscript x indicates bending stress, and superscripts b and t denote the bottom and top of a beam element of a layer. u_i is the nodal displacement as shown in Fig. 8.3.3.

The average shear stress at an element is computed from

$$\tau_{avg} = \frac{G}{2}\left[\frac{(u_1^t + u_2^t) - (u_1^b + u_2^b)}{2} + \frac{2(v_2 - v_1)}{l}\right] \tag{12.5.5}$$

The transverse normal stress is computed from the transverse shear stress

$$\sigma_y = \frac{\partial}{\partial x}\int \tau_{xy}\,dy \tag{12.5.6}$$

where the integration is through the thickness of the top layer for σ_y on the upper interface and the thickness of the bottom layer for σ_y on the lower interface if there are three layers. Interface normal and shear stresses of tri-layer materials are computed in Ref. [19].

12.6 Buckling Analysis

When a beam is subjected to an axial compression with lateral loading as shown in Fig. 12.6.1, the equilibrium of force and moment are given as

$$\frac{dV}{dx} = -q \tag{12.6.1}$$

$$-\frac{dM}{dx} - P\frac{dv}{dx} = V \tag{12.6.2}$$

where V is the shear force, q is the lateral load, M is the bending moment, P is the axial compressive load, and v is the lateral deflection of the beam.

Substitution of Eq. (12.6.2) into Eq. (12.6.1) results in

$$\frac{d^2M}{dx^2} + P\frac{d^2V}{dx^2} = q \qquad (12.6.3)$$

Using the moment-curvature relation $M = EId^2v/dx^2$, Eq. (12.6.3) becomes

$$EI\frac{d^4v}{dx^4} + P\frac{d^2v}{dx^2} = q \qquad (12.6.4)$$

When the lateral load vanishes, the equation becomes

$$EI\frac{d^4v}{dx^4} + P\frac{d^2v}{dx^2} = 0 \qquad (12.6.5)$$

This is a homogeneous differential equation. When boundary condtions are also homogeneous, the problem becomes an *eigenvalue* problem.

Applying the Galerkin method with weak formulation to Eq. (12.6.5) will result in the element stiffness matrix as discussed in Sec. 8.1 for the first term. The second term in Eq. (12.6.5) yields the following matrix expression.

$$P[K_G]\{d\} = P\int_0^l [H']^T[H']dx\{d\} \qquad (12.6.6)$$

where

$$[H'] = \begin{bmatrix} \frac{\partial H_1}{\partial x} & \frac{\partial H_2}{\partial x} & \frac{\partial H_3}{\partial x} & \frac{\partial H_4}{\partial x} \end{bmatrix} \qquad (12.6.7)$$

and H_i is the shape function as defined in Eq. (8.1.8).

Undertaking the integration of Eq. (12.6.6) gives

$$[K_G] = \frac{1}{30l}\begin{bmatrix} 36 & 3l & -36 & 3l \\ 3l & 4l^2 & -3l & -l^2 \\ -36 & -3l & 36 & -3l \\ 3l & -l^2 & -3l & 4l^2 \end{bmatrix} \qquad (12.6.8)$$

This is called the *geometric stiffness matrix*. Finally, the resultant finite element matrix equation for Eq. (12.6.5) is

$$[K]\{d\} - P[K_G]\{d\} = ([K] - P[K_G])\{d\} = 0 \qquad (12.6.9)$$

This form is similar to the free vibration equation of the beam as given in Eq. (8.1.26). The *eigenvalue* of Eq. (12.6.9) is the critical buckling load, and *eigenvector* indicates the buckling mode shape.

Dynamic Buckling. The equation of beam bending including the axial compressive load is written as

$$m\frac{\partial^2 v}{\partial t^2} + EI\frac{\partial^4 v}{\partial x^4} + P\frac{\partial^2 v}{\partial x^2} + kv = 0 \qquad (12.6.10)$$

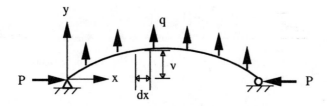

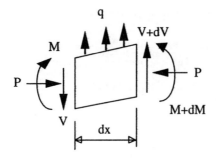

Figure 12.6.1 Beam Subjected to Axial Compression with Lateral Deflection

Here, all the parameters were defined previously except for k which is the spring constant of elastic foundation. Equation (12.6.10) is obtained from Eq. (12.6.4) by including the inertia term and substituting $q = -kv$.

Applying the finite element technique to Eq. (12.6.10) yields

$$[M]\{\ddot{v}\} + [K]\{v\} - P[K_G]\{v\} + [K_k]\{v\} = 0 \qquad (12.6.11)$$

in which the last term comes from the elastic foundation. When the first and the last terms of Eq. (12.6.11) vanish, it is the same as Eq. (12.6.9). When the first term drops out, it becomes

$$[[K] + [K_k] - P[K_G]]\{v\} = 0 \qquad (12.6.12)$$

Here, $[K_k]$ is given below using the Hermitian element.

$$[K_k] = \frac{kl}{420} \begin{bmatrix} 156 & 22l & 54 & -13l \\ 22l & 4l^2 & 13l & -3l^2 \\ 54 & 13l & 156 & -22l \\ -13l & -3l^2 & -22l & 4l^2 \end{bmatrix} \qquad (12.6.13)$$

For dynamic instability, the axial force is assumed to be

$$P = \bar{P}cos(\theta t) \qquad (12.6.14)$$

where $\bar{P}$ is the magnitude of the pulsating in-plane force and θ is the circular frequency of the pulsation. Let the transverse displacement vector be expressed in terms of trigonometric forms

$$\{v\} = \sum_i \{v_i\} sin\frac{i\theta t}{2} + \sum_i \{\hat{v}_i\} cos\frac{i\theta t}{2} \qquad (12.6.15)$$

where $\{v_i\}$ and $\{\hat{v}_i\}$ are time-independent vectors of nodal displacement. Substituting Eqs. (12.6.14) and (12.6.15) into Eq. (12.6.11), and considering the first term (i.e. $i = 1$) of Eq. (12.6.15) as the first-order approximation, yield the following eigenvalue matrix equation

$$det\left[[K] + [K_k] \pm \frac{\bar{P}}{2}[K_G] - \frac{\theta^2}{4}[M]\right] = 0 \qquad (12.6.16)$$

The eigenvalues of θ obtained from Eq. (12.6.16) determines the boundary of domain of principal instability. An example of the dynamic instability of beams and the extension to plate bending are found in Ref. [62].

12.7 Nonlinear Analysis

Let the differential equation be linear in the derivative of highest order and have the general form

$$u^{(n)} + \Re(u, u', ..., u^{(n-1)}; x) + \aleph(u, u', ..., u^{(n-1)}; x) = f(x) \qquad (12.7.1)$$

where the superscript notation $u^{(n)}$ denotes the nth derivative of u. $\Re$ is the linear differential operator and $\aleph$ indicates the nonlinear differential operator.

To fix ideas, applications to three second order two-point boundary value problems will be discussed. The three linearization schemes are presented in order of ascending ability. In all cases, each nonlinear operator is approximated by a linear operator

$$\aleph \approx \ell^*(u) + \hbar^* \qquad (12.7.2)$$

where both linear differential operators ℓ^* and $\hbar^*$ are functions of previous values of u which change during the iteration process. Substitution of Eq. (12.7.2) into (12.7.1) leads to a differential equation with operator ℓ^* on the left hand side, and function $\hbar^*$ on the right hand side with excitation function f. Depending on the linearization technique utilized, either $\hbar^*$ or ℓ^* may not present [63].

The simplest linearization, called *constant linearization*, of the nonlinear term in Eq. (12.7.1) treats the term as a constant, that is, with $\ell^*(u) = 0$, and where $\hbar^*$ is evaluated at previous iterations, thereby bringing the term to the right-hand side of Eq. (12.7.1) with the $f(x)$ excitation term. That is,

$$\aleph(u, u', ..., u^{(n-1)}; x) \approx \hbar^* = \aleph^*(u, u', ..., u^{(n-1)}; x) \qquad (12.7.3)$$

where the asterisk superscript on $\aleph$ denotes that it is evaluated from known values of the arguments from previous iterations. For example, for the three nonlinear operators considered in this study, u^2, $u'u$, and $(u')^2$, their constant linearization approximations are

$$u^2 \approx (u^*)^2 = \hbar^* \qquad (12.7.4)$$

$$u'u \approx (u')^*u^* = \hbar^* \qquad (12.7.5)$$

and

$$(u')^2 \approx (u'^*)^2 = \hbar^* \qquad (12.7.6)$$

respectively.

Substituting Eq. (12.7.3) into Eq. (12.7.1) yields the linear differential equation,

$$u^{(n)} + \Re(u) = f(x) - \hbar^* \qquad (12.7.7)$$

where $\hbar^*$ is given by Eq. (12.7.3).

The finite element matrix equation resulting from constant linearization has the form

$$[K]\{u\} = \{F\} - \{H^*\} \qquad (12.7.8)$$

where $[K]$ is the matrix arising from the finite element approximation of the linear differential operators $u^{(n)}$ and $\Re$, $\{F\}$ is the vector from the finite element approximation of the excitation function f, and $\{H^*\}$ is the vector from the constant linearization of the nonlinear operator, $\aleph$. Matrix $[K]$ and vector $\{F\}$ do not change during the iteration process, while the vector representation of the nonlinear term $\{H^*\}$ is updated during the iteration process. A solution of Eq. (12.7.8) is obtained when successive iterations yield solutions that satisfy a convergence criterion.

Another linearization technique is called *classical linearization*. The basis for classical linearization is the desire to linearize the nonlinear differential equations as linear operators acting on u or its derivatives. This should provide a better linearization than the constant linearization scheme. The classical linearization of $\aleph$ is obtained by the approximation

$$\aleph \approx \ell^* = \left(\frac{\aleph^*}{\Im^*}\right)\Im \qquad (12.7.9)$$

where $\Im$ is a linear opeartor acting on u or any of its derivatives up to $u^{(n-1)}$.

The result of classical linearization is to reduce the nonlinear differential equation, Eq. (12.7.1), to the linear differential equation

$$u^{(n)} + \Re(u) + \ell^*(u) = f(x) \qquad (12.7.10)$$

Linear operator $\Im$ may not be unique and therefore there may be alternate forms of classical linearization. For the three applications studied here, their classical linearization is given below.

$$u^2 \approx u^* u = \Im^* \qquad (12.7.11)$$

$$u'u \approx (u')^* u = \Im^* \qquad (12.7.12)$$

$$u'u \approx u^*(u') = \Im^* \qquad (12.7.13)$$

$$(u')^2 \approx (u')^* u' \qquad (12.7.14)$$

As shown in Eqs. (12.7.12) and (12.7.13), there are two forms of classical linearization for $u'u$. There is no reason to expect that alternate forms provide the same solutions. A study showed that the Eq. (12.7.12) approximation was far superior to the Eq. (12.7.13) approximation. The final form of the finite element analysis resulting from classical linearization is

$$\left([K] + [M]^*\right)\{u\} = \{F\} \qquad (12.7.15)$$

where $[M]^*$ is the finite element approximation of $\Im^*$ given by Eq. (12.7.9) and needs to be updated with each iteration.

The third linearization scheme is called a *quasilinearization* scheme developed in Ref. [63]. Here, a nonlinear operator $\aleph$ is expanded into a Taylor series with respect to u and each of the derivatives of u which appear as arguments of $\aleph$. The series is terminated after the linear terms. That is

$$\aleph(u^{(n-1)}, ..., u', u; x) \approx \ell^* + \hbar^* = \aleph^*_{u^{(n-1)}}\left\{u^{(n-1)} - \left(u^{(n-1)}\right)^*\right\} + ... + \aleph^*_u\{u - u^*\} \qquad (12.7.16)$$

The subscripts on $\aleph$ denote a partial derivative of $\aleph$ with respect to the subscript. The asterisk subscript on a quantity denotes that the quantity is evaluated from previous iteration values. Equation (12.7.16) provides a unique approximation to any nonlinear operator $\aleph$. Quasilinearization leads to finite element algebraic equations of the form

$$\left([K] + [M]^*\right)\{u\} = \{F\} - \{H\}^* \qquad (12.7.17)$$

Using Eq. (12.7.16), the quasilinear approximations of u^2, $u'u$, and u'^2 operators are

$$u^2 \approx (u^*)^2 + 2u^*(u - u^*) = 2u^*u - (u^*)^2 \qquad (12.7.18)$$

$$u'u \approx (u')^*u^* + u^*\{u' - (u')^*\} + (u')^*\{u - u^*\} = -u^*u'^* + u^*(u') + (u')^*u \qquad (12.7.19)$$

and

$$(u')^2 \approx (u'^*)^2 + 2u'^*(u' - u'^*) = -(u'^*)^2 + 2u'^*u' \qquad (12.7.20)$$

respectively.

12.8 MATLAB Application to Buckling Problem

This section shows an example of the beam buckling problem.

♣ **Example 12.8.1** This solves buckling loads of a simply supported beam. The beam has elastic modulus of 12, and moment inertia of cross-section of 1/12. The length of the beam is 1. All are in proper units. The following program computes the buckling loads using 4 beam elements. The lowest buckling load is 9.87 which agrees with the exact solution.

```
%------------------------------------------------
% Example 12.8.1
```

```
% to find the critical buckling loads of of a simply supported
% beam using Hermitian beam elements %
% Variable descriptions
% k = element stiffness matrix
% m = element mass matrix
% kk = system stiffness matrix
% mm = system mass matrix
% index = a vector containing system dofs associated with each element
% bcdof = a vector containing dofs associated with boundary conditions
% bcval = a vector containing boundary condition values associated with
%          the dofs in 'bcdof'
%---------------------------------------------------------------
%
clear
nel=4;                          % number of elements
nnel=2;                         % number of nodes per element
ndof=2;                         % number of dofs per node
nnode=(nnel-1)*nel+1;           % total number of nodes in system
sdof=nnode*ndof;                % total system dofs
%
el=12;                          % elastic modulus
xi=1/12;                        % moment of inertia of cross-section
tleng=1;                        % total length of the beam
leng=tleng/nel;                 % uniform mesh (equal size of elements)
%
kk=zeros(sdof,sdof);            % initialization of system stiffness matrix
kkg=zeros(sdof,sdof);           % initialization of system geometric matrix
index=zeros(nel*ndof,1);        % initialization of index vector
%
bcdof(1)=1;                     % deflection at node 1 is constrained
bcdof(2)=sdof-1;                % deflection at the last node is constrained
%
for iel=1:nel                   % loop for the total number of elements
%
index=feeldof1(iel,nnel,ndof);  % extract system dofs
                                % associated with the element
%
[k,m]=febeam1(el,xi,leng,0,0,1);  % compute element stiffness
                                  % and mass matrix
%
[kg,kef]=febeambk(leng,0);      % compute geometric stiffness matrix
%
kk=feasmbl1(kk,k,index);        % assemble element stiffness matrices
                                % into the system matrix
%
kkg=feasmbl1(kkg,kg,index);     % assemble geometric matrices
                                % into system matrix
%
```

```
end
%
[kk,kkg]=feaplycs(kk,kkg,bcdof);              % apply constraints
%
fsol=eig(kk,kkg)                              % solve the eigenvalue problem
%
%―――――――――――――――――――――――――――――――――――――――――――――
```

‡

12.9 MATLAB Application to Nonlinear Problem

This section shows some examples for nonlinear problems using different linearization techniques as discussed in Sec. 12.7. In addition, the first example illustrates the convergence of a linear finite element solution as the element size decreases in the problem domain.

♣ **Example 12.9.1** This example solves the linear ODE using different sizes of elements and computes the finite element errors compared to the exact solution. The equation and boundary conditions are

$$\frac{d^2u}{dx^2} = 1 \quad 0 < x < 1 \tag{12.9.1}$$

$$u(0) = 0 \text{ and } u(1) = 0 \tag{12.9.2}$$

The mean-square norm of the error is computed as below:

$$||e||_0 = \left(\int_0^1 \left(u_{exact} - u_{fem}\right)^2 dx\right)^{1/2} \tag{12.9.3}$$

The log-log plot of the error vs. element size is shown in Fig. 12.9.1.

```
%―――――――――――――――――――――――――――――――――――――――――――
% Example 12.9.1
% to solve the ordinary differential equation given as
% u" = 1, 0 < x < 1
% u(0) = 0 and u(1) = 0
% and to compute the mean-square norm for different element sizes.
%
% Variable descriptions
% k = element matrix
```

```
% f = element vector
% kk = system matrix
% ff = system vector
% index = a vector containing system dofs associated with each element
% bcdof = a vector containing dofs associated with boundary conditions
% bcval = a vector containing boundary condition values associated with
%              the dofs in 'bcdof'
%----------------------------------------------------------------
%
%--------------------------------------
% input data for control parameters
%--------------------------------------
%
clear
for int=1:7                           % loop for the different element numbers
%
nel=2^int;                                       % number of elements
nnel=2;                                          % number of nodes per element
ndof=1;                                          % number of dofs per node
nnode=nel+1;                                     % total number of nodes in system
sdof=nnode*ndof;                                 % total system dofs
%
%--------------------------------------
% input data for nodal coordinate values
%--------------------------------------
%
elemsize(int)=1.0/nel;
for i=1:nnode
gcoord(i)=elemsize(int)*(i-1);
end
%
%--------------------------------------
% input data for nodal connectivity for each element
%--------------------------------------
%
for i=1:nel
nodes(i,1)=i;
nodes(i,2)=i+1;
end
%
%--------------------------------------
% input data for coefficients of the ODE
%--------------------------------------
%
acoef=1;                                         % coefficient 'a' of the diff eqn
bcoef=0;                                         % coefficient 'b' of the diff eqn
ccoef=0;                                         % coefficient 'c' of the diff eqn
%
```

```
%------------------------------------------------
% input data for boundary conditions
%------------------------------------------------
%
bcdof(1)=1;                          % first node is constrained
bcval(1)=0;                          % whose described value is 0
bcdof(2)=nnode;                      % 4th node is constrained
bcval(2)=0;                          % whose described value is 0
%
%------------------------------------------------
% initialization of matrices and vectors
%------------------------------------------------
%
ff=zeros(sdof,1);                    % initialization of system force vector
kk=zeros(sdof,sdof);                 % initialization of system matrix
index=zeros(nnel*ndof,1);            % initialization of index vector
%
%------------------------------------------------
% computation of element matrices and vectors and their assembly
%------------------------------------------------
%
for iel=1:nel                        % loop for the total number of elements
%
nl=nodes(iel,1); nr=nodes(iel,2);    % extract nodes for (iel)-th element
xl=gcoord(nl); xr=gcoord(nr);        % extract nodal coord values
eleng=xr-xl;                         % element length
index=feeldof1(iel,nnel,ndof);       % extract dofs associated with element
%
k=feode2l(acoef,bcoef,ccoef,eleng);  % compute element matrix
f=fef1l(xl,xr);                      % compute element vector
[kk,ff]=feasmbl2(kk,ff,k,f,index);   % assemble element matrices and vectors
%
end                                  % end of loop for elements
%
%------------------------------------------------
% apply boundary conditions
%------------------------------------------------
%
[kk,ff]=feaplyc2(kk,ff,bcdof,bcval);
%
%------------------------------------------------
% solve the matrix equation
%------------------------------------------------
%
fsol=kk\ff;
%
%------------------------------------------------
% compute mean-square error
```

```
%------------------------------
%
error(int)=0.0;
%
for iel=1:nel
%
nl=nodes(iel,1); nr=nodes(iel,2);     % extract nodes for (iel)-th element
xl=gcoord(nl); xr=gcoord(nr);         % extract nodal coord values
eleng=xr-xl;                          % element length
soll=fsol(nl); solr=fsol(nr);         % extract fem solution
%
co1=(soll-solr)/eleng-0.5;
co2=(xl*solr-xr*soll)/eleng;
%
error(int)=error(int)+(xr^5-xl^5)/20.0+co1*(xr^4-xl^4)/4.0 ...
+(co1^2+co2)*(xr^3-xl^3)/3.0+co1*co2*(xr^2-xl^2)+co2^2*eleng;
%
end
%
error(int)=sqrt(error(int));
%
end                     % end of loop of different element numbers
%
%------------------------------------------------
% log-log plot of the mean-square error vs element size
%------------------------------------------------
%
loglog(elemsize,error);
xlabel('Element Size')
ylabel('Mean-Square Error')
%
%------------------------------------------------
```

‡

♣ **Example 12.9.2** This example solves the nonlinear ODE using the quasilinearization technique. The ODE equation and boundary conditions are

$$\frac{d^2u}{dx^2} - \alpha u \frac{du}{dx} = 0 \quad 0 < x < 1 \tag{12.9.4}$$

$$u(0) = 0 \text{ and } u(1) = 1 \tag{12.9.5}$$

where $\alpha=100$ is chosen in the following program. The error tolerance to stop the iteration is also selected to be 0.0001. The converged finite element solution is plotted in Fig. 12.9.2. The number of iterations for convergency is 7.

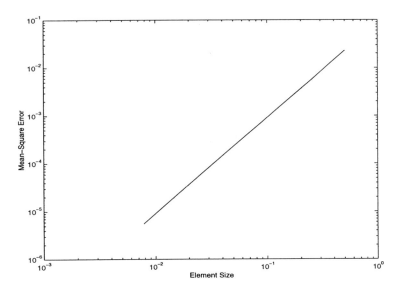

Figure 12.9.1 Log-Log Plot of Error vs. Element Size

```
%------------------------------------------------------------
% Example 12.9.2
% to solve the ordinary differential equation given as
% u" - alpa u u'= 0, 0 < x < 1
% u(0) = 0 and u(1) = 1
% using quasilinearization.
%
% Variable descriptions
% k = element matrix
% f = element vector
% kk = system matrix
% ff = system vector
% index = a vector containing system dofs associated with each element
% bcdof = a vector containing dofs associated with boundary conditions
% bcval = a vector containing boundary condition values associated with
%         the dofs in 'bcdof'
%------------------------------------------------------------
%
%------------------------------------
% input data for control parameters
%------------------------------------
%
clear
%
nel=100;                            % number of elements
nnel=2;                             % number of nodes per element
```

```
ndof=1;                          % number of dofs per node
nnode=nel+1;                     % total number of nodes in system
sdof=nnode*ndof;                 % total system dofs
alpa=100;                        % coefficient of the nonlinear term
toler=0.0001;                    % error tolerance to terminate iterations
%
%----------------------------------
% input data for nodal coordinate values
%----------------------------------
%
elemsize=1.0/nel;
for i=1:nnode
gcoord(i)=elemsize*(i-1);
end
%
%----------------------------------
% input data for nodal connectivity for each element
%----------------------------------
%
for i=1:nel
nodes(i,1)=i;
nodes(i,2)=i+1;
end
%
%----------------------------------
% input data for boundary conditions
%----------------------------------
%
bcdof(1)=1;                      % first node is constrained
bcval(1)=0;                      % whose described value is 0
bcdof(2)=nnode;                  % 4th node is constrained
bcval(2)=1;                      % whose described value is 0
%
%----------------------------------
% loop for iteration
%----------------------------------
%
error=1;                         % error is set to 1 arbitrarily
solold=gcoord';                  % assume a linear function initially
%
it=0;
while error > toler
it=it+1;                         % iteration counter
%
%----------------------------------
% initialization of matrices and vectors
%----------------------------------
%
```

```
ff=zeros(sdof,1);                   % initialization of system force vector
kk=zeros(sdof,sdof);                % initialization of system matrix
index=zeros(nnel*ndof,1);           % initialization of index vector
%
%-----------------------------------------------------------------
% computation of element matrices and vectors and their assembly
%-----------------------------------------------------------------
%
for iel=1:nel                       % loop for the total number of elements
%
nl=nodes(iel,1); nr=nodes(iel,2);   % extract nodes for (iel)-th element
xl=gcoord(nl); xr=gcoord(nr);       % extract nodal coord values
eleng=xr-xl;                        % element length
index=feeldof1(iel,nnel,ndof);      % extract dofs associated with element
soll=solold(nl); solr=solold(nr);   % extract old solutions
%
%-----------------------------------------------------------------
% element matrix and vector for quasilinearization
%-----------------------------------------------------------------
%
% element matrix
%
k(1,1)=1/eleng-alpa*(2*soll+solr)/6+alpa*(solr-soll)/3;
k(1,2)=-1/eleng+alpa*(2*soll+solr)/6+alpa*(solr-soll)/6;
k(2,1)=-1/eleng-alpa*(soll+2*solr)/6+alpa*(solr-soll)/6;
k(2,2)=1/eleng+alpa*(soll+2*solr)/6+alpa*(solr-soll)/3;
%
% element vector
%
f(1)=alpa*(solr-soll)*(2*soll+solr)/6;
f(2)=alpa*(solr-soll)*(soll+2*solr)/6;
%
% assemble element matrices and vectors
%
[kk,ff]=feasmbl2(kk,ff,k,f,index);
%
end                                 % end of loop for elements
%
%-----------------------------------------------------------------
% apply boundary conditions
%-----------------------------------------------------------------
%
[kk,ff]=feaplyc2(kk,ff,bcdof,bcval);
%
%-----------------------------------------------------------------
% solve the matrix equation
%-----------------------------------------------------------------
%
```

```
fsol=kk\ff;
%
%------------------------
% check the error
%------------------------
%
error=0;
for i=2:(nnode-1)
error=error+(fsol(i)-solold(i))^2/fsol(i)^2;
end
%
error=sqrt(error);
%
solold=fsol;                              % assign previous solution
%
end                                       % end of iteration loop
%
%------------------------
% plot of the solution
%------------------------
%
IterationNo=it            % print number of iteration for convergency
%
plot(gcoord',fsol);
xlabel('x-axis')
ylabel('Solution')
%
%---------------------------------------------------------------
```

‡

♣ **Example 12.9.3** This example solves the same nonlinear ODE as given in Eqs. (12.9.4) and (12.9.5) using the classical linearization. The nonlienar term is linearized as u^*u' where ()* denotes the term assumed to be known for linearization. Here, $\alpha=10$ is chosen with the error tolerance of 0.0001. The converged finite element solution is plotted in Fig. 12.9.3. The number of iterations for convergency is 92.

```
%---------------------------------------------------------------
% Example 12.9.3
% to solve the ordinary differential equation given as
% u" - alpa u u'= 0, 0 < x < 1
% u(0) = 0 and u(1) = 1
% using classical linearization such as (u*)u'.
%
% Variable descriptions
```

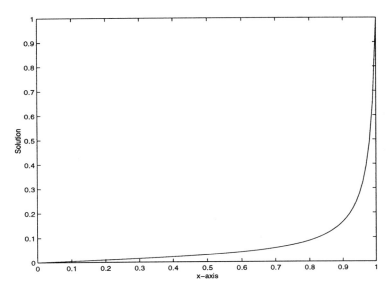

Figure 12.9.2 Finite Element Solution Using Quasilinearization

```
% k = element matrix
% f = element vector
% kk = system matrix
% ff = system vector
% index = a vector containing system dofs associated with each element
% bcdof = a vector containing dofs associated with boundary conditions
% bcval = a vector containing boundary condition values associated with
%              the dofs in 'bcdof'
%---------------------------------------------------------------
%
%---------------------------------
% input data for control parameters
%---------------------------------
%
clear
%
nel=50;                       % number of elements
nnel=2;                       % number of nodes per element
ndof=1;                       % number of dofs per node
nnode=nel+1;                  % total number of nodes in system
sdof=nnode*ndof;              % total system dofs
alpa=10;                      % coefficient of the nonlinear term
toler=0.0001;                 % error tolerance to terminate iterations
%
%-------------------------------
% input data for nodal coordinate values
```

```
%-----------------------------------------------
%
elemsize=1.0/nel;
for i=1:nnode
gcoord(i)=elemsize*(i-1);
end
%
%-----------------------------------------------
% input data for nodal connectivity for each element
%-----------------------------------------------
%
for i=1:nel
nodes(i,1)=i;
nodes(i,2)=i+1;
end
%
%-----------------------------------------------
% input data for boundary conditions
%-----------------------------------------------
%
bcdof(1)=1;                        % first node is constrained
bcval(1)=0;                        % whose described value is 0
bcdof(2)=nnode;                    % 4th node is constrained
bcval(2)=1;                        % whose described value is 0
%
%-----------------------------------------------
% loop for iteration
%-----------------------------------------------
%
error=1;                           % error is set to 1 arbitrarily
solold=gcoord';                    % assume a linear function initially
%
it=0;
while error > toler
it=it+1;                           % iteration counter
%
%-----------------------------------------------
% initialization of matrices and vectors
%-----------------------------------------------
%
ff=zeros(sdof,1);                  % initialization of system force vector
kk=zeros(sdof,sdof);               % initialization of system matrix
index=zeros(nnel*ndof,1);          % initialization of index vector
%
%-----------------------------------------------
% computation of element matrices and vectors and their assembly
%-----------------------------------------------
%
```

```
for iel=1:nel                              % loop for the total number of elements
%
nl=nodes(iel,1); nr=nodes(iel,2);          % extract nodes for (iel)-th element
xl=gcoord(nl); xr=gcoord(nr);              % extract nodal coord values
eleng=xr-xl;                               % element length
index=feeldof1(iel,nnel,ndof);             % extract dofs associated with element
soll=solold(nl); solr=solold(nr);          % extract old solutions
%
%-------------------------------------------------
% element matrix and vector for quasilinearization
%-------------------------------------------------
%
% element matrix
%
k(1,1)=1/eleng-alpa*(2*soll+solr)/6;
k(1,2)=-1/eleng+alpa*(2*soll+solr)/6;
k(2,1)=-1/eleng-alpa*(soll+2*solr)/6;
k(2,2)=1/eleng+alpa*(soll+2*solr)/6;
%
% element vector
%
f(1)=0;
f(2)=0;
%
% assemble element matrices and vectors
%
[kk,ff]=feasmbl2(kk,ff,k,f,index);
%
end                                        % end of loop for elements
%
%-------------------------
% apply boundary conditions
%-------------------------
%
[kk,ff]=feaplyc2(kk,ff,bcdof,bcval);
%
%-------------------------
% solve the matrix equation
%-------------------------
%
fsol=kk\ff;
%
%-------------------
% check the error
%-------------------
%
error=0;
for i=2:(nnode-1)
```

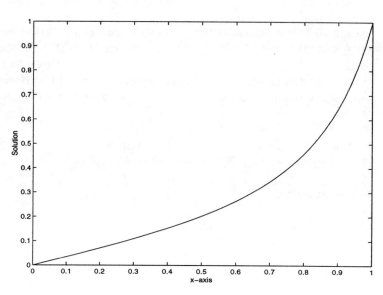

Figure 12.9.3 FE Solution Using Classical Linearization $(u^*)u'$

```
error=error+(fsol(i)-solold(i))^2/fsol(i)^2;
end
%
error=sqrt(error);
%
solold=fsol;                            % assign previous solution
%
end %
%--------------------------------
% plot of the solution
%--------------------------------
%
InterationNo=it                         % number of iteration for convergency
%
plot(gcoord',fsol);
xlabel('x-axis')
ylabel('Solution')
%
%------------------------------------------------------
```

♣ **Example 12.9.4** This example solves the same nonlinear ODE as given in Eqs. (12.9.4) and (12.9.5) using the classical linearization. The nonlinear term is linearized as $u(u')^*$ where $()^*$ denotes the term assumed to be known for linearization. Here, $\alpha=100$ is chosen with the error tolerance of 0.0001. The converged finite element solution is plotted in Fig. 12.9.4. The number of iterations for convergency is 21.

```
%----------------------------------------------------
% Example 12.9.4
% to solve the ordinary differential equation given as
%   u" - alpa*u*u'= 0, 0 < x < 1
%   u(0) = 0 and u(1) = 1
% using classical linearization (u')*u.
%
% Variable descriptions
%   k = element matrix
%   f = element vector
%   kk = system matrix
%   ff = system vector
%   index = a vector containing system dofs associated with each element
%   bcdof = a vector containing dofs associated with boundary conditions
%   bcval = a vector containing boundary condition values associated with
%           the dofs in 'bcdof'
%----------------------------------------------------
%
%----------------------------------------------------
% input data for control parameters
%----------------------------------------------------
%
clear
%
nel=100;                    % number of elements
nnel=2;                     % number of nodes per element
ndof=1;                     % number of dofs per node
nnode=nel+1;                % total number of nodes in system
sdof=nnode*ndof;            % total system dofs
alpa=100;                   % coefficient of the nonlinear term
toler=0.0001;               % error tolerance to terminate iterations
%
%----------------------------------------------------
% input data for nodal coordinate values
%----------------------------------------------------
%
elemsize=1.0/nel;
for i=1:nnode
gcoord(i)=elemsize*(i-1);
end
%
%----------------------------------------------------
% input data for nodal connectivity for each element
```

```
%------------------------------------------------
%
for i=1:nel
nodes(i,1)=i;
nodes(i,2)=i+1;
end
%
%------------------------------
% input data for boundary conditions
%------------------------------
%
bcdof(1)=1;                          % first node is constrained
bcval(1)=0;                          % whose described value is 0
bcdof(2)=nnode;                      % 4th node is constrained
bcval(2)=1;                          % whose described value is 0
%
%------------------------------
% loop for iteration
%------------------------------
%
error=1;                             % error is set to 1 arbitrarily
solold=gcoord';                      % assume a linear function initially
%
it=0;
while error > toler
it=it+1;                             % iteration counter
%
%------------------------------------------
% initialization of matrices and vectors
%------------------------------------------
%
ff=zeros(sdof,1);                    % initialization of system force vector
kk=zeros(sdof,sdof);                 % initialization of system matrix
index=zeros(nnel*ndof,1);            % initialization of index vector
%
%--------------------------------------------------------
% computation of element matrices and vectors and their assembly
%--------------------------------------------------------
%
for iel=1:nel                        % loop for the total number of elements
%
nl=nodes(iel,1); nr=nodes(iel,2);    % extract nodes for (iel)-th element
xl=gcoord(nl); xr=gcoord(nr);        % extract nodal coord values
eleng=xr-xl;                         % element length
index=feeldof1(iel,nnel,ndof);       % extract dofs associated with element
soll=solold(nl); solr=solold(nr);    % extract old solutions
%
%--------------------------------------------------
```

```
% element matrix and vector for quasilinearization
%------------------------------------------------
%
% element matrix
%
k(1,1)=1/eleng+alpa*(solr-soll)/3;
k(1,2)=-1/eleng+alpa*(solr-soll)/6;
k(2,1)=-1/eleng+alpa*(solr-soll)/6;
k(2,2)=1/eleng+alpa*(solr-soll)/3;
%
% element vector
%
f(1)=0;
f(2)=0;
%
% assemble element matrices and vectors
%
[kk,ff]=feasmbl2(kk,ff,k,f,index);
%
end                                      % end of loop for elements
%
%------------------------------------------------
% apply boundary conditions
%------------------------------------------------
%
[kk,ff]=feaplyc2(kk,ff,bcdof,bcval);
%
%------------------------------------------------
% solve the matrix equation
%------------------------------------------------
%
fsol=kk\ff;
%
%------------------------------
% check the error
%------------------------------
%
error=0;
for i=2:(nnode-1)
error=error+(fsol(i)-solold(i))^2/fsol(i)^2;
end
%
error=sqrt(error);
%
solold=fsol;                             % assign previous solution
%
end                                      % end of iteration loop
%
```

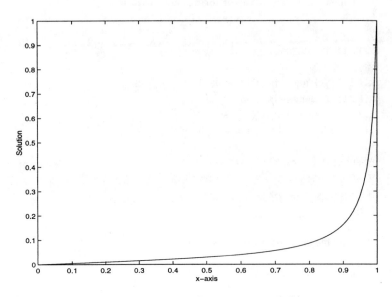

Figure 12.9.4 FE Solution Using Classical Linearization u(u')*

```
%————————————————————-
% plot of the solution
%————————————————————-
%
IterationNo=it            % print number of iteration for convergency
%
plot(gcoord',fsol);
xlabel('x-axis')
ylabel('Solution')
%
%————————————————————————————-
```

‡

REFERENCES

1. Crandall, S. H., *Engineering Analysis, A Survey of Numerical Procedures*, McGraw-Hill, New York, 1956.
2. Finlayson, B. A., *The Method of Weighted Residuals and Variational Principles*, Academic Press, New York, 1972.
3. Cook, R. D., *Concepts and Applications of Finite Element Analysis*, 2nd ed., John Wiley & Sons, New York, 1981.
4. Langhaar, H. L., *Energy Methods in Applied Mechanics*, Krieger, Malabar, FL, 1989.
5. Boresi, A. P., Schmidt, R. J., and Sidebottom, O. M., *Advanced Mechanics of Materials*, 5th ed., John Wiley & Sons, New York, 1993.
6. Mikhlin, S. G., *Variational Methods in Mathematical Physics*, Pergamon, New York, 1964.
7. Washizu, K., *Variational Methods in Elasticity and Plasticity*, Pergamon, New York, 1975.
8. Reddy, J. N., *Applied Functional Analysis and Variational Methods in Engineering*, McGraw-Hill, New York, 1986.
9. Akin, J. E., *Finite Element Analysis for Undergraduates*, Academic Press, London, 1986.
10. Irons, B. M., "Engineering Applications of Numerical Integration in Stiffness Methods", *Journal of the American Institute of Aeronautics and Astronautics*, Vol. 4, No. 11, 1966, pp. 2035-2037.
11. Archer, J. S., "Consistent Mass Matrix for Distributed Systems", *Proceedings of American Society of Civil Engineers*, Vol. 89, ST4, 1963, pp. 161-178.
12. Archer, J. S., "Consistent Matrix Formulations for Structural Analysis Using Finite-Element Techniques", *Journal of American Institute of Aeronautics and Astronautics*, Vol. 3, No. 10, 1965, pp. 1910-1918.
13. Hinton, E., Rock, T., and Zienkiewicz, "A Note on Mass Lumping and Related Processes in the Finite Element Method", *International Journal of Earthquake Engineering and Structural Engineering*, Vol. 4, No. 3, 1976, pp. 245-249.
14. Tong, P., Pian, T. H. H., and Bociovelli, L. L., "Mode Shapes and Frequencies by the Finite Element Method Using Consistent and Lumped Matrices", *Computers and Structures*, Vol. 1, 1971, pp. 623-638.
15. Fried, I. and Malkus, D. S., "Finite Element Mass Matrix Lumping by Numerical Integration With the Convergence Rate Loss", *International Journal of Solids and Structures*, Vol. 11, 1975, pp. 461-465.
16. Craig, Roy R., Jr., *Structural Dynamics, An Introduction to Computer Methods*, John Wiley & Sons, New York, 1981.
17. Zienkiewicz, O. C., Taylor, R. L., and Too, J. M., "Reduced Integration Technique in General Analysis of Plates and Shells", *International Journal for Numerical Methods in Engineering*, Vol. 3, 1971, pp. 275-290.
18. Hughes, T. J. R., Taylor, R. L., and Kanoknukulchai, W., "A Simple and Efficient Element for Plate Bending", *International Journal for Numerical Methods in Engineering*, Vol. 11, 1977, pp. 1529-1543.
19. Kwon, Y. W., Salinas, D., and Neibert, M. J., "Thermally Induced Stresses in a Trilayered System", *Journal of Thermal Stresses*, Vol. 17, 1994, pp. 489-506.

20. Kwon, Y. W. and Akin, J. E., "A Simple Efficient Algorithm for Elasto-plastic Plate Bending", *Engineering Computations*, Vol. 3, No. 4, 1986, pp. 283-286.
21. Kwon, Y. W. and Akin, J. E., "Linear Elastic and Non-linear Elastoplastic Plate Bending Analysis Using a Mixed Galerkin Finite Element Technique", *Engineering Computations*, Vol. 1, No. 3, 1984, pp. 268-272.
22. Akin, J. E. and Kwon, Y. W., "A Mixed Finite Element Method for Layered Composite Plates", *Computational Mechanics'86: Theory and Application* (edited by Yagawa et al.), Springer, Tokyo, 1986, pp. I95-I102.
23. Akin, J. E. and Kwon, Y. W., "Analysis of Plates With Through-Wall Cracks Using a Hybrid-Strain Element", *The Mathematics of Finite Elements and Applications VI* (edited by Whiteman, J. R.), Academic Press, London, 1988, pp. 115-122.
24. Yang, P. C., Horris, C. H., and Stavsky, Y., "Elastic Wave Propagation in Heterogeneous Plates", *International Journal of Solids and Structures*, Vol. 2, 1966, pp. 665-684.
25. Whitney, J. M. and Pagano, N. J., "Shear Deformation in Heterogeneous Anisotropic Plates", *Journal of Applied Mechanics*, Vol. 37, 1970, pp. 1031-1036.
26. Bathe, K. J., *Finite Element Procedures in Engineering Analysis*, Prentice-Hall, Englewood Cliffs, NJ, 1982.
27. Hughes, T. J. R., *The Finite Element Method, Linear Static and Dynamic Finite Element Analysis*, Prentice-Hall, Englewood Cliffs, NJ, 1987.
28. Zienkiewicz, O. C. and Taylor, R. L., *The Finite Element Method, Vol. 2 Solid and Fluid Mechanics Dynamics and Non-linearity*, 4th Ed., McGraw-Hill, London, 1991.
29. Park, K. C., "Practical Aspects of Numerical Time Integration", *Computers and Structures*, Vol. 7, 1977, pp. 343-353.
30. Kwon, Y. W. and Christy, C., "An Efficient Finite Element Modeling of Dynamic Crack Propagation Using a Moving Node Element", *Structural Engineering and Mechanics*, Vol. 2, No. 2, 1994, pp. 173-184.
31. Meirovitch, L., *Computational Methods in Structural Dynamics*, Kluwer Academic Publishers, The Netherlands, 1980.
32. Inman, D. J., *Vibration with Control, Measurement, and Stability*, Prentice-Hall, Englewood Cliffs, NJ, 1989.
33. Press, W. H., Flaneery, B. P., Teukolsky, S. A., and Vetterling, W.T., "Numerical Recipes", Cambridge University Press, New York, 1986.
34. Tocher, J. L., *Analysis of Plate Bending Using Triangular Elements*, Ph.D. Dissertation, University of California, Berkeley, 1962.
35. Owen, D. R. J. and Li, Z. H., "A Refined Analysis of Laminated Plates by Finite Element Displacement Methods - I. Fundamentals and Static Analysis", *Computers and Structures*, Vol. 26, No. 6, 1987, pp. 907-914.
36. Kwon, Y. W., "Finite Element Analysis of Crack Closure in Plate Bending", *Computers and Structures*, Vol. 32, No. 6, 1989, pp. 1439-1445.
37. Ashwell, D. G., "Strain Elements, with Applications to Arches, Rings and Cylindrical Shells", *Finite Elements for Thin Shells and Curved Members* (edited by Ashwell, D. G. and Gallagher, R. H.), Vol. 6, 1970, pp. 1133-1156.
38. Brogan, W. L., *Modern Control Theory*, Prentice-Hall, Englewood Cliffs, NJ, 1985.

39. Ogata, K., *Modern Control Engineering*, Prentice-Hall, Englewood Cliffs, NJ, 1990.
40. Chen, C. T., *Linear System Theory and Design*, Holt, Rinehart and Winston, New York, 1984.
41. Kailath, T., *Linear Systems*, Prentice-Hall, Englewood Cliffs, NJ, 1980.
42. Friedland, B., *Control System Design*, McGraw-Hill, New York, 1986.
43. Ogata, K., *Solving Control Engineering Problems with MATLAB*, Prentice-Hall, Englewood Cliffs, NJ, 1994.
44. Kuo, B. C. and Hanselman, D. C., "MATLAB Tools for Control System Analysis and Design", Prentice-Hall, Englewood Cliffs, NJ, 1994.
45. Athans, M. and Falb, P. L., *Optimal Control, and Introduction to Theory and Its Applications*, McGraw-Hill, New York, 1966.
46. Kirk, D. E., *Optimal Control Theory*, Prentice-Hall, Englewood Cliffs, NJ, 1970.
47. Laub, A., "A Schur Method for Solving Algebraic Riccati Equations", *IEEE Transactions on Automatic Control*, Vol. AC-24, Jun. 1979, pp. 913-925.
48. Potter, J. E., "Matrix Quadratic Solutions," *Journal of SIAM*, Applied Mathematics, Vol. 14, No. 3, 1966, pp. 496-501.
49. Meirovitch, L., *Introduction to Dynamics and Control*, John Wiley & Sons, New York, 1985.
50. Meirovitch, L., *Elements of Vibration Analysis*, McGraw-Hill, Singapore, 1986.
51. Junkins, J. L. and Kim, Y. D., *Introduction to Dynamics and Control of Flexible Structures*, AIAA Education Series, AIAA, Washington, DC, 1993.
52. Bayo E., "A Finite-Element Approach to Control the End-Point Motion of a Single-Link Flexible Robot", *Journal of Robotic Systems*, Vol. 4, No. 1, 1987, pp. 63-75.
53. Kim, Y., Junkins, J. L., and Kurdila, A. J., "On the Consequences of Certain Modeling Approximations in Dynamics and Control of Flexible Space Structures", Preprint, *Proceedings of the 33rd SDM Conference*, AIAA, Dallas, TX, April 1992, pp. 1173-1184.
54. Fujii, H. and Ishijima, S., "Mission Function Control for Slew Maneuver of a Flexible Space Structure", *Journal of Guidance, Control, and Dynamics*, Vol. 12, No. 6, November-December 1989, pp. 858-865.
55. Junkins, J. L., Rahman, Z. H., and Bang, H., "Near-Minimum-Time Control of Distributed Parameter Systems: Analytical and Experimental Results", *Journal of Guidance, Control, and Dynamics*, Vol. 14, No. 2, Mar.-Apr. 1991, pp. 406-415.
56. Junkins, J. L., Rahman, Z. H., and Bang, H., "Near-Minimum-Time Maneuvers of Flexible Vehicles: A Lyapunov Control Law Design Method", *Mechanics and Control of Large Flexible Structures*, Chapter 22, Progress in Astronautics and Aeronautics (edited by J. L. Junkins), Vol. 129, AIAA, New York, 1990.
57. Bang, H. and Kwon, Y. W., "Boundary Force Feedback for Flexible Structure Maneuver and Vibration Control", *Proceedings for 1994 ASME Winter Meeting*, Chicago, IL, November 6-11, pp. 59-70.
58. Kwon, Y. W., "Development of Finite Element Shape Functions with Derivative Singularity", *Computers and Structures*, Vol. 30, No. 5, 1988, pp. 1159-1163.
59. Kwon, Y.W., "Fracture Analysis of Plate Bending Using the Finite Element Method", *International Journal of Fracture*, Vol. 35, 1987, pp. R79-R81.

60. Kwon, Y.W. and Akin, J.E., "Development of a Derivative Singular Element for Application to Crack Propagation Problems", *Computers and Structures*, Vol. 31, No. 3, 1989, pp. 467-471.
61. Bettess, P., "Infinite Elements", *Internal Journal for Numerical Methods in Engineering*, Vol. 11, 1977, pp. 53-64.
62. Kwon, Y.W., "Finite Element Analysis of Dynamic Instability of Layered Composite Plates Using a High-Order Bending Theory", *Computers and Structures*, Vol. 38, No. 1, 1991, pp. 57-62.
63. Salinas, D., Kwon, Y.W., and Ritter, B.S., "A Study of Linearization Techniques in Nonlinear FEM", *Mathematical Modeling and Scientific Computing*, Vol. 1, No. 5, 1993, pp. 480-493.

FOR ADDITIONAL READING

Akin, J. E., *Finite Elements for Analysis and Design*, Academic Press, London, 1993.

Becker, E. B., Carey, G. F., and Oden, J. T., *Finite Elements: Vol. I An Introduction*, Prentice-Hall, Englewood Cliffs, NJ, 1981.

Bickford, W. B., *A First Course in the Finite Element Method*, Irwin, Homewood, IL, 1990.

Carey, G. F. and Oden, J. T., *Finite Elements: Vol. II A Second Course*, Prentice-Hall, Englewood Cliffs, NJ, 1983.

Craig, Jr., R. R., *Structural Dynamics - An Introduction to Computer Methods*, John Wiley & Sons, 1981.

Gibson, R. F., *Principles of Composite Material Mechanics*, McGraw-Hill, New York, 1994.

Jones, R. M., *Mechanics of Composite Materials*, McGraw-Hill, New York, 1975.

Reddy, J. N., *An Introduction to the Finite Element Method*, 2nd Ed., McGraw-Hill, New York, 1993.

Skelton, R. E., *Dynamic System Control - Linear System Analysis and Synthesis*, John Wiley & Sons, New York, 1988.

Zienkiewicz, O. C. and Taylor, R. L., *The Finite Element Method, Vol. 1 Basic Formulation and Linear Problems*, 4th Ed., McGraw-Hill, London, 1989.

APPENDIX A:
FEM MATLAB FUNCTION FILES

Appendix A   MATLAB Function Files   565

function [kk,mm]=feaplycs(kk,mm,bcdof)

Purpose:
    Apply constraints to eigenvalue matrix equation.
    [kk]{x}=lamda[mm]{x}

Synopsis:
    [kk,mm]=feaplycs(kk,mm,bcdof)

Variable Description:
    kk - system stiffness matrix before applying constraints
    mm - system mass matrix before applying constraints
    bcdof - a vector containing constrained dof

function [kk,ff]=feaplyc2(kk,ff,bcdof,bcval)

Purpose:
    Apply constraints to matrix equation [kk]{x}={ff}.

Synopsis:
    [kk,ff]=feaplybc(kk,ff,bcdof,bcval)

Variable Description:
    kk - system matrix before applying constraints
    ff - system vector before applying constraints
    bcdof - a vector containing constrained dof
    bcval - a vector containing contained value

Note:
    For example, there are constraints at dof=2 and 10
    and their constrained values are 0.0 and 2.5,
    respectively. Then, bcdof(1)=2 and bcdof(2)=10; and
    bcval(1)=1.0 and bcval(2)=2.5.

function [kk]=feasmbl1(kk,k,index)

Purpose:
    Assembly of element matrices into the system matrix.

Synopsis:
    [kk]=feasmbl1(kk,k,index)

Variable Description:
    kk - system matrix

k - element matrix
index - dof vector associated with an element

---

function [kk,ff]=feasmbl2(kk,ff,k,f,index)

Purpose:
   Assembly of element matrices into the system matrix and
   assembly of element vectors into the system vector.

Synopsis:
   [kk,ff]=feasmbl2(kk,ff,k,f,index)

Variable Description:
   kk - system matrix
   ff - system vector
   k - element matrix
   f - element vector
   index - dof vector associated with an element

---

function [g]=febasgr(A,B,dc)

Purpose:
   Calculate a feedback gain for a single input system by
   Bass-Gura formula.
   System equation : xdot = Ax + bu

Synopsis:
   [g]=fbasgur(A,B,dc)

Variable Description:
   dc - a vector consisting of desired closed-loop poles
   g - a feedback gain vector.

---

function [kg,kef]=febeambk(leng,elfd)

Purpose:
   Geometric stiffness matrix and stiffness matrix for
   elastic foundation using Hermitian beam element
   nodal dof $\{v\_1\ theta\_1\ v\_2\ theta\_2\}$

Appendix A   MATLAB Function Files

Synopsis:
    [kg,kef]=febeambk(el,xi,leng,area,rho,ipt)

Variable Description:
    kg - element geometric stiffness matrix (size 4x4)
    kef - stiffness matrix for elastic foundation(size 4x4)
    leng - element length
    elfd - spring constant of elastic foundation (force/length^2)

---

function [k,m]=febeam1(el,xi,leng,area,rho,ipt)

Purpose:
    Stiffness and mass matrices for Hermitian beam element.
    nodal dof $\{v_1 \; theta_1 \; v_2 \; theta_2\}$

Synopsis:
    [k,m]=febeam1(el,xi,leng,area,rho,ipt)

Variable Description:
    k - element stiffness matrix (size 4x4)
    m - element mass matrix (size 4x4)
    el - elastic modulus
    xi - second moment of inertia of cross-section
    leng - element length
    area - area of beam cross-section
    rho - mass density (mass per unit volume)
    ipt = 1: consistent mass matrix
           2: lumped mass matrix
           otherwise: diagonal mass matrix

---

function [k,m]=febeam2(el,xi,leng,sh,area,rho,ipt)

Purpose:
    Stiffness and mass matrices for the Timoshenko beam element.
    nodal dof $\{v_1 \; theta_1 \; v_2 \; theta_2\}$

Synopsis:
    [k,m]=febeam2(el,xi,leng,sh,area,rho,ipt)

Variable Description:
    k - element stiffness matrix (size 4x4)
    m - element mass matrix (size 4x4)
    el - elastic modulus

xi - second moment of inertia of cross-section
leng - length of the beam element
rho - mass density of the beam element (mass per unit volume)
sh - shear modulus
area - area of cross-section
ipt = 1: consistent mass matrix
    2: lumped mass matrix
    otherwise: diagonal mass matrix

---

function [k,m]=febeam3(el,sh,leng,heig,width,rho)

Purpose:
Stiffness and mass matrices for beam element with displacement degrees of freedom only.
nodal dof $\{u_1\hat{\ }b\ u_1\hat{\ }t\ v_1\ u_2\hat{\ }b\ u_2\hat{\ }t\ v_2\}$

Synopsis:
[k,m]=febeam1(el,sh,leng,heig,rho,area,ipt)

Variable Description:
k - element stiffness matrix (size 6x6)
m - element mass matrix (size 6x6)
el - elastic modulus
sh - shear modulus
leng - element length
heig - element thickness
width - width of the beam element
rho - mass density of the beam element (mass per unit volume)
    lumped mass matrix only

---

function [k,m]=febeam4(el,xi,leng,sh,heig,rho,ipt)

Purpose:
Stiffness and mass matrices for mixed beam element.
Bending moment and deflection as nodal degrees of freedom.
nodal dof $\{M_1\ v_1\ M_2\ v_2\}$

Synopsis:
[k,m]=febeam4(el,xi,leng,sh,heig,rho,ipt)

Variable Description:
k - element stiffness matrix (size 4x4)
m - element mass matrix (size 4x4)

el - elastic modulus
xi - second moment of inertia of cross-section
leng - length of the beam element
sh - shear modulus
heig - beam thickness
rho - mass density of the beam element (mass per unit volume)
ipt = 1 - lumped mass matrix
    = otherwise - diagonalized mass matrix

---

function [Ctobty,rrank,ccond]=fectobt(A,B)

Purpose:
 Calculate controllability matrix and/or observability of
 a system described in state space form
  xdot = Ax + Bu

Synopsis:
 [Ctobty,rrank,ccond]=fectobt(A,B)

Variable Description:
 Ctobty - controllability or observability matrix
 rrank - rank of Ctobty which determines yes/no type answer
 ccond - condition number of Ctobty

Note:
 For controllability test, the input argument should follow as
 fctobty(A,B).
 For observability test, we should provide the input argument as
 fctobty(A^T, C^T). ( )^T is transpose of ( ).

---

function [dhdx,dhdy]=federiv2(nnel,dhdr,dhds,invjacob)

Purpose:
 Determine derivatives of 2-D isoparametric shape functions with
 respect to physical coordinate system.

Synopsis:
 [dhdx,dhdy]=federiv2(nnel,dhdr,dhds,invjacob)

Variable Description:
 dhdx - derivative of shape function w.r.t. physical coordinate x
 dhdy - derivative of shape function w.r.t. physical coordinate y
 nnel - number of nodes per element

dhdr - derivative of shape functions w.r.t. natural coordinate r
dhds - derivative of shape functions w.r.t. natural coordinate s
invjacob - inverse of 2-D Jacobian matrix

---

function [dhdx,dhdy,dhdz]=federiv3(nnel,dhdr,dhds,dhdt,invjacob)

Purpose:
    Determine derivatives of 3-D isoparametric shape functions with respect to physical coordinate system.

Synopsis:
    [dhdx,dhdy,dhdz]=federiv3(nnel,dhdr,dhds,dhdt,invjacob)

Variable Description:
    dhdx - derivative of shape function w.r.t. physical coordinate x
    dhdy - derivative of shape function w.r.t. physical coordinate y
    dhdz - derivative of shape function w.r.t. physical coordinate z
    nnel - number of nodes per element
    dhdr - derivative of shape functions w.r.t. natural coordinate r
    dhds - derivative of shape functions w.r.t. natural coordinate s
    dhdt - derivative of shape functions w.r.t. natural coordinate t
    invjacob - inverse of 3-D Jacobian matrix

---

function [eta,yim]=fediresp(M,K,F,u,t,C,q0,dq0,a,b)

Purpose:
    Calculate impulse response for a damped structural system using modal analysis. It uses modal coordinate equations to evaluate modal responses analytically, then convert modal coordinates into physical responses.

Synopsis:
    [eta,yim]=fediresp(M,K,F,u,t,C,q0,dq0,a,b)

Variable Description:
    M, K - mass and stiffness matrices
    F - input or forcing influence matrix
    u - index for excitation
    t - time of evaluation
    u - index for the excitation
    C - output matrix
    q0, dq0 - initial conditions
    a, b - parameters for proportional damping [C]=a[M]+b[K]

Appendix A   MATLAB Function Files   571

eta - modal coordinate response
yim - physical coordinate response

---

function [index]=feeldof(nd,nnel,ndof)

Purpose:
   Compute system dofs associated with each element.

Synopsis:
   [index]=feeldof(nd,nnel,ndof)

Variable Description:
   index - system dof vector associated with element "iel"
   iel - element number whose system dofs are to be determined
   nnel - number of nodes per element
   ndof - number of dofs per node

---

function [index]=feeldof1(iel,nnel,ndof)

Purpose:
   Compute system dofs associated with each element in one-dimensional problem.

Synopsis:
   [index]=feeldof1(iel,nnel,ndof)

Variable Description:
   index - system dof vector associated with element "iel"
   nd - element node numbers whose system dofs are to be determined
   nnel - number of nodes per element
   ndof - number of dofs per node

---

function [yfft,freq]=fefft(y,t)

Purpose:
   Calculate Fast Fourier Transform (FFT) using the time domain signal. The time domain data are provided with corresponding time interval.

Synopsis:

[yf, freq]=feﬀt(y,t)

Variable Description:
    y - time domain data n by 1
    t - time interval for y of n by 1 size
    yf - absolute value of FFT of y
    freq - frequency axis values

Notes:
    The number of data points for y should be power of 2, and truncation is needed to achieve the requirement.

---

function [k]=feflxl2(eleng)

Purpose:
    Element matrix for Cauchy-type boundary such as $du/dn=a(u-b)$ using linear element where a and b are known constants.

Synopsis:
    [k]=feflxl2(eleng)

Variable Description:
    k - element vector (size 2x2)
    eleng - length of element side with given flux

---

function [k,m]=feframe2(el,xi,leng,area,rho,beta,ipt)

Purpose:
    Stiffness and mass matrices for the 2-D frame element.
    nodal dof u_1 v_1 $theta\_1$ u_2 v_2 $theta\_2$

Synopsis:
    [k,m]=feframe2(el,xi,leng,area,rho,beta,ipt)

Variable Description:
    k - element stiffness matrix (size 6x6)
    m - element mass matrix (size 6x6)
    el - elastic modulus
    xi - second moment of inertia of cross-section
    leng - element length
    area - area of beam cross-section
    rho - mass density (mass per unit volume)
    beta - angle between the local and global axes
        is positive if the local axis is in the ccw direction from

Appendix A   MATLAB Function Files

   the global axis
ipt = 1 - consistent mass matrix
    = 2 - lumped mass matrix
    = 3 - diagonal mass matrix

---

function [f]=fefxl(xl,xr)

Purpose:
   Element vector for f(x)=x using linear element.

Synopsis:
   [f]=fefxl(xl,xr)

Variable Description:
   f - element vector (size 2x1)
   xl - coordinate value of the left node
   xr - coordinate value of the right node

---

function [f]=fefx2l(xl,xr)

Purpose:
   Element vector for f(x)=x^2 using linear element.

Synopsis:
   [f]=fefx2l(xl,xr)

Variable Description:
   f - element vector (size 2x1)
   xl - coordinate value of the left node
   xr - coordinate value of the right node

---

function [f]=fef1l(xl,xr)

Purpose:
   Element vector for f(x)=1 using linear element.

Synopsis:
   [f]=fef1l(xl,xr)

Variable Description:

    f - element vector (size 2x1)
    xl - coordinate value of the left node
    xr - coordinate value of the right node

---

function [point1,weight1]=feglqd1(ngl)

Purpose:
   Determine the integration points and weighting coefficients
   of Gauss-Legendre quadrature for one-dimensional integration.

Synopsis:
   [point1,weight1]=feglqd1(ngl)

Variable Description:
   ngl - number of integration points
   point1 - vector containing integration points
   weight1 - vector containing weighting coefficients

---

function [point2,weight2]=feglqd2(nglx,ngly)

Purpose:
   Determine the integration points and weighting coefficients
   of Gauss-Legendre quadrature for two-dimensional integration.

Synopsis:
   [point2,weight2]=feglqd2(nglx,ngly)

Variable Description:
   nglx - number of integration points in the x-axis
   ngly - number of integration points in the y-axis
   point2 - vector containing integration points
   weight2 - vector containing weighting coefficients

---

function [point3,weight3]=feglqd3(nglx,ngly,nglz)

Purpose:
   Determine the integration points and weighting coefficients
   of Gauss-Legendre quadrature for three-dimensional integration.

Synopsis:

Appendix A   MATLAB Function Files   575

    [point3,weight3]=feglqd3(nglx,ngly,nglz)

Variable Description:
    nglx - number of integration points in the x-axis
    ngly - number of integration points in the y-axis
    nglz - number of integration points in the z-axis
    point3 - vector containing integration points
    weight3 - vector containing weighting coefficients

---

function [eta,yim]=feiresp(M,K,F,u,t,C,q0,dq0)

Purpose:
    Calculate impulse response for a given structural system
    using modal analysis. It uses modal coordinate equations
    to evaluate modal responses analytically, then convert modal
    coordinates into physical responses.

Synopsis:
    [eta,yim]=impresp(M,K,F,u,t,C,q0,dq0)

Variable Description:
    M, K - mass and stiffness matrices
    F - input or forcing function
    u - index for excitation
    t - vector of time duration
    C - output matrix
    q0, dq0 - initial conditions
    eta - modal coordinate response
    yim - physical coordinate response

---

function [shape,dhdr]=feisol2(rvalue)

Purpose
    Compute isoparametric 2-node shape functions
    and their derivatives at the selected
    point in terms of the natural coordinate.

Synopsis:
    [shape,dhdr]=kwisol2(rvalue)

Variable Description: shape - shape functions for the linear element
    dhdr - derivatives of shape functions
    rvalue - r coordinate value of the selected point

Notes:
> 1st node at rvalue=-1
> 2nd node at rvalue=1

---

function [shapeq4,dhdrq4,dhdsq4]=feisoq4(rvalue,svalue)

Purpose:
> Compute isoparametric four-node quadilateral shape functions
> and their derivatives at the selected (integration) point
> in terms of the natural coordinate.

Synopsis:
> [shapeq4,dhdrq4,dhdsq4]=feisoq4(rvalue,svalue)

Variable Description:
> shapeq4 - shape functions for four-node element
> dhdrq4 - derivatives of the shape functions w.r.t. r
> dhdsq4 - derivatives of the shape functions w.r.t. s
> rvalue - r coordinate value of the selected point
> svalue - s coordinate value of the selected point

Notes:
> 1st node at (-1,-1), 2nd node at (1,-1)
> 3rd node at (1,1), 4th node at (-1,1)

---

function [shapes8,dhdrs8,dhdss8,dhdts8]=feisos8(rvalue,svalue,tvalue)

Purpose:
> Compute isoparametric eight-node solid shape functions
> and their derivatives at the selected (integration) point
> in terms of the natural coordinate.

Synopsis:
> [shapes8,dhdrs8,dhdss8,dhdts8]=feisos8(rvalue,svalue,tvalue)

Variable Description:
> shapes8 - shape functions for four-node element
> dhdrs8 - derivatives of the shape functions w.r.t. r
> dhdss8 - derivatives of the shape functions w.r.t. s
> dhdts8 - derivatives of the shape functions w.r.t. t
> rvalue - r coordinate value of the selected point
> svalue - s coordinate value of the selected point
> tvalue - t coordinate value of the selected point

Notes:
> 1st node at (-1,-1,-1), 2nd node at (1,-1,-1)
> 3rd node at (1,1,-1), 4th node at (-1,1,-1)
> 5th node at (-1,-1,1), 6th node at (1,-1,1)
> 7th node at (1,1,1), 8th node at (-1,1,1)

function [shapet3,dhdrt3,dhdst3]=feisot3(rvalue,svalue)

Purpose:
> Compute isoparametric three-node triangular shape functions
> and their derivatives at the selected (integration) point
> in terms of the natural coordinate.

Synopsis:
> [shapet3,dhdrt3,dhdst3]=feisot3(rvalue,svalue)

Variable Description:
> shapet3 - shape functions for three-node element
> dhdrt3 - derivatives of the shape functions w.r.t. r
> dhdst3 - derivatives of the shape functions w.r.t. s
> rvalue - r coordinate value of the selected point
> svalue - s coordinate value of the selected point

Notes:
> 1st node at (0,0), 2nd node at (1,0), 3rd node at (0,1)

function [jacob1]=fejacob1(nnel,dhdr,xcoord)

Purpose:
> Determine the Jacobian for one-dimensional mapping.

Synopsis:
> [jacob1]=fejacob1(nnel,dhdr,xcoord)

Variable Description:
> jacob1 - Jacobian for one-dimension
> nnel - number of nodes per element
> dhdr - derivative of shape functions w.r.t. natural coordinate
> xcoord - x axis coordinate values of nodes

function [jacob2]=fejacob2(nnel,dhdr,dhds,xcoord,ycoord)

Purpose:
    Determine the Jacobian for two-dimensional mapping.

Synopsis:
    [jacob2]=fejacob2(nnel,dhdr,dhds,xcoord,ycoord)

Variable Description:
    jacob2 - Jacobian for one-dimension
    nnel - number of nodes per element
    dhdr - derivative of shape functions w.r.t. natural coordinate r
    dhds - derivative of shape functions w.r.t. natural coordinate s
    xcoord - x axis coordinate values of nodes
    ycoord - y axis coordinate values of nodes

function [jacob3]=fejacob3(nnel,dhdr,dhds,dhdt,xcoord,ycoord,zcoord)

Purpose:
    Determine the Jacobian for three-dimensional mapping.

Synopsis:
    [jacob3]=fejacob3(nnel,dhdr,dhds,dhdt,xcoord,ycoord,zcoord)

Variable Description:
    jacob3 - Jacobian for one-dimension
    nnel - number of nodes per element
    dhdr - derivative of shape functions w.r.t. natural coordinate r
    dhds - derivative of shape functions w.r.t. natural coordinate s
    dhdt - derivative of shape functions w.r.t. natural coordinate t
    xcoord - x axis coordinate values of nodes
    ycoord - y axis coordinate values of nodes
    zcoord - z axis coordinate values of nodes

function [kinmtxax]=fekineax(nnel,dhdx,dhdy,shape,radist)

Purpose:
    Determine kinematic equations between strains and displacements for axisymmetric solids.

Synopsis:
    [kinmtxax]=fekineax(nnel,dhdx,dhdy,shape,radist)

Variable Description:

nnel - number of nodes per element
shape - shape functions
dhdx - derivatives of shape functions with respect to x
dhdy - derivatives of shape functions with respect to y
radist - radial distance of integration point or central point
    for hoop strain component

---

function [kinmtpb]=fekinepb(nnel,dhdx,dhdy)

Purpose:
   Determine the kinematic matrix expression relating bending curvatures
   to rotations and displacements for shear deformable plate bending.

Synopsis:
   [kinmtpb]=fekinepb(nnel,dhdx,dhdy)

Variable Description:
   nnel - number of nodes per element
   dhdx - derivatives of shape functions with respect to x
   dhdy - derivatives of shape functions with respect to y

---

function [kinmtps]=fekineps(nnel,dhdx,dhdy,shape)

Purpose:
   Determine the kinematic matrix expression relating shear strains
   to rotations and displacements for shear deformable plate bending.

Synopsis:
   [kinmtps]=fekineps(nnel,dhdx,dhdy,shape)

Variable Description:
   nnel - number of nodes per element
   dhdx - derivatives of shape functions with respect to x
   dhdy - derivatives of shape functions with respect to y
   shape - shape function

function [kinmtsb]=fekinesb(nnel,dhdx,dhdy)

Purpose:
    Determine the kinematic matrix expression relating bending curvatures
    to rotations and displacements for shear deformable plate bending

Synopsis:
    [kinmtsb]=fekinesb(nnel,dhdx,dhdy)

Variable Description:
    nnel - number of nodes per element
    dhdx - derivatives of shape functions with respect to x
    dhdy - derivatives of shape functions with respect to y

---

function [kinmtsm]=fekinesm(nnel,dhdx,dhdy)

Purpose:
    Determine the kinematic equation between strains and displacements
    for two-dimensional solids

Synopsis:
    [kinmtsm]=fekinesm(nnel,dhdx,dhdy)

Variable Description:
    nnel - number of nodes per element
    dhdx - derivatives of shape functions with respect to x
    dhdy - derivatives of shape functions with respect to y

---

function [kinmtss]=fekiness(nnel,dhdx,dhdy,shape)

Purpose:
    Determine the kinematic matrix expression relating shear strains
    to rotations and displacements for shear deformable plate bending

Synopsis:
    [kinmtss]=fekiness(nnel,dhdx,dhdy,shape)

Variable Description:
    nnel - number of nodes per element
    dhdx - derivatives of shape functions with respect to x
    dhdy - derivatives of shape functions with respect to y
    shape - shape function

Appendix A   MATLAB Function Files   581

function [kinmtx2]=fekine2d(nnel,dhdx,dhdy)

Purpose:
    Determine the kinematic equation between strains and displacements for two-dimensional solids.

Synopsis:
    [kinmtx2]=fekine2d(nnel,dhdx,dhdy)

Variable Description:
    nnel - number of nodes per element
    dhdx - derivatives of shape functions with respect to x
    dhdy - derivatives of shape functions with respect to y

---

function [kinmtx3]=fekine3d(nnel,dhdx,dhdy,dhdz)

Purpose:
    Determine the kinematic equation between strains and displacements for three-dimensional solids.

Synopsis:
    [kinmtx3]=fekine3d(nnel,dhdx,dhdy,dhdz)

Variable Description:
    nnel - number of nodes per element
    dhdx - derivatives of shape functions with respect to x
    dhdy - derivatives of shape functions with respect to y
    dhdz - derivatives of shape functions with respect to z

---

function [k]=felpaxt3(r1,z1,r2,z2,r3,z3)

Purpose:
    Element matrix for axisymmetric Laplace's equation using three-node linear triangular element.

Synopsis:
    [k]=felpaxt3(r1,z1,r2,z2,r3,z3)

Variable Description:
    k - element stiffness matrix (size 3x3)
    r1, z1 - r and z coordinate values of the first node of element
    r2, z2 - r and z coordinate values of the second node of element
    r3, z3 - r and z coordinate values of the third node of element

---

function [m]=felpt2r4(xleng,yleng)

Purpose:
    Element matrix of transient term for two-dimensional Laplace's equation using four-node bilinear rectangular element.

Synopsis:
    [m]=felpt2r4(xleng,yleng)

Variable Description:
    m - element stiffness matrix (size 4x4)
    xleng - element size in the x-axis
    yleng - element size in the y-axis

function [m]=felpt2t3(x1,y1,x2,y2,x3,y3)

Purpose:
    Element matrix for transient term of two-dimensional Laplace's equation using linear triangular element.

Synopsis:
    [m]=felpt2t3(x1,y1,x2,y2,x3,y3)

Variable Description:
    m - element stiffness matrix (size 3x3)
    x1, y1 - x and y coordinate values of the first node of element
    x2, y2 - x and y coordinate values of the second node of element
    x3, y3 - x and y coordinate values of the third node of element

function [m]=felpt3t4(x,y,z)

Purpose:
    Element matrix of transient term for three-dimensional Laplace's equation using four-node tetrahedral element.

Synopsis:
    [m]=felpt3t4(x,y,z)

Variable Description:
    m - element stiffness matrix (size 4x4)
    xleng - element size in the x-axis
    yleng - element size in the y-axis

Appendix A MATLAB Function Files 583

function [k]=felp2dr4(xleng,yleng)

Purpose:
    Element matrix for two-dimensional Laplace's equation
    using four-node bilinear rectangular element.

Synopsis:
    [k]=felp2dr4(xleng,yleng)

Variable Description:
    k - element stiffness matrix (size 4x4)
    xleng - element size in the x-axis
    yleng - element size in the y-axis

function [k]=felp2dt3(x1,y1,x2,y2,x3,y3)

Purpose:
    Element matrix for two-dimensional Laplace's equation
    using three-node linear triangular element.

Synopsis:
    [k]=felp2dt3(x1,y1,x2,y2,x3,y3)

Variable Description:
    k - element stiffness matrix (size 3x3)
    x1, y1 - x and y coordinate values of the first node of element
    x2, y2 - x and y coordinate values of the second node of element
    x3, y3 - x and y coordinate values of the third node of element

function [k]=felp3dt4(x,y,z)

Purpose:
    Element matrix for three-dimensional Laplace's equation
    using four-node tetrahedral element.

Synopsis:
    [k]=felp3dt4(x,y,z)

Variable Description:
    k - element matrix (size 4x4)
    x - x coordinate values of the four nodes
    y - y coordinate values of the four nodes
    z - z coordinate values of the four nodes

function [G,S]=felqr(A,B,Q,R)

Purpose:
    Calculate the feedback gain matrix by Linear Quadratic
    Regulator(LQR) technique. The given system is
        xdot = Ax + Bu, u = -Gx
    and the cost function to be minimized is defined as
        J=(1/2)integral(x'Qx+u'Ru)dt

Synopsis:
    [G,S]=felqr(A,B,Q,R)

Variable Description:
    A, B, Q, R - input arguments
    $G = R^{-1}G'S$ - feedback gain matrix
    S - solution of the Algebraic Ricatti Equation (ARE)
        $AS+A'S-SBR^{-1}S+Q=0$

Notes:
    (A,B) should be controllable.
    Q is at least positive semidefinite.
    R is at least positive definite.

---

function [x,y]=felresp(A,B,C,D,x0,u,t)

Purpose:
    Find the time response of a linear system driven by an initial condition
    and an external input. The numerical algorithm used in this program is
    a zero order hold approximation for control input for discretized
    system.

Synopsis:
    [x,y]=felresp(A,B,C,D,x0,u,t)

Variable Description:
    A, B, C, D - system matrices in, xdot = Ax + Bu, y = Cx + Du
    x0 - initial condition vector for the state variables
    t - integration time at equal distance as t=0:dt:tf
    dt - time step, tf - final time
    u - control input vector with as many rows as the size of t
    x(y) - state(output) vector

Notes:
    The control input vector must have as many columns as the number of input.

Appendix A   MATLAB Function Files   585

function [matmtrx]=fematiso(iopt,elastic,poisson)

Purpose:
    determine the constitutive equation for isotropic material.

Synopsis:
    [matmtrx]=fematiso(iopt,elastic,poisson)

Variable Description:
    elastic - elastic modulus
    poisson - Poisson's ratio
    iopt=1 - plane stress analysis
    iopt=2 - plane strain analysis
    iopt=3 - axisymmetric analysis
    iopt=4 - three dimensional analysis

function [Omega,Phi,ModF]=femodal(M,K,F)

Purpose:
    Calculate modal parameters for a given structural system.
    It calculates natural frequency and eigenvector.
    The eigenvectors are normalized so that the modal mass matrix
    becomes an identity matrix.

Synopsis:
    [Omega, Phi, ModF]=femodal(M,K,F)

Variable Description:
    M, K - mass and stiffness matrices
    F - input or forcing function
    Omega - natural frequency (rad/sec)
    Phi - modal matrix with each column corresponding to
        the eigenvector
    ModF - modal input matrices.

function [k]=feodex2l(xl,xr)

Purpose:
    Element matrix for ($x^2 u'' - 2x u' - 4 u$) using linear element.

Synopsis:
    [k]=feodex2l(xl,xr)

Variable Description:

k - element matrix (size 2x2)
xl - coordinate value of the left node of the linear element
xr - coordinate value of the right node of the linear element

---

function [k]=feode2l(acoef,bcoef,ccoef,eleng)

Purpose:
Element matrix for (a u" + b u' + c u) using linear element.

Synopsis:
[k]=feode2l(acoef,bcoef,ccoef,eleng)

Variable Description:
k - element matrix (size 2x2)
acoef - coefficient of the second order derivative term
bcoef - coefficient of the first order derivative term
ccoef - coefficient of the zero-th order derivative term
eleng - element length

---

function [y]=ferbsim(M,K,F,g1,g2,g3,EI,h,l0,thf,tf)

Purpose:
Simulate a rotating flexible beam attached to a base. The mathematical model is created from *frobfem.m* as system mass and stiffness matrices.

Synopsis:
[y] = ferbsim(M,K,F,g1,g2,g3,EI,h,l0,thf,tf)

Variable Description:
M, K, F - system matrices
g1, g2, g3 - feedback gains
EI, h, l0 - parameters for boundary force calculation
thf, tf - final angle and final simulation time
y - output parameter

Appendix A   MATLAB Function Files

function [w,M,K]=ferobem(N,EI,rho,I_c,I_t,m_t,l_0,L)

Purpose:
    Produce a finite element modeling of a rotating beam attached to a rigid base.

Synopsis:
    [w,M,K]=ferobem(N,EI,rho,I_c,I_t,m_t,l_0,L)

Variable Description:
    N - number of elements
    EI - elastic rigidity
    rho - linear mass density
    I_c(I_t) - moment of inertia of center body(tip mass)
    m_t - tip mass
    l_0 - radius of the center body
    L - beam length
    M, K - system mass, stiffness matrices
    w - natural frequency

---

function [t_p, t_r, t_s, M_p]=fesecnd(zeta, w_n)

Purpose:
    Calculate dynamic characteristics of a typical standard second order system. Transfer function is

$$H(s) = \frac{w_n^2}{s^2 + 2*zeta*w_n*s + w_n^2}$$

Synopsis:
    [t_p, t_r, M_p, t_s]=fsecond(zeta, w_n)

Variable Description:
    zeta - damping ratio
    w_n - natural frequency (rad/sec)
    t_p - peak time
    t_r - rise time
    t_s - settling time
    M_p - maximum overshoot

function [num, den] = festotf(A,B,C,D,iu)

Purpose:
> Convert a state space form of system into a transfer function form for the given system
> $$\dot{x} = Ax + Bu$$
> $$y = Cx + Du$$
> The transfer function becomes
> $$H(s) = \frac{N(s)}{D(s)} = C(sI-A)^{-1}B + D$$

Synopsis:
> [num,den]=festotf(A,B,C,D,iu)

Variable Description:
> A, B, C, D - system matrix
> iu - index for control input (iu-th input)
> D(s) - vector of coefficients of the denominator polynomials
> N(s) - vector of coefficients of the numerator polynomials

Note:
> There is same number of rows in N(s) as the number of output.

function [tr3d,xprime,yprime]=fetransh(xcoord,ycoord,zcoord,n)

Purpose:
> Compute direction cosines between three-dimensional local and global coordinate axes

Synopsis:
> [tr3d,xprime,yprime]=fetransh(xcoord,ycoord,zcoord,n)

Variable Description:
> xcoord - nodal x coordinates (4x1)
> ycoord - nodal y coordinates (4x1)
> zcoord - nodal z coordinates (4x1)
> n - number of nodes per element
> tr3d - 3-D transformation matrix from local to global axes
> xprime - coordinate in terms of the local axes (4x1)
> yprime - coordinate in terms of the global axes (4x1)

Note:
> The local x-axis is defined in the direction from the first node to the second node. Nodes 1, 2, and 4 define the local xy-plane. The local z-axis is defined normal to the local xy-plane. The local y-axis is defined normal to the x and z axes.

Appendix A MATLAB Function Files

function [k,m]=fetruss1(el,leng,area,rho,ipt)

Purpose:
    Stiffness and mass matrices for the 1-d truss element.
    nodal dof u_1 u_2

Synopsis:
    [k,m]=fetruss1(el,leng,area,rho,ipt)

Variable Description:
    k - element stiffness matrix (size 4x4)
    m - element mass matrix (size 4x4)
    el - elastic modulus
    leng - element length
    area - area of truss cross-section
    rho - mass density (mass per unit volume)
    ipt = 1 - consistent mass matrix
        = 2 - lumped mass matrix

function [k,m]=fetruss2(el,leng,area,rho,beta,ipt)

Purpose:
    Stiffness and mass matrices for the 2-d truss element.
    nodal dof u_1 v_1 u_2 v_2

Synopsis:
    [k,m]=fetruss2(el,leng,area,rho,beta,ipt)

Variable Description:
    k - element stiffness matrix (size 4x4)
    m - element mass matrix (size 4x4)
    el - elastic modulus
    leng - element length
    area - area of truss cross-section
    rho - mass density (mass per unit volume)
    beta - angle between the local and global axes
    ipt = 1: consistent mass matrix
        positive if the local axis is in the ccw direction from
        the global axis
    ipt = 1 - consistent mass matrix
        = 2 - lumped mass matrix

APPENDIX B:
EXAMPLES OF PRE- AND POST-PROCESSOR

For finite element analysis, pre-process and post-process are highly important parts of the whole process. They are usually very labor-intensive. As a result, graphic programs are used to ease the processes.

Mesh generation in a complex domain is not easy even with a mesh generation program. However, a simple domain can be meshed using a simple algorithm. For example, a rectangular shape of 2-D domain can be broken into a number of rectangular elements easily. An example is given below to illustrate how to mesh a rectangular domain with elements of the same size.

♣ **Example B.1** A rectangular domain of size 2 by 4 is divided into a mesh with 4 elements by 8 elements in the x- and y-axis, respectively. Each element has four nodes and nodal connectivity is chosen in the counterclockwise direction as shown in Fig. 6.2.1. As a result, the total number of elements is 32 and the total number of nodes is 45.

```
%----------------------------------------------------------------
%
% generate uniform 2-D rectangular mesh for a rectangular
% shape of domain
%
clear
xnode=5;                                    % number of nodes in x-axis
ynode=9;                                    % number of nodes in y-axis
%
xzero=0.0;                                  % x-coord of bottom left corner
yzero=0.0;                                  % y-coord of bottom left corner
%
xlength=2.0;                                % size of domain in x-axis
ylength=4.0;                                % size of domain in y-axis
%
xnel=xnode-1;                               % no of elements in x-axis
ynel=ynode-1;                               % no of elements in y-axis
%
nel=xnel*ynel;                              % total number of elements
nnode=xnode*ynode;                          % total number of nodes
%
delx=xlength/xnel;                          % element size in x-axis
dely=ylength/ynel;                          % element size in y-axis
%
nodes=zeros(nel,4);
gcoord=zeros(nnode,2);
%
% generate nodal coordinate values
%
for iy=1:ynode
for ix=1:xnode
gcoord((iy-1)*xnode+ix,1)=xzero+(ix-1)*delx;
```

```
      gcoord((iy-1)*xnode+ix,2)=yzero+(iy-1)*dely;
      end
      end
      %
      % generate nodal connectivity in ccw direction
      %
      for iiy=1:ynel
      for iix=1:xnel
      nodes((iiy-1)*xnel+iix,1)=(iiy-1)*xnode+iix;
      nodes((iiy-1)*xnel+iix,2)=(iiy-1)*xnode+iix+1;
      nodes((iiy-1)*xnel+iix,3)=iiy*xnode+iix+1;
      nodes((iiy-1)*xnel+iix,4)=iiy*xnode+iix;
      end
      end
      %
      %————————————————————————————————
```

‡

♣ **Example B.2** This example plots the mesh generated from the previous example. It shows elements and nodes in the mesh. (See Fig. B.1.)

```
      %————————————————————————————————
      %
      % x and y coordinates
      %
      xcoord=gcoord(:,1);
      ycoord=gcoord(:,2);
      %
      % extract coordinates for each element
      %
      for i=1:nel
      for j=1:4
      x(j)=xcoord(nodes(i,j));
      y(j)=ycoord(nodes(i,j));
      end;                                                %j loop
      %
      xvec=[x(1),x(2),x(3),x(4),x(1)];
      yvec=[y(1),y(2),y(3),y(4),y(1)];
      plot(xvec,yvec);                                    % plot element
      hold on;
      %
      % place element number
      %
      midx=mean(xvec(1:4));
      midy=mean(yvec(1:4));
      text(midx,midy,num2str(i));
```

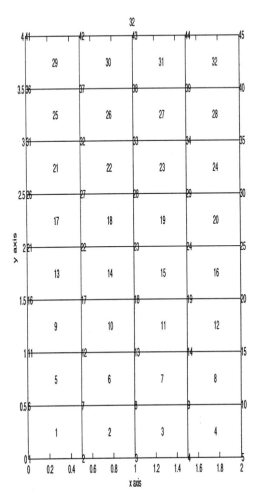

Figure B.1 Mesh with 4 by 8 elements

```
%
end;                                              % i loop
%
xlabel('x axis');
ylabel('y axis');
title(num2str(i));
%
% put node numbers
%
for jj=1:nnode
text(gcoord(jj,1),gcoord(jj,2),num2str(jj));
end;
%
```

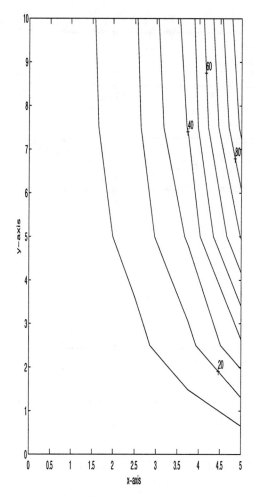

Figure B.2 Solution Contour of Example 5.9.2

For a rectangular domain, contour plot of the nodal solution is simple. The following examples show the contour plots in a rectangular domain

♣ **Example B.3** This example plots 2-D contour lines of the solution obtained in Example 5.9.2. Thus, the example must be run before the following program is executed. Figure B.2 shows the contour plot.

```
%————————————————————————
%
% Plot contours of the solution to Ex. 5.9.2
% See Fig. 5.9.2 for the problem mesh
%
xx=0.0:1.25:5.0;                            % x-axis grid
yy=0.0:2.5:10.0;                            % y-axis grid
%
for j=1:5
for i=1:5
sol(i,j)=fsol((j-1)*5+i);                   % nodal values
end
end
v=[20 40 60 80];                            % values for contour label
c=contour(xx,yy,sol);                       % plot contour
xlabel('x-axis');
ylabel('y-axis');
clabel(c,v);                                % label contours
%
%————————————————————————
```

‡

♣ **Example B.4**   This example plots 3-D contour lines of the solution obtained in Example 5.9.2. The MATLAB command *contour* is replaced by *contour3*. (See Fig. B.3.)

‡

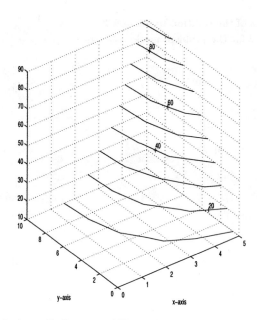

Figure B.3 Solution 3-D Contour of Example 5.9.2

INDEX

A

Acceleration, 288, 289
Active control, 299, 428
Actuator, 428, 437, 479, 481, 495, 496, 498
Adjoint, 454
Admissible solution, 43
Algebraic Ricatti Equation (ARE), 476, 477
Angle of twist, 77
Angular displacement, 504
  velocity, 504
Approximate solution, 35
Asymptotically stable, 429, 430, 431, 435
Automatic mesh generation, 52
Autonomous system, 442
Averaged weighted residual, 37, 238
Axial axis, 103
Axial bar (member), 74, 75, 199, 201, see Truss
Axial displacement (deformation), 199, 244, 248, 262
Axial force (load), 71, 74, 201, 261, 265
Axial member, 74, 75, 201
Axial stiffness, 262
Axial strain, 245, 250
Axisymmetric, 103-105, 121, 322, 323

B

Backward difference, 100, 101, 126, 136, 141
Bar structure, see Truss
Bass-Gura formula, 468, 470, 511
Beam, see Chapter 8
  element, 240, 242, 247-249, 253, 260, 463
  linear element, 247
  rigidity, 237, 240, 444, 502
Bending deformation, 201, 259, 262
Bending energy, 373
Bending moment, 241, 252-254
Bending stiffness, 246, 251, 257
  matrix, 248
Bending strain, 246, 258, 372, 376, 384, 386
Bending stress, 252, 259, 338, 371
BIBO stability, 432, 456
Biharmonic governing equation, 367
Bilinear elements, 93, 116, 163
Bilinear isoparametric element, 163, 164, 373
Bilinear isoparametric shape function, 372
Bilinear plate element, 386
Bilinear rectangular element, 92, 93, 99, 109, 116, 117, 120, 121, 131, 135, 141
Bilinear shape functions, 93, 165
Biquadratic isoparametric shape function, 373
Body coordinate system, 205
Body force, 315, 319
Boundary
  force, 506
  integral, 87, 94, 105, 315
  segment, 316
  traction, 316
Buckling analysis, 534

C

C^0 compatibility (*or* continuity), 257, 371, 383
C^0 type beam elements, 262
C^1 type, 239
Cantilever beam, 241, 289, 333

Central difference, 101, 223, 288, 289
Centroid, 324
Characteristic equation, 437, 454, 455, 457-459, 468, 471, 492
Characteristic root, 437, 454
Circumferential axis, 103
Circumferential direction, 322
Circumferential strain, 323
Clamped, 253
Classical linearization, 538
Closed-loop, 459
  eigenvalue, 468, 481, 492
  poles, 472
  system, 457, 458, 459, 467, 468, 471, 477, 481, 483, 488, 496
  transfer function, 457
Collocation method, 32, 33
Collocated sensor and actuator, 494, 495, 504
Command input, 457
Compatibility, 72
Compensator, 490
Concentrated force, 241
Condensation technique, 260
Condition number, 464
Conditionally stable, 130, 288
Consistent mass matrix, 200, 205, 206, 243, 247, 294, 325
Constant linearization, 537
Constitutive equation, 311, 313, 318, 319, 366, 371, 377, 379
Constitutive matrix, 370, 376
Constraint, 54, 55, 73, 283
Continuity, 238, 249
Control system design, 427, 428, 451
Controllability, 461-463, 485
  Grammian, 462, 463
  matrix, 463, 464, 469
  test, 486, 487
Convection coefficient, 140
Convolution integral, 296, 445, 456
Coordinate transformation, 203, 266
Correction factor for shear energy, 245
Crank-Nicolson, 101, 102, 126

Critical time step, 288
Critically damped case, 437
Current flow, 78
Curvature, 366
Cylindrical coordinate, 103

D

Damped natural frequency, 300
Damped response, 300
Damping coefficient, 435
Damping matrix, 287, 299, 434
Damping ratio, 436, 437, 459, 484
Determinant of Jacobian, 167, 168
Diagonalized mass matrix, 243, 247
Dirac delta function, 32, 241, 455, 498
Direct Velocity Feedback (DVF), 499
Direction cosine, 206, 312, 378
Dirichlet boundary condition, 41, 43
Discrete Fourier transform (DFT), 303, 304
Distributed parameter system, 428
Distributed pressure loading, 367
Disturbance, 457, 458, 460, 466
Duality, 486
Dynamic analysis, 244, 324
Dynamic buckling, 535
Dynamic characteristics, 454

E

Effective force vector, 288
Effective mass matrix, 288
Eigenvalue, 218, 244, 283, 292, 302, 442, 454, 476 488
Eigenvector, 292, 302, 442-444, 476
Eight-node element, 169, 170
Elasticity, see Chapter 9
Electric circuit, 78
Element boundary, 95-98, 108, 315
Element domain, 90, 91
Element force vector, 72
Element mass matrix, 200, 243, 500

Element (stiffness) matrix, 51, 54, 72, 91, 108, 186, 200, 202, 240, 245, 246, 251, 252, 262, 314, 316, 319, 320, 322, 370-372, 377, 500
Element topology, 53
Embedded crack, 248
Energy method, 245, 312, 317
Equation of motion, 226, 448, 486, 497, 499, 510
Equilibrium, 71, 299, 366, 367, 377, 384, 429-431
point 434, 435, 495, 505, 506, 509
Equivalent beam rigidity, 259
Equivalent spring system, 78
Error signal, 457
Error vector, 488
Essential boundary condition, 43, 86, 312, 317
Estimated state variable, 490-492, 497
Euler-Bernoulli beam, 237, 244, 294, 365
Euler's formula, 444
Extended Hamilton's principle, 499
External load, 317
External work, 318

100, 121, 130, 136, 154, 218, 221, 241, 258, 271, 283, 285, 327, 376, 377, 387, 427, 433, 440, 495, 499, 500, 511
formulation, 71, 75, 76, 237, 312, 324
First order equation, 441
First order system, 441, 443, 447
Flexual rigidity of plate, 377
Flow rate, 78, 79
Fluid flow, 78, 79
Fluid viscosity, 78
Flux, 95, 105, 108, 109, 142
boundary condition, 85, 108, 144
Force equilibrium, 366
Forward difference, 100, 101, 126, 127, 136
Fourier transform, 303
Frame, *see* Chapter 8
element, 262, 263, 268, 286
Free body diagram, 199, 311, 366
Frequency domain, 303, 427, 456
Full state feedback, 467, 471, 477, 481, 485, 487, 491
Functional, 42-44

F

Factorial term, 446
Fast Fourier Transform (FFT), 303, 304
Feedback control, 456, 466, 490-492, 506, 510
Feedback gain, 457, 467, 469-471, 476, 477, 480, 492, 495-498
Fibrous composite, 375
Fictitious displacement, 288
Fictitious time step, 289
Fictitious velocity, 289
Final value theorem, 458-460
Finite difference method, 100
Finite dimensional approximation, 497
Finite element, 37, 43, 51, 52, 61, 64,

G

Galerkin's method, 32-35, 36, 40, 43, 65, 200, 237, 239, 244, 252, 255, 312, 314, 319, 324, 377, 380
Gauss-Legendre quadrature, 175, 176, 177, 179, 180, 181-186, 334, 373
Generalized coordinate vector, 433
Generalized strain, 257, 258, 384, 386
Geometric compatibility, 76
Global axes, 215, 232, 279
Global beam displacement, 261
Global coordinate system, 202, 262
Global derivative, 164-166
Global displacement, 263
Global matrix, 204
Global node number, 96, 316

Global truncation error, 100, 101
Green's theorem, 87

H

Hamiltonian, 472, 475, 476
Harmonic motion, 218
Heat conduction, 78, 85, 87, 95, 98
    coefficient, 140
Heat flux, 78, 87
Heat sink, 87
Heat source, 87
Heat transfer, 97
Hermitian beam element, 241, 246,
    262, 268, 289
Hermitian shape function, 239, 242
Hooke's law, 199
Hybrid beam element, 255, 258
Hybrid plate element, 383

I

Impulse response, 297, 304, 455, 456
Inertia, 200
    force, 241, 324
Infinite dimensional system, 428, 434
Initial solution, 289
Inplane axis, 366
Inplane deformation
    (*or* displacement), 259, 371, 373
Inplane strain, 370
Input influence matrix, 295, 433, 452,
    494, 496
Integral operator, 42
Integration by parts, 34, 42, 86,
    104, 238
Inter-element compatibility, 246
Interlaminar delamination, 248
Internal bending moment, 505
Internal energy, 317, 370
Internal forces, 78
Internal layers, 260
Internal pressure, 339

Internal shear force, 505
Internal strain energy, 376
Interpolation, 372, 374, 382
    function, 92, 106
Inverse Fourier transform, 303
Inverse Laplace transform technique,
    296, 300, 437, 445, 455
Isoparametric elements, *see* Chapter 6,
    247, 327, 333, 345, 374, 378, 382,
    383
Isoparametric quadrilateral elements,
    339
Isoparametric shape function,
    371, 373, 382
Isoparametric solid element, 351
Isotropic material, 312, 319, 375, 376

J

Jacobian, 161, 162, 165-169, 172

K

Kinematic equation, 312, 313, 320,
    322, 323, 374, 379
Kinematic matrix, 376
Kinetic energy, 288, 495
Kirchhoff plate bending theory, 365
Kronecker delta, 89

L

Lagrange multiplier, 472
Lagrange shape functions, 93
Lagrangian, 499
Laminar flow, 78
Laminated (composite) beam, 248,
    259, 263, 264
Laplace transform, 300, 437, 445, 451
    455, 456, 458
Laplace's equation, *see* Chapter 5

Least squares method, 32, 33
Line integral, 87
Linear elements, 52, 56, 65, 159
Linear frame element, 265
Linear Quadratic Regulator (LQR), 467, 472, 477, 480, 497
Linear shape functions, 44, 52, 92, 98 200, 246
Linear spring, 71, 76
   equivalency, 79
Linear triangular element,
   see triangular element
Local axis, 265
Local coordinate system, 268
Local derivative, 164
Local node number, 316
Local truncation error, 100
Longitudinal vibration, 511, 512
Lower triangular Toeplitz matrix, 469
L-shape frame, 279
Lumped mass matrix, 200, 205, 206, 243, 247, 251, 255, 325
Lyapunov
   equation, 462, 463, 485
   function, 430-435, 495, 498, 505, 509, 512
   instability theorem, 430
   stability theory, 429

M

Mass matrix, 262, 287, 293, 299, 433, 443, 463, 483, 496
Matrix differential equation, 446, 475
Maximum overshoot, 439, 509
Measurement noise, 458
Mechanical force, 79
Midplane axis, 259
Mindlin/Reissner plate theory, 370
Mixed beam element, 252, 268
Mixed formulation, 382
Mixed plate bending formulation, 378
Modal
   analysis, 292, 299
   control, 482
   coordinate, 294, 296, 300, 302, 400, 482-484
   input force vector, 483
   matrix, 294, 300, 483
   testing, 455
   truncation, 464
Mode shape, 218, 284, 292
Modes, 292
Moment, 366, 382
   equilibrium, 366
   of inertia, 245, 268
Multi-input system, 471

N

Natural boundary condition, 42, 56, 62, 86, 312, 313
Natural coordinate, 159-163, 171, 172
Natural frequency, 218, 244, 283-286, 292, 300, 436, 459, 484, 502-504
Neuman boundary condition 41, 42
Neutral axis, 238, 245
Neutrally stable, 459
Newton's second law, 199
Newton's third law, 78
Nine-node element, 170, 171
Nodal connectivity, 53
Nodal degrees of freedom, 53, 204
Nodal displacement, 203, 248, 314, 316, 324, 374, 377
Nodal flux, 96, 97, 122
Nodal sequence, 89
Nodal variable, 38, 41, 45, 88, 98, 99, 164, 238, 239, 241, 243, 324, 368, 371
Nodal vector, 45
Nonlinear analysis, 537
Nonlinear function, 429
Nontrivial solution, 442
Normal strain, 374
Normalization, 293

Normalized eigenvector, 293
Numerical integration, 92, 161, 173, 177-179, 446
Nyquist, 304

O

Observability, 485
  Grammina, 485
  matrix, 486
  test, 486, 487
Observer, 485, 491, 492
  gain, 488, 489, 492
One-dimensional truss, 199
One-point Gaussian quadrature, 247
Open-loop, 452
  eigenvalues, 480, 511
Optimal control, 472, 474, 475
Optimal trajectory, 473
Optimality condition, 474, 475
Ordinary differential equation, 101
Orthogonal, 293
Oscillation, 229, 291
Output distribution matrix, 494
Output feedback, 467, 495
Overdamped case, 438

P

Parabolic differential equation, 100
Parabolic type disturbance, 460
Partial fraction expansion, 455
Peak time, 439, 509
Performance index, 472-474, 512
Perturbation, 429, 447
Physical coordinate, 159-162, 302
Physical element, 167, 168
Piecewise continuous function, 37
Piecewise linear boundary, 90
Piecewise linear functions, 35
Pin joints, 201
Planar geometry, 262
Planar frame structure, 265

Planar transformation, 262, 263
Plane
  strain, 311
  stress, 248, 311, 316, 326, 370
  truss, 202, 205
Plate bending, see Chapter 10
  three-node element, 368
Plate rigidity, 367
Poisson's equation, see Chapter 5
Polar moment of inertia, 76, 268
Pole, 452
Pole placement technique, 468, 471, 477
Positive definite, 293, 432, 434, 462, 474, 495
  matrix, 431, 474
Positive semidefinite, 293, 474
Potential energy, 42, 257, 376, 383, 499
Potential flow, 85
Potter's algorithm, 477
Pre-processor, 52
Pressure difference, 78
Pressure loading, 377, 382, 386
Primary variable, 252
Principle of minimum potential energy, 318
Proportional damping, 299, 300
Proportional plus Derivative (PD) control, 458-460
Proportional plus Integral (PI) control, 460

Q

Quadratic isoparametric element, 162
Quadratic triangular element, 171
Quadrature point, 244
Quadrature rule, 178
Quadrilateral
  isoparametric element, 169
  shape, 163, 378
  element, 168, 190, 326, 385
Quasilinearization, 539

R

Radial axis, 103
Radial direction, 322
Radial displacement, 323
Rank deficient, 247
Rank test, 486
Rayleigh damping, 299, 400
Rayleigh-Ritz method, 42, 43
Reaction forces, 75
Reciprocal relation, 376
Rectangular element, 166
Reduced integration, 246, 251, 259, 373
Residual, 31
Resistance, 78
Ricatti matrix, 475-477
Rigid body motions, 55, 442
Rise time, 438, 509
Robustness, 472, 497
Rotating beam, 499, 500, 502, 504, 508, 512
Rotational degree of freedom, 268
Rotational motion, 442

S

Sampling period, 304, 448
Sampling point, 173-175
Second order system, 292, 430, 436, 444, 482
Secondary variable, 51, 252
Selective integration, 373, 395
Self-adjoint operator, 35
Semi-infinite element, 528
Sensor, 428, 457, 485, 495
  noise, 457
  output, 485, 488
Separation principle, 492
Settling time, 439
Shape function, 39, 89, 99, 92, 99, 107, 159-166, 238, 246, 247, 248, 253, 255, 313, 320, 321, 323, 324, 369, 372, 373, 384

Shear correction factor, 255, 379
Shear deformable, 370, 394
Shear deformation, 259
Shear energy, 247, 250, 373
Shear force, 254, 367, 378-380
Shear locking, 246, 251, 373, 377
Shear stiffness, 246, 257
Shell, *see* Chapter 10
Similarity transformation, 443
Simply supported, 253, 268
  plates, 383, 385, 394
Simpson's rule, 174
Single input system, 471
Singular, 55
  element, 513, 519, 522
Sinusoidal motion, 433, 438
Slope, 238, 241, 246
Space truss, 205
Spatial coordinate system, 34
Spatial frame, 265
Spatial variables, 324
Specific heat, 99
Stability, *see* Chapter 11
State space form, 441, 453, 461, 466
State transition matrix, 446, 447
State variables, 466, 467, 487
State vector, 441, 447, 466, 472
Static analysis, 206, 244, 268
Static equilibrium, 75
Statically determinate, 72
Statically indeterminate, 72
  system, 76, 77
  torsional members, 77
Stationary value, 258, 318, 377, 384, 385
Steady state, 99, 102, 109, 437-439, 456, 459, 474, 475
  error, 458, 460
  solution, 146
Step function, 229
Step input, 456, 459
Step response, 437, 439
Stiffness matrix, 202, 218, 239, 241, 261, 263, 293, 294, 299, 300, 433,

443, 463, 483
Strain analysis, *see* Chapter 9
Strain-displacement relation, 199, 250, 257, 384
Strain energy, 203, 204, 245, 250, 318
   bending strain energy, 245
   shear strain energy, 245
Stress analysis, *see* Chapter 9
Strong formulation, 34, 35
Structural damping, 299
Subdomains, 38
Substructure's energy, 505
Surface traction, 319

T

Taylor series, 445
Temporal axis, 199
Temporal derivative, 243, 324
Test function, 32, 33, 37, 40, 98, 161, 200, 238
Tetrahedral element, 106-109, 150
Thermal stress, 325, 531
Thick beam, 255, 256, 259
Thick plate theory, 379, 382
Thin beam, 255, 256, 259
Thin plate theory, 379, 382
Three-dimensional
   elasticity, 319
   truss element, 207
Three point integration, 174
Three point quadrature, 177, 179
Time
   constant, 438
   domain approach, 302, 427
   step size, 291
Timoshenko beam, 244, 268, 370
Torque, 76
Torsion of noncircular members, 85
Torsional load, 265
Torsional members, 76
Total potential energy, 317
Total system energy, 433

Traction, 311, 315, 316
   boundary condition, 320
   surface, 320
Transfer function, 437, 452, 453, 455, 457, 458, 509
Transformation matrix, 261
Transformed stiffness matrix, 204
Transient
   analysis, 287
   heat conduction, 98
   problem, 99
   response, 229, 289, 438, 439, 459
Translational degree of freedom, 243
Translational motion, 442
Transverse
   direction, 376
   shear deformation, 244, 254, 255, 366, 370, 371, 378
   shear energy, 370, 376, 377
   shear force, 378
   shear strain, 284
   stiffness matrix, 251
Trapezoidal rule, 173
Trapezoidal shape, 168
Trial function, 31-37, 98, 161
Triangular domain, 91, 175
Triangular elements, 99, 105, 135, 172, 313, 339
   linear, 88-92, 104, 109, 121, 316, 323, 325
   six-node, 172
Triangular isoparametric element, 171
Truss, *see* Chapter 7
   1-D truss element, 199
   plane truss element, 201
Twisting moment, 76, 268
Two point boundary value problem, 474
Two-point quadrature rule, 177, 178

U

Unconditionally stable, 101, 136

Underdamped case, 437
Under-integrated, 247, 377
Unidirectional composite, 375, 376
Unit impulsive input, 455
Unit step input, 437, 509

V

Variable interpolation, 106
Variational
   method, 42
   operator, 42
   principle, 472, 473
Vibration control, 428

W

Weak formulation, 34, 52, 65, 86, 104,
   200, 238, 378
Weighted average, 32
Weighted residual, 31, 39, 86, 99, 104,
   238
Weighting coefficient, 173-176,
   177, 178
Weighting function, 313, 314
Weighting matrix, 474, 477, 480

Z

Zero order hold, 447